化工原理实验

（双语版）

叶长燊　李玲　施小芳　侯琳熙　等编

Experiments of Chemical Engineering Principles

化学工业出版社

·北京·

内容简介

《化工原理实验（双语版）》共分10章，内容包括绪论、化工原理实验研究方法、实验数据的误差分析、实验数据处理、Excel与Origin软件的应用、实验室常用测量仪表、化工原理虚拟仿真实验、化工原理演示实验、化工原理“三型”实验、化工原理自组装实验、化工原理提高型实验与实训等内容。全书从基础知识到专业知识，从演示实验、虚拟仿真实验到实物实验，从“三型”实验到自组装实验，从单一的单元操作实验到复杂的综合实训实验，由浅入深、由易到难进行介绍，旨在从多方面、多层次实现学生实验设计能力、实践动手能力、工程研究能力、工程问题的综合分析与解决能力的系统性和个性化训练与培养。

本书为中英对照，可作为高等学校化工及相关专业的化工原理实验教材或教学参考书，也可以作为石油、化工、环境、轻工、医药等行业科研和生产技术人员的参考书。

图书在版编目（CIP）数据

化工原理实验：双语版：中文、英文/叶长燊等编. —北京：化学工业出版社，2021.2 (2023.8重印)

ISBN 978-7-122-38073-9

Ⅰ.①化… Ⅱ.①叶… Ⅲ.①化工原理-实验-高等学校-教材-汉、英 Ⅳ.①TQ02-33

中国版本图书馆CIP数据核字（2020）第244058号

责任编辑：刘兴春 刘 婧　　文字编辑：王云霞 陈小滔
责任校对：边 涛　　装帧设计：李子姮

出版发行：化学工业出版社（北京市东城区青年湖南街13号 邮政编码100011）
印 装：北京天宇星印刷厂
787mm×1092mm 1/16 印张27½ 字数656千字 2023年8月北京第1版第4次印刷

购书咨询：010-64518888　　售后服务：010-64518899
网 址：http://www.cip.com.cn
凡购买本书，如有缺损质量问题，本社销售中心负责调换。

定 价：86.00元

前 言

化工原理是化学工程学科中形成最早、基础性最强、应用面最广的学科分支。化工原理紧密联系化工生产实际，它来源于实践，又面向实践，应用于实践，是一门实践性很强的专业基础课程。“化工原理”、“化工原理实验”及“化工原理课程设计”三大课程构成了化工原理课程体系，其中化工原理实验则是树立工程意识、培养工程研究能力和解决工程问题能力必不可少的重要实践教学环节。

福州大学从二十世纪九十年代开始，在化工原理实验课程的教学内容、教学方法等方面一直紧跟人才培养的时代要求开展持续的教学改革，实验设备也经过多次的改进和更新换代。经过三十年的努力，已逐步形成演示实验、虚拟仿真实验、“设计型和研究型以及综合型”实验、自组装实验、提高型实验和综合实训实验等系统完善的实验教学内容，以学生自主学习为主、教师指导为辅的启发式、交互式、探究式等多样的实验教学方法；从实验研究目标确立、实验研究方案制订、实验装置组装、实验步骤设计、实验现象观察、参数测定、实验数据处理以及实验结果分析与讨论等环节，有效地培养学生的工程思维、实践动手能力和创新能力。

本书的特点如下。

(1) 教学内容系统完善、循序渐进。

全书包括绪论、化工原理实验研究方法、实验数据的误差分析、实验数据处理、 Excel 与 Origin 软件的应用、实验室常用测量仪表、化工原理虚拟仿真实验、化工原理演示实验、化工原理“三型”实验、化工原理自组装实验、化工原理提高型实验与实训等内容。

(2) 研究型、设计型和综合型实验进一步完善。

在实验装置改进和更新换代的基础上，“三型”实验的实验装置和教学内容进一步完善，学生可根据任务书独立完成实验研究并撰写研究小论文，本书可引导学生熟练应用所学知识、开展探究性学习、掌握论文撰写的基本要求和技能。

(3) 自组装实验操作性强

在“三型”实验基础上，开发建设了流体流动阻力、离心泵、传热过程强化、列管换热器等自组装实验。学生可根据自身学习需求和兴趣设置实验研究课题，自主搭建实验装置，制订研究计划和目标，独立开展实验研究，并获得相关结果和结论，充分发挥学习的自主性和主动性，强化实践动手能力。

(4) 设置提高型和实训实验。

在单元操作实验的基础上，进一步增强实验研究的综合性和复杂性，培养学生综合应用知识的能力和应对复杂工程问题的能力。

(5) 思考题更加丰富，针对性强、富有启发性。

各实验均提供了大量针对本实验研究内容的思考题，有利于启发学生针对实验的理论依据、实验过程现象、实验与理论的结合等各个方面逐步进行深入、全面的思考与探讨。

本书由叶长燊、李玲、施小芳、侯琳熙等编写，其中英文由吴乃昕、张孟佳翻译编写，全书由叶长燊统稿。本书在编写过程中借鉴参考了其他院校的有关教材，在此向相关教材的作者表示诚挚的谢意。同时，感谢天津大学化工基础实验中心、北京欧倍尔软件技术开发有限公司、北京恒久实验设备有限公司、福州大学石油化工学院、福州大学石油化工虚拟仿真实验教学中心、福州大学石油化工学院化工原理与实验教学团队全体老师的大力支持。

鉴于编者的水平及经验有限，且编写时间仓促，书中不妥和疏漏之处在所难免。在此，真诚希望广大读者和同行不吝赐教，使本教材日臻完善。

编者

2020 年 7 月 1 日于福州大学

目 录

绪论

1

0.1 化工原理实验的教学目的 1
0.2 化工原理实验的教学特点 1
0.3 化工原理实验的教学内容与方法 2
0.4 化工原理实验的教学环节和要求 2
0.5 化工原理实验的基本安全知识 6
0.5.1 实验守则 7
0.5.2 基本安全知识 7
0.6 任务书 12
0.6.1 阻力实验 12
0.6.2 离心泵实验 13
0.6.3 过滤实验 14
0.6.4 传热实验 15
0.6.5 吸收实验 16
0.6.6 精馏实验 18
0.6.7 干燥实验 19
0.6.8 液液萃取实验 20

第1章
化工原理实验研究方法

22

1.1 直接实验法 22
1.2 量纲分析法 22
1.2.1 量纲分析法具体步骤 23
1.2.2 量纲分析法举例说明 23
1.3 数学模型法 25
1.3.1 数学模型法主要步骤 25
1.3.2 数学模型法举例说明 25
1.3.3 数学模型法和量纲分析法的比较 26

第2章
实验数据的误差分析

28

2.1 真值与平均值 28
2.2 误差的分类 29
2.3 误差的表示方法 30
2.4 精密度、正确度和精确度 31
2.5 实验数据的有效数字 31
2.6 误差的处理方法 33

2.7 直接测量的误差估算 36
2.8 间接测量中的误差传递 38
2.9 误差分析的具体应用 39

第 3 章
实验数据处理
42

3.1 列表法 42
3.1.1 数据表的分类 42
3.1.2 设计实验数据表应注意的事项 43
3.2 图示法 43
3.2.1 坐标系选择的基本原则 44
3.2.2 坐标分度的确定 44
3.2.3 设计实验数据图应注意的事项 45
3.3 数学方程表示法 46
3.3.1 图解分析法 46
3.3.2 回归分析法 49

第 4 章
Excel 与 Origin 软件的应用
54

4.1 Excel 在实验数据处理中的应用 54
4.2 Origin 在实验数据回归与绘图中的应用 60

第 5 章
实验室常用测量仪表
69

5.1 温度测量 69
5.1.1 膨胀式温度计 70
5.1.2 热电偶温度计 71
5.1.3 热电阻温度计 74
5.1.4 温度计的选择和使用原则 75
5.2 压力（差）测量 75
5.2.1 液柱式压力计 75
5.2.2 弹性式压力计 77
5.2.3 电气式压力计 79
5.2.4 活塞式压力计 81
5.2.5 压力（差）计的校验 81
5.3 流量测量 81

5.3.1 速度式流量计 81
5.3.2 容积式流量计 85
5.3.3 质量流量计 85
5.3.4 流量计的检验和标定 86
5.4 液位测量 86
5.4.1 直读式液位计 86
5.4.2 差压式液位计 87
5.4.3 浮力式液位计 88

第6章
化工原理虚拟仿真实验
91

6.1 虚拟仿真实验概述 91
6.1.1 基本操作 91
6.1.2 菜单选择项功能说明 93
6.1.3 仪表和阀门调节说明 95
6.1.4 注意事项 97
6.2 雷诺演示仿真实验 97
6.2.1 仿真主界面 97
6.2.2 雷诺演示实验装置 97
6.2.3 实验项目 98
6.3 化工流动过程综合实验 98
6.3.1 仿真主界面 98
6.3.2 化工流动过程综合实验装置 99
6.3.3 实验项目 99
6.4 离心泵综合性能测定实验 99
6.4.1 仿真主界面 99
6.4.2 离心泵实验装置 100
6.4.3 实验项目 100
6.5 恒压过滤实验 101
6.5.1 仿真主界面 101
6.5.2 恒压过滤实验装置 101
6.5.3 实验项目 102
6.5.4 可变换的实验条件 102
6.5.5 数据处理注意事项 102
6.6 传热综合实验 102
6.6.1 仿真主界面 102
6.6.2 传热综合实验装置 103
6.6.3 实验项目 103
6.6.4 可变换的实验条件 103

6.7 二氧化碳吸收与解吸实验 104
6.7.1 仿真主界面 104
6.7.2 二氧化碳吸收与解吸实验装置 104
6.7.3 实验项目 105
6.7.4 可变换的实验条件 105
6.7.5 注意事项 105
6.8 精馏综合拓展实验 105
6.8.1 仿真主界面 105
6.8.2 精馏综合拓展实验装置 106
6.8.3 实验项目 107
6.8.4 注意事项 107
6.9 液液萃取实验 108
6.9.1 仿真主界面 108
6.9.2 液液萃取实验装置 108
6.9.3 实验项目 109
6.9.4 可变换的实验条件 109
6.9.5 注意事项 109
6.10 洞道干燥实验 109
6.10.1 仿真主界面 109
6.10.2 洞道干燥实验装置 109
6.10.3 实验项目 110
6.10.4 可变换的实验条件 110
6.10.5 注意事项 110

第7章
化工原理演示实验
111

7.1 雷诺实验 111
7.1.1 实验目的 111
7.1.2 实验原理 111
7.1.3 实验装置 111
7.1.4 实验步骤 111
7.1.5 注意事项 113
7.1.6 思考题 113
7.2 能量转化 (流体机械能转化) 演示实验 113
7.2.1 实验目的 113
7.2.2 实验原理 113
7.2.3 实验装置 113
7.2.4 实验步骤 114
7.2.5 注意事项 115
7.2.6 思考题 115

7.3　流线演示实验　115
7.3.1　实验目的　115
7.3.2　实验原理　115
7.3.3　实验装置　115
7.3.4　实验步骤　116
7.3.5　思考题　117
7.4　非均相气固分离演示实验　117
7.4.1　实验目的　117
7.4.2　实验原理　117
7.4.3　实验装置　118
7.4.4　实验步骤　118
7.4.5　注意事项　120
7.4.6　思考题　120
7.5　板式塔流体力学性能演示实验　120
7.5.1　实验目的　120
7.5.2　实验原理　120
7.5.3　实验装置　121
7.5.4　实验步骤　122
7.5.5　注意事项　122
7.5.6　思考题　122

第8章
化工原理“三型”实验
123

8.1　流体流动阻力实验　123
8.1.1　实验目的　123
8.1.2　实验原理　123
8.1.3　实验装置与流程　125
8.1.4　实验步骤及注意事项　126
8.1.5　思考题　127
8.2　离心泵实验　128
8.2.1　实验目的　128
8.2.2　实验原理　128
8.2.3　实验装置与流程　129
8.2.4　实验步骤及注意事项　130
8.2.5　思考题　131
8.3　过滤实验　132
8.3.1　实验目的　132
8.3.2　实验原理　132

8.3.3 实验装置与流程 133
8.3.4 实验步骤及注意事项 134
8.3.5 思考题 135
8.4 传热实验 136
8.4.1 实验目的 136
8.4.2 实验原理 136
8.4.3 实验装置与流程 137
8.4.4 实验步骤与注意事项 140
8.4.5 思考题 142
8.5 吸收实验 142
8.5.1 实验目的 142
8.5.2 实验原理 143
8.5.3 实验装置与流程 144
8.5.4 实验步骤与注意事项 146
8.5.5 思考题 147
8.6 精馏实验 148
8.6.1 实验目的 148
8.6.2 实验原理 148
8.6.3 实验装置与流程 151
8.6.4 实验步骤与注意事项 154
8.6.5 思考题 155
8.7 干燥实验 156
8.7.1 实验目的 156
8.7.2 实验原理 156
8.7.3 实验装置与流程 158
8.7.4 实验步骤与注意事项 159
8.7.5 思考题 160

第 9 章
化工原理自组装实验
161

9.1 自组装实验目的 161
9.2 自组装实验要求 161
9.3 自组装实验主要设备、仪表与配件 162
9.3.1 流体流动阻力自组装实验 162
9.3.2 泵性能测试自组装实验 163
9.3.3 传热过程强化自组装实验 164
9.3.4 列管换热自组装实验 164

第 10 章
化工原理提高型实验与实训

166

10.1 液液萃取提高型实验 166
10.1.1 实验目的 166
10.1.2 实验原理 166
10.1.3 实验装置与流程 167
10.1.4 实验步骤与注意事项 167
10.1.5 思考题 171
10.2 多效蒸发综合实训 172
10.2.1 实训目的 172
10.2.2 基本原理 172
10.2.3 工艺流程、主要设备及仪表 172
10.2.4 实训项目 174
10.3 吸收解吸综合实训 175
10.3.1 实训目的 175
10.3.2 基本原理 176
10.3.3 工艺流程、主要设备及仪表 176
10.3.4 实训项目 176

Introduction

180

0.1 Teaching objectives 180
0.2 Teaching characteristics 181
0.3 Teaching content and methods 181
0.4 Teaching requirements 182
0.5 Basic safety knowledge of the experiment 189
0.5.1 Experimental rules 189
0.5.2 Basic safety knowledge 190
0.6 Tasks 196
0.6.1 Resistance experiment 196
0.6.2 Centrifugal pump experiment 197
0.6.3 Filtration experiment 199
0.6.4 Heat transfer experiment 201
0.6.5 Absorption experiment 202
0.6.6 Distillation experiment 205
0.6.7 Drying experiment 207
0.6.8 Liquid-liquid extraction experiment 208

Chapter 1
Research Methods of Chemical Engineering Principle Experiment

211

1.1 Direct experiment method 211
1.2 Dimensional analysis method 211
1.2.1 Specific steps of dimensional analysis 212
1.2.2 Dimensional analysis examples 213
1.3 Mathematical model method 215
1.3.1 Main steps of mathematical model method 215
1.3.2 Mathematical model method example 216
1.3.3 Comparison of mathematical model method and dimensional analysis method 217

Chapter 2
Error Analysis of Experimental Data

219

2.1 True values and the mean 219
2.2 Error classification 220
2.3 Error representation methods 221
2.4 Precision, trueness and accuracy 223
2.5 Effective digits of experimental data 224
2.6 Error handling methods 227
2.7 Error estimation of directly measured values 230
2.8 Error propagation in indirect measurements 233
2.9 Application examples of error analysis 234

Chapter 3
Experimental Data Processing

238

3.1 Tabular method 238
3.1.1 Classification of data tables 238
3.1.2 Considerations for designing experimental data tables 239
3.2 Graphic method 240
3.2.1 Basic principles of coordinate system selection 240
3.2.2 Determination of coordinate division 241
3.2.3 Considerations for designing experimental data graphs 242
3.3 Mathematical equation representation method 243
3.3.1 Graphic analysis method 244
3.3.2 Regression analysis method 247

Chapter 4
Applications of Excel and Origin Software

253

4.1 Application of Excel in the experimental data processing 253
4.2 Application of Origin in experimental data regression and plotting 260

Chapter 5
Commonly Used Measuring Instruments in the Laboratory

269

5.1 Temperature measurement 269
5.1.1 Expansion thermometer 270
5.1.2 Thermocouple thermometer 273
5.1.3 Thermistor thermometer 276
5.1.4 Principles of thermometer selection and use 277
5.2 Pressure (pressure drop) measurement 278
5.2.1 Liquid column pressure gauge 278
5.2.2 Elastic pressure gauge 280
5.2.3 Electric pressure gauge 283
5.2.4 Piston pressure gauge 285
5.2.5 Calibration of pressure gauge 285
5.3 Flow measurement 286
5.3.1 Velocity flowmeter 286
5.3.2 Volumetric flowmeter 291
5.3.3 Mass flowmeter 292
5.3.4 Flowmeter inspection and calibration 292
5.4 Liquid level measurement 293
5.4.1 Direct reading level gauge 293
5.4.2 Differential pressure level gauge 295
5.4.3 Buoyancy liquid level gauge 296

Chapter 6
Virtual Simulation Experiments of Chemical Engineering Principle

299

6.1 Overview of virtual simulation experiments 299
6.1.1 Basic operations 299
6.1.2 Description of menu selection functions 301
6.1.3 Instrument and valve adjustment instructions 304
6.1.4 Precautions 306
6.2 Simulation experiment of Reynolds demonstration 306

6.2.1 Main interface of simulation 306
6.2.2 Reynolds demonstration experimental device 307
6.2.3 Experiment projects 308
6.3 Comprehensive experiment of chemical flow processes 308
6.3.1 Main interface of simulation 308
6.3.2 Experimental device for the comprehensive experiment of chemical flow processes 308
6.3.3 Experiment projects 309
6.4 Experiment of centrifugal pump comprehensive performance measurement 309
6.4.1 Main interface of simulation 309
6.4.2 Centrifugal pump experimental device 310
6.4.3 Experiment projects 311
6.5 Filtration experiment under constant pressure 311
6.5.1 Main interface of simulation 311
6.5.2 Experimental device of filtration under constant pressure 311
6.5.3 Experiment projects 312
6.5.4 Adjustable experimental conditions 312
6.5.5 Notes for data processing 312
6.6 Comprehensive experiment of heat transfer 313
6.6.1 Main interface of simulation 313
6.6.2 Experimental device for comprehensive experiment of heat transfer 313
6.6.3 Experiment projects 314
6.6.4 Adjustable experimental conditions 315
6.7 Carbon dioxide absorption and desorption experiment 315
6.7.1 Main interface of simulation 315
6.7.2 Experimental device of carbon dioxide absorption and desorption 315
6.7.3 Experiment projects 316
6.7.4 Adjustable experimental conditions 317
6.7.5 Precautions 317
6.8 Comprehensive experiment of rectification 317
6.8.1 Main interface of simulation 317

6.8.2 Experimental device for comprehensive experiment of rectification 318
6.8.3 Experiment projects 319
6.8.4 Precautions 319
6.9 Liquid-liquid extraction experiment 320
6.9.1 Main interface of simulation 320
6.9.2 Experimental device for liquid-liquid extraction 320
6.9.3 Experiment projects 321
6.9.4 Adjustable experimental conditions 321
6.9.5 Precautions 322
6.10 Tunnel drying experiment 322
6.10.1 Main interface of simulation 322
6.10.2 Tunnel drying experimental device 323
6.10.3 Experiment projects 323
6.10.4 Adjustable experimental conditions 323
6.10.5 Precautions 324

Chapter 7 **Demonstration Experiment of Chemical Engineering Principle** 325

7.1 Reynolds experiment 325
7.1.1 Experimental objectives 325
7.1.2 Experimental principles 325
7.1.3 Experimental device 325
7.1.4 Experimental procedures 326
7.1.5 Precautions 327
7.1.6 Questions 328
7.2 Demonstration experiment of energy conversion (fluid mechanical energy conversion) 328
7.2.1 Experimental objectives 328
7.2.2 Experimental principles 328
7.2.3 Experimental device 329
7.2.4 Experimental procedures 330
7.2.5 Precautions 330
7.2.6 Questions 330
7.3 Streamline demonstration experiment 331
7.3.1 Experimental objectives 331
7.3.2 Experimental principles 331
7.3.3 Experimental device 332

7.3.4 Experimental procedures 333
7.3.5 Questions 333
7.4 Demonstration experiment of heterogeneous gas-solid separation 334
7.4.1 Experimental objectives 334
7.4.2 Experimental principles 334
7.4.3 Experimental device 336
7.4.4 Experimental procedures 337
7.4.5 Precautions 337
7.4.6 Questions 337
7.5 Demonstration experiment of hydrodynamic performance of plate column 338
7.5.1 Experimental objectives 338
7.5.2 Experimental principles 338
7.5.3 Experimental device 339
7.5.4 Experimental procedures 340
7.5.5 Precautions 340
7.5.6 Questions 341

Chapter 8
Chemical Engineering Principle "Three Type" Experiment
342

8.1 Fluid flow resistance experiment 342
8.1.1 Experimental objective 342
8.1.2 Experimental principles 342
8.1.3 Experimental device 344
8.1.4 Experimental procedures and precautions 344
8.1.5 Questions 347
8.2 Centrifugal pump experiment 348
8.2.1 Experimental objective 348
8.2.2 Experimental principles 348
8.2.3 Experimental device 350
8.2.4 Experimental procedures and precautions 351
8.2.5 Questions 352
8.3 Filtration experiment 353
8.3.1 Experimental objective 353
8.3.2 Experimental principles 353
8.3.3 Experimental device 355
8.3.4 Experimental procedures and precautions 356
8.3.5 Questions 358

8.4 Heat transfer experiment 359
8.4.1 Experimental objectives 359
8.4.2 Experimental principles 359
8.4.3 Experimental device 360
8.4.4 Experimental procedures and precautions 363
8.4.5 Questions 366
8.5 Absorption experiment 367
8.5.1 Experimental objectives 367
8.5.2 Experimental principles 367
8.5.3 Experimental device 369
8.5.4 Experimental procedures and precautions 371
8.5.5 Questions 373
8.6 Distillation experiment 374
8.6.1 Experimental objectives 374
8.6.2 Experimental principles 375
8.6.3 Experimental device 378
8.6.4 Experimental procedures and precautions 381
8.6.5 Questions 383
8.7 Drying experiment 384
8.7.1 Experimental objectives 384
8.7.2 Experimental principles 385
8.7.3 Experimental device 387
8.7.4 Experimental procedures and precautions 387
8.7.5 Questions 389

Chapter 9
Self-designed Experiments of Chemical Engineering Principle
391

9.1 Purpose of self-designed experiments 391
9.2 Requirements of self-designed experiments 392
9.3 Equipment for self-designed experiments 392
9.3.1 Self-designed experiment of fluid flow resistance 392
9.3.2 Self-designed experiment of pump performance testing 393
9.3.3 Self-designed experiment of enhanced heat transfer 394
9.3.4 Self-designed experiment of shell-and-tube heat exchanger 395

Chapter 10
Advanced Experiment and Practical Training

397

10.1 Advanced experiment of liquid-liquid extraction 397
10.1.1 Experimental objectives 397
10.1.2 Experimental principles 397
10.1.3 Experimental device 399
10.1.4 Experimental procedures and precautions 399
10.1.5 Questions 404
10.2 Multi-effect evaporation comprehensive training 404
10.2.1 Training objectives 404
10.2.2 Basic principles 405
10.2.3 Process, main devices and instruments 406
10.2.4 Training projects 407
10.3 Absorption-desorption comprehensive training 410
10.3.1 Training objectives 410
10.3.2 Basic principles 410
10.3.3 Process, main equipment and instruments 411
10.3.4 Training projects 411

附录

416

参考文献

421

绪　论

化工原理实验是化工、制药、环境、食品、生物工程等院系或专业的一门必修课程，是化工原理课程体系中一个非常重要的实践教学环节，是理论与实践互通的桥梁。化工原理实验属于工程实验范畴，每个实验项目都相当于化工生产中的一个单元操作，学生通过实验能建立起一定的工程概念。同时，随着实验课的进行，会遇到大量的工程实际问题，因此可以在实验过程中更实际、更有效地学到更多工程实验方面的原理及测试手段；可以初步认识化工各单元操作设备的结构，掌握其开停车、稳定运行等操作方法；可以发现复杂的真实设备与工艺过程同描述这一过程的数学模型之间的关系；可以学会使用现代工具如先进的仪器仪表和计算机软件来分析测试和处理数据；还可以掌握应用因次分析法、参数合并法、当量法等工程研究方法解决复杂工程问题的基本技能。因此，在化工原理实验课的全过程中，学生可在基础理论、工程素质、问题分析、实验研究、思维方法和创新能力等方面都得到培养和提高，为今后的学习和工作打下坚实的基础。

0.1　化工原理实验的教学目的

化工原理实验的教学目的主要有以下几点。

(1) 巩固和深化理论知识

在学习化工原理基本理论的基础上，通过一些比较典型的、已被或将被广泛应用的化工过程与设备操作，开展相关实验研究，进一步巩固和深化所学的化工原理理论知识。

(2) 提供一个理论联系实际的机会

每套实验装置都是小型化的工厂操作单元，通过实验接触工程装置，应用所学的化工原理等化学化工的理论知识去解决实验中遇到的各种实际问题，去领悟工程实验的精髓。

(3) 培养从事工程实验研究的能力

① 为完成特定的研究课题，应用工程研究方法设计实验方案的能力；

② 在实验过程中，观察和分析实验现象的能力及解决实验问题的能力；

③ 正确选择和使用测量仪表的能力，以及团队协作自主完成实验的能力；

④ 利用实验的原始数据，选择和应用恰当的方法、软件进行数据处理以获得实验结果的能力；

⑤ 运用文字、图表等正确完整地撰写实验报告或研究报告的能力等。

(4) 提高自身综合素质和水平

通过一定数量和不同层次的实验训练，掌握各种实验技能，学会通过实验获取新的知识和信息的方法，提高自身综合素质和水平。

0.2　化工原理实验的教学特点

本课程内容强调实践性和工程观念，并将能力和素质培养贯穿于实验课的全过程。围绕化工原理课程的基本理论，开设有演示实验，虚拟仿真实验以及设计型、研究型、综合型“三型”实验，培养学生掌握实验研究的方法，训练其独立思考、综合分析问题和解决问题的能力。

除完成实验教学基本内容外，还可为有学习潜力的学生提供实验场所，提供基础零部件，供学生进行自组装实验，学生也可以预约参与提高型实验或实训实验。以上实验可培养学生独立自主完成工程实验研究、实践动手和探究学习的能力。

本课程的部分实验报告采用小论文形式撰写，这种类型实验报告的撰写是提高学生写作能力、综合应用知识能力和科研能力的一个重要手段，可为毕业论文环节、今后工作所需的科学研究和科学论文的撰写打下坚实的基础。

0.3 化工原理实验的教学内容与方法

工程实验不仅是一项技术工作，而且是一门重要的技术学科，有其自身的特点和系统。因此，我们将化工原理实验单独设课。化工原理实验教学内容主要包括实验理论教学、演示实验及虚拟仿真实验和单元操作实验三大部分。

(1) 实验理论教学

主要讲述化工原理实验教学的目的、要求和方法，化工原理实验的特点，化工原理实验的研究方法，实验数据的误差分析，实验数据的处理方法以及实验的安全操作规程等。

(2) 演示实验及虚拟仿真实验

演示实验和虚拟仿真实验采用开放式教学，由学生在课外时间独立完成。虚拟仿真实验包括仿真运行、数据处理和实验测评三部分。

(3) 单元操作实验

单元操作实验分为必修实验和选修实验。必修实验项目主要包括流体流动阻力实验、离心泵实验、过滤实验、传热实验、吸收实验、精馏实验、干燥实验等“三型”实验和液液萃取提高型实验。每个实验均安排预习和实验操作两个环节。化工原理实验工程性较强，有许多问题需要事先考虑、分析，并做好必要的准备。因此，必须在实验操作前进行预习和完成演示实验及虚拟仿真实验。

选修实验供学有余力或感兴趣的学生选择，包括流体流动阻力、流体输送机械特性、传热过程强化、列管换热器自组装实验和吸收解吸、多效蒸发综合实训实验。学生自行设置实验研究目标、内容、操作规程，经过教师审核后独立完成实验项目。

0.4 化工原理实验的教学环节和要求

化工原理实验的教学环节主要有：实验预习（包括演示实验和虚拟仿真实验）、实验操作、数据测试、数据处理和实验报告撰写。各个环节的具体要求如下。

(1) 实验预习环节

要完成实验任务，仅仅掌握实验原理是不够的，还必须做到以下几点。

① 认真阅读实验讲义，复习《化工原理》教材以及参考书的有关内容，撰写实验预习报告，为培养能力，应试图对每个实验提出问题，带着问题上预习课。

② 到实验室现场熟悉设备装置的结构和流程。

③ 明确操作程序与所要测定的参数，了解相关仪表的类型和使用方法，以及参数的调整、实验测试点的分布等。

④ 完成演示实验和虚拟仿真实验。

(2) 实验操作环节

一般以3～4人为一小组合作进行实验。进入实验室之前必须了解相关的用电、防火、防爆和防毒等安全知识，进入实验室后要观察配电箱、灭火器、防护用品和应急出口的位置。组员之间要做到既分工又合作，每个组员要各负其责，并且要在适当的时候进行轮换工作，这样既能保证质量，又能获得全面的训练。实验操作注意事项如下。

① 实验设备的开车操作应按教材说明的程序逐项进行。设备开车前必须检查设备和管道上各个阀门的开、关状态是否合乎要求，检查合格并征得实验指导教师同意方可开始操作。

② 操作过程中设备及仪表有异常情况时，应立即按停车步骤停车并报告指导教师，同时应利用这个时机研究产生异常情况的原因，因为这是分析问题和处理问题的极好机会。

③ 操作过程中应随时观察仪表指示值的变动，确保操作在稳定条件下进行。出现不符合规律的现象时，应注意观察研究，分析其原因，不要轻易放过。

④ 实验过程注意观察各种现象和所测量数据的变化，认真思考其内在规律和变化原因，并正确、完整、规范地做好数据记录。

⑤ 停车时应严格按操作说明依次关闭有关的气源、水源、电源等，并将各阀门恢复至实验前所处的位置（开或关）。

(3) 数据测试环节

1) 确定要测试哪些数据　凡是与实验结果有关或是整理数据时必需的参数都应一一测定。原始数据记录表的设计应在实验前完成。原始数据应包括工作介质性质、操作条件、设备几何尺寸及环境条件等。

2) 实验数据的分割　一般来说，实验要测的数据尽管有许多个，但常常选择其中一个数据作为自变量来控制，而把其他受其影响或控制的并随之而变的数据作为因变量，如离心泵特性曲线就把流量作为自变量，而把其他同流量有关的扬程、轴功率、效率等作为因变量。在数据处理时，又往往把这些所测的数据标绘在各种坐标系上，为了使数据点在图上分布均匀，这就涉及实验数据点均匀分割的问题。化工原理实验最常用的有直角坐标系和双对数坐标系，坐标系不同所采用的分割方法也不同。其分割值x与实验预定的测定次数n，及其最大、最小的控制量x_{max}和x_{min}之间的关系如下。

① 对于直角坐标系

$$x_i=x_{min} \qquad \Delta x=\frac{x_{max}-x_{min}}{n-1} \qquad \Delta x_{i+1}=x_i+\Delta x$$

② 对于双对数坐标系

$$x_i=x_{min} \qquad \lg\Delta x=\frac{\lg x_{max}-\lg x_{min}}{n-1}$$

$$\Delta x=\left(\frac{x_{max}}{x_{min}}\right)^{\frac{1}{n-1}} \qquad x_{i+1}=x_i \cdot \Delta x$$

3) 读数与记录

① 待设备各部分运转正常，操作稳定后才能读取数据。如何判断是否已达稳定？一般经两次测定，其读数相同或十分相近，即可认为已达稳定。当变更操作条件后，各项参

数达到稳定需要一定的时间，因此也要待其稳定后读数，否则易造成实验结果无规律，甚至反常。

② 同一操作条件下，不同数据最好由数人同时读取，若操作者同时兼读几个数据，应尽可能快速、准确地读取各个数据。

③ 每次读数都应与其他有关数据及前一点数据对照，看看相互关系是否合理。如不合理应查找原因，重新测试，并在实验记录上注明。

④ 所记录的数据应是直接读取的原始数值，不要经过运算后记录，如秒表读数 1 分 23 秒应记为 1′23″，不要记为 83″。

⑤ 读取数据应根据仪表的精度，读至仪表最小分度以下一位数，这个数为估读值。如水银温度计最小分度为 0.1℃，若水银柱恰指 22.4℃时，应记为 22.40℃。注意，过多取估读值的位数是毫无意义的。

有些参数在读数过程中波动较大，首先要设法减小其波动。在波动不能完全消除的情况下，可取波动的最高点与最低点两个数据，然后取平均值。

⑥ 不能凭主观臆测修改数据，也不要随意舍弃数据。对可疑数据，除有明显原因，如读错、误记等情况使数据不正常可以舍弃之外，一般应在数据处理时检查并处理。

⑦ 数据记录完毕要仔细检查一遍，看有无漏记或记错处，特别要注意仪表上的计量单位。实验完毕，须将原始数据记录表格交实验指导教师检查并签字，确认准确无误后方可结束实验。

(4) 数据处理环节

原始记录只可整理，绝不可以随便修改，经判断确实为过失误差造成的不正确数据须注明后，方可剔除不计入结果。数据处理过程应有计算示例，数据处理结果可以用列表法、图示法或回归分析法来说明，但均要标明实验条件。列表法、图示法和回归分析法详见第 3 章实验数据处理。

(5) 实验报告撰写

实验报告是实验工作的全面总结和系统概括，是实验环节中不可缺少的一个重要组成部分。化工原理实验具有显著的工程性，属于工程技术科学的范畴，它研究的对象是复杂的实际问题和工程问题，因此化工原理的实验报告可以按传统实验报告格式或小论文格式撰写。

1) 传统实验报告格式

① 封面：主要内容有实验名称，报告人姓名、学号、班级及同组实验人姓名，实验地点，指导教师，实验日期等。

② 实验目的：简明扼要地说明为什么要进行本实验，实验要解决什么问题。

③ 实验原理：即实验的理论依据。简要说明实验所依据的基本原理，包括实验涉及的主要概念，实验依据的重要定律、公式及据此推算的重要结果。实验的理论依据要求准确、充分。

④ 实验装置与流程图：简单地画出实验装置流程示意图和测试点、控制点的具体位置，标出设备、仪器仪表及调节阀等的标号，在流程图的下方写出图名及与标号相对应的设备、仪器等的名称。

⑤ 实验步骤：根据实际操作顺序将实验过程划分为几个步骤，并在每个操作步骤前面加上序号，以使条理更为清晰。对于操作过程的说明应简单明了。

⑥ 注意事项：对于容易引起设备或仪器仪表损坏，容易发生危险，以及一些对实验结果影响比较大的操作，应在注意事项中注明，以引起注意。

⑦ 原始数据记录：通常采用列表法记录实验过程中从测量仪表上所读取的数值。

⑧ 数据处理：数据处理是实验报告的重点内容之一，要求将实验原始数据进行整理、计算。数据处理计算过程要举例说明，例如，以某一组原始数据为例，把各项计算过程列出（以此说明数据结果表中的数字是如何得到的），将计算结果制作出便于分析讨论的表格、图或回归出变量之间的关系式。表格要易于显示数据的变化规律及各参数的相关性，图要能直观地表达变量间的相互关系。

⑨ 实验结果的分析与讨论：实验结果的分析与讨论是实验人员理论水平的具体体现，也是对实验方法和结果进行的综合分析与研究，是工程实验报告的重要内容之一。其主要内容包括以下几方面。

ⅰ. 从理论上对实验所得结果进行分析和解释，说明其必然性；

ⅱ. 对实验中的现象进行分析讨论，说明其内在原因以及主要影响因素；

ⅲ. 分析误差的大小和原因，指出提高实验质量的途径；

ⅳ. 将实验结果与前人和他人的结果对比，说明结果的异同，并解释这种异同；

ⅴ. 本实验结果在生产实践中的价值和意义、推广和应用效果的预测等；

ⅵ. 由实验结果提出进一步的研究方向或对实验方法及装置提出改进建议等。

⑩ 实验结论：结论是根据实验结果所作出的最后判断，得出的结论要从实际出发，要有理论依据，简明扼要。

⑪ 参考文献：同以下小论文格式部分。

2）小论文格式　科学论文有其独特的写作格式，其构成常包括以下部分：标题、作者、单位、摘要、关键词、前言（或引言、序言）、正文、结论、致谢、参考文献等。具体可参考国家标准《学位论文编写规则》(GB/T 7713.1—2006)。

① 标题：标题又叫题目，它是论文的总纲，是文献检索的依据，是全篇文章的实质与精华，也是引导读者判断是否阅读该文的一个依据，因此要求标题能准确地反映论文的中心内容。

② 作者和单位：署名作者只限于那些选定研究课题和制订研究方案，直接参加全部或主要部分研究工作并作出主要贡献，以及参加撰写论文并能对内容负责的人，按其贡献大小排列名次。工作单位写在作者名下。

③ 摘要：撰写摘要的目的是让读者一目了然地了解本文研究了什么问题，用什么方法，得到什么结果，这些结果有什么重要意义。摘要是对论文内容不加注解和评论的概括性陈述，是全文的高度浓缩。摘要一般以几十个字至三百字为宜。

④ 关键词：关键词是从论文中选出起关键作用的、最说明问题的、代表论文内容特征的或最有意义的单词或术语，以便于检索。可选 3～8 个关键词。

⑤ 前言：前言是论文主体部分的开端。简要说明研究工作的目的、范围、相关领域的前人工作和知识空白、理论基础和分析、研究设想、研究方法（前言中提及方法的名称即可，无须展开细述）、预期结果和意义等。前言应言简意赅，不要与摘要雷同。比较短的论文用一小段文字作简要说明，则不用“引言”或“前言”两字。

⑥ 正文：这是论文的核心部分，占主要篇幅，主要包括实验方法、仪器设备、材料原料、实验和观测结果、计算方法、数据资料、经过加工整理的图表、形成的实验结果以

及对实验结果的分析讨论等。这一部分的形式主要根据作者意图和文章内容决定，不可能也不应该规定一个统一的形式，但必须实事求是、客观真切、准确完备、合乎逻辑、层次分明、简练可读。本部分可根据论文内容分成若干个标题来叙述。

需要强调的是实验结果与分析讨论是正文的重点，是结论赖以产生的基础。需对数据处理的实验结果进一步加以整理，从中选出最能反映事物本质的数据或现象，并将其制成便于分析、讨论的图或表。在结果与分析中既要包含所取得的结果，还要说明结果的可信度、再现性、误差、与理论或分析结果的比较，以及经验公式的建立等。

⑦ 结论：结论是在理论分析和计算结果（实验结果）中分析和归纳出的观点，它是以结果和讨论（或实验验证）为前提，经过严密的逻辑推理做出的最后判断，是整个研究过程的结晶，是全篇论文的精髓，据此可以看出研究成果的水平。可以在结论或讨论中提出建议、研究设想、仪器设备改进意见、尚待解决的问题等。

⑧ 致谢：致谢的作用主要是为了表示尊重所有合作者的劳动。致谢对象包括除作者以外所有对研究工作和论文写作有贡献、有帮助的人，如指导过论文的专家、教授，帮助搜集和整理过资料的人，对研究工作和论文写作提过建议的人等。

⑨ 参考文献：参考文献反映作者的科学态度和研究工作的依据，也反映出该论文的起点和深度。可提示读者查阅原始文献，同时也表示作者对他人成果的尊重。参考文献的著录方法采用顺序编码制。顺序编码制是指作者在论文中所引用的文献按它们在文中出现的先后顺序，用阿拉伯数字加方括号连续编码，视具体情况把序号作为上角或作为语句的组成部分进行标注，并在文后参考文献表中，各条文献按在论文中出现的文献序号顺序依次排列。具体可参考国家标准《信息与文献 参考文献著录规则》（GB/T 7714—2015）。

例如：被引用的文献为期刊论文的单篇文献时，著录格式为：“［序号］主要责任者. 文献题名［J］. 刊名，年，卷（期）：起止页码.”。

被引用的文献为图书时，著录格式为：“［序号］主要责任者. 文献题名［M］. 版本（第一版本不标注）. 出版地：出版者，出版年：起止页码.”。

被引用的文献为电子文献时，著录格式为：“［序号］主要责任者. 电子文献题名［电子文献及载体类型标识］. 电子文献的出版或获得地址，发表更新日期/引用日期.”。

⑩ 附录：附录是在论文末尾作为正文主体的补充项目，并不是必需的。对于某些数量较大的重要原始数据、计算程序、篇幅过大不便于作为正文的材料、对专业同行有参考价值的资料等，可作为附录放在论文的最后（参考文献之后）。每一附录均应另页。

⑪ 外文摘要：对于正式发表的论文，有些刊物要求要有外文摘要。通常是将中文标题（title）、作者（author）、摘要（abstract）及关键词（key words）译为英文。其排放位置因刊物而异。

用论文形式撰写化工原理实验实验报告，是一种综合素质和能力培养的重要手段，应提倡这种形式的实验报告，尤其对于研究型实验、自组装实验和实训实验。但无论何种形式的实验报告，均应体现出它的学术性、科学性、理论性、规范性、创造性和探索性。

0.5 化工原理实验的基本安全知识

化工原理实验装置集电器、仪表和机械传动设备等于一体，且大部分化工原理实验还需要使用化学药品，因此，进入化工原理实验室之前必须了解相关的水、电、气的安全使用以及防火、防爆和防毒等安全知识，严格遵守实验守则。

0.5.1 实验守则

① 学生必须按时到指定实验室做实验，无正当理由不得迟到、早退和旷课。

② 实验前学生应做好预习，明确实验的目的、要求，掌握实验的内容、方法和步骤，了解实验设备的基本情况，接受指导教师提问检查。经检查合格后，方可进行实验。

③ 进入实验室前，学生必须穿好实验服，着装规范（例如不得穿背心、短裤、裙子和拖鞋等），留长发者必须将头发盘起，实验期间不得进行与实验无关的其他活动，不得随意触碰与实验无关的设备或仪表开关，实验过程必须保持安静。

④ 要按实验规程正确操作，仔细观察，真实记录实验数据，不得修改原始记录，不抄袭他人数据，不得擅自离开操作岗位或干扰他人实验。

⑤ 实验过程中要注意安全，应根据实验性质选择佩戴相应的防护用品（例如护目镜、手套等），出现意外事故时要保持镇静，并迅速采取措施（如切断电源、气源，化学灼伤时迅速用水冲洗等），防止事故扩大，并及时向指导教师报告。

⑥ 实验过程中，如发现仪器设备发生故障或损坏，应及时报告指导教师处理，不得让有问题的设备继续运转，以免发生事故。

⑦ 学生进入开放实验室做实验前，应事先和有关实验室负责人联系（上网预约或直接与实验室负责人预约），报告自己的实验目的、内容和所需实验仪器。经审核同意后，方可在实验室负责人安排的时间内进行。

⑧ 要爱护仪器设备，节约用水、用电、用气和实验材料。实验完成后，必须将实验数据交给指导教师签字确认，并整理好所使用的仪器设备，清理实验场地，关闭电源、气源等。经指导教师检查合格后，方可离开实验室。

0.5.2 基本安全知识

化工原理实验过程中都会涉及水、电、气和化学药品等。每个环节都可能存在不安全因素，任何疏漏和差错，均可能酿成安全事故。因此实验室全体人员均须熟悉与掌握有关安全知识和事故的防范技术，并具备一定的安全事故处理技能。本节列出的是与化工原理实验相关的最基本的安全知识，由于每个实验的性质不同，对有特殊安全要求的实验，将在每个实验的注意事项中列出。

0.5.2.1 安全用电

(1) 安全用电规定

化工原理实验中电气设备较多，安全用电十分重要，在实验开始前、实验过程中和实验结束后都要严格遵守以下规定。

1）在实验开始之前必须做到如下几点。

① 了解实验室内总电闸的位置，以便出现用电事故时能及时切断各电源。

② 检查电气设备和电路是否符合规定要求。检查用电导线有无裸露，电气设备是否漏电，是否有保护接地或保护接零措施。

③ 在接通电源之前，必须十分清楚实验装置的启动和停车操作顺序，以及紧急停车的方法。

2）在实验过程中必须做到如下几点。

① 电气设备要保持干燥清洁。所有电气设备带电时不可以用湿布擦拭，更不能有水落于其上。操作者的双手也必须是干燥的。

② 实验过程中若出现跳闸现象或发现保险丝熔断，应立即检查实验设备是否有问题，切忌不经检查便换上保险丝或保险管就再次合闸，否则可能会造成设备的损坏。

③ 电源或实验设备上的保险丝或保险管，都应按规定电流标准使用。严禁私自加粗保险丝或用其他金属丝代替。

④ 若实验设备发生过热现象或有煳焦味时，应立即切断电源。

⑤ 设备维修时必须停电作业。

3）实验结束后必须做到如下几点。

① 把实验设备上的加热开关、泵或风机开关、仪表开关、总电源开关等按顺序关闭，尤其要注意切断加热电气设备的电源开关。

② 离开实验室前必须把本实验室的总电闸拉下断电。

③ 若发生停电情况，必须拉下所有的电闸，防止操作人员离开现场后，因突然供电而导致实验设备在无人监视下运行。

（2）触电急救原则

1）迅速脱离电源　使伤者脱离电源的方法：a. 切断电源开关；b. 若电源开关较远，可用干燥的木棍、竹竿等挑开触电者身上的电线或带电设备；c. 可用几层干燥的衣服将手包住，或者站在干燥的木板上，拉触电者的衣服，使其脱离电源。注意：触电者未脱离电源前，救护人员不得用手直接触及触电者。

2）就地急救处理　触电者脱离电源后，应观察其神志是否清醒。神志清醒者，应使其就地躺平，严密观察，暂时不要站立或走动；如神志不清，应就地仰面躺平，且确保气道通畅，并以 5s 时间间隔呼叫触电者或轻拍其肩膀，以判定触电者是否意识丧失。禁止摇动触电者头部呼叫触电者。联系医疗部门。

3）准确地使用人工呼吸　检查触电者的呼吸和心跳情况，呼吸停止或心脏停搏时应立即实施心肺复苏的正确抢救方法（心脏按压和人工呼吸），并尽快联系医疗部门救治。

4）坚持抢救　坚持就是触电者复生的希望，百分之一的希望也要尽百分之百的努力。

0.5.2.2　化学品安全

1）在进入实验室实验之前，要知道本次实验需要用到的化学品名称，根据名称，上网或查相关手册，了解该化学品的性质和使用注意事项等。例如查 MSDS（Material Safety Data Sheet），即化学品安全数据说明书。它提供化学品的理化参数、燃爆性能、对健康的危害、安全使用和贮存、泄漏处置、急救措施以及有关的法律法规等十六项内容。通过学习相关资料掌握化学品的正确使用方法和急救措施。

2）每间实验室均配备有洗眼器和紧急喷淋装置，实验前要学会其正确的使用方法。

3）实验室要保持通风。

4）在实验过程中根据化学品的性质，选择合适的防护用品，例如护目镜和防护手套等。但要注意没有一种材质的手套能够防护所有的化学物质，天然橡胶对于稀的水溶液具有很好的防护作用，但是却容易被油、脂和许多有机溶剂穿透。丁腈橡胶手套可以用来防护油、脂类的物质，但却不能防护芳香族物质或含卤素的溶剂，所以应合理选择。

5）针对化工原理实验所用到的化学药品，若遇紧急情况按下列方法处置。

① 强酸、强碱及其他一些化学物质，具有强烈的刺激性和腐蚀性，被这类物质灼伤

时，应用大量流动清水冲洗，再分别用低浓度的（2%～5%）弱碱（强酸引起的）、弱酸（强碱引起的）进行中和，而后就医。

② 若酸（或碱）不慎溅入眼内，应立即就近用大量的清水或生理盐水冲洗。冲洗时，眼睛置于洗眼装置上方，水向上冲洗眼睛，冲洗时间应不少于15min，切不可因疼痛而紧闭眼睛。处理后，再送眼科医院治疗。

③ 实验中若感觉咽喉灼痛，可能为吸入刺激性气体，应立即将患者转移到安全地带，解开领扣，使其呼吸通畅并呼吸到新鲜空气，并及时就医。

0.5.2.3 高压气瓶

高压气瓶是一种贮存各种压缩气体或液化气体的高压容器，标准高压气瓶按国家标准制造，并经有关部门严格检验方可使用。气瓶盛装的气体通常分为易燃气体（包括氢气、甲烷、乙烯、丙烯、乙炔、甲醚、液态烃、氯甲烷、一氧化碳等）、助燃气体（包括氧气、压缩空气、氯气等）、不燃气体（包括氮气、二氧化碳、氖、氩等）和有毒气体（包括氨气、氯气、硫化氢等）。各类气瓶外表面均涂敷有一定颜色的涂料，主要是能从颜色上迅速辨别气瓶中所贮存气体的种类，以免混淆。《气瓶颜色标志》（GB/T 7144—2016）规定了气瓶外表面的涂敷颜色、字样、字色、色环、色带和检验色标等要求。表0-1列出了化工原理实验常用气瓶的颜色及标识。

表0-1 常用气瓶的颜色及标识

气体种类	气瓶颜色	字样	字色	阀门出口螺纹
氧	淡(酞)蓝	氧	黑	正扣
氢	淡绿	氢	大红	反扣
氮	黑	液氮	白	正扣
氨	银灰	液氨	深绿	正扣
空气	黑	空气	白	正扣
二氧化碳	铝白	液化二氧化碳	黑	正扣
氨	淡黄	液氨	黑	正扣

使用气瓶的主要危险是气瓶可能爆炸和漏气，已充气的气瓶爆炸的主要原因是气体受热膨胀，压力超过气瓶的最大负荷；或是瓶颈螺纹损坏，当内部压力升高时，冲脱瓶颈，在这种情况下，气瓶会向放出气体的相反方向高速飞行；另外气瓶坠落或撞击坚硬物时就会发生爆炸，这些均可造成很大的破坏和伤亡事故。因此，使用时应注意以下事项。

① 不同种类的气瓶要严格按照国家标准或行业内部标准分类存放。例如易燃和助燃气瓶必须隔间存放，严禁混放一处（如氢气钢瓶和氧气钢瓶）；氢气瓶应单独存放，最好放置在室外专用的小屋内，若因需要在实验室内使用氢气瓶，其数量不得超过2瓶，且保证空气中氢气最高含量不得超过1%（体积分数）。气瓶放置地点应避免曝晒和强烈震动，不得靠近热源和明火，并应保证气瓶瓶体干燥。

② 涉及有毒、易燃易爆气体的场所，必须配有通风设施和监控报警装置，张贴安全警示标识。易燃、助燃气瓶与明火距离不得小于10m。

③ 各种气瓶必须定期进行检验。在气瓶肩部，有用钢印打出的标记：制造厂、制造日期、气瓶型号、工作压力、气压试验压力、气压试验日期及下次送验日期、气体容积、

气瓶重量等，如在使用中发现有严重腐蚀或严重损伤的，应提前进行检验。瓶阀发生故障时，不要擅自拆卸瓶阀或瓶阀上的零件。此外，投入使用的气瓶还要有充气厂家出示的合格证书。

④ 移动气瓶时，必须佩戴好气瓶瓶帽和防震圈，最好用特制的担架或小推车搬运，也可以用手抬或垂直转动，但绝不允许手握气瓶瓶阀移动。

⑤ 气瓶直立放置时要固定稳妥，根据气瓶形状采用适当的安全装置和防倾倒装置。

⑥ 所有气瓶必须装有减压阀，安装时螺扣要旋紧，防止泄漏。选用的减压阀要分类专用，不得混用。减压阀的高压腔与气瓶连接，低压腔为气体出口，并通往使用系统。高压表的示值为气瓶内贮存气体的压力，低压表的出口压力可由调节螺杆控制。顺时针转动低压表压力调节螺杆，使其压缩主弹簧并传动薄膜、弹簧垫块和顶杆而将活门打开，转动调节螺杆，改变活门开启的高度，从而调节高压气体的通过量并达到所需的压力值，这样进口的高压气体由高压室经节流减压后进入低压室，并经出口通往使用系统。

⑦ 在打开气瓶总开关阀前切记：要先将减压阀的调节螺杆沿逆时针方向旋松。使用时应先旋动气瓶总开关阀，后开减压阀；使用完毕后，先关闭总开关阀，放尽余气后，再关闭减压阀。切不可只关减压阀，不关总开关阀。

⑧ 开高压气瓶总开关阀或调减压阀时，操作人员不能站在气体出口的前方，应站在与气瓶接口处垂直的位置上，以防阀门或减压表冲出伤人。操作时严禁敲打撞击，并经常检查有无漏气，应注意压力表读数。

⑨ 氧气瓶或氢气瓶等严禁与油类接触，并绝对避免让其他可燃性气体混入氧气瓶；操作人员不能穿戴沾有各种油脂或易感应产生静电的服装、手套操作，以免引起燃烧或爆炸。

⑩ 用后的气瓶，应按规定留 0.05MPa 以上的残余压力。可燃性气体应剩余 0.2～0.3MPa，氢气应保留 2MPa，以防重新充气时发生危险，不可用完用尽。需返回厂家充气的气瓶要贴上空瓶的标志。

0.5.2.4 消防知识

（1）常用消防器材

实验室按规定配备有一定数量的消防器材，实验者应熟悉本实验室消防器材的存放位置、种类和使用方法，还应当熟悉几条不同方向的逃生路线。需要特别强调的是，在同一大类的消防器材中，由于型号、式样、内容物等的不同，其使用方法也会有所不同。例如化学泡沫灭火器使用时要倒置，空气泡沫灭火器使用时不能倒置，干粉灭火器有外挂式贮压式、内置式贮气瓶式、开启有提环式、手轮式、压把式等，所以进入实验室前要有针对性地学习其使用方法。

1）消防沙箱　沙子能隔断火焰和空气并降低火焰温度，从而灭火。易燃液体和其他不能用水灭火的危险品着火可用沙子来扑灭。需要注意的是，沙子中不能混有可燃性杂物，并且要干燥，潮湿的沙子遇火后可能因水分蒸发，而使燃着的液体飞溅。因为沙箱中存沙量有限，所以只能扑灭局部小规模的火灾。

2）石棉、灭火毯或湿布　灭火原理是隔绝空气以达到灭火的目的。适用于迅速扑灭火源面积不大的火灾，也是扑灭衣服着火的常用方法。

3）二氧化碳灭火器　二氧化碳灭火器筒内充装有液态二氧化碳。二氧化碳气体可以排除空气而包围在燃烧物体的表面或分布于较密闭的空间中，降低可燃物周围或防护空间内的氧浓度，产生窒息作用而灭火。另外，二氧化碳从贮存容器中喷出时，会由液体迅速汽化成气体，而从周围吸收部分热量，起到冷却的作用。

使用时，先拔出保险销，再压合压把，将喷嘴对准火焰根部喷射。注意：手要放在钢瓶的木柄上，防止冻伤，还要防止现场人员窒息。

主要用于扑救贵重设备、档案资料、仪器仪表、600V 以下电气设备及油类的初期火灾。

4）泡沫灭火器　实验室多用手提式化学泡沫灭火器，其外壳是铁皮制成的，内装碳酸氢钠与发泡剂的混合溶液，另有一玻璃瓶内胆，装有硫酸铝水溶液。使用时将灭火器筒身倒置，碳酸氢钠和硫酸铝两溶液混合后发生化学作用，产生二氧化碳气体泡沫。喷射出的大量二氧化碳气体泡沫，它们能黏附在可燃物上，使可燃物与空气隔绝，并降低温度，达到灭火的目的。

使用时，不要提早倒置以防提前喷出，当距离着火点 10m 左右时，即可将筒体颠倒过来，一只手紧握提环，另一只手扶住筒体的底圈，用力摇晃几下，将射流对准燃烧物。灭火器应始终保持倒置状态，否则会中断喷射。

可用于扑灭固体物质类火灾，如木材、棉、纸张等；液体类火灾，如汽油、柴油等；但不可用于可燃、易燃液体的火灾，如醇、酯、醚、酮等。不可用于气体、金属、精密仪器火灾，也不可用于扑灭带电设备的火灾，因为泡沫本身是导电的，会造成灭火人的触电事故，应切断电源后再灭火。

5）干粉灭火器　干粉灭火器内部装有磷酸铵盐等干粉灭火剂，这种干粉灭火剂具有易流动性、干燥性，由无机盐和粉碎干燥的添加剂组成。干粉灭火器是利用二氧化碳气体或氮气气体作动力，将筒内的干粉喷出灭火的。灭火原理为干粉中无机盐的挥发性分解物与燃烧过程中燃料所产生的自由基或活性基团发生化学抑制和负催化作用，使燃烧的链反应中断而灭火，干粉落在可燃物表面，发生化学反应，并在高温作用下形成一层玻璃状覆盖层，从而隔绝氧，进而窒息灭火，另外，还有部分稀释氧和冷却的作用。

干粉灭火器最常用的开启方法为压把法，将灭火器提到距火源适当位置后，先上下颠倒几次，使筒内的干粉松动，然后让喷嘴对准燃烧最猛烈处，拔去保险销，压下压把，灭火剂便会喷出灭火。若使用外装式手提灭火器时，一只手握住喷嘴，另一只手向上提起提环，干粉即可喷出。

干粉灭火器可扑灭一般火灾，还可扑灭石油、有机溶剂等易燃液体和可燃气体以及电气设备的初期火灾。

(2) 火灾应急处理

① 局部起火，根据起火的原因，立即选用正确的灭火器材灭火；发生大面积火灾时，实验人员应通知所有人员沿消防通道紧急疏散并立即向消防部门报警，同时向学院领导报告。

② 有人员受伤时，立即向医疗部门报告，请求支援。人员撤离到预定地点后，实验指导教师、实验室工作人员、学生干部立即组织清点人数，尽快确认未到人员所在位置。

③ 逃离火场时若遇浓烟，应尽量用多层湿布捂住口鼻，并放低身体或是爬行，千万不要直立行走，以免被浓烟熏呛而窒息。

④ 逃离火场时切勿使用电梯，应尽快沿着安全出口方向离开火场，并到空旷处汇合，

未得到许可不得擅自返回火场。

0.6 任务书

0.6.1 阻力实验

实验任务书（1）
直管摩擦阻力损失随 *Re* 变化规律的研究

任务要求

① 设计实验方案，测定水在圆形直管（包括光滑管与粗糙管）中层流流动、湍流流动时的阻力损失。

② 根据实验结果，探讨直管摩擦阻力损失、摩擦系数 λ 随 Re 变化的规律。

③ 将实验结果与莫狄（Moody）摩擦系数图比较，通过误差分析，探讨其合理性。

④ 分析直管摩擦阻力损失产生的原因，结合实验结果提出工程上减少阻力损失的意义与方法。

⑤ 测定流体流过管件的局部阻力损失，了解局部阻力产生的原因，掌握局部阻力损失的测定和计算方法。

实验任务书（2）
局部阻力损失机理及减少局部阻力损失若干问题的探讨

任务要求

① 设计实验方案，测定管件在某一状态下不同 Re 时的局部阻力损失。

② 分析局部阻力产生的原因。结合实验结果，探讨工程上减少局部阻力损失的意义与方法。

③ 将本实验的计算方法和结果与有关手册查得的管件与阀门的阻力系数、当量长度数据等进行比较，分析误差的原因。

④ 测定光滑管或粗糙管的直管摩擦阻力损失，了解直管阻力产生的原因，掌握直管阻力损失的测定和计算方法。

实验任务书（3）
间接法确定管壁粗糙度的研究

任务要求

① 设计实验方案，测定水在圆形直管（包括光滑管与粗糙管）中层流流动、湍流流动时的阻力损失。

② 分析直管摩擦阻力损失产生的原因及直管摩擦阻力损失随 Re 变化的规律。

③ 探讨测定管壁粗糙度的意义，提出间接测定管壁粗糙度的依据与方法。

④ 将间接测定的管壁粗糙度与实际管壁粗糙度比较，说明该方法的可行性。

⑤ 测定管件局部阻力损失，了解局部阻力产生的原因，掌握局部阻力损失的测定和计算方法。

0.6.2 离心泵实验

实验任务书（1）
离心泵流量调节方式与能耗分析

化工过程中常由于生产任务、工艺条件等发生变化，需要对泵的流量进行调节。离心泵流量调节的方式主要有：出口阀门调节和电机变速调节，以及串、并联调节等。各种调节方式，除了有它们各自的优缺点外，其能量的损耗也不一样。本实验拟通过如下实验任务，寻求适宜的、节能的流量调节方式。

任务要求

① 设计实验方案，测定单台离心泵的特性曲线和高、低阻管路的管路特性曲线。

② 设计实验方案，测定两台同型号离心泵在相同转速下的串联操作和并联操作的扬程曲线。

③ 分析出口阀门调节和电机变速调节，以及串、并联调节适用的场合和优缺点。

④ 根据实验结果，通过作图和计算，对能耗进行分析，得出结论。

实验任务书（2）
离心泵工作点的确定与调节

当离心泵的实际工作点不在高效区时或流量不能满足要求时，需要调整离心泵的工作点。在生产实际中调整工作点通常采用以下两种方式：一是改变管路特性曲线，如调节出口阀门开度等；二是改变离心泵本身的特性曲线，如改变泵的转速、将泵并联或串联等。拟通过实验，完成以下任务。

任务要求

① 设计实验方案，测定单台离心泵的特性曲线和高、低阻管路的管路特性曲线。

② 设计实验方案，测定两台同型号离心泵在相同转速下的串联操作和并联操作的扬程曲线。

③ 通过实验，分析比较出口阀门调节和电机变速调节，以及串、并联调节工作点的特点。

④ 以本实验装置为例，若管路一定，离心泵的实际工作点不在高效区时，你认为采取哪种方式调节较为合理？为什么？

实验任务书（3）
离心泵串、并联组合操作特性研究

在实际生产中用一台离心泵不能满足流量或扬程需求时，常采用两台或两台以上的泵组合操作。通过实验进行离心泵串、并联组合操作特性的研究，掌握双泵组合操作的特点。

任务要求

① 需要对离心泵串、并联操作特性曲线进行分析研究，掌握串、并联泵特性曲线的特点及其与单泵特性曲线的联系，掌握串、并联泵对管路流量和扬程的影响等。

② 根据研究内容，确定需要测定哪些曲线，并制订实验方案，完成实验。

③ 现需将20℃的清水从贮水池输送至水塔，已知塔内水面高于贮水池水面5m。水塔及贮水池水面恒定不变，均与大气相通。输送管为$\phi 32mm \times 2mm$的钢管，总长为10m（或40m）（均包括局部阻力在内，计算时λ取0.02）。试分析（应写出详细的计算过程、分析步骤、列出相关的图表）：

ⅰ. 若采用单泵进行输送，则流量为多少？该泵在运行时的功率、效率为多少？从经济角度来看是否合理？

ⅱ. 若要提高流量，采用哪种操作方式（单泵、串联、并联）更为合理？为什么？

ⅲ. 通过实验研究和对本实例的分析可以得到什么结论？

0.6.3 过滤实验

实验任务书（1）

板框过滤机最佳过滤时间和最大生产能力的研究

在板框过滤机的一个操作循环中，过滤、洗涤、卸渣、重装等操作依次进行。其中卸渣和重装时间是固定的，而过滤和洗涤时间是可以人为选择的。过滤时间过长或过短，都会使生产能力变小，因此，存在着一个使生产能力最大的最佳操作周期。请你根据实验室的设备，完成该课题的研究。

任务要求

① 根据研究主题，设计实验方案。

② 通过实验确定板框过滤机最佳过滤时间和最大生产能力研究所必需的参数，如过滤常数、压缩性指数等。

③ 结合实验现象和数据图表对实验结果进行必要的分析。

④ 利用微分法或积分法处理实验数据，用图解法或解析法确定板框过滤机的最佳过滤时间、生产能力和最大生产能力，并对最佳过滤时间、生产能力和最大生产能力的影响因素进行探讨。

实验任务书（2）

过滤压力对$CaCO_3$悬浮液过滤过程的影响研究

过滤压力是过滤过程中一个十分重要的操作参数，它对过滤速率、过滤常数、过滤机的生产能力以及滤饼的结构都有着重要的影响。因此，利用实验室板框过滤机，探究过滤压力对过滤过程的影响。

任务要求

① 根据研究主题，设计实验方案。

② 通过实验确定过滤常数、压缩性指数等过滤参数。

③ 结合实验现象对不同过滤压力下滤饼空隙率、含水量等参数以及滤饼内部结构进行必要的分析。

④ 利用积分法或微分法处理实验数据，结合图表探讨过滤压力对过滤速率、过滤常数、生产能力等方面的影响。

实验任务书（3）

工业转筒真空过滤机的设计研究

设计工业用过滤机必须先测定有关的过滤参数，这项工作一般是用同一悬浮液在小型过滤实验设备中进行。

现有某工厂需过滤含 $CaCO_3$ 5.0%～5.5% 的水悬浮液，过滤温度为25℃，固体 $CaCO_3$ 的密度为 $2930kg/m^3$。要求工业转筒真空过滤机的操作真空度为0.08MPa，以滤液计的生产能力为 $0.001m^3/s$。

请你利用实验室的小型板框压滤机进行实验，测定有关的过滤参数，确定转筒真空过滤机的转速 n、转筒的浸没度 φ、转筒的直径 D 和长度 L。

完成上述设计任务时可认为滤布阻力不随过滤压力变化，每获得 $1m^3$ 滤液所生成的滤饼体积也无显著变化。

0.6.4 传热实验

实验任务书（1）

螺旋线圈强化传热过程研究

在管内插入螺旋线圈是强化管内流体对流传热的重要措施，管内插入螺旋线圈对传热的强化效果如何？对管内流体的流动阻力又产生怎样的影响？利用套管换热器，完成螺旋线圈强化传热过程研究。

任务要求

① 查阅资料，对管内插入螺旋线圈强化传热的现状与发展作出简要评述。

② 设计实验方案，测定普通套管换热器和管内插入螺旋线圈的套管换热器的传热系数 K 及空气对管壁的对流给热系数 α_1。

③ 用作图法或最小二乘法关联出上述两种换热管中空气侧对流给热系数关联式 $Nu=ARe^mPr^{0.4}$，并确定常数 A、m 的值。

④ 根据实验数据处理结果，比较普通传热管与管内插入螺旋线圈传热管的传热效果，探讨螺旋线圈强化传热的机理。

⑤ 分析管内螺旋线圈的哪些结构尺寸会对传热及流体流动阻力产生影响？分析使用螺旋线圈强化传热的优缺点。

实验任务书（2）

管内流体对流给热系数的测定与列管换热器设计

列管换热器是工业生产中应用广泛的一种换热器，现需要设计一列管换热器，用

110℃ 的水蒸气将空气加热至 90℃，空气来自环境，流量分别为 $5000m^3/h$ 、$8000m^3/h$ 和 $10000m^3/h$ 。请结合实验测量结果实现换热器的准确设计。

任务要求

① 选择实验室中适宜的换热器，设计实验方案，测定空气在管内传热的对流给热系数。

② 根据实验结果，确定管内对流给热系数经验关联式。

③ 根据当地气候条件确定空气进口温度和适宜的空气流速，以上述测定的对流给热系数关联式为基础进行列管换热器的设计，确定列管换热器的主要结构参数，如管数、管径、壁厚、管长等，以及换热器具体型号，并校核空气流动阻力。

实验任务书（3）
管内流体对流给热系数的测定与套管换热器设计

套管换热器具有结构简单、传热推动力大等特点，但其传热面积较小。为提高其传热效果，在管内插入螺旋线圈以强化传热，将此换热器用于预热空气，加热介质为 120℃ 的水蒸气，需将空气加热至 110℃ ，空气来自环境，流量分别为 $80m^3/h$ 、$90m^3/h$ 和 $100m^3/h$ 。

任务要求

① 选择实验室中适宜的换热器，设计实验方案，测定内插螺旋线圈管内空气的对流给热系数。

② 根据实验结果，确定管内对流给热系数经验关联式。

③ 根据当地气候条件确定空气进口温度和适宜的空气流速，以上述测定的对流给热系数关联式为基础进行套管换热器的设计，确定套管换热器的主要结构参数，如内管直径、内管壁厚、管长等，并校核管程流动阻力。

0.6.5 吸收实验

实验任务书（1）
填料吸收塔流体力学性能及传质性能研究

吸收实验装置由陶瓷拉西环吸收塔和不锈钢鲍尔环解吸塔构成，利用该装置完成以下流体力学性能及传质性能的测定。

任务要求

① 在解吸塔中，测定干塔填料层压降曲线，即空塔气速 u 与每米填料压降 $\Delta p/z$ 之间的关系，关联出 u-$\Delta p/z$ 函数关系并与理论值比较。

② 在解吸塔中，测定喷淋量 $L=100L/h$ 时的湿塔填料层压降曲线，确定该喷淋量下载点、泛点的位置及 3 个流动区域的划分，并深入探讨干塔和湿塔的流体力学性能差异及测定填料层压降曲线的工程意义。

③ 在吸收塔中，控制吸收剂流量为 80～120L/h，进塔混合气中二氧化碳浓度控制在 10%～20%（体积分数），在吸收塔可正常操作前提下，取两个差别较大的液相流量。分

别测定吸收塔的传质单元高度 H_{OL}、传质单元数 N_{OL}、液相总体积吸收系数 $K_x a$、吸收因子 A、气相出口组成 Y_a 和回收率 η，并对吸收剂不同流量下测定的参数以及物料衡算的结果进行分析讨论。

实验任务书（2）
填料塔流体力学特性——压降规律与液泛规律的研究

利用实验室现有的不锈钢鲍尔环填料解吸塔，完成填料塔流体力学特性——压降规律与液泛规律的研究。

任务要求

① 在解吸塔中，测定干塔填料层压降曲线，即空塔气速 u 与每米填料压降 $\Delta p/z$ 之间的关系，关联出 u-$\Delta p/z$ 函数关系并与理论值比较。

② 在解吸塔中，测定喷淋量 $L=0\sim200\text{L/h}$ 范围内若干条湿塔填料层压降曲线，确定各喷淋量下载点、泛点的位置及3个流动区域的划分，并深入探讨干塔和湿塔的流体力学性能差异以及气、液流量对流体力学性能的影响规律。

③ 将液泛数据与埃克特填料塔液泛速度通用关联图比较，分析其误差产生的主要原因。

④ 完成在一定操作条件下，吸收塔总传质系数 $K_x a$ 和回收率 η 的测定。

实验任务书（3）
吸收解吸填料塔适宜操作条件的确定及传质性能研究

利用实验室吸收解吸装置，在吸收塔入塔气体体积流量比为 $V_{CO_2}/V_{空气}=0.2$，吸收率不低于5%及适宜填料润湿率的条件下，确定适宜的操作条件，测定吸收塔液相总传质单元高度、液相总体积传质系数和吸收率。

任务要求

① 通过理论分析，初步确定吸收塔、解吸塔操作条件。

② 制订解吸塔、吸收塔水力学性能实验方案，测定各塔压降曲线，判断吸收塔和解吸塔适宜气、液操作范围，为传质实验奠定基础。

③ 根据上述理论分析和水力学性能实验结果，确定传质性能实验的适宜操作条件。

④ 在适宜操作条件下，测定吸收塔 $K_x a$、H_{OL}、N_{OL}、A、Y_a、η 等参数，并对吸收塔的传质性能，以及是否实现预期目标，未实现目标又该如何调整参数等问题进行讨论分析。

实验任务书（4）
二氧化碳在水中吸收传质速率特征的研究

利用陶瓷拉西环填料吸收塔，测定不同气、液流量下，水吸收空气中二氧化碳这一传质过程传质系数、回收率等参数，分析探讨水吸收空气中二氧化碳传质过程的基本特征和强化传质过程的主要途径。

任务要求

① 混合气流量为 0.8m^3/h、二氧化碳体积分数为 20%，吸收剂流量分别为 80L/h、100L/h 和 120L/h 时，完成吸收塔传质单元高度 H_{OL}、传质单元数 N_{OL}、液相总体积吸收系数 K_xa、吸收因子 A、气相出口组成 Y_a 和回收率 η 的测定与计算。关联回归出 K_xa 与 L 的函数方程，并分析探讨吸收剂流量对吸收传质过程的影响规律。

② 吸收剂流量为 100L/h，二氧化碳体积分数为 20%，混合气流量分别为 0.8m^3/h、1.0m^3/h、1.2m^3/h 时，完成吸收塔传质单元高度 H_{OL}、传质单元数 N_{OL}、液相总体积吸收系数 K_xa、吸收因子 A、气相出口组成 Y_a 和回收率 η 的测定与计算。分析探讨混合气流量对吸收传质过程的影响规律。

③ 结合上述结果，探讨用水吸收空气中二氧化碳这一过程传质速率的特征和速率控制步骤。

④ 探讨强化二氧化碳在水中吸收传质速率的有效途径。

0.6.6 精馏实验

实验任务书（1）
原料浓度对精馏塔操作条件和分离能力的影响研究

对于一给定的精馏塔，冷液进料，由于前段工序的原因，进料浓度发生了变化，直接影响精馏操作。请你根据实验室的设备和物料，完成下列任务。

任务要求

① 从理论上分析，对于已给定的精馏塔，当进料浓度发生变化时，若不改变操作条件，对塔顶和塔釜产品质量有何影响。

② 根据实验室现有条件，拟订改变进料浓度的方法，制订出实验方案（包括实验操作条件、实验操作方法和注意事项等）。

③ 在全回流、稳定操作条件下进行实验，测定全塔理论塔板数、单板效率和全塔效率。

④ 在某一回流比下连续精馏、稳定操作条件下进行实验，测定全塔理论塔板数、单板效率和全塔效率。

⑤ 其他操作条件不变，只改变进料浓度，测定全塔理论塔板数、单板效率和全塔效率。

⑥ 根据实验结果，探讨进料浓度变化对全塔效率和单板效率的影响，以及在进料浓度发生变化时，若要保证塔顶和塔釜产品的质量，可采取的措施。

实验任务书（2）
回流比对精馏塔操作条件和分离能力的影响研究

对于一给定的精馏塔，回流比是一个对产品质量和产量有重大影响而又便于调节的参数。请你根据实验室提供的设备和物料，完成下列实验任务。

任务要求

① 从理论上分析，对于已给定的精馏塔，回流比的改变对精馏操作和分离能力的

影响。

② 根据实验室现有条件，拟订改变回流比的方法，制订出实验方案（包括实验操作条件、实验操作方法和注意事项等）。

③ 在全回流、稳定操作条件下进行实验，测定全塔理论塔板数、单板效率和全塔效率。

④ 在某一回流比下连续精馏、稳定操作条件下进行实验，测定全塔理论塔板数、单板效率和全塔效率。

⑤ 其他操作条件不变，只改变回流比，测定全塔理论塔板数、单板效率和全塔效率。

⑥ 探讨不同回流比对全塔效率和单板效率的影响，以及不同回流比时浓度曲线分布有何不同。当回流比变小时，若要保证塔顶馏出液的质量，可采取哪些措施。

⑦ 确定其中一组操作条件下的最小回流比，并计算最小回流比与实际回流比的关系。

实验任务书（3）

进料位置对精馏塔操作条件和分离能力的影响研究

最适宜进料板的位置是指在相同的理论板数和同样的操作条件下，具有最大分离能力的进料板位置，或在同一操作条件下所需理论板数最少的进料板位置。

在化学工业中，多数精馏塔都设有两个以上的进料板，调节进料板的位置是以进料组分发生变化为依据的。请你根据实验室提供的设备和物料，完成下列实验任务。

任务要求

① 从理论上分析，改变进料位置对精馏操作和分离能力的影响。

② 根据实验室现有条件，拟订改变进料位置的方法，制订出实验方案（包括实验操作条件、实验操作方法和注意事项等）。

③ 在全回流、稳定操作条件下进行实验，测定全塔理论塔板数、单板效率和全塔效率。

④ 在某一回流比下连续精馏、稳定操作条件下进行实验，测定全塔理论塔板数、单板效率和全塔效率。

⑤ 其他操作条件不变，只改变进料位置，测定全塔理论塔板数、单板效率和全塔效率。

⑥ 探讨不同进料位置对全塔效率和单板效率的影响，以及不同进料位置的浓度曲线分布有何不同。

⑦ 在本实验的进料浓度下，你认为最佳进料位置应该在哪里？进料位置发生变化后，为保证塔顶馏出液产品质量，应如何调整精馏塔操作参数？

0.6.7 干燥实验

实验任务书（1）

干燥条件对干燥特性曲线的影响研究

利用实验室洞道式干燥器，完成湿含量、干燥面积、绝干质量一定的纸浆板在不同干燥条件下的恒定干燥速率的比较实验，进行不同干燥条件对干燥速率曲线的影响研究。

任务要求

① 制订出测定恒定干燥速率曲线的实验方案，包括实验步骤、实验原始数据记录表、实验注意事项等。

② 不同的干燥条件分别是：

ⅰ. 相同介质流速、不同介质温度（两种温度）。

ⅱ. 相同介质温度、不同介质流速（两种流速）。

ⅲ. 相同介质温度、介质流速，不同纸浆板厚度。

③ 根据实验结果（图、表）探讨不同的干燥条件对恒速干燥速率、降速干燥速率以及临界湿含量和平衡湿含量等干燥特性参数的影响，以及测定干燥速率曲线的工程意义。

实验任务书（2）

恒定干燥条件下干燥时间的确定

某造纸厂欲在洞道干燥器中干燥每批质量为200kg的纸浆板，湿含量由1.76kg水/kg绝干干燥至0.22kg水/kg绝干，干燥表面积为0.025m^2/kg绝干。经生产现场测定，洞道中的气速为2.6m/s，温度为75℃。实验室现有该纸浆板的试样，试利用小型洞道干燥器及其他相关测试仪器仪表，测定该物料的恒定干燥特性参数以确定生产中每批纸浆板在洞道干燥器中所需要的停留时间。

任务要求

① 制订出实验方案，包括实验步骤、实验原始数据记录表、实验注意事项等。

② 利用实验所测数据，用公式、图表等说明你是如何确定干燥时间的。

③ 对实验结果进行讨论并以此说明测定干燥速率曲线的工程意义。

0.6.8 液液萃取实验

实验任务书（1）

萃取塔流体力学性能及传质性能研究

萃取塔具有显著的放大效应，因此，在放大过程中无论小试、中试，其流体力学性能和传质性能都是十分重要的研究对象。利用实验室的喷洒萃取塔、机械搅拌萃取塔、填料萃取塔等装置完成本项目的研究。

任务要求

① 选择某一类型萃取塔，制订实验方案（包括实验步骤、实验原始数据记录表、实验注意事项等）。

② 测定不同轻、重两相流量条件下，萃取塔的滞留分数和泛点。

③ 测定不同轻、重两相流量条件下，萃取塔的传质单元高度、传质单元数、传质系数等参数。

④ 根据实验结果探究轻、重两相流量或相比对萃取塔滞留分数、泛点、传质系数的影响，总结萃取塔流体力学性能和传质性能的影响因素及其规律。

实验任务书（2）
外加机械能对萃取塔性能的影响研究

为强化萃取塔的传质，可以通过外加机械能来实现，常见的外加机械能有脉冲、机械搅拌、振动等。由此也就有了机械搅拌萃取塔、脉冲填料萃取塔、振动筛板萃取塔等。因此，外加机械能的方式、强度等因素自然对萃取过程产生不容忽视的影响。在实验室具备的条件下，完成外加机械能对萃取塔性能的影响研究。

任务要求

① 在机械搅拌萃取塔、喷洒萃取塔和填料萃取塔中选择某一类型萃取塔作为研究对象，制订实验方案（包括实验步骤、实验原始数据记录表、实验注意事项等）。

② 在没有外加机械能（无搅拌、无脉冲）的条件下，测定萃取塔的滞留分数、泛点及传质单元高度、传质单元数、传质系数等参数。

③ 在不同搅拌速度（对机械搅拌萃取塔）、不同脉冲频率（对喷洒萃取塔和填料萃取塔）条件下，测定萃取塔的传质单元高度、传质单元数、传质系数等参数。

④ 根据实验结果探究有无外加机械能对萃取塔流体力学性能和传质性能的影响，探究机械搅拌或脉冲强度对萃取塔流体力学性能和传质性能的影响规律。

实验任务书（3）
颗粒型填料结构、尺寸对萃取塔性能的影响研究

填料塔结构简单、操作方便，是萃取过程重要的传质设备之一，但填料的材质和结构尺寸分别影响萃取过程连续相对填料的润湿性能以及分散相的破碎和聚并行为。以实验室填料萃取塔为基础，研究不同材质、不同结构尺寸的填料对萃取塔流体力学性能和传质性能的影响。

任务要求

① 在实验室拥有的拉西环、鲍尔环、θ环、阶梯环、花环等颗粒填料中，选择若干不同材质、不同结构、不同尺寸的填料，制订适当的实验方案。

② 在填料萃取塔中分别填装各种选定的填料，分别进行流体力学性能和传质性能实验。

③ 根据实验结果分析不同材质、不同结构、不同尺寸的填料对实验体系液液萃取流体力学性能和传质性能的影响。

以上任务书仅供“三型”实验和提高型实验使用，各实验列出几例典型任务书供实验者参考，但不仅限于这些任务书，教师和实验者可以根据实际需要制订相应的实验任务书。

第 1 章 化工原理实验研究方法

化工原理实验属于工程实验范畴，工程实验不同于基础课程的实验，后者采用的方法是理论的、严密的，研究的对象通常是简单的、基本的，甚至是理想的，而工程实验面对的是复杂的实验问题和工程问题，困难在于变量多，涉及的物料千变万化，设备结构、大小悬殊等。因此，它同其他工程学科一样，除了生产经验总结以外，实验研究是学科建立和发展的重要基础。多年来，化工原理实验在发展过程中形成的研究方法有直接实验法、量纲分析法和数学模型法三种。

1.1 直接实验法

直接实验法就是对研究对象进行直接实验，以获取其相关参数间的关系和规律。这是一种解决工程实际问题最基本的方法，针对性强，得到的结果较为可靠。但是，此研究方法得出的通常是变量较少的部分参数间的规律性关系，得出的实验结果只能应用于特定的实验条件和实验设备中，或只能推广到实验条件完全相同的场合，因此具有较大的局限性。

1.2 量纲分析法

对于一个多变量影响的工程问题，为研究过程的规律，往往采用网格法规划实验，即依次固定其他变量，改变某一变量测定目标值。例如，影响流体流动阻力的主要因素有管径 d、管长 l、平均流速 u、流体密度 ρ、流体黏度 μ 及管壁粗糙度 ε，变量数为 6 个，如果每个变量改变条件次数为 10 次，则需要做 10^6 次实验。不难看出变量数出现在幂上，涉及变量越多，所需实验次数将会剧增。因此，需要寻找一种方法以减少工作量，并使得到的结果具有一定的普遍性。量纲分析法就是一种能解决上述问题并在工程研究中广泛使用的实验研究方法。

量纲分析法所依据的基本理论是量纲一致性原则和白金汉（Buckingham）的 π 定理。量纲一致性原则是：凡是根据物理规律导出的物理方程，其中各项的量纲必然相同。π 定理是：用量纲分析所得到的独立的量纲数群个数 N，等于变量数 n 与基本量纲数 m 之差，即 $N=n-m$。

量纲分析法是将多变量函数整理为简单的若干无量纲数群构成的准数关系式，然后通过实验获得准数关系式中的系数与指数，从而大大减少实验工作量，同时也容易将实验结果应用到工程计算和设计中。

使用量纲分析法时，应明确量纲与单位是不同的。量纲又称因次，是指物理量的种类，而单位是比较同一种类物理量大小所采用的标准，例如，长度可以用米（m）、厘米（cm）、毫米（mm）等来表示。但单位的种类同属长度类，如以 L、M、T 分别表示长度、质量、时间，则 [L]、[M]、[T] 分别表示长度、质量、时间的量纲。

量纲有两类，一类是基本量纲，它们是彼此独立的，不能相互导出；另一类是导出量纲，由基本量纲导出。例如，在国际单位制中基本量纲有长度、质量、时间、热力学温度、物质的量、电流强度、发光强度 7 个，常用的长度、质量、时间的量纲可分别用 [L]、[M]、[T] 表示，其他物理量的量纲都可以由这 7 个基本量纲导出，并可写成幂指数乘积的形式。

现设某个物理量的导出量纲为 Q，$[Q]=[M^a L^b \theta^c]$，式中 a、b、c 为常数。如果基本量纲的指数均为零，这个物理量称为无量纲数（或无量纲数群），如反映流体流动状态的雷诺数就是无量纲数群。

$$Re=du\rho/\mu \tag{1-1}$$

管径 d，基本量纲：$[d]=[L]$；

流速 u，导出量纲：$[u]=[LT^{-1}]$；

密度 ρ，导出量纲：$[\rho]=[ML^{-3}]$；

黏度 μ，导出量纲：$[\mu]=[ML^{-1}T^{-1}]$。

$$[Re]=[d][u][\rho]/[\mu]=[L][LT^{-1}][ML^{-3}]/[ML^{-1}T^{-1}]=[L^0M^0T^0]$$

1.2.1 量纲分析法具体步骤

① 找出影响过程的独立变量；

② 确定独立变量所涉及的基本量纲；

③ 构造因变量和自变量的函数式，通常以指数方程的形式表示；

④ 用基本量纲表示所有独立变量的量纲，并写出包含各独立变量的量纲式；

⑤ 依据物理方程的量纲一致性原则和 π 定理得到准数方程；

⑥ 通过实验归纳总结准数方程的具体函数式。

1.2.2 量纲分析法举例说明

以获得流体在管内流动的阻力和摩擦系数 λ 的关系式为例。根据摩擦阻力的性质和有关实验研究，得知由于流体内摩擦而出现的单位体积流体的沿程损失 Δp_f 与 6 个因素有关，其函数关系式为：

$$\Delta p_f=f(d,l,u,\rho,\mu,\varepsilon) \tag{1-2}$$

这个函数具体是什么形式并不知道，但从数学上讲，任何非周期性函数，用幂函数的形式逼近是可取的，所以化工上一般将其改为下列幂函数的形式：

$$\Delta p_f=Kd^a l^b u^c \rho^d \mu^e \varepsilon^f \tag{1-3}$$

尽管上式中各物理量上的幂指数是未知的，但根据量纲一致性原则可知，方程式等号右侧的量纲必须与 Δp_f 的量纲相同，那么组合成几个无量纲数群才能满足要求呢？由式(1-2) 分析，变量数 $n=7$（包括 Δp_f），表示这些物理量的基本量纲 $m=3$（长度 [L]、时间 [T]、质量 [M]），因此根据白金汉的 π 定理可知，组成的无量纲数群的个数为 $N=n-m=4$。

通过量纲分析，将变量无量纲化。式(1-3) 中各物理量的量纲分别是：

$[\Delta p_f]=[ML^{-1}T^{-2}]$　　$[d]=[l]=[L]$　　$[u]=[LT^{-1}]$

$[\rho]=[ML^{-3}]$　　$[\mu]=[ML^{-1}T^{-1}]$　　$[\varepsilon]=[L]$

将各物理量的量纲代入式(1-3)，则两端量纲为：

$$[ML^{-1}T^{-2}]=K[L]^a[L]^b[LT^{-1}]^c[ML^{-3}]^d[ML^{-1}T^{-1}]^e[L]^f$$

根据量纲一致性原则，上式等号两边各基本量的量纲的指数必然相等，可得方程组：

对基本量纲［M］：$d+e=1$

对基本量纲［L］：$a+b+c-3d-e+f=-1$

对基本量纲［T］：$-c-e=-2$

此方程组包括3个方程，却有6个未知数，用其中3个未知数b、e、f来表示a、d、c，解此方程组，可得：

$$\begin{cases}a=-b-c+3d+e-f-1\\d=1-e\\c=2-e\end{cases}\qquad\begin{cases}a=-b-e-f\\d=1-e\\c=2-e\end{cases}$$

将求得的a、d、c代入式(1-3)，即得：

$$\Delta p_f=Kd^{-b-e-f}l^bu^{2-e}\rho^{1-e}\mu^e\varepsilon^f \tag{1-4}$$

将指数相同的各物理量归并在一起得：

$$\frac{\Delta p_f}{u^2\rho}=K\left(\frac{l}{d}\right)^b\left(\frac{du\rho}{\mu}\right)^{-e}\left(\frac{\varepsilon}{d}\right)^f \tag{1-5}$$

$$\Delta p_f=2K\left(\frac{l}{d}\right)^b\left(\frac{du\rho}{\mu}\right)^{-e}\left(\frac{\varepsilon}{d}\right)^f\left(\frac{u^2\rho}{2}\right) \tag{1-6}$$

由于摩擦损失Δp_f应与管长l成正比，故式中$b=1$（实验也证实这一点），因此将此式与计算流体在管内摩擦阻力的公式（范宁公式）

$$\Delta p_f=\lambda\frac{l}{d}\left(\frac{u^2\rho}{2}\right) \tag{1-7}$$

相比较，整理得到研究摩擦系数λ的关系式，即得：

$$\lambda=2K\left(\frac{du\rho}{\mu}\right)^{-e}\left(\frac{\varepsilon}{d}\right)^f \tag{1-8}$$

或

$$\lambda=\Phi\left(Re,\frac{\varepsilon}{d}\right) \tag{1-9}$$

由以上分析可以看出：在量纲分析法的指导下，将一个复杂的多变量的管内流体流动阻力的计算问题，简化为摩擦系数λ的研究和确定。它是建立在正确判断过程影响因素的基础上，进行了逻辑加工而归纳出的数群。上面的例子只能告诉我们：λ是Re与ε/d的函数，至于它们之间的具体形式，归根到底还得靠实验来确定。通过实验获得经验公式或算图并用以指导工程计算和工程设计。著名的莫狄（Moody）摩擦系数图即“摩擦系数λ与Re、ε/d的关系曲线”，就是这种实验的结果。许多实验研究了各种具体条件下的摩擦系数λ的计算公式，其中较著名的适用于光滑管的柏拉修斯（Blasius）公式：

$$\lambda=\frac{0.3164}{Re^{0.25}} \tag{1-10}$$

其他研究结果可以参看有关教科书及手册。

量纲分析法有两点需要注意。

① 最终所得数群的形式与求解联立方程组的方法有关。如何合并变量为有用的准数，这是研究者必须注意的问题。在前例中如果不以b、e、f来表示a、d、c，而改为以d、

e、f 表示 a、b、c，整理得到的数群形式也就不同。不过，这些形式不同的数群可以通过互相乘除，仍然可以变换成前例中所求得的 4 个数群。

② 必须对所研究的过程问题有本质的了解，如果有一个重要的变量被遗漏或者引进一个无关的变量，就会得出不正确的结果，甚至导致谬误的结论。所以应用量纲分析法必须持谨慎的态度。

从以上分析可知：量纲分析法是通过将变量组合成无量纲数群，从而减少实验自变量的个数，大幅度地减少实验次数。另一个优点是，若按式(1-2) 进行实验时，为改变 ρ 和 μ，实验中必须更换多种流体；为改变 d，必须改变实验装置（管径）。而应用量纲分析所得的式(1-6) 指导实验时，要改变 $du\rho/\mu$ 只需改变流速；要改变 l/d，只需改变测量段的距离，即两测压点的距离。从而可以将水、空气等的实验结果推广应用于其他流体，将小尺寸模型的实验结果应用于大型实验装置。因此，实验前的无量纲化工作是规划一个实验的一种有效手段，并在化工实验研究中广为应用。

1.3　数学模型法

1.3.1　数学模型法主要步骤

数学模型法是在对研究的问题有充分认识的基础上，按以下主要步骤进行工作。

① 将复杂问题作合理又不过于失真的简化，提出一个近似实际过程又易于用数学方程式描述的物理模型；

② 对所得到的物理模型进行数学描述，即建立数学模型，然后确定该方程的初始条件和边界条件，并求解方程；

③ 通过实验对数学模型的合理性进行检验，并测定模型参数。

1.3.2　数学模型法举例说明

以求取流体通过固定床的压降为例进行说明。固定床中颗粒间的空隙形成许多可供流体通过的细小通道，这些通道是曲折而且互相交联的。同时，这些通道的截面大小和形状又是很不规则的，流体通过如此复杂的通道时的压降自然很难进行理论计算，但可以用数学模型法来解决。

(1) 物理模型

流体通过颗粒层的流动多呈爬流状态，单位体积床层所具有的表面积对流动阻力有决定性作用。这样，为解决压降问题，可在保证单位体积表面积相等的前提下，将颗粒层内的实际流动过程作如下大幅度的简化，使之可以用数学方程式加以描述。

将床层中的不规则通道简化成长度为 L_e 的一组平行细管，并规定：细管的内表面积等于床层颗粒的全部表面积，细管的全部流动空间等于颗粒床层的空隙容积。

根据上述假定，可求得这些虚拟细管的当量直径 d_e

$$d_e=\frac{4\times\text{通道的截面积}}{\text{润湿周边}} \tag{1-11}$$

分子、分母同乘 L_e，则有

$$d_e=\frac{4\times\text{床层的流动空间}}{\text{细管的全部内表面积}} \tag{1-12}$$

以 $1m^3$ 床层体积为基准，则床层的流动空间为 ε，$1m^3$ 床层的颗粒全部表面积即为床层的比表面 α_B。如果忽略因颗粒相互接触而使裸露的颗粒表面积减小，则 α_B 与颗粒的比表面 α 之间关系为 $\alpha_B=\alpha(1-\varepsilon)$，因此

$$d_e=\frac{4\varepsilon}{\alpha_B}=\frac{4\varepsilon}{\alpha(1-\varepsilon)} \tag{1-13}$$

按此简化的物理模型，流体通过固定床的压降即可等同于流体通过一组当量直径为 d_e、长度为 L_e 的细管的压降。

(2) 数学模型

上述简化的物理模型，已将流体通过具有复杂几何边界的床层的压降简化为通过均匀圆管的压降。对此，可用现有的理论作出如下数学描述：

$$h_f=\frac{\Delta p}{\rho}=\lambda\frac{L_e}{d_e}\frac{u_1^2}{2} \tag{1-14}$$

式中，u_1 为流体在细管内的流速。u_1 可取实际填充床中颗粒空隙间的流速，它与空床流速（表观流速）u 的关系为：

$$u=\varepsilon u_1 \tag{1-15}$$

将式(1-13)、式(1-15) 代入式(1-14) 得

$$\frac{\Delta p}{L}=\left(\lambda\frac{L_e}{8L}\right)\frac{(1-\varepsilon)\alpha}{\varepsilon^3}\rho u^2 \tag{1-16}$$

细管长度 L_e 与实际长度 L 不等，但可以认为 L_e 与实际床层高度 L 成正比，即 $L_e=kL$，并将系数 k 并入摩擦系数中，于是

$$\frac{\Delta p}{L}=\lambda'\frac{(1-\varepsilon)\alpha}{\varepsilon^3}\rho u^2 \tag{1-17}$$

式中，$\lambda'=\frac{\lambda}{8}\frac{L_e}{L}$。

上式即为流体通过固定床压降的数学模型，其中包括一个未知的待定系数 λ'。λ'称为模型参数，就其物理意义而言，也可称为固定床的流动摩擦系数。

(3) 模型的检验和模型参数的估值

上述床层的简化处理只是一种假定，其有效性必须经过实验检验，其中模型参数 λ'也必须由实验测定。康采尼（Kozeny）和欧根（Ergun）等均对此进行了实验研究，获得了不同实验条件下不同范围的 λ'与 Re'的关联式。由于篇幅所限，详细内容请查阅其他有关书籍。

1.3.3 数学模型法和量纲分析法的比较

对于数学模型法，决定成败的关键是对复杂过程的合理简化，即能否得到一个足够简单的既可用数学方程式表示而又不失真的物理模型。只有对过程的内在规律，特别是过程的特殊性，有着深刻的理解并根据特定的研究目的加以利用，才有可能对真实的复杂过程进行大幅度的合理简化，同时在指定的某一侧面保持等效。上述例子在进行简化时，只是物理模型与实际过程在阻力损失这一侧面保持等效。

对于量纲分析法，决定成败的关键在于如数列出影响过程的主要因素。它无须对过程本身的规律深入理解，只要做若干量纲分析实验，考察每个变量对实验结果的影响程度即可。在量纲分析法指导下的实验研究只能得到过程的外部联系，对过程的内部规律则不甚了然。然而，这正是量纲分析法的一大特点，它使量纲分析法成为对各种研究对象原则上皆适用的一般方法。

无论是数学模型法还是量纲分析法，最后都要通过实验解决问题，但实验的目的大相径庭。数学模型法的实验目的是为了检验物理模型的合理性，并测定为数较少的模型参数；而量纲分析法的实验目的是为了寻找各无量纲变量之间的函数关系。

第2章 实验数据的误差分析

任何一项实验和测量均存在误差。产生误差的原因是极其复杂的，不同因素产生的误差性质也不同，所以误差分析的目的就是评价实验数据的精确性，通过误差分析认清误差的来源及其影响，并设法消除或减小误差，以提高实验的质量。本章就化工原理实验中遇到的一些误差的基本概念与估算方法作一扼要介绍。

2.1 真值与平均值

真值是指某物理量客观存在的确定值，也称理论值或定义值。严格来讲，由于测量仪器、测定方法、环境、人的观察力、测量的程序等都不可能是完美无缺的，故真值是无法测得的，是一个理想值。科学实验中真值的定义是：设在测量中观察的次数为无限多，则根据误差分布定律正负误差出现的概率相等，故将各观察值相加，加以平均，在无系统误差情况下，可能获得极近于真值的数值。故“真值”在现实中是指观察次数无限多时，所求得的平均值（或是写入文献手册中所谓的“公认值”）。然而对工程实验而言，观察的次数都是有限的，故用有限观察次数求出的平均值近似真值，或称为最佳值。一般称这一最佳值为平均值。常用的平均值有下列几种。

(1) 算术平均值 $\overline{x}$

这种平均值最常用。测量值的分布服从正态分布时，用最小二乘法原理可以证明：在一组等精度的测量中，算术平均值为最佳值或最可信赖值。

$$\overline{x}=\frac{x_1+x_2+\cdots+x_n}{n}=\frac{1}{n}\sum_{i=1}^{n}x_i \tag{2-1}$$

式中，x_1，x_2，…，x_n 为各次观测值；n 为观察的次数。

(2) 均方根平均值 $\overline{x}_{均}$

$$\overline{x}_{均}=\sqrt{\frac{x_1^2+x_2^2+\cdots+x_n^2}{n}}=\sqrt{\frac{1}{n}\sum_{i=1}^{n}x_i^2} \tag{2-2}$$

(3) 加权平均值 $\overline{w}$

设对同一物理量用不同方法测定，或对同一物理量由不同人测定，计算平均值时，常对比较可靠的数值予以加权重平均，称为加权平均。

$$\overline{w}=\frac{w_1x_1+w_2x_2+\cdots+w_nx_n}{w_1+w_2+\cdots+w_n}=\frac{\sum_{i=1}^{n}w_ix_i}{\sum_{i=1}^{n}w_i} \tag{2-3}$$

式中，x_1，x_2，…，x_n 为各次观测值；w_1，w_2，…，w_n 为各测量值的对应权重，一般凭经验确定。

(4) 几何平均值 $\bar{x}_G$

$$\bar{x}_G=\sqrt[n]{x_1x_2x_3\cdots x_n} \tag{2-4}$$

(5) 对数平均值 x_m

$$x_m=\frac{x_1-x_2}{\ln x_1-\ln x_2}=\frac{x_1-x_2}{\ln\frac{x_1}{x_2}} \tag{2-5}$$

若 $1<\frac{x_1}{x_2}<2$，可用算术平均值代替对数平均值，引起的误差在 4%以内。

介绍以上各种平均值，目的是要从一组测定值中找出最接近真值的那个值。平均值的选择主要由一组观测值的分布类型决定，在化工原理实验研究中，数据分布多属于正态分布，故通常采用算术平均值。

2.2　误差的分类

在任何一种测量中，无论所用仪器多么精密，方法多么完善，实验者多么细心，不同时间所测得的结果不一定完全相同，而有一定的误差和偏差。严格来讲，误差是指实验测量值（包括直接和间接测量值）与真值（客观存在的确定值）之差，偏差是指实验测量值与平均值之差，但习惯上通常对两者不加以区别。根据误差的性质及其产生的原因，可将误差分为系统误差、随机误差和过失误差 3 种。

(1) 系统误差

系统误差是指在一定条件下，对同一物理量进行多次测量时，误差的数字保持恒定，或随条件改变按一定的规律变化。可以用正确度一词来表征系统误差的大小，系统误差越小，正确度越高，反之亦然。

产生系统误差的原因通常有以下几点。

1）测量仪器　仪器的精度不能满足要求或仪器存在零点偏差等。

2）试剂质量　试剂不纯或质量不符合要求等。

3）测量方法　以近似的测量方法测量或利用简化的计算公式进行计算。

4）环境因素　如温度、压力和湿度等外界因素的改变。

5）个人习惯　如读取数据常偏高或偏低，滴定分析时，判定滴定终点的颜色偏深或偏浅。

系统误差是误差的重要组成部分，在测量时，应尽力消除其影响。一般系统误差是有规律的，其产生的原因也往往是可知或找出原因后可以消除的，至于不能消除的系统误差也应设法确定或估计出来。

(2) 随机误差

随机误差又称偶然误差，是由某些不易控制的随机因素造成的。它主要表现在测量结果的分散性，但完全服从统计规律。研究随机误差可以采用概率统计的方法。随着测量次数的增加，平均值的随机误差可以减小，但不会消除。在误差理论中，常用精密度一词来表征随机误差的大小，随机误差越大，精密度越低，反之亦然。

(3) 过失误差

过失误差又称粗大误差，即与实际明显不符的误差，主要是由实验人员粗心大意所致，读错、测错、记错等都会带来过失误差。含有过失误差的测量值称为坏值，应在整理数据时依据相关的准则加以剔除。

综上所述，可以认为系统误差和过失误差总是可以设法避免的，而随机误差是不可避免的，因此最好的实验结果应该只含有随机误差。

2.3 误差的表示方法

(1) 绝对误差 $D(x)$

测量值 x 与其真值 A 之差的绝对值称为绝对误差，其表达式为：

$$D(x)=|x-A| \tag{2-6}$$

在工程计算中，真值常用算术平均值 $\bar{x}$ 或相对真值代替。相对真值是指使用高精度级的标准仪器所测量的值。故绝对误差又可表示为：

$$D(x)=|x-\bar{x}| \tag{2-7}$$

(2) 相对误差 $E_r(x)$

绝对误差与真值绝对值之比称为相对误差，其表达式为：

$$E_r(x)=\frac{D(x)}{|A|} \tag{2-8}$$

与绝对误差一样，其真值也常用算术平均值 $\bar{x}$ 代替，故相对误差又可表示为：

$$E_r(x)=\frac{D(x)}{|\bar{x}|} \tag{2-9}$$

相对误差反映测量的准确程度，常用百分数或千分数表示。因此，不同物理量的相对误差可以互相比较，相对误差与被测量值的大小及绝对误差的数值都有关系。

(3) 算术平均误差 δ

n 次测量值的算术平均误差为：

$$\delta=\frac{\sum_{i=1}^{n} d_i}{n}=\frac{\sum_{i=1}^{n}|x_i-\bar{x}|}{n},\quad i=1,2,\cdots,n \tag{2-10}$$

式中，n 为观测次数；d_i 为测量值与平均值的偏差，$d_i=|x_i-\bar{x}|$。

算术平均误差的缺点是无法表示出各次测量间彼此符合的情况。

(4) 标准误差 σ

标准误差也称为均方根误差，其表达式为：

$$\sigma=\sqrt{\frac{\sum_{i=1}^{n} d_i^2}{n}}=\sqrt{\frac{\sum_{i=1}^{n}(x_i-\bar{x})^2}{n}} \tag{2-11}$$

标准误差对一组测量中的较大误差或较小误差比较敏感，故标准误差可较好地表示精确度。上式适用于无限次测量的场合。实际测量中，测量次数是有限的，式(2-11) 可改写为：

$$\sigma=\sqrt{\frac{\sum_{i=1}^{n}(x_i-\bar{x})^2}{n-1}} \tag{2-12}$$

标准误差不是一个具体的误差，σ 的大小只说明在一定条件下等精度测量集合所属的任一次测量值对其算术平均值的分散程度。如果 σ 的值小，说明该测量集合中相应小的误差就占优势，任一次测量值对其算术平均值的分散度就小，测量的可靠性就高。

上述的各种误差表示方法中，不论是比较各种测量的精度或是评定测量结果的质量，均以相对误差和标准误差表示为佳，而在文献中标准误差更常被采用。

2.4　精密度、正确度和精确度

测量的质量和水平，可用误差的概念来描述，也可用准确度的概念来描述。为了指明误差的来源和性质，通常用以下 3 个概念。

(1) 精密度

精密度可以衡量某些物理量几次测量之间的一致性，即重复性，它可以反映随机误差大小的程度，精密度高表示随机误差小。

(2) 正确度

正确度指在规定条件下，测量中所有系统误差的综合，它可以反映系统误差大小的程度，正确度高表示系统误差小。

(3) 精确度

精确度又称准确度，指测量结果与真值的逼近程度，它可以反映系统误差和随机误差综合大小的程度，精确度高表示系统误差和随机误差都小。

为说明它们间的区别，往往用打靶来做比喻。如图 2-1 所示，A 的系统误差小而随机误差大，即正确度高而精密度低；B 的系统误差大而随机误差小，即正确度低而精密度高；C 的系统误差和随机误差都小，表示精确度高。当然实验测量中没有像靶心那样的明确真值，而是设法去测定这个未知的真值。

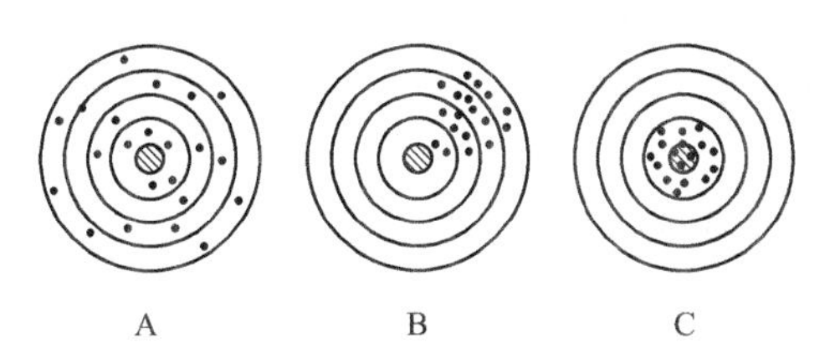

图 2-1　精密度、正确度、精确度含义示意图

对于实验测量来说，精密度高，正确度不一定高。正确度高，精密度也不一定高。但精确度高，必然是精密度与正确度都高。

2.5　实验数据的有效数字

在实验中，无论是直接测量的数据还是计算结果，都会涉及有效数字的问题。并不是小数点后面的数字越多就越准确，或者运算结果保留位数越多就越准确。一是因为在记录测量数据时包括最后一位估读数字，不确定的数字。通过直接读取获得的准确数字叫做可靠数字，通过估读得到的数字叫做存疑数字，测量结果中能够反映被测量大小的带有一位存疑数字的全部数字叫有效数字。数据中小数点的位置在前或在后仅与所用的测量单位有关，例如 28.3mm 与 0.0283m 这两个数据，其准确度相同，有效数字相同，但小数点的位置不同。二是因为在实验测量中所使用的仪器仪表只能达到一定的准确度，因此测量或计算的结果不可能也不应该超越仪器仪表所允许的准确度范围，如上述的长度测量中，标尺最小分度 1mm，其读数可以到 0.1mm（上述的 3 即为估读值），故数据的有效数字是 3 位。

（1）“0”在有效数字中的作用

数字前面所有的“0”只起定值作用，数字中间的“0”和末尾的“0”都是有效数字，但以“0”结尾的正整数除外。以“0”结尾的正整数的有效数字的位数不确定。例如 350 克不一定是 350.0 克，前者可能是 2 位也可能是 3 位有效数字，后者是 4 位，而 0.0350 克虽然有 5 位数字，但有效数字仅为 3 位。

在科学研究与计算中，为了清楚地表示出数据的准确度，可采用科学计数法表示。其方法为：先将有效数字写出，并在第一个有效数字后面加上小数点，并用 10 的整数幂表示数据的数量级。例如要表示 35800 的有效数字为 4 位，可以写成 3.580×10^4，若其只有 3 位有效数字，则可以写成 3.58×10^4。

（2）有效数字的舍入规则

数字的舍入，习惯上用四舍五入的方法，但是在一些精度要求较高的场合，则应按照《数字修约规则与极限数值的表示和判定》(GB/T 8170—2008) 中的方法进行修约。也就是说当有效数字确定后，在书写时一般只保留一位存疑数字，多余数字按数字修约规则处理。国标中规定的进舍规则如下。

① 拟舍弃数字的最左一位数字小于 5，则舍去，保留其余各位数字不变。

② 拟舍弃数字的最左一位数字大于 5，则进一，即保留数字的末位数字加 1。

③ 拟舍弃数字的最左一位数字是 5，且其后有非 0 数字时进一，即保留数字的末位数字加 1。

④ 拟舍弃数字的最左一位数字是 5，且其后无数字或皆为 0 时，若保留的末位数字为奇数（1，3，5，7，9）则进一，即保留数字的末位数字加 1；若保留的末位数字为偶数（0，2，4，6，8），则舍去。

⑤ 负数修约时，先将它的绝对值按上面规定进行修约，然后在所得值前面加上负号。

例如将下面的几个数保留 3 位有效数字：

12.1498 ⟶12.1　　规则①

12.1698 ⟶12.2　　规则②

12.2598 ⟶12.3　　规则③

12.2508 ⟶12.3　　规则③

12.2500 ⟶12.2　　规则④

12.3500 ⟶12.4　　规则④

（3）有效数字的运算规则

1）加减法　各数所保留的小数点后的位数，以各数中小数点后的位数最少的数据为基准，其他数据修约至与其相同，再进行加减计算。

例如：将 12.2508、17.4、0.083 三个数相加，应写为

$12.3+17.4+0.1=29.8$

2）乘除法　各数所保留的位数，以原来各数中有效数字最少的数据为基准，其他有效数字修约至相同，再进行乘除运算，计算结果仍保留最少的有效数字。

例如：将 13.786、1.034、0.128 三个数相乘，应写为

$13.8\times1.03\times0.128=1.82$

3）对数运算 对数的有效数字位数应与真数相同。

例如：lg12.34=1.091 ln12.34=2.513

(4) 直接测量值的有效数字

直接测量的有效数字的位数取决于测量仪器的精确度。一般来说有效数字的位数可保留到测量仪器最小刻度的后一位，即包含一位估读数字。例如图2-2(a)的数据可记为1.71，最后一位为估读值。估读值可随实验者的读取习惯不同而略有差异。但是对于测量仪器的最小刻度不以1或$1\times10^{\pm n}$为单位，则估读数字为测量仪器的最小刻度位即可。图2-2(b)的最小刻度位为0.2，因此数据可记为1.7。

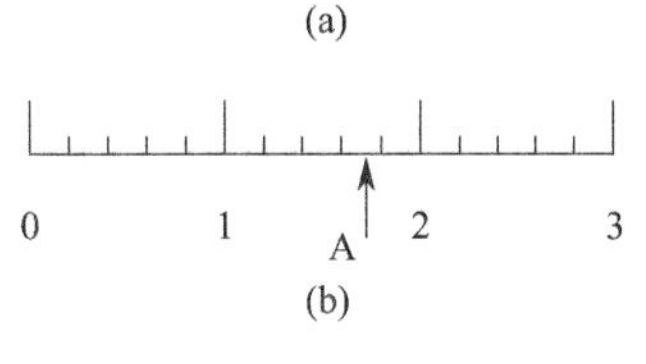

图2-2 不同坐标刻度的读数情况

(5) 非直接测量值的有效数字

在实验中，有些数据不是用仪器直接测量得到的，而是需要通过计算获得。化工原理实验属于工程实验，在计算过程中原则上要遵循有效数字的运算规则，但也需要根据实际情况综合考虑。

① 参加运算的常数π、g等的数值以及某些因子如$\sqrt{2}$、1/3等的有效数字，取几位为宜，原则上取决于计算所用的原始数据的有效数字的位数。假设参与计算的原始数据中，位数最多的有效数字是n位，则引用上述常数时宜$n+2$位，目的是避免常数的引入造成更大的误差。

② 在数据运算过程中，为兼顾结果的精度和运算的方便，所有的中间运算结果，工程上，一般宜取5或6位有效数字。

③ 表示误差大小的数据一般宜取1或2位有效数字。由于误差是用来为数据提供准确程度的信息，为避免过于乐观，并提供必要的保险，故在确定误差的有效数字时，也用截断的办法，然后将保留数字末位加1，以使给出的误差值大一些，而无须考虑前面所说的数字舍入规则。如误差为0.5612，可写成0.6或0.57。

2.6 误差的处理方法

(1) 系统误差

系统误差的修正是不能依靠误差理论来解决的，而是要在掌握物理量的测定方法、测定原理的基础上来校正系统误差。分析和处理系统误差的关键，首先在于如何发现测量数据中存在显著的系统误差，从而进一步设法将它消除或加以校正。

系统误差可分为恒定系统误差和可变系统误差两大类。其简易判别方法有观察法、比较法等。常用的消除或减小系统误差的方法有根源消除法、修正消除法、代替消除法、异号消除法、交换消除法、对称消除法、半周期消除法、回归消除法等。对系统误差能否处理得当，在很大程度上取决于对测量技术掌握的熟悉程度，以及分析各种测量技术问题的丰富经验。关于系统误差的消除、校正以及消除程度的判别准则，请参考有关专业书籍。

(2) 随机误差

在实验测量过程中，随机误差是不可避免的。如何从含有随机误差的实验数据中确定出最可靠的测量结果，这是实验数据处理的一个基本问题。

随机误差大多数情况下服从正态分布，具有单峰性、有界性、对称性和抵偿性。抵偿性指的是随测量次数无限增加，误差平均趋向于零，这是随机误差最本质的统计特性。换言之，凡是有抵偿性的误差，原则上均按随机误差处理。根据随机误差的统计规律，常用算术平均值及其标准偏差等来表示测量结果。

(3) 过失误差

在整理实验数据时，往往会发现几个偏差特别大的数据，若保留它，则对平均值及随机误差都有很大的影响，但是随意舍弃这些数据，以获得实验结果的一致性，会有丢失有用信息的危险，这显然是不恰当的。如果这些数据是由于测量中的过失误差产生的，通常称其为可疑值或坏值，必须将其删除，但这些数据也有可能是由随机误差产生的，并不属于坏值，则不能将其删除。因此，若是从技术上和物理上找不出产生异常值的原因，不能确定数据中是否含有过失误差或随机误差，这时可采用下述的统计方法判断与删除。

(4) 可疑值的判断与删除

统计方法基本思想是：给定一个显著性水平，按一定分布确定一个临界值，凡是超过这个临界值的误差，就认为它不属于随机误差的范围，而是过失误差，该数据应予以剔除。以下介绍 3 种常用的统计判断准则。

1）拉依达准则　拉依达准则又称 3σ 准则，它是以测量次数充分多为前提。在一般情况下，测量次数都比较少，因此 3σ 准则只能是一个近似准则。

对于某个测量列 $x_i(i=1\sim n)$，若各测量值 x_i 只含有随机误差，根据随机误差正态分布规律，其偏差 d_i 落在 $\pm3\sigma$ 以外的概率约为 0.3%，σ 的计算见式(2-12)。如果在测量列中发现某测量值的偏差大于 3σ，亦即

$$|d_i|>3\sigma \tag{2-13}$$

则可以认为它含有过失误差，应该剔除。

当使用 3σ 准则时，允许一次将偏差大于 3σ 的所有数据剔除，然后再将剩余各个数据重新计算 σ，并再次用 3σ 准则继续剔除超差数据。

这种方法的最大优点是计算简单，应用方便，但当实验点数较少时，则很难将坏点剔除。在 $n\leqslant10$ 情况下，不可能出现 $|d_i|>3\sigma$ 的情况。为此在测量次数较少时，最好不要选用 3σ 准则。

2）格拉布斯准则　设对某量作多次独立测量，得一组测量数列 $x_i(i=1\sim n)$，当 x_i 服从正态分布时，可按式(2-1) 和式(2-12) 计算得 $\bar{x}$、σ。为了检验数列 $x_i(i=1\sim n)$ 中是否存在过失误差，将 x_i 按从小到大顺序排列，即：

$$x_{(1)}\leqslant x_{(2)}\leqslant\cdots\leqslant x_{(n)}$$

格拉布斯给出了 $g_{(1)}$ 和 $g_{(n)}$ 分布：

$$g_{(1)}=\frac{\bar{x}-x_{(1)}}{\sigma} \tag{2-14}$$

$$g_{(n)}=\frac{x_{(n)}-\bar{x}}{\sigma} \tag{2-15}$$

当选定了显著性水平 α，根据实验次数 n，可由表 2-1 查得相应的临界值 $g_{(0)}(n，\alpha)$。若有

$$g_{(i)}\geqslant g_{(0)}(n,\alpha) \tag{2-16}$$

即判别该测量值含有过失误差，应当剔除。

表 2-1　格拉布斯临界值 $g_{(0)}(n, \alpha)$

n	显著性水平 α			n	显著性水平 α		
	0.05	0.025	0.01		0.05	0.025	0.01
3	1.15	1.16	1.16	17	2.47	2.62	2.79
4	1.46	1.48	1.49	18	2.50	2.65	2.82
5	1.67	1.71	1.75	19	2.53	2.68	2.85
6	1.82	1.89	1.94	20	2.56	2.71	2.88
7	1.94	2.02	2.10	21	2.58	2.73	2.91
8	2.03	2.13	2.22	22	2.60	2.76	2.94
9	2.11	2.21	2.32	23	2.62	2.78	2.96
10	2.18	2.29	2.41	24	2.64	2.80	2.99
11	2.23	2.36	2.48	25	2.66	2.82	3.01
12	2.28	2.41	2.55	30	2.74	2.91	3.10
13	2.33	2.46	2.61	35	2.81	2.98	3.18
14	2.37	2.51	2.66	40	2.87	3.04	3.24
15	2.41	2.55	2.70	45	2.92	3.09	3.29
16	2.44	2.59	2.75	50	2.96	3.13	3.34

3）t 检验准则　由数学统计理论证明，在测量次数较少时，随机变量服从 t 分布，t 分布不仅与测量值有关，还与测量次数 n 有关，当 $n>10$ 时 t 分布就很接近正态分布了。所以当测量次数较少时，依据 t 分布原理的 t 检验准则来判别过失误差较为合理。t 检验准则的特点是先剔除一个可疑的测量值，而后再按 t 分布检验准则确定该测量值是否应该被删除。

设对某一物理量作多次测量，得测量列 $x_i(i=1\sim n)$，若认为其中测量值 x_j 为可疑数据，将它剔除后计算平均值为（计算时不包括 x_j）：

$$\bar{x}=\frac{1}{n-1}\sum_{\substack{i=1\\i\neq j}}^{n}x_i \tag{2-17}$$

求得的测量列的标准误差 σ（不包括 $d_j=x_j-\bar{x}$）为：

$$\sigma=\sqrt{\frac{1}{n-2}\sum_{\substack{i=1\\i\neq j}}^{n}d_i^2} \tag{2-18}$$

根据测量次数 n 和选取的显著性水平 α，即可由表 2-2 查得 t 检验系数 $K(n，\alpha)$，若有

$$|x_j-\bar{x}|>K(n,\alpha)\sigma \tag{2-19}$$

则认为测量值 x_j 含有过失误差，剔除 x_j 是正确的，否则，认为 x_j 不含有过失误差，应当保留。

表 2-2　t 检验系数 $K(n, \alpha)$ 表

n	显著性水平 α		n	显著性水平 α	
	0.05	0.01		0.05	0.01
4	4.97	11.46	6	3.04	5.04
5	3.56	6.53	7	2.78	4.36

续表

n	显著性水平 α		n	显著性水平 α	
	0.05	0.01		0.05	0.01
8	2.62	3.96	20	2.16	2.95
9	2.51	3.71	21	2.15	2.93
10	2.43	3.54	22	2.14	2.91
11	2.37	3.41	23	2.13	2.90
12	2.33	3.31	24	2.12	2.88
13	2.29	3.23	25	2.11	2.86
14	2.26	3.17	26	2.10	2.85
15	2.24	3.12	27	2.10	2.84
16	2.22	3.08	28	2.09	2.83
17	2.20	3.04	29	2.09	2.82
18	2.18	3.01	30	2.08	2.81
19	2.17	3.00			

在上述 3 个准则中，t 检验准则一般用于测量次数很少的场合；3σ 准则适用于测量次数较多的场合，但由于它使用简便，又不需要查表，所以对测量次数比较少但要求又不高的场合，还是经常使用的。

2.7 直接测量的误差估算

（1）一次测量值的误差估算

在实验中，由于条件不许可或要求不高等原因，对一个物理量的直接测量只进行一次，这时可以根据具体的实际情况，对测量值的误差进行合理的估计。

下面介绍如何根据所使用的仪表估算一次测量值的误差。

1）给出精确度等级的仪表与测量误差的估算（如电工仪表、转子流量计等） 这些仪表的精确度常采用仪表的最大引用误差和精确度的等级来表示。仪表最大引用误差的定义为

$$\text{最大引用误差}=\frac{\text{仪表显示值的绝对误差}}{\text{该仪表相应档次量程的绝对值}}\times 100\% \tag{2-20}$$

式中，仪表显示值的绝对误差是指在规定的情况下，被测参数的测量值与被测参数的标准值之差的绝对值的最大值。对于多档仪表，不同档次显示值的绝对误差和量程范围均不相同。

式(2-20) 表明，若仪表显示值的绝对误差相同，则量程范围越大，最大引用误差越小。

例如我国电工仪表的精确度等级有 7 种：0.1，0.2，0.5，1.0，1.5，2.5 和 5.0，金属管转子流量计的精确度等级有 3 种：1.5，2.5，4。数字越小表示精确度越高。如某仪表的精确度等级为 2.5 级，则说明此仪表的最大引用误差为 2.5%。

在使用仪表时，如何估算某一次测量值的绝对误差和相对误差？

设仪表的精确度等级为 P 级，其最大引用误差为 P%。若仪表的测量范围为 x_n，仪表的示值为 x_i，则由式(2-20) 得该示值的绝对误差和相对误差分别为：

$$D(x) \leqslant x_n \times P\% \tag{2-21}$$

$$E_r(x) = \frac{D(x)}{x_i} \leqslant \frac{x_n}{x_i} \times P\% \tag{2-22}$$

式(2-21) 和式(2-22) 表明：

① 若仪表的精确度等级 P 和测量范围 x_n 已固定，则测量的示值 x_i 越大，测量的相对误差越小。

② 选用仪表时，不能盲目地追求仪表的精确度等级，因为测量的相对误差还与$\frac{x_n}{x_i}$有关，而应兼顾仪表的精确度等级和$\frac{x_n}{x_i}$两者。

2）未给出精确度等级的仪表与测量误差的估算（如电子天平类）　例如国标《电子天平》(GB/T 26497—2011) 将电子天平划分为 4 个准确度等级，分别为特种准确度级、高准确度级、中准确度级和普通准确度级，分别用符号①、Ⅱ、Ⅲ、Ⅲ表示。这类仪器的精确度可用以下公式来表示：

$$\text{仪器的精确度} = \frac{0.5 \times \text{名义分度值}}{\text{量程的范围}} \tag{2-23}$$

式中名义分度值指测量仪表最小分度所代表的数值。

若有一电子天平，其名义分度值为 0.1mg，测量范围为 0～200g，则其精确度为：

$$\text{精确度} = \frac{0.5 \times 0.1}{(200-0) \times 10^3} = 2.5 \times 10^{-7}$$

若仪器的精确度已知，也可用式(2-23) 求得其名义分度值。

使用这些仪表时，测量的绝对误差和相对误差可分别由式(2-24) 和式(2-25) 确定。

$$D(x) \leqslant 0.5 \times \text{名义分度值} \tag{2-24}$$

$$E_r(x) \leqslant \frac{0.5 \times \text{名义分度值}}{\text{测量值}} \tag{2-25}$$

(2) 多次测量值的误差估算

如果一个物理量的值是通过多次测量得出的，那么该测量值的误差可通过标准误差来估算。

设某一量重复测量了 n 次，各次测量值为 x_1，x_2，…，x_n，该组数据的平均值 $\bar{x} = (x_1 + x_2 + \cdots + x_n)/n$，标准误差 $\sigma = \sqrt{\sum (x_i - \bar{x})^2/(n-1)}$，则 $\bar{x}$ 值的绝对误差和相对误差可分别用式(2-26) 和式(2-27) 计算。

$$D(\bar{x}) = \frac{\sigma}{\sqrt{n}} \tag{2-26}$$

$$E_r(\bar{x}) = \frac{D(\bar{x})}{|\bar{x}|} \tag{2-27}$$

(3) 测量值的实际误差

用上述方法所确定的测量误差，一般总是比测量值的实际误差小得多。这是因为仪器没有调整到理想状态，如不垂直、不水平、零位没有调整好等，会引起误差；仪表的实际工作条件不符合规定的正常工作条件，会引起附加误差；仪器经过长期使用后，零件发生磨损，装配状况发生变化等，会引起误差；操作者的习惯和偏向会引起误差；仪表所感受的信号实际上可能并不等于待测的信号，仪表电路可能会受到干扰等也会引起误差。

总之，测量值实际误差大小的影响因素是很多的。为了获得较准确的测量结果，需要有较好的仪器，也需要有科学的态度和方法，以及扎实的理论知识和实践经验。

2.8 间接测量中的误差传递

在许多实验和研究中，所得到的结果有时不是用仪器直接测量得到的，而是要把实验现场直接测量的值代入一定的关系式中，通过计算才能求得所需要的结果，即间接测量值。例如 $Re=du\rho/\mu$ 就是间接测量值。由于直接测量值总有一定的误差，因此它们必然引起间接测量值也有一定的误差，也就是说直接测量误差不可避免地传递到间接测量值中去，而产生间接测量误差。下面介绍误差传递的基本方程。

若间接测量值 y 与直接测量值 x_1，x_2，…，x_n 的函数关系为

$$y=f(x_1,x_2,\cdots,x_n) \tag{2-28}$$

则其微分式为：

$$\mathrm{d}y=\frac{\partial y}{\partial x_1}\mathrm{d}x_1+\frac{\partial y}{\partial x_2}\mathrm{d}x_2+\cdots+\frac{\partial y}{\partial x_n}\mathrm{d}x_n \tag{2-29}$$

如果以 Δy、Δx_1、Δx_2、…、Δx_n 分别代替上式中的 $\mathrm{d}y$、$\mathrm{d}x_1$、$\mathrm{d}x_2$、…、$\mathrm{d}x_n$，且当直接测量值的误差（Δx_1、Δx_2、…、Δx_n）很小，并且考虑到最不利的情况，应是误差累积和取绝对值，则可出求间接测量值的最大绝对误差和相对误差的传递公式：

最大绝对误差传递公式：

$$\Delta y=\left|\frac{\partial y}{\partial x_1}\right|\cdot|\Delta x_1|+\left|\frac{\partial y}{\partial x_2}\right|\cdot|\Delta x_2|+\cdots+\left|\frac{\partial y}{\partial x_n}\right|\cdot|\Delta x_n| \tag{2-30}$$

或

$$\Delta y=\sum_{i=1}^{n}\left|\frac{\partial y}{\partial x_i}\cdot\Delta x_i\right| \tag{2-31}$$

式中，$\frac{\partial y}{\partial x_i}$为误差传递系数；$\Delta x_i$ 为直接测量值的绝对误差；Δy 为间接测量值的最大绝对误差。

最大相对误差传递公式：

$$E_\mathrm{r}=\frac{\Delta y}{y}=\frac{1}{f(x_1,x_2,\cdots,x_n)}\left(\left|\frac{\partial y}{\partial x_1}\right|\cdot|\Delta x_1|+\left|\frac{\partial y}{\partial x_2}\right|\cdot|\Delta x_2|+\cdots+\left|\frac{\partial y}{\partial x_n}\right|\cdot|\Delta x_n|\right) \tag{2-32}$$

或

$$E_\mathrm{r}(y)=\frac{\Delta y}{y}=\sum_{i=1}^{n}\left|\frac{\partial y}{\partial x_i}\cdot\frac{\Delta x_i}{y}\right| \tag{2-33}$$

式(2-30)～式(2-33) 是相加合成法的一般公式，是误差的最大值，它近似等于误差实际值的概率是极小的。根据概率论，采用几何合成法则较符合事物固有的规律。以下式(2-34)～式(2-36) 为几何合成法的一般公式。

间接测量值 y 的绝对误差：

$$\Delta y=\sqrt{\left(\frac{\partial y}{\partial x_1}\Delta x_1\right)^2+\left(\frac{\partial y}{\partial x_2}\Delta x_2\right)^2+\cdots+\left(\frac{\partial y}{\partial x_n}\Delta x_n\right)^2} \tag{2-34}$$

间接测量值 y 的相对误差：

$$E_{\mathrm{r}}(\lambda)=\frac{\Delta y}{y}=\sqrt{\left(\frac{\partial y}{\partial x_1}\cdot\frac{\Delta x_1}{y}\right)^2+\left(\frac{\partial y}{\partial x_2}\cdot\frac{\Delta x_2}{y}\right)^2+\cdots+\left(\frac{\partial y}{\partial x_n}\cdot\frac{\Delta x_n}{y}\right)^2} \tag{2-35}$$

间接测量 y 值的标准误差：

$$\sigma_y=\sqrt{\left(\frac{\partial y}{\partial x_1}\sigma_{x_1}\right)^2+\left(\frac{\partial y}{\partial x_2}\sigma_{x_2}\right)^2+\cdots+\left(\frac{\partial y}{\partial x_n}\sigma_{x_n}\right)^2} \tag{2-36}$$

式中，σ_{x_1}，σ_{x_2}，…，σ_{x_n} 为直接测量值的标准误差；σ_y 为间接测量值的标准误差。

一些常用函数的相加合成法和几何合成法的简便公式请查阅相关书籍。

2.9 误差分析的具体应用

误差分析除用于计算测量结果的精确度外，还可以对具体的实验设计先进行误差分析，在找到误差的主要来源及每一个因素所引起的误差大小后，对实验方案和选用仪器仪表提出有益的建议。下面举 3 个例子进行说明。

【例 2-1】 现需在 $DN6$（公称直径为 6mm）的小铜管中进行阻力实验，因铜管内径太小，不能采用一般的游标卡尺测量，而是采用体积法进行直径间接测量。截取高度为 400mm（绝对误差±0.5mm）的管子，测量这段管子中水的容积，从而计算管子的平均内径。测量的量具为移液管，其体积刻度线准确，而且它的系统误差可以忽略。体积测量 3 次，分别为 11.31mL、11.26mL 和 11.30mL。求体积的算术平均值 $\bar{x}$、平均绝对误差 D（x）、相对误差 E_{r}（x）。

解： 算术平均值 $\bar{x}=\dfrac{\sum x_i}{n}=\dfrac{11.31+11.26+11.30}{3}\mathrm{m}=11.29\mathrm{mL}$

平均绝对误差　$D(x)=\dfrac{|11.29-11.31|+|11.29-11.26|+|11.29-11.30|}{3}\mathrm{mL}=0.02\mathrm{mL}$

相对误差　$E_{\mathrm{r}}(x)=\dfrac{D(x)}{|\bar{x}|}=\dfrac{0.02}{11.29}\times100\%=0.18\%$

【例 2-2】 流体流动阻力实验中测定摩擦系数 λ 使用的管为直径 8mm 的不锈钢管，测压点间距 $l=1.7\mathrm{m}$，现准备将层流管改为例 2-1 所述的 $DN6$ 的小铜管，测压点间距不变，采用 500mL 的量筒测其流量。希望在 $Re=2000$ 时，摩擦系数 λ 的精确度不低于 4.5%，问改用铜管后，是否能满足 λ 的精确度要求？测压点间距是否需要调整？采用 500mL 的量筒测其流量误差是否在合理范围内？

解： λ 的函数形式是：　　$\lambda=\dfrac{2g\pi^2}{16}\cdot\dfrac{d^5(R_1-R_2)}{lV_{\mathrm{s}}^2}$

式中，R_1，R_2 为被测量段两点间的表压（液柱）读数值，m；V_{s} 为流量，$\mathrm{m^3/s}$；l 为被测量段长度，m。

相对误差：

$$E_{\mathrm{r}}(\lambda)=\frac{\Delta\lambda}{\lambda}=\pm\sqrt{\left[5\left(\frac{\Delta d}{d}\right)\right]^2+\left[2\left(\frac{\Delta V_{\mathrm{s}}}{V_{\mathrm{s}}}\right)\right]^2+\left(\frac{\Delta l}{l}\right)^2+\left(\frac{\Delta R_1+\Delta R_2}{R_1-R_2}\right)^2}\times100\%$$

要求 $E_{\mathrm{r}}(\lambda)<4.5\%$，由于 $\dfrac{\Delta l}{l}$ 所引起的误差小于 $\dfrac{E_{\mathrm{r}}(\lambda)}{10}$，故可以略去不考虑。剩下 3 项分

误差，可按等效法进行分配，每项分误差和总误差的关系为：

$$E_r(\lambda)=\sqrt{3m_i^{\ 2}}=4.5\%$$

每项分误差　$m_i=\frac{4.5}{\sqrt{3}}\%=2.6\%$。

（1）流量项的分误差 m_1 估计

首先确定 V_s 值：

$$V_s=Re\frac{d\mu\pi}{4\rho}=2000\times\frac{0.008\times10^{-3}\times\pi}{4\times1000}\mathrm{m^3/s}=1\times10^{-5}\mathrm{m^3/s}=10\mathrm{mL/s}$$

这么小的流量可以采用 500mL 的量筒测量，量筒系统误差很小，可以忽略，读数误差为 $\Delta V=\pm5\mathrm{mL}$，计时用的秒表系统误差也可忽略，开停秒表的随机误差估计为 $\Delta\tau=\pm0.1\mathrm{s}$，当 $Re=2000$ 时，若每次测量水体积 V 约为 450mL，则需时间 τ 为 48s 左右。故流量测量最大误差为：

$$\frac{\Delta V_s}{V_s}=\pm\left(\frac{\Delta V}{V}+\frac{\Delta\tau}{\tau}\right)=\pm\left(\frac{5}{450}+\frac{0.1}{48}\right)=\pm(0.011+2.08\times10^{-3})$$

式中具体数字说明$\frac{\Delta V}{V}$的误差较大，$\frac{\Delta\tau}{\tau}$很小可以忽略。因此流量项的分误差为：

$$m_1=2\frac{\Delta V_s}{V_s}=2\times0.011\times100\%=2.2\%$$

没有超过每项分误差范围。

（2）管径项的分误差 m_2

由例 2-1 知道管径 d 可由体积法进行间接测量：

$$V=\frac{\pi}{4}d^2h\text{，则 }d=\sqrt{\frac{V}{h}\times\frac{4}{\pi}}$$

已知管高度 $h=400\mathrm{mm}$，绝对误差为$\pm0.5\mathrm{mm}$。为保险起见，仍采用几何合成法计算 d 的相对误差：

$$\frac{\Delta d}{d}=\frac{1}{2}\left(\frac{\Delta V}{V}+\frac{\Delta h}{h}\right)$$

由例 2-1 计算出$\frac{\Delta V}{V}$的相对误差为 0.18%。再代入具体数值，可得

$$m_2=5\frac{\Delta d}{d}=\frac{5}{2}\left(\frac{\Delta V}{V}+\frac{\Delta h}{h}\times100\%\right)=\frac{5}{2}\times\left(0.18\%+\frac{0.5}{400}\times100\%\right)=0.8\%$$

也没有超过每项分误差范围。

（3）压差项的分误差 m_3

单管式压差计用分度为 1mm 的尺子测量，系统误差可以忽略，读数随机绝对误差 ΔR 为$\pm0.0005\mathrm{m}$。

$$m_3=\frac{\Delta R_1+\Delta R_2}{R_1-R_2}=\frac{2\Delta R_1}{R_1-R_2}=\frac{2\times0.0005}{R_1-R_2}$$

压差测量值 R_1-R_2 与两测压点间的距离 l 成正比，则有

$$R_1-R_2=\frac{64}{Re}\cdot\frac{l}{d}\cdot\frac{u^2}{2g}=\frac{64}{2000}\cdot\frac{l}{0.006}\cdot\frac{\left(\frac{9.4\times10^{-6}}{0.785\times0.006^2}\right)^2}{2g}=0.03l$$

式中，u 为平均流速，m/s。

由上式可算出 l 的变化对压差项分误差的影响，见表 2-3。

表 2-3 l 的变化对压差项分误差的影响

l/m	R_1-R_2/m	$\frac{2\Delta R_1}{R_1-R_2}$/%
0.500	0.015	6.7
1.000	0.030	3.3
1.500	0.045	2.2
2.000	0.060	1.6

由表 2-3 中可见，选用 $l \geqslant 1.500$m 可满足要求。若实验采用 $l=1.500$m，其压差项的分误差 m_3 为：$m_3=\frac{\Delta R_1+\Delta R_2}{R_1-R_2}=\frac{2\Delta R_1}{R_1-R_2}=\frac{2\times 0.0005}{0.03\times 1.500}\times 100\%=2.2\%$

总误差为

$$E_r(\lambda)=\frac{\Delta\lambda}{\lambda}=\pm\sqrt{m_1^2+m_2^2+m_3^2}\times 100\%=\pm\sqrt{(2.2\%)^2+(0.8\%)^2+(2.2\%)^2}\times 100\%=\pm 3.2\%$$

通过以上误差分析可知：

① 改用铜管后，λ 的精确度能满足要求。

② 测压点的间距不需要调整即可满足要求。若 $l>1.500$m，可以使误差进一步减小。

③ 直径项的分误差，虽传递系数较大（等于 5），对总误差影响较大，但因选择体积法进行直径间接测量的方案合理，这项测量精确度高，对总误差的影响反而减小了。

④ 流量项的分误差在合理的误差范围内，即采用 500mL 的量筒测流量是合适的，但若改用精确度更高一级的量筒，使读数误差减小，则可以提高实验结果的精确度。

【例 2-3】 试求例 2-2 实验中，当 $Re=300$ 时，所测 λ 的相对误差为多少？（l 选用 1.7m，水温 20℃，$R_1-R_2=0.0068$m，当出水量为 450mL 时，所需时间为 319s）

解： 由例 2-2 知 $m_1=2.2\%$，$m_2=0.8\%$

$$m_3=\frac{2\Delta R_1}{R_1-R_2}=\frac{2\times 0.0005}{0.0068}\times 100\%=14.7\%$$

$$E_r(\lambda)=\pm\sqrt{m_1^2+m_2^2+m_3^2}\times 100\%=\pm\sqrt{(2.2\%)^2+(0.8\%)^2+(14.7\%)^2}\times 100\%=\pm 14.9\%$$

结果表明，由于压差下降，压差测量的相对误差上升，致使 λ 测量的相对误差增大。当 $Re=300$ 时，λ 的理论值为 $\frac{64}{Re}=0.213$，如果实验结果与此值有差异（例如 $\lambda=0.181$ 或 $\lambda=0.245$），并不一定说明 λ 的测量值与理论值不符，还要看偏差多少。像括号中的这种偏差是由测量精密度不高引起的，如果提高压差测量精度或者增加测量次数并取平均值，就有可能与理论值相符。以上例子充分说明了误差分析在实验中的重要作用。

第3章 实验数据处理

化工原理实验的重要目的之一就是要将所测试的实验原始数据进行数据处理，获得各变量之间的定量关系，在此基础上进一步分析实验现象、总结规律，验证实验结果与理论的一致性，并为生产与设计提供基础数据和理论指导。实验数据处理就是以测量为手段，以研究对象的概念、状态为基础，以数学运算为工具推断出某测量值的真值，并导出某些具有规律性结论的整个过程。

常用的实验数据处理方法有列表法、图示法和数学方程表示法三种。

3.1 列表法

将实验数据按自变量和因变量的关系，以一定的顺序列出数据表，即为列表法。列表法有许多优点，如原始数据记录表会给数据处理带来方便，使数据易于比较，形式紧凑，同一表格内可以表示几个变量间的关系等。列表通常是整理数据的第一步，为曲线图的标绘或数学公式的整理打下基础。

3.1.1 数据表的分类

实验数据表一般分为原始数据记录表和实验数据处理结果表两大类。现以流体阻力实验测定 λ-Re 关系为例进行说明。

(1) 原始数据记录表

原始数据记录表是根据实验的具体内容设计的，以清楚地记录所有待测数据。该表必须在实验前完成。原始数据记录表如表 3-1 所列。

表 3-1 流体阻力实验原始数据记录表

实验装置 第________套 实验起始温度________℃ 实验结束温度________℃ 实验时间____________

光滑管内径 $d=$______m 测压点间距 $l=$______m 粗糙管内径 $d=$______m 测压点间距 $l=$______m

序号	流量 V_s/(L/h)	光滑管直管压差 Δp_f			流量 V_s/(L/h)	粗糙管直管压差 Δp_f		
		kPa	cmH_2O			kPa	cmH_2O	
			左	右			左	右
1								
2								
⋮								
n								

(2) 实验数据处理结果表

实验数据处理结果表可细分为中间计算结果表（体现出实验过程主要变量的计算结果）、综合结果表（表达实验过程中得出的结论）和误差分析表（表达实验值与参照值或理论值的误差范围）等。实验报告中要用到哪些表，应根据实验具体情况而定。流体流动阻力实验数据处理结果表如表 3-2 所列，误差分析结果表如表 3-3 所列。

表 3-2　流体流动阻力实验数据处理结果表（光滑管）

序号	流量V_s/(m^3/s)	平均流速 u/(m/s)	压力损失值 Δp_f/kPa	Re	λ	λ-Re 关系式
1						
2						
⋮						
n						

表 3-3　流体流动阻力实验误差分析结果表

序号	Re	实验 λ-Re 关系式	$\lambda_{实验}$	$\lambda_{理论}$(或 $\lambda_{经验}$)公式	$\lambda_{理论}$(或 $\lambda_{经验}$)	相对误差/%
1						
2						
⋮						
n						

3.1.2　设计实验数据表应注意的事项

① 表格设计力求简明扼要，一目了然，便于阅读和使用。记录、计算项目要满足实验需要，各种实验条件可以写在表的名称和表格之间。

② 表头列出物理量的名称、符号和计量单位。符号与计量单位之间用斜线“/”隔开。斜线不能重叠使用，可以根据情况使用“()”。计量单位不宜混在数字之中，造成分辨不清。

③ 注意有效数字位数，即记录的数字应与测量仪表的准确度相匹配，不可过多或过少。

④ 物理量的数值较大或较小时，要用科学计数法表示。以“物理量的符号$\times 10^{\pm n}$/计量单位”的形式记入表头。注意：表头中的 $10^{\pm n}$ 与表中的数据应服从下式。

$$物理量的实际值\times 10^{\pm n}=表中数据$$

⑤ 为便于引用，每一个数据表都应在表的上方写明表的序号和表的名称。表的序号应按其在报告中出现的顺序编写，并在正文中有所交代。同一个表尽量不跨页，必须跨页时，在后页的表上须注“续表”。

⑥ 数据书写要清楚整齐。修改时宜用单线将错误的画掉，将正确的写在下面。记录者的姓名以及要备注的问题可作为“表注”，写在表的下方。

3.2　图示法

图示法是表示实验中各变量之间关系最常用的方法。它是将整理得到的实验数据或计算结果标绘在适宜的坐标上，然后将数据点连成光滑的曲线或直线。该法的优点是直观清晰，便于比较，容易看出数据中的极值点、转折点、周期性、变化率以及其他特性，准确的图形还可以在不知数学表达式的情况下进行微积分运算，因此得到广泛的应用。

在工程实验中正确的作图必须遵循以下基本原则，才能得到与实验点位置偏差最小且光滑的曲线（或直线）图形。

3.2.1 坐标系选择的基本原则

化工原理实验中常用的坐标系为直角坐标系、单对数坐标系和双对数坐标系。单对数坐标系如图 3-1 所示，一个轴是分度均匀的普通坐标轴，另一个轴是分度不均匀的对数坐标轴。双对数坐标系如图 3-2 所示，两个轴都是对数坐标轴。

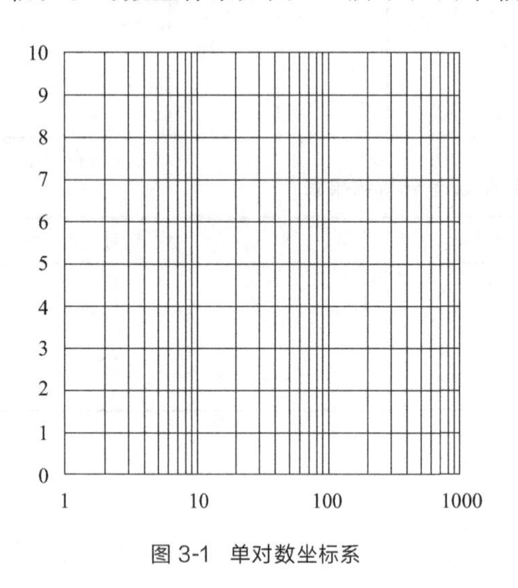

图 3-1 单对数坐标系

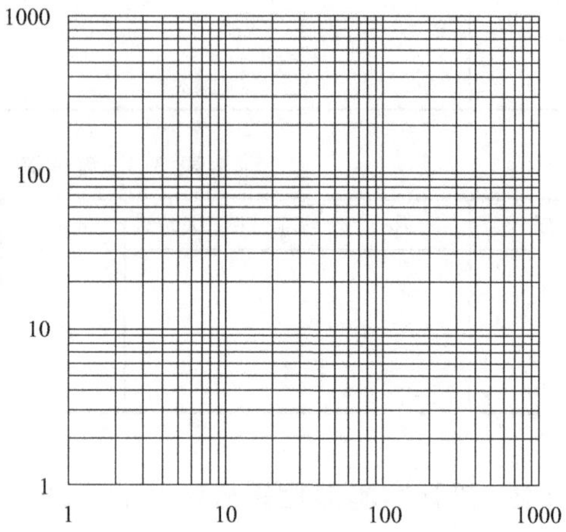

图 3-2 双对数坐标系

(1) 直角坐标系

变量 x、y 间的函数关系式为：

$$y=a+bx$$

此为直线函数，将变量 x、y 标绘在直角坐标纸上得到一直线图，即可求出截距 a 和斜率 b。

(2) 单对数坐标系

在下列情况下，建议使用单对数坐标系。

① 变量之一在所研究的范围内发生了几个数量级的变化。

② 在自变量由零开始逐渐增大的初始阶段，当自变量的少许变化引起因变量极大变化时，采用单对数坐标可使曲线最大变化范围伸长，使图形轮廓更清楚。

③ 当需要变换某种非线性关系为线性关系时，可用单对数坐标。例如指数函数 $y=a\mathrm{e}^{bx}$、对数函数 $y=a+b\lg x$。a、b 为待定系数。详见 3.3.1 部分。

(3) 双对数坐标系

在下列情况下，建议使用双对数坐标系。

① 变量 x、y 在数值上均发生了几个数量级的变化。

② 需要将曲线开始部分划分成展开的形式。

③ 当需要变换某种非线性关系为线性关系时，可用双对数坐标。例如幂函数 $y=ax^b$，式中，a、b 为待定系数。详见 3.3.1 部分。

3.2.2 坐标分度的确定

坐标分度是按每条坐标轴所能代表的物理量大小来定的，即选择适当的坐标比例尺。坐标分度的确定方法如下。

① 为了得到良好的图形，在 x、y 的误差 $D(x)$、$D(y)$ 已知的情况下，比例尺的取法应使实验“点”的边长为 $2D(x)$、$2D(y)$（近似于正方形），而且使 $2D(x)=2D(y)=1\sim2\text{mm}$，若 $2D(x)=2D(y)=2\text{mm}$，则它们的比例尺应为：

$$M_y=\frac{2\text{mm}}{2\Delta y}=\frac{1}{\Delta y}\text{mm} \tag{3-1}$$

$$M_x=\frac{2\text{mm}}{2\Delta x}=\frac{1}{\Delta x}\text{mm} \tag{3-2}$$

如已知温度误差 $D(\text{t})=0.05℃$，则

$$M=\frac{1\text{mm}}{0.05℃}=20\text{mm}/℃$$

此时温度 1℃ 的坐标为 20mm 长。若感觉太长，可取 $2D(x)=2D(y)=1\text{mm}$，此时 1℃ 的坐标为 10mm 长。

② 若测量数据的误差未知，那么坐标的分度应与实验数据的有效数字相匹配，即最适合的分度是使实验曲线坐标读数和实验数据具有同样的有效数字位数。其次，横、纵坐标之间的比例不一定一致，应根据具体情况选择，使实验曲线的坡度处于 30°～60°，这样的曲线坐标读数准确度较高。

③ 推荐使用坐标轴的比例常数 $M=(1、2、5)\times10^{\pm n}$（$n$ 为正整数），不使用 3、4、6、7、8、9 等的比例常数，因为后者的比例常数会引起绘制图形的麻烦，容易导致错误。

3.2.3 设计实验数据图应注意的事项

① 对于两个变量的系统，习惯上选横轴为自变量，纵轴为因变量。在两轴侧要标明变量名称、符号和单位，如离心泵特性曲线的横轴须标明：流量 $Q/(\text{m}^3/\text{h})$。尤其是单位，初学者往往容易忽略。

② 坐标分度要适当，使变量的函数关系表现清楚。直角坐标的原点不一定选为零点，应根据所标绘数据范围而定，以使图形匀称居中为原则。

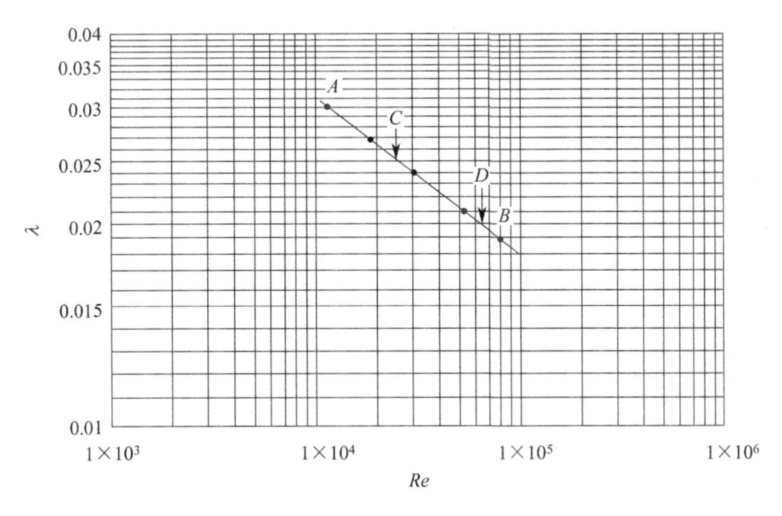

图 3-3　摩擦系数 λ 与 Re 关系图

对数坐标的原点不是零。标在对数坐标轴上的值是真值，而不是对数值。例如某一次

阻力实验计算得到$Re_1=1.1\times10^4$，$\lambda_1=0.031$，$Re_2=7.8\times10^4$，$\lambda_1=0.019$，将其标绘在图3-3的双对数坐标上，即为A点和B点。由于1、10、100、1000等数的对数分别为0、1、2、3等，所以在对数坐标纸上每一数量级的距离是相等的，但在同一数量级内的刻度并不是等分的，使用时应严格遵循图纸的坐标系。

③ 若在同一张坐标纸上同时标绘几组测量值，则各组要用不同符号（如：o、Δ、×等）以示区别。若n组不同函数同绘在一张坐标纸上，则在曲线上要标明函数关系名称。

④ 图必须有图的序号和图的名称，图号应按报告中出现的顺序编写，并在正文中有所交代，必要时还应有图注。

⑤ 图线应光滑。利用曲线板等工具将各离散点连接成光滑曲线或直线，并使线条尽可能通过较多的实验点，或者使线条以外的点尽可能位于线条的附近，并使线条两侧的点数大致相等。

3.3 数学方程表示法

在实验研究中，除了用表格和图形描述变量间的关系外，还常常把实验数据整理成方程式，以描述过程或现象的自变量和因变量之间的关系，即建立过程的数学模型。其方法是将实验数据绘制成曲线，与已知的函数关系式的典型曲线（线性方程、幂函数方程、指数函数方程、抛物线函数方程、双曲线函数方程等）进行对照并确定适宜的函数关系，然后用图解法或者回归分析法确定函数式中的各种常数。运用计算机将实验数据结果回归为数学方程已成为实验数据处理的主要手段，例如Excel和Origin软件就是化工原理实验中常用的数据处理和绘图软件，详见第4章。所得函数表达式是否能准确地反映实验数据所存在的关系，应通过检验加以确认。

数学方程式选择的原则是：既要求形式简单，所含常数较少，同时也希望能准确地表达实验数据之间的关系。但要同时满足两者的条件往往难以做到，通常是在保证必要的准确度的前提下，尽可能选择简单的线性关系或者经过适当方法转换成线性关系的形式，例如将幂函数和指数函数通过取对数转化成线性方程，使数据处理工作得到简化。

3.3.1 图解分析法

当公式选定后，可用图解法求方程式中的常数，本小节以幂函数和指数函数、对数函数为例进行说明。

(1) 幂函数的线性图解

幂函数$y=ax^b$中a、b为需要求解的系数。若直接在直角坐标系上作图必为曲线，为此把等式两边同时取对数，则：

$$\lg y=\lg a+b\lg x \tag{3-3}$$

令$\lg y=Y$，$\lg x=X$，则式(3-3)变换为：

$$Y=\lg a+bX \tag{3-4}$$

式(3-4)为一直线方程，在直角坐标纸上作图为一条直线，通过其截距和斜率可求得a、b值。但为了避免对每一个实验数据x、y都求对数值的麻烦，所以我们可以采用双对数坐标纸作图。双对数坐标中求系数a、b的方法如下。

1）系数b的求法　系数b即为直线的斜率，如图3-3所示。在对数坐标上求取斜率的方法与直角坐标上的求法不同。双对数坐标纸上直线的斜率需要用对数值来计算。在直线上任意取两个点（注意：一定要在线上取点，不可以直接用实验点）代入式(3-5)计算，则：

$$b=\frac{\lg y_2-\lg y_1}{\lg x_2-\lg x_1} \tag{3-5}$$

2）系数 a 的求法　在直线上任取一个点的数值和已求出的斜率 b，代入原方程 $y=ax^b$ 中，通过计算求得 a 值。

3）联立方程法　a、b 值也可通过在直线上任意取两个点，用联立方程的方法求解。

【例 3-1】 在流体流动阻力实验中，使用光滑管，在湍流区测定数据。现已将根据实验数据计算出的摩擦系数 λ 与 Re 的值标绘在图 3-3 上，并将实验点拟合成了一条直线(此处具体数字略)。请确定摩擦系数 λ 与 Re 的关系式 $\lambda=aRe^b$ 中的 a、b 值。

解：方法一

在图 3-3 的直线上任意取两点：C 点（2.6×10^4，0.025）和 D 点（6.3×10^4，0.020），根据式(3-5) 计算 b 值

$$b=\frac{\lg y_2-\lg y_1}{\lg x_2-\lg x_1}=\frac{\lg 0.020-\lg 0.025}{\lg(6.3\times10^4)-\lg(2.6\times10^4)}=-0.2521$$

$$a=\frac{\lambda}{Re^b}=\frac{0.025}{(2.6\times10^4)^{-0.2521}}=0.3243$$

方法二

将 C 点和 D 点的值分别代入方程 $\lambda=aRe^b$ 联立求解：

$$\begin{cases}0.025=a\times(2.6\times10^4)^b\\0.020=a\times(6.3\times10^4)^b\end{cases}$$

解得：$a=0.3243$，$b=-0.2521$。

按有效数字规则写成 $a=0.32$，$b=-0.25$，则实验所得摩擦系数 λ 与 Re 的关系式为：

$$\lambda=0.32Re^{-0.25} \quad 或 \quad \lambda=\frac{0.32}{Re^{0.25}}$$

（2）指数或对数函数的线性图解

当所研究的函数关系呈指数函数 $y=ae^{bx}$ 或对数函数 $y=a+b\lg x$ 时，将实验数据标绘在单对数坐标纸上的图形是一条直线。线性化方法如下：

对 $y=ae^{bx}$ 上式两边同时取对数，则

$$\lg y=\lg a+bx\lg e \tag{3-6}$$

令 $\lg y=Y$，$b\lg e=k$，则式(3-6) 变为

$$Y=\lg a+kx \tag{3-7}$$

对于对数函数 $y=a+b\lg x$，令 $\lg x=X$，则有：

$$y=a+bX \tag{3-8}$$

系数的求法如下。

1）系数 b 的求法　对于 $y=ae^{bx}$，纵轴为对数坐标，斜率为：

$$k=\frac{\lg y_2-\lg y_1}{x_2-x_1} \tag{3-9}$$

$$b=\frac{k}{\lg e} \tag{3-10}$$

对 $y=a+b\lg x$，横轴为对数坐标，斜率为：

$$b=\frac{y_2-y_1}{\lg x_2-\lg x_1} \tag{3-11}$$

2）系数 a 的求法　系数 a 的求法与幂函数中所述方法基本相同，可将直线上任一点处的坐标值和已经求出的系数 b 代入函数关系式后求解。

3）联立方程法　a、b 值也可通过在直线上任意取两个点，用联立方程的方法求解。

（3）二元线性方程的图解

若实验研究中，所研究对象的物理量是一个因变量与两个自变量，它们若成线性关系，则可采用以下函数式表示：

$$y=a+bx_1+cx_2 \tag{3-12}$$

在图解此类函数式时，应首先令其中一自变量恒定不变，例如使 x_1 为常数，则上式可改写成：

$$y=d+cx_2 \tag{3-13}$$

式中，$d=a+bx_1=$常数。

由 y 与 x_2 的数据可在直角坐标中标绘出一条直线，如图 3-4(a) 所示。采用上述图解法即可确定 x_2 的系数 c。

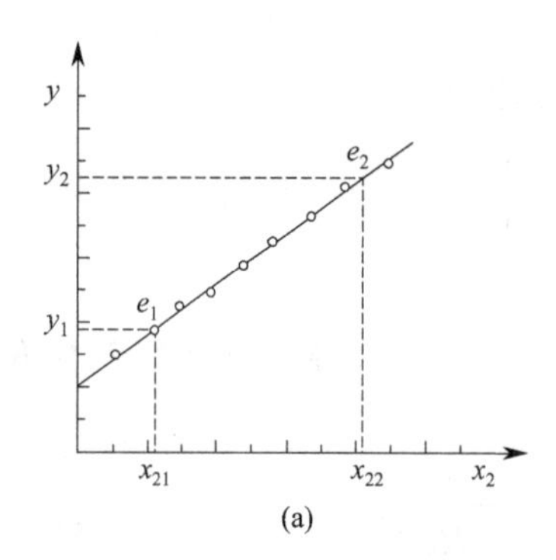

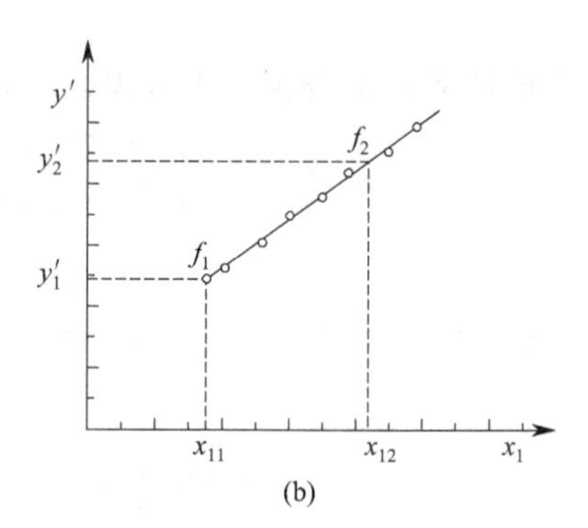

图 3-4　二元线性方程图解示意

在图 3-4(a) 中直线上任取两点 $e_1(x_{21}, y_1)$，$e_2(x_{22}, y_2)$，则有

$$c=\frac{y_2-y_1}{x_{22}-x_{21}} \tag{3-14}$$

当求得 c 后，将其代入式(3-12) 中，并将式(3-12) 重新改写成以下形式：

$$y-cx_2=a+bx_1 \tag{3-15}$$

令 $y'=y-cx_2$ 于是可得一新的线性方程：

$$y'=a+bx_1 \tag{3-16}$$

由实验数据 y、x_2 和 c 计算得 y'，由 y' 与 x_1 在图 3-4(b) 中标绘其直线，并在该直线上任取 $f_1(x_{11}, y'_1)$ 及 $f_2(x_{12}, y'_2)$ 两点，由 f_1、f_2 两点即可确定 a、b 两个常数。

$$b=\frac{y'_2-y'_1}{x_{12}-x_{11}} \tag{3-17}$$

$$a=\frac{y'_1x_{12}-y'_2x_{11}}{x_{12}-x_{11}} \tag{3-18}$$

应该指出的是，在确定 b、a 时，其自变量 x_1，x_2 应同时改变，才能使其结果覆盖整个实验范围。

施伍德（Sherwood）利用7种不同流体对流过圆形直管的强制对流传热进行研究，并取得大量数据，采用幂函数形式进行处理，其函数形式为：

$$Nu=CRe^mPr^n \tag{3-19}$$

式中，Nu 为努塞特（Nusselt）数；Re 为雷诺（Reynolds）数；Pr 为普朗特（Prandtl）数；C、m 和 n 为待定常数。

Nu 随 Re 及 Pr 的变化而变化，将上式两边取对数，采用变量代换，使之化为二元线性方程形式：

$$\lg Nu=\lg C+m\lg Re+n\lg Pr \tag{3-20}$$

令 $y=\lg Nu$，$x_1=\lg Re$，$x_2=\lg Pr$，$a=\lg C$，上式即可表示为二元线性方程式：

$$y=a+mx_1+nx_2 \tag{3-21}$$

现将式(3-20）改写为以下形式，确定常数 n（固定变量 Re 值，使 Re=常数，自变量减少一个）。

$$\lg Nu=(\lg C+m\lg Re)+n\lg Pr \tag{3-22}$$

施伍德固定 $Re=10^4$，将7种不同流体的实验数据在双对数坐标纸上标绘 Nu 和 Pr 之间的关系，得出 Pr 准数的指数 n，然后按下式图解法求解：

$$\lg(Nu/Pr^n)=\lg C+m\lg Re \tag{3-23}$$

以 Nu/Pr^n 为纵坐标，以 Re 为横坐标，在双对数坐标纸上作图，即可由斜率和截距求出 C 和 m 值。这样，经验公式中的所有待定常数 C、m 和 n 均被确定。

3.3.2 回归分析法

在3.3.1部分中介绍了用图解法获得经验公式的过程。尽管图解法有很多优点，但是由于作图过程中是通过离散点画出的拟合直线，任意性较大，且是从坐标图上读数，读数也会带来误差。那么要得到这些实验数据变量之间的定量关系式，使之尽可能地符合实际情况，应用最广泛的一种数理统计方法就是回归分析法，用这种数学方法可以从大量观测的散点数据中寻找到能反映事物内部的一些统计规律，并可以用数学模型形式表达出来。

回归也称拟合。对具有相关关系的两个变量，若用一条直线描述，则称一元线性回归，用一条曲线描述，则称一元非线性回归。对具有相关关系的3个变量，其中一个因变量、两个自变量，若用平面描述，则称二元线性回归，用曲面描述，则称二元非线性回归。依次类推，可以延伸到 n 维空间进行回归，则称多元线性回归或多元非线性回归。处理实验问题时，往往将非线性问题转化为线性问题来处理，建立线性回归方程的最有效方法为线性最小二乘法。虽然化工原理实验过程中大量遇到的回归问题多为二元以上回归，但因为一元线性回归的概念容易理解。因此，本小节主要讲述一元线性回归方程的求法，为其他的回归分析学习和用Origin软件进行回归分析打下基础。

3.3.2.1 一元线性回归方程的求法

在科学实验的数据统计方法中，通常要从获得的实验数据（x_i，y_i，$i=1, 2, \cdots$，

n）中，寻找其自变量 x_i 与因变量 y_i 之间函数关系 $y=f(x)$。由于实验测定数据一般都存在误差，因此，不能要求所有的实验点均在 $y=f(x)$ 所表示的曲线上，只需满足实验点（x_i，y_i）与 $f(x_i)$ 的残差 $d_i=y_i-f(x_i)$ 小于给定的误差即可。此类寻求实验数据关系近似函数表达式 $y=f(x)$ 的问题称为曲线拟合。

曲线拟合首先应针对实验数据的特点，选择适宜的函数形式，确定拟合的目标函数。例如在取得两个变量的实验数据之后，若在普通直角坐标纸上标出各个数据点，如果各点的分布近似于一条直线，则可考虑采用线性回归求其表达式。

设给定 n 个实验点 (x_1, y_1)，(x_2, y_2)，…，(x_n, y_n)，其离散点如图 3-5 所示，于是可以利用如下一条直线来代表它们之间的关系

$$y'=a+bx \tag{3-24}$$

式中，y'为由回归式算出的值，称回归值；a、b 为回归系数。

对每一测量值 x_i 可由式(3-24) 求出一回归值 y'。回归值 y'与实测值 y_i 之差的绝对值 $d_i=|y_i-y_i'|=|y_i-(a+bx_i)|$，它表明 y_i 与回归直线的偏离程度。两者偏离程度越小，说明直线与实验数据点拟合越好。$|y_i-y_i'|$ 值代表点（x_i，y_i）沿平行于 y 轴方向到回归直线的距离，如图 3-6 上各竖直线 d_i 所示。

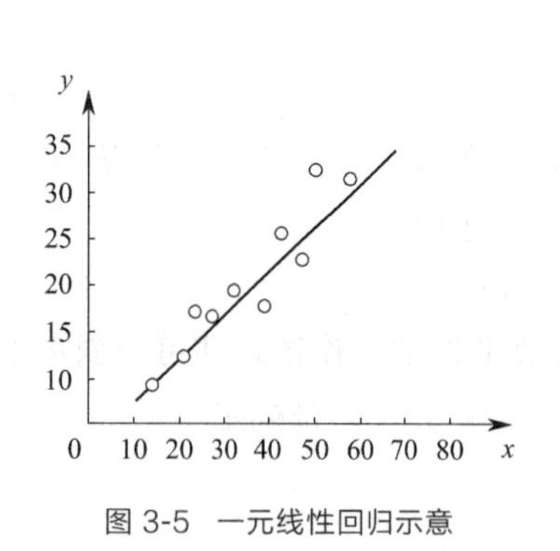

图 3-5　一元线性回归示意

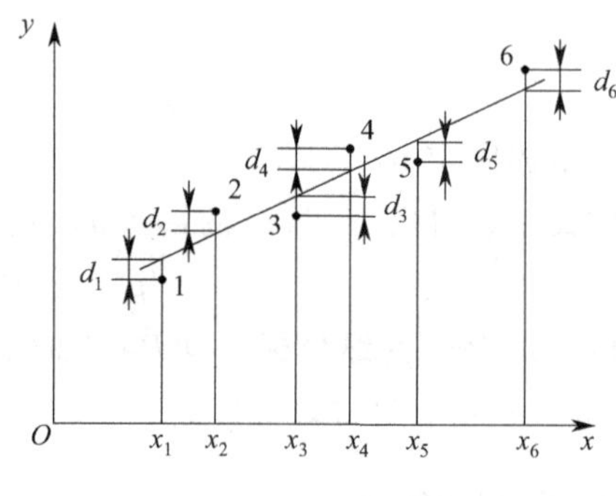

图 3-6　实验曲线示意

曲线拟合时应确定拟合的目标函数。选择残差平方和为目标函数的处理方法即为最小二乘法。此法是寻求实验数据近似函数表达式的更为严格而有效的方法。其定义为：最理想的曲线就是能使各点同曲线的残差平方和为最小。

设残差平方和 Q 为：

$$Q=\sum_{i=1}^{n} d_i^2=\sum_{i=1}^{n}\left[y_i-(a+bx_i)\right]^2 \tag{3-25}$$

式中，x_i、y_i 是已知值，故 Q 为 a 和 b 的函数。为使 Q 值达到最小，根据极值原理，只要将式(3-25) 分别对 a、b 求偏导数 $\frac{\partial Q}{\partial a}$、$\frac{\partial Q}{\partial b}$，并令其等于零，即可求 a、b 值，这就是最小二乘法原理，即

$$\begin{cases}\dfrac{\partial Q}{\partial a}=-2\sum\limits_{i=1}^{n}(y_i-a-bx_i)=0\\[2ex]\dfrac{\partial Q}{\partial b}=-2\sum\limits_{i=1}^{n}(y_i-a-bx_i)x_i=0\end{cases} \tag{3-26}$$

由式(3-26) 可得正规方程：

$$\begin{cases} a+\bar{x}b=\bar{y} \\ n\bar{x}a+(\sum_{i=1}^{n}x_i^2)b=\sum_{i=1}^{n}x_iy_i \end{cases} \tag{3-27}$$

其中

$$\bar{x}=\frac{1}{n}\sum_{i=1}^{n}x_i \quad \bar{y}=\frac{1}{n}\sum_{i=1}^{n}y_i \tag{3-28}$$

解方程（3-27），可得到回归式中的 a（截距）和 b（斜率）

$$b=\frac{\sum(x_i \cdot y_i)-n\bar{x}\bar{y}}{\sum x_i^2-n(\bar{x})^2} \tag{3-29}$$

$$a=\bar{y}-b\bar{x} \tag{3-30}$$

【例 3-2】 转子流量计标定时得到的读数与流量数据如表 3-4 所列，用最小二乘法求实验方程。

表 3-4　转子流量计标定时得到的读数与流量数据

读数 x/格	0	2	4	6	8	10	12	14	16
流量 y/(m^3/h)	30.00	31.25	32.58	33.71	35.01	36.20	37.31	38.79	40.04

解： $\sum(x_iy_i)=2668.58, \bar{x}=8, \bar{y}=34.9878, \sum x_i^2=816$

$$b=\frac{\sum(x_iy_i)-n\bar{x}\bar{y}}{\sum x_i^2-n(\bar{x})^2}=\frac{2668.58-9\times8\times34.9878}{816-9\times8^2}=0.623$$

$$a=\bar{y}-b\bar{x}=34.9878-0.623\times8=30.0$$

所以，回归方程为 $y=30.0+0.623x$。

3.3.2.2　回归效果的检验

实验数据变量之间的关系具有不确定性，一个变量的每一个值对应的是整个集合值。当 x 改变时，y 的分布也以一定的方式改变。在这种情况下，变量 x 和 y 间的关系就称为相关关系。

在以上求回归方程的计算过程中，并不需要事先假定两个变量之间一定有某种相关关系。就方法本身而论，即使平面图上是一群完全杂乱无章的离散点，也能用最小二乘法给其配一条直线来表示 x 和 y 之间的关系，但显然这是毫无意义的。实际上只有两变量是线性关系时进行线性回归才有意义，因此，必须对回归效果进行检验。

(1) 相关系数

我们可引入相关系数 r 对回归效果进行检验，相关系数 r 是表明两个变量线性关系密切程度的一个数量性指标。若回归所得线性方程为 $y'=a+bx$，则相关系数 r 的计算式为（推导过程略）：

$$r=\frac{\sum(x_i-\bar{x})(y_i-\bar{y})}{\sqrt{\sum(x_i-\bar{x})^2\sum(y_i-\bar{y})^2}} \tag{3-31}$$

r 的变化范围为 $-1\leqslant r\leqslant1$，其正、负号取决于 $\sum(x_i-\bar{x})(y_i-\bar{y})$，与回归直线方程的斜率 b 一致。r 的几何意义可用图 3-7 来说明。

当 $r=\pm1$ 时，即 n 组实验值（x_i，y_i）全部落在直线 $y=a+bx$ 上，此时称完全相关，如图 3-7(d) 和(e) 所示。

当 $0<|r|<1$ 时，代表绝大多数的情况，这时 x 与 y 存在着一定的线性关系。当 $r>0$ 时，散点图的分布是 y 随 x 增大而增大，此时称 x 与 y 正相关，如图 3-7(b) 所示；当 $r<0$ 时，散点图的分布是 y 随 x 增大而减小，此时称 x 与 y 负相关，如图 3-7(c) 所示；$|r|$ 越小，散点离回归线越远，越分散。当 $|r|$ 越接近 1 时，即 n 组实验值 (x_i, y_i) 越靠近 $y=a+bx$，变量 x 与 y 之间的关系越接近于线性关系。当 $r=0$ 时，变量之间就完全没有线性关系了，如图 3-7(a) 所示。应该指出，没有线性关系，并不等于不存在其他函数关系，如图 3-7(f) 所示。

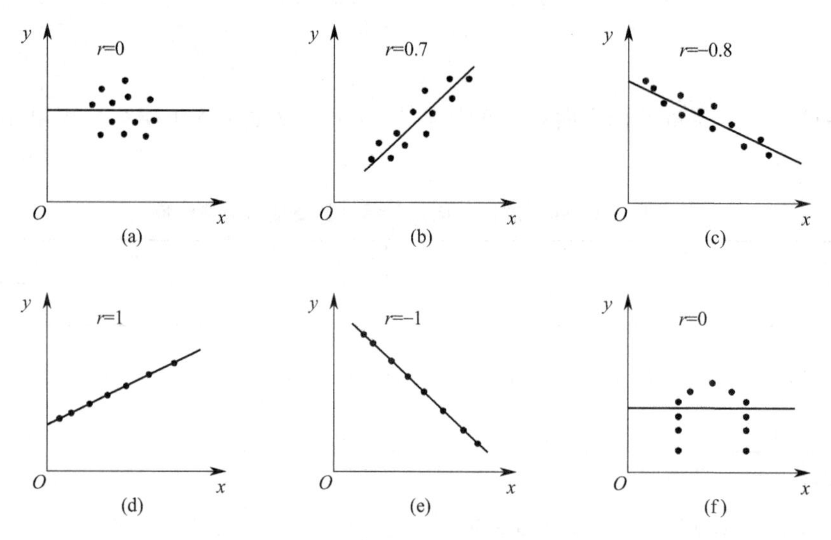

图 3-7 相关系数的几何意义图

(2) 显著性检验

如上所述，相关系数 r 的绝对值越接近 1，x、y 间越线性相关。但究竟 $|r|$ 接近到什么程度才能说明 x 与 y 之间存在线性相关关系呢？这就有必要对相关系数进行显著性检验。只有当 $|r|$ 达到一定程度才可以采用回归直线来近似地表示 x、y 之间的关系，此时可以说明相关关系显著。一般来说，相关系数 r 要达到相应显著性水平对应的 $r_{\min}$ 值与实验数据的个数 n 有关。因此只有 $|r|>r_{\min}$ 时，才能采用线性回归方程来描述其变量之间的关系。$r_{\min}$ 值可以从表 3-5 中查出。利用该表可根据实验点个数 n 及显著水平系数 α 查出相应的 $r_{\min}$。显著水平系数 α 一般可取 1%或 5%。α 越小，显著程度越高。

【例 3-3】 求例 3-2 中转子流量计标定实验的实际相关系数 r。

解： $n=9$，$n-2=7$，查表 3-5 得

$\alpha=0.01$ 时，$r_{\min}=0.798$；$\alpha=0.05$ 时，$r_{\min}=0.666$

$\bar{x}=8$，$\bar{y}=34.9878$

$\sum(x_i-\bar{x})(y_i-\bar{y})=149.46$，$\sum(x_i-\bar{x})^2=240$，$\sum(y_i-\bar{y})^2=93.12$

$$r=\frac{\sum(x_i-\bar{x})(y_i-\bar{y})}{\sqrt{\sum(x_i-\bar{x})^2\sum(y_i-\bar{y})^2}}=\frac{149.46}{\sqrt{240\times93.12}}=0.99976>0.798$$

上述计算结果说明此例的相关系数在 $\alpha=0.01$ 的水平是高度显著的。

表 3-5　相关系数 r_{min} 检验表

$n-2$	α		$n-2$	α	
	0.05	0.01		0.05	0.01
1	0.997	1.000	21	0.413	0.526
2	0.950	0.990	22	0.404	0.515
3	0.878	0.959	23	0.396	0.505
4	0.811	0.917	24	0.388	0.496
5	0.754	0.874	25	0.381	0.487
6	0.707	0.834	26	0.374	0.478
7	0.666	0.798	27	0.367	0.470
8	0.632	0.765	28	0.361	0.463
9	0.602	0.735	29	0.355	0.456
10	0.576	0.708	30	0.349	0.449
11	0.553	0.684	35	0.325	0.418
12	0.532	0.661	40	0.304	0.393
13	0.514	0.641	45	0.288	0.272
14	0.497	0.623	50	0.273	0.354
15	0.482	0.606	60	0.250	0.325
16	0.468	0.590	70	0.232	0.302
17	0.456	0.575	80	0.217	0.283
18	0.444	0.561	90	0.205	0.267
19	0.433	0.549	100	0.195	0.254
20	0.423	0.537	200	0.138	0.181

第4章 Excel与Origin软件的应用

4.1 Excel在实验数据处理中的应用

Microsoft Excel是一款个人计算机数据处理软件，拥有友好的操作界面、功能齐全的计算功能和丰富的图表工具，是化工原理实验数据处理的高效工具。在化工原理实验数据处理过程中主要涉及算术运算和函数计算，其中算术运算符如表4-1所列，其运算优先顺序为乘方、乘除法、加减法，相同优先级运算符（乘法和除法、加法和减法）的计算顺序则按照计算公式的顺序进行，若需改变运算顺序，则可以通过添加括号“（）”来实现，若“－”作为负号，则其优先于其他算术运算符。

表4-1 算术运算符及优先级

算术运算符	含义	优先级
+	加法	3
—	减法	3
*	乘法	2
/	除法	2
∧	乘方	1

常用的计算函数如表4-2所列。

表4-2 常用计算函数

函数	含义
SUM(number1:number2)	单元格number1与number2之间求和
AVERAGE(number1:number2)	单元格number1与number2之间求平均值
ABS(number)	求单元格number绝对值
MAX(number1:number2)	求单元格number1与number2之间最大值
MIN(number1:number2)	求单元格number1与number2之间最小值
LOG(number，base)	求单元格number以base为底的对数值
LN(number)	求单元格number自然对数值
SQRT(number)	求单元格number开方
PI()	圆周率π

运用上述运算和函数即可完成化工原理实验数据处理，例如在离心泵单泵特性曲线测定实验中，根据实验测定的流量、真空度、出口压力、电机功率等原始数据，计算流速、扬程、有效功率、效率等参数，以进口流速计算为例，如图4-1所示，在单元格I5输入“=B5/3600/PI（）/0.035∧2*4”（其中B5单元格为流量，0.035为进口管管径），根据

实验数据计算得到扬程 H，通过拟合得到扬程 H 和流量 Q 的关系式为：

$$H=18.8354-0.1712Q^2$$

因此 P5 单元输入“=18.8354－0.1712*B5^2”即可得到上述拟合式的扬程计算值，将其与实验测定的扬程值（即单元格 L5）对比，在单元格 Q5 中输入“=ABS((L5－P5)/L5*100)”，即可获得扬程与流量拟合关联式与实验的误差。其他需要计算的物理量同样处理，在完成数据点 1（对应 Excel 表中第 5 行数据）所有物理量的计算后，如图 4-2 所示，选择 I5～Q5，通过填充柄拖放即可获得各个物理量的计算结果（图 4-3）。

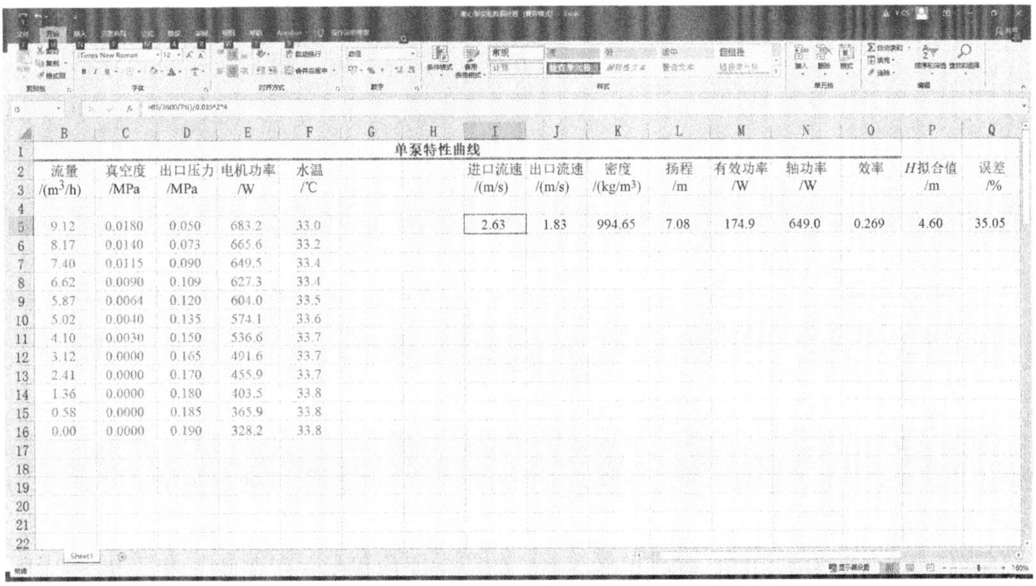

单泵特性曲线

流量/(m³/h)	真空度/MPa	出口压力/MPa	电机功率/W	水温/℃	进口流速/(m/s)	出口流速/(m/s)	密度/(kg/m³)	扬程/m	有效功率/W	轴功率/W	效率	H拟合值/m	误差/%
9.12	0.0180	0.050	683.2	33.0	2.63	1.83	994.65	7.08	174.9	649.0	0.269	4.60	35.05
8.17	0.0140	0.073	665.6	33.2									
7.40	0.0115	0.090	649.5	33.4									
6.62	0.0090	0.109	627.3	33.4									
5.87	0.0064	0.120	604.0	33.5									
5.02	0.0040	0.135	574.1	33.6									
4.10	0.0030	0.150	536.6	33.7									
3.12	0.0000	0.165	491.6	33.7									
2.41	0.0000	0.170	455.9	33.7									
1.36	0.0000	0.180	403.5	33.8									
0.58	0.0000	0.185	365.9	33.8									
0.00	0.0000	0.190	328.2	33.8									

图 4-1　算术运算示例

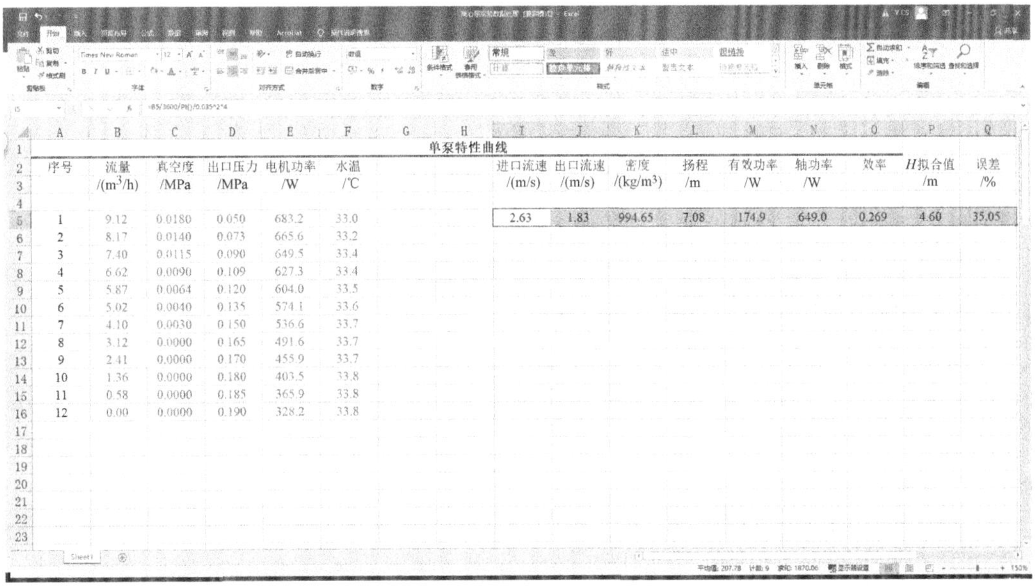

单泵特性曲线

序号	流量/(m³/h)	真空度/MPa	出口压力/MPa	电机功率/W	水温/℃	进口流速/(m/s)	出口流速/(m/s)	密度/(kg/m³)	扬程/m	有效功率/W	轴功率/W	效率	H拟合值/m	误差/%
1	9.12	0.0180	0.050	683.2	33.0	2.63	1.83	994.65	7.08	174.9	649.0	0.269	4.60	35.05
2	8.17	0.0140	0.073	665.6	33.2									
3	7.40	0.0115	0.090	649.5	33.4									
4	6.62	0.0090	0.109	627.3	33.4									
5	5.87	0.0064	0.120	604.0	33.5									
6	5.02	0.0040	0.135	574.1	33.6									
7	4.10	0.0030	0.150	536.6	33.7									
8	3.12	0.0000	0.165	491.6	33.7									
9	2.41	0.0000	0.170	455.9	33.7									
10	1.36	0.0000	0.180	403.5	33.8									
11	0.58	0.0000	0.185	365.9	33.8									
12	0.00	0.0000	0.190	328.2	33.8									

图 4-2　第 1 组数据计算设置

单泵特性曲线

序号	流量 /(m³/h)	真空度 /MPa	出口压力 /MPa	电机功率 /W	水温 /℃
1	9.12	0.0180	0.050	683.2	33.0
2	8.17	0.0140	0.073	665.6	33.2
3	7.40	0.0115	0.090	649.5	33.4
4	6.62	0.0090	0.109	627.3	33.4
5	5.87	0.0064	0.120	604.0	33.5
6	5.02	0.0040	0.135	574.1	33.6
7	4.10	0.0030	0.150	536.6	33.7
8	3.12	0.0000	0.165	491.6	33.7
9	2.41	0.0000	0.170	455.9	33.7
10	1.36	0.0000	0.180	403.5	33.8
11	0.58	0.0000	0.185	365.9	33.8
12	0.00	0.0000	0.190	328.2	33.8

序号	进口流速 /(m/s)	出口流速 /(m/s)	密度 /(kg/m³)	扬程 /m	有效功率 /W	轴功率 /W	效率	H拟合值 /m	误差 /%
1	2.63	1.83	994.65	7.08	174.9	649.0	0.269	4.60	35.05
2	2.36	1.64	994.59	9.06	200.6	632.3	0.317	7.41	18.23
3	2.14	1.48	994.52	10.57	212.0	617.0	0.344	9.46	10.52
4	1.91	1.33	994.52	12.29	220.5	595.9	0.370	11.33	7.78
5	1.69	1.18	994.49	13.17	209.5	573.8	0.365	12.94	1.78
6	1.45	1.01	994.46	14.48	197.0	545.4	0.361	14.52	0.26
7	1.18	0.82	994.42	15.94	177.1	509.8	0.347	15.96	0.13
8	0.90	0.63	994.42	17.18	145.3	467.0	0.311	17.17	0.08
9	0.70	0.48	994.42	17.70	115.6	433.1	0.267	17.84	0.78
10	0.39	0.27	994.39	18.74	69.1	383.3	0.180	18.52	1.17
11	0.17	0.12	994.39	19.25	30.3	347.6	0.087	18.78	2.47
12	0.00	0.00	994.39	19.77	0.0	311.8	0.000	18.84	4.71

图 4-3　数据处理结果

上述方法可实现化工原理实验数据处理所需的大部分功能。除此之外，在化工原理实验中还涉及传质单元数的数值积分计算，下面就以萃取实验的数据处理为例说明如何应用 Excel 工具进行数值积分。

萃取实验中虽然溶质在原溶剂和萃取剂中的溶解度均很低，但是由于相平衡方程中萃取相与萃余相间呈现非线性关系，如 25℃时苯甲酸在水（萃取剂）和煤油（原溶剂）中的相平衡方程为

$$y=4.443060\times10^{-6}+1.247730x-4.440948\times10^{2}x^{2}+5.048579\times10^{4}x^{3}+6.414260\times10^{6}x^{4}$$

式中，x、y 分别为萃余相和萃取相中溶质的组成。因此，在计算传质单元数

$$N_{\mathrm{OE}}=\int_{y_a}^{y_b}\frac{\mathrm{d}y}{y^{*}-y}$$

时，通常不易获得解析积分；而需要采用图解积分法或数值积分法求取传质单元数。积分上、下限即萃取相的出塔和进塔组成，可由实验组成分析获得，现以 $y_a=0$、$y_b=0.0010$ 为例说明如何利用 Excel 软件对上式进行积分。

首先，如图 4-4 所示，在 Excel 表中选择 A 列设置积分步长为 0.0001，用 B 列记录 y 值，在 B 列从 $y=0$ 开始，下一行单元格在其相邻的上一行单元格基础上递增一个步长，并利用单元格的填充柄下拉，直至 B 列最下方单元格数值达到 0.0010，即 $y_b=0.0010$ 时，停止填充。

这样就在积分的上、下限之间划分了若干区间，从实例可见，由 0 至 0.0010 间分割为 10 个区间，而这些分割点也必然与萃取塔内某一截面的萃取相组成 y 一一对应，那么在这些截面上的萃余相组成 x 则可通过全塔的物料衡算获得，而后由萃余相组成 x 根据相平衡方程即可求得 y^{*}，得到图 4-5 所示结果。

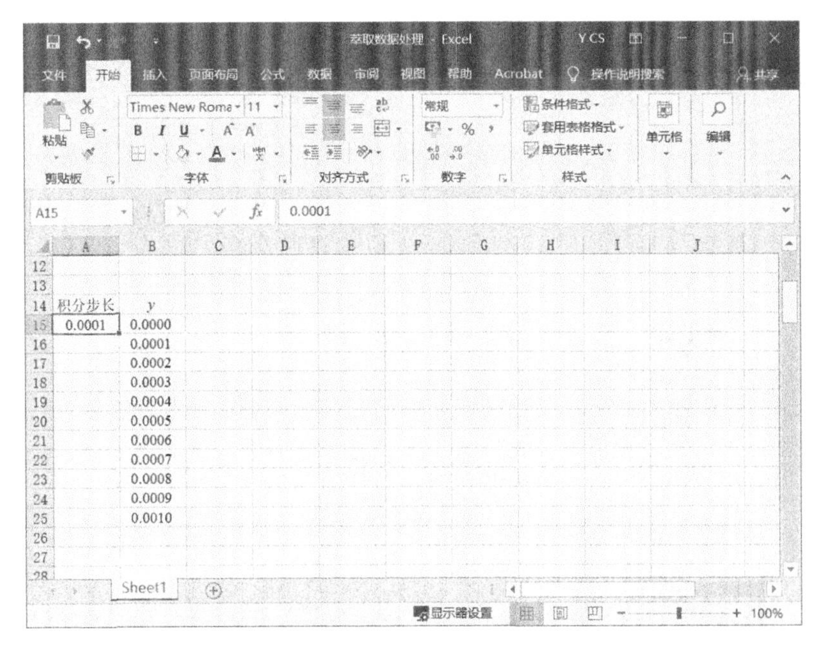

图 4-4　积分区间设置

积分步长	y	y	y*
0.0001	0.0000	8.2000E-04	7.5971E-04
	0.0001	9.0333E-04	8.1066E-04
	0.0002	9.8667E-04	8.5778E-04
	0.0003	1.0700E-03	9.0133E-04
	0.0004	1.1533E-03	9.4157E-04
	0.0005	1.2367E-03	9.7878E-04
	0.0006	1.3200E-03	1.0132E-03
	0.0007	1.4033E-03	1.0452E-03
	0.0008	1.4867E-03	1.0751E-03
	0.0009	1.5700E-03	1.1031E-03
	0.0010	1.6533E-03	1.1295E-03

图 4-5　物料衡算与相平衡计算

根据上述结果，即可获得各个区间分割点的被积函数值 $1/(y^*-y)$，如图 4-6 所示。

将 y 与 $1/(y^*-y)$ 作图，得到被积函数图，如图 4-7 所示。很明显要求解上述积分值 N_{OE}，就是求取 $y_a=0$、$y_b=0.0010$ 之间函数 $1/(y^*-y)$ 与横轴所围成图形的面积。

该图形虽然不规则，但是如图 4-8 所示，其面积可分解为各个区间的面积加和，而各个区间的面积可近似视为一个矩形，矩形的面积是容易获得的，例如（0.0001，0.0002）区间的面积，可利用 $y=0.0001$ 这个区间分割点的被积函数值 $1/(y^*-y)\big|_{y=0.0001}$ 与区间

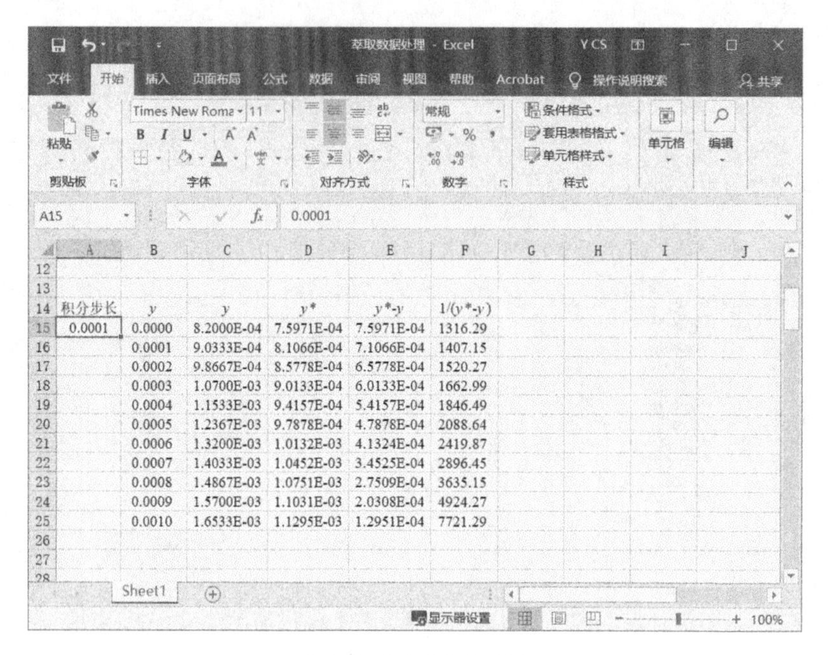

积分步长	y	y	y*	y*-y	1/(y*-y)
0.0001	0.0000	8.2000E-04	7.5971E-04	7.5971E-04	1316.29
	0.0001	9.0333E-04	8.1066E-04	7.1066E-04	1407.15
	0.0002	9.8667E-04	8.5778E-04	6.5778E-04	1520.27
	0.0003	1.0700E-03	9.0133E-04	6.0133E-04	1662.99
	0.0004	1.1533E-03	9.4157E-04	5.4157E-04	1846.49
	0.0005	1.2367E-03	9.7878E-04	4.7878E-04	2088.64
	0.0006	1.3200E-03	1.0132E-03	4.1324E-04	2419.87
	0.0007	1.4033E-03	1.0452E-03	3.4525E-04	2896.45
	0.0008	1.4867E-03	1.0751E-03	2.7509E-04	3635.15
	0.0009	1.5700E-03	1.1031E-03	2.0308E-04	4924.27
	0.0010	1.6533E-03	1.1295E-03	1.2951E-04	7721.29

图 4-6　被积函数的计算

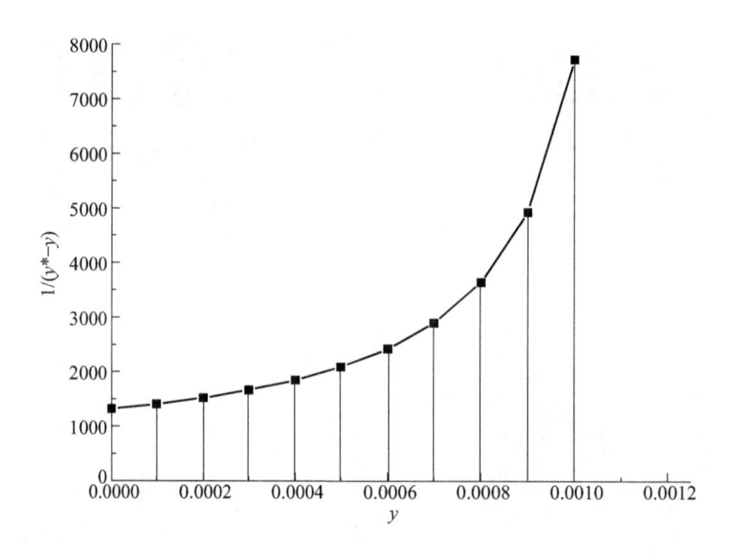

图 4-7　被积函数图

步长的乘积求得该矩形面积。由图 4-8 可见，虽然该矩形面积与实际面积之间存在误差，但是，只要区间步长足够小，那么矩形面积就无限接近实际面积。因此，理论上只要区间步长足够小，就可获得实际面积。

除了矩形积分外，还可采用梯形积分，与上述方法类似，梯形积分就是在各个区间内用如图 4-9 所示的梯形面积近似实际面积。同样以(0.0001，0.0002)区间为例，其梯形面积为

$$\frac{[1/(y^*-y)|_{y=0.0001}+1/(y^*-y)|_{y=0.0002}]\times 0.0001}{2}$$

很明显采用梯形积分在同样条件下，其结果更符合实际情况。

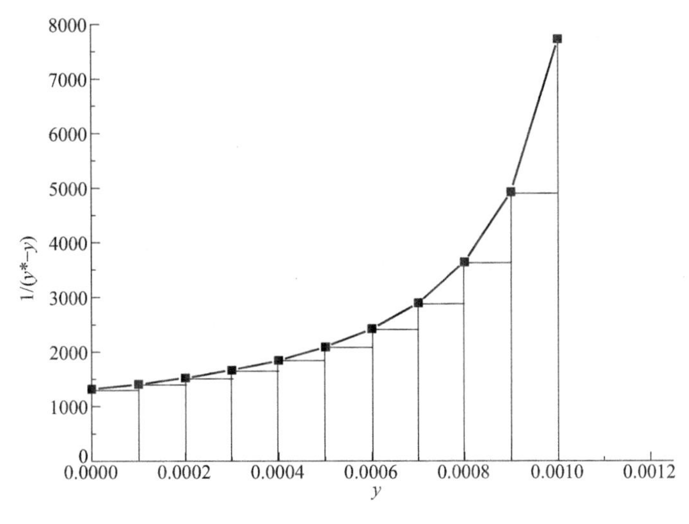

图 4-8　矩形积分

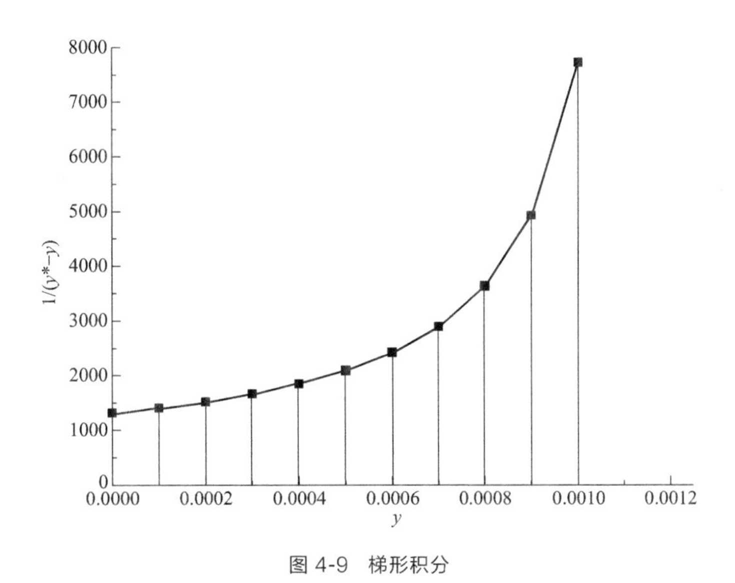

图 4-9　梯形积分

在 Excel 中，同样以（0.0001，0.0002）区间为例，如图 4-10 所示，其矩形积分为 F14＊＄A＄15，梯形积分为（F14＋F15）＊＄A＄15/2，其他区间的面积计算，通过填充柄复制即可，结果如图 4-10 所示。

将各个区间的积分结果累加即可得到 N_{OE} 积分结果，梯形积分与矩形积分值分别如 G26 和 H26 所示。然而该结果是否是正确结果呢？这需要缩小区间步长重复上述计算过程，若前后不同步长的积分结果相近或误差小于允许值，那么该积分结果既为最终结果，否则，需要继续缩小步长重复上述计算过程。根据上述介绍的方法，设置步长为 0.0001、0.00005、0.00002、0.00001，分别计算得到梯形积分的结果分别为 2.692、2.666、2.658、2.657，矩形积分的结果分别为 2.372、2.506、2.594、2.625。可见，梯形积分在

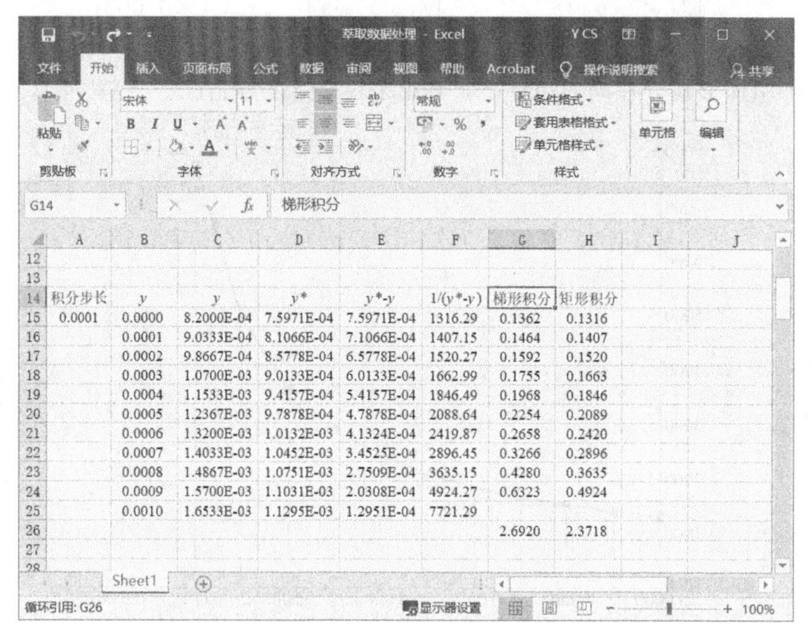

积分步长	y	y	y*	y*-y	1/(y*-y)	梯形积分	矩形积分
0.0001	0.0000	8.2000E-04	7.5971E-04	7.5971E-04	1316.29	0.1362	0.1316
	0.0001	9.0333E-04	8.1066E-04	7.1066E-04	1407.15	0.1464	0.1407
	0.0002	9.8667E-04	8.5778E-04	6.5778E-04	1520.27	0.1592	0.1520
	0.0003	1.0700E-03	9.0133E-04	6.0133E-04	1662.99	0.1755	0.1663
	0.0004	1.1533E-03	9.4157E-04	5.4157E-04	1846.49	0.1968	0.1846
	0.0005	1.2367E-03	9.7878E-04	4.7878E-04	2088.64	0.2254	0.2089
	0.0006	1.3200E-03	1.0132E-03	4.1324E-04	2419.87	0.2658	0.2420
	0.0007	1.4033E-03	1.0452E-03	3.4525E-04	2896.45	0.3266	0.2896
	0.0008	1.4867E-03	1.0751E-03	2.7509E-04	3635.15	0.4280	0.3635
	0.0009	1.5700E-03	1.1031E-03	2.0308E-04	4924.27	0.6323	0.4924
	0.0010	1.6533E-03	1.1295E-03	1.2951E-04	7721.29		
						2.6920	2.3718

图 4-10　积分结果

缩小步长的过程中相比于矩形积分迅速收敛，当步长由 0.00002 缩小为 0.00001 时，两次积分误差为 0.04%，而矩形积分的误差为 1.2%。因此，$N_{OE}=2.657$ 可视为最终的准确解。

上述数值积分方法是利用 Excel 表格进行计算的，需要不断改变步长重复计算，直至获得准确解，计算过程较为烦琐。所以，也可在图 4-7 基础上，对被积函数进行多项式回归，得到

$$\frac{1}{y^*-y}=1261.11846+3.096720\times10^{6}y-1.945360\times10^{10}y^{2}+6.816080\times10^{13}y^{3}-9.377030\times10^{16}y^{4}+4.828130\times10^{19}y^{5}+1.448440\times10^{17}y^{6}$$

因此

$$N_{OE}=\int_{0}^{0.001}\frac{dy}{y^*-y}=1261.11846y+1.54836\times10^{6}y^{2}-6.48453\times10^{9}y^{3}+1.70402\times10^{13}y^{4}-1.87541\times10^{16}y^{5}+8.04688\times10^{18}y^{6}+2.0692\times10^{16}y^{7}\Big|_{0}^{0.001}$$

$$=1.2611185y+1.54836-6.48453+17.0402-18.75406+8.0468833+2.069\times10^{5}$$

$$=2.658$$

该结果与梯形积分结果相近。

4.2　Origin 在实验数据回归与绘图中的应用

Origin 被公认为最快、最灵活、使用最方便的工程绘图软件，与其他绘图软件相比，具有简洁的界面和强大的科技绘图及数据处理功能，是科技工作者和工程技术人员首选的绘图工具。Origin 功能众多，现以离心泵特性曲线图的绘制和处理为例说明其在化工原理实验数据绘图和处理中的应用。打开 Origin 主界面如图 4-11 所示，菜单栏“Column”中选择“Add New Column…”，在对话框中输入 2，新增 2 列，得到 A、B、C、D 四列数据

表 Book1，在其中填写或复制离心泵特性曲线数据：流量、扬程、功率和效率（如图 4-12 所示）。

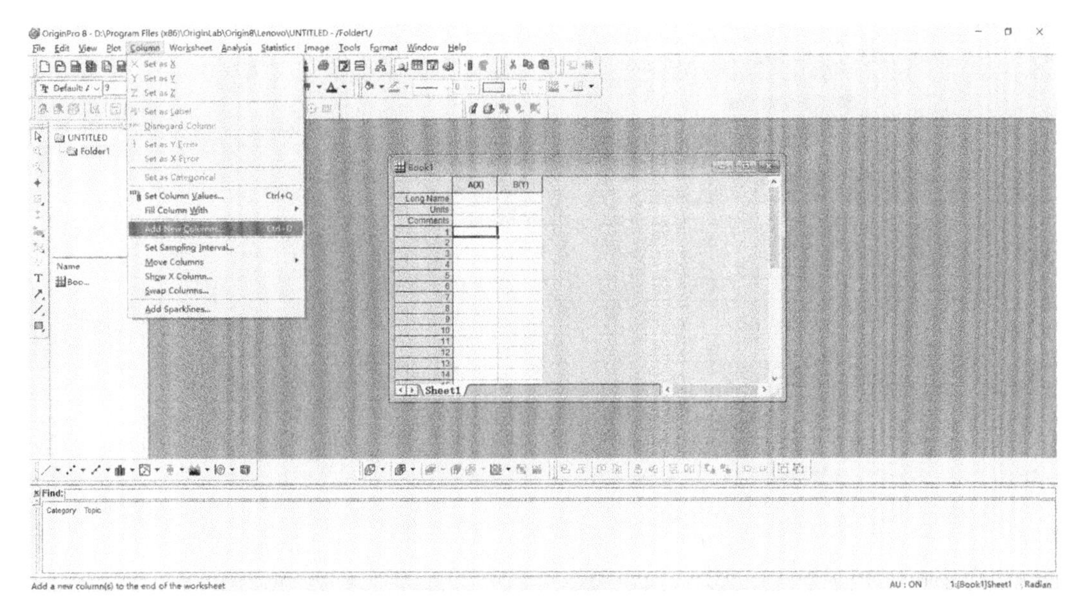

图 4-11　Origin 主界面

Book1

	A(X)	B(Y)	C(Y)	D(Y)
Long Name	Q	H	N	η
Units	m³/s	m	W	
Comments				
1	9.12	7.08	649.04	0.2695
2	8.17	9.06	632.32	0.3173
3	7.4	10.57	617.03	0.3437
4	6.62	12.29	595.94	0.3699
5	5.87	13.17	573.8	0.3651
6	5.02	14.82	545.4	0.3612
7	4.1	15.94	509.77	0.3473
8	3.12	17.18	467.02	0.3111
9	2.41	17.7	433.11	0.267
10	1.36	18.74	383.33	0.1801
11	0.58	19.25	347.61	0.0871
12	0	19.77	311.79	0
13				

Sheet1

图 4-12　数据输入

首先，选择菜单栏“Plot”中“symbol”—“scatter”，打开对话框如图 4-13，将数据列 A 设置为 x 轴，数据列 B 设置为 y 轴，点击“Add”按钮，添加该数据图至下方对话框中，默认为图层 layer1。确认后可生成图 4-14 所示的默认数据图（该窗口默认名称为 Graph1），该图还比较粗糙，图中任何元素都可以通过双击或右击鼠标选择“Properties”打开相应的属性设置对话框（图 4-15），在这些对话框的“Scale”“Tick Labels”“Title & Format”等标签中可对图 4-14 的数据图中的坐标、坐标刻度、坐标名称与格式、坐标起始值、坐标刻度形式、数据点形式、图例等进行相应的设置，设置后的扬程与流量关系图如图 4-16 所示。

离心泵特性曲线中扬程线、功率线、效率线的数值差异很大，若将这三条关系线画在统一的坐标系中，不仅影响美观，也会使个别关系线因为坐标范围不合适而无法正确体现。因此，需要对不同的关系线设置不同的坐标轴。

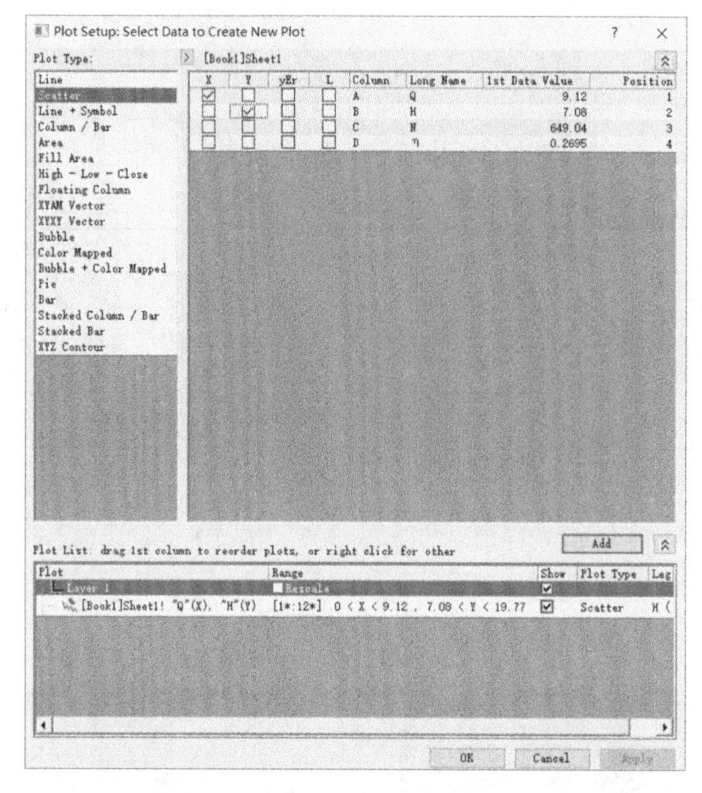

图 4-13　数据图设置

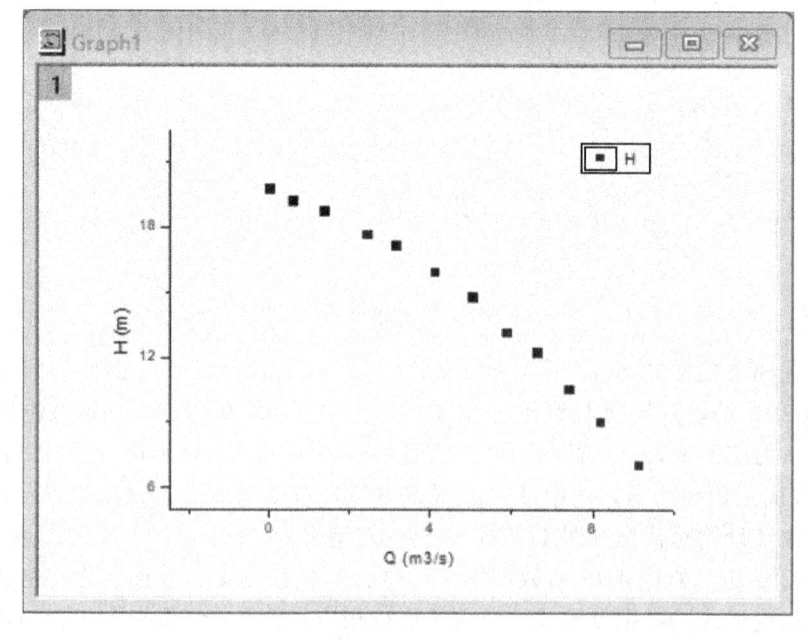

图 4-14　数据图

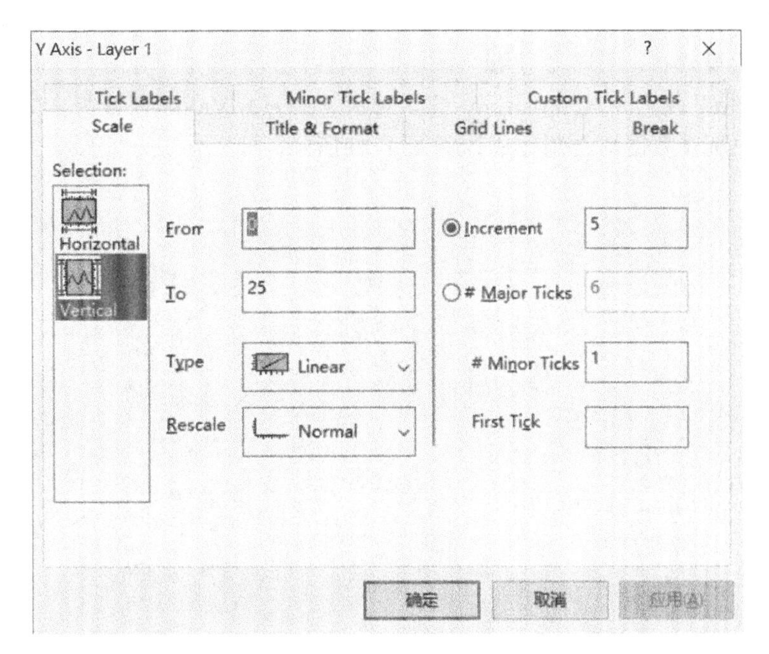

图 4-15　属性设置对话框

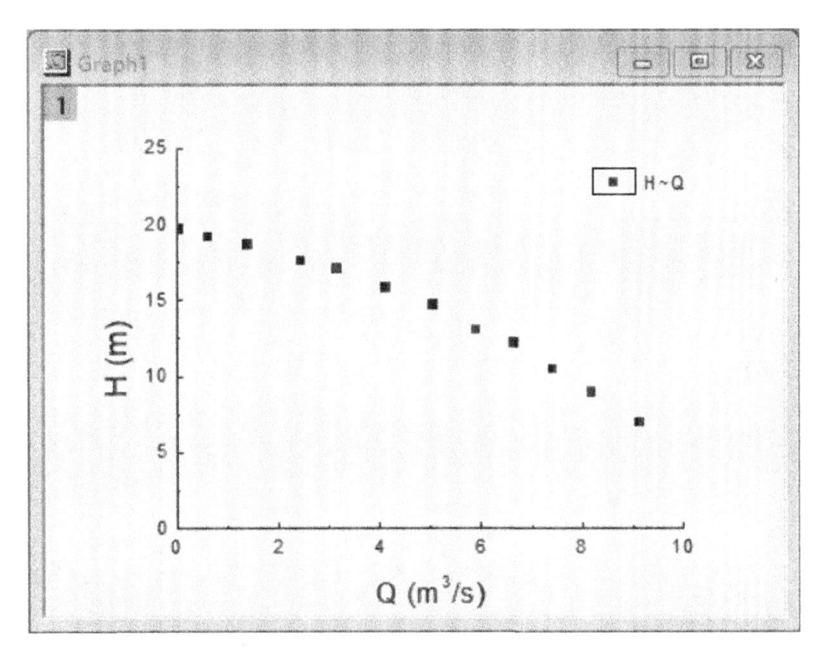

图 4-16　扬程与流量关系图

在激活图形窗口 Graph1 条件下，选择菜单中“Graph”中的“New Layer (Axes)”—“(Linked) Right Y”，新增一图层 Layer2（图层标识显示于 Graph1 窗口左上角），并在图形右侧设置一新的 y 轴，图层 2 与图层 1 共用同一 x 轴。而后选择“Graph”—“Add Plot to Layer”—“Scatter”，在弹出的对话框（图 4-17）中，依次选择绘图所需的数据表 Book1，将数据列 A 和 C 分别设置为 x 轴和 y 轴，点击“Add”和“OK”按钮，绘制功率与流量的数据图，经过适当设置得到图 4-18。同样的方法，设置图层 3，并在图形右侧新建一 y 轴用于表示效率，为避免其与图层 2 的功率 y 轴重合，将图层 3 的 y 轴位置设置在 $x=12.5$ 位置（图 4-15 属性设置窗口中“Title & Format”标签中 Axis 值由“Right”改为“At Position=”，对应 Percent/Value 设置为 12.5）。由此离心泵特性曲线图中所有数据均绘制完成，如图 4-19 所示。

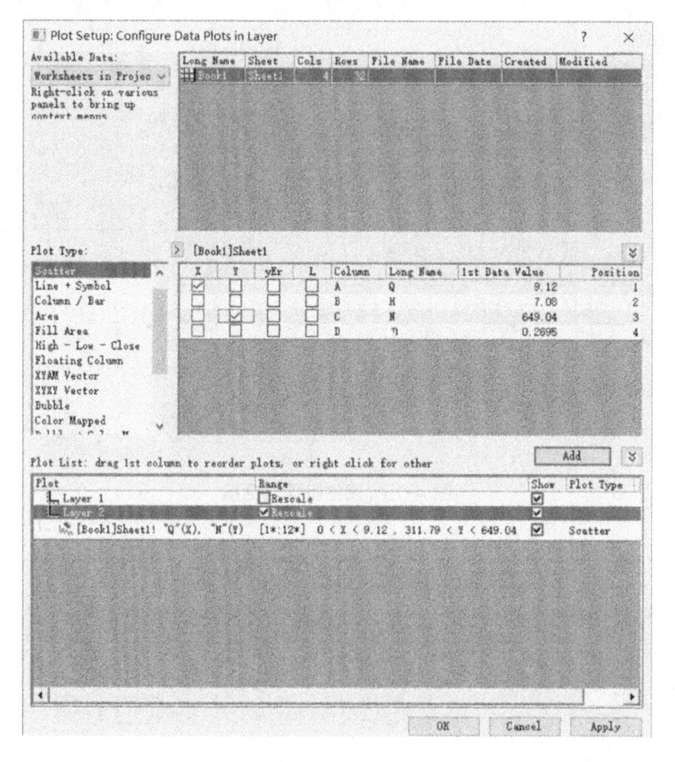

图 4-17　Plot 设置对话框

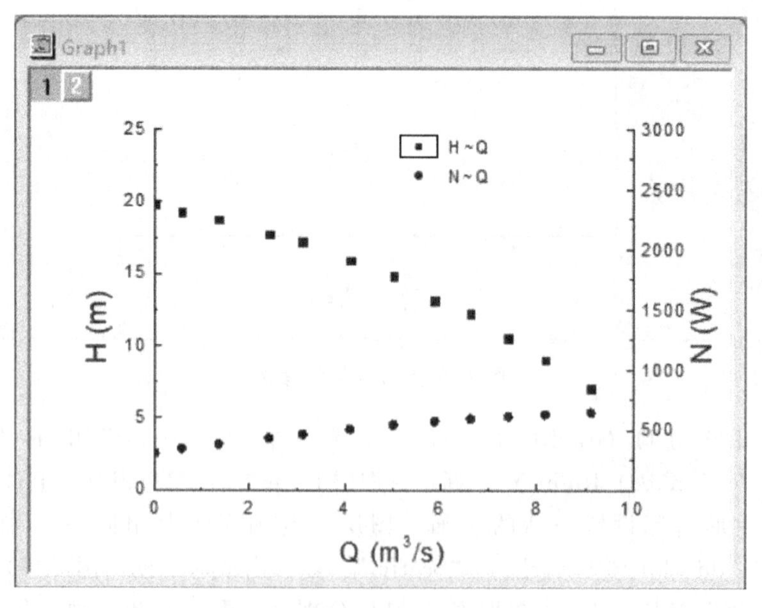

图 4-18　扬程、功率与流量关系图

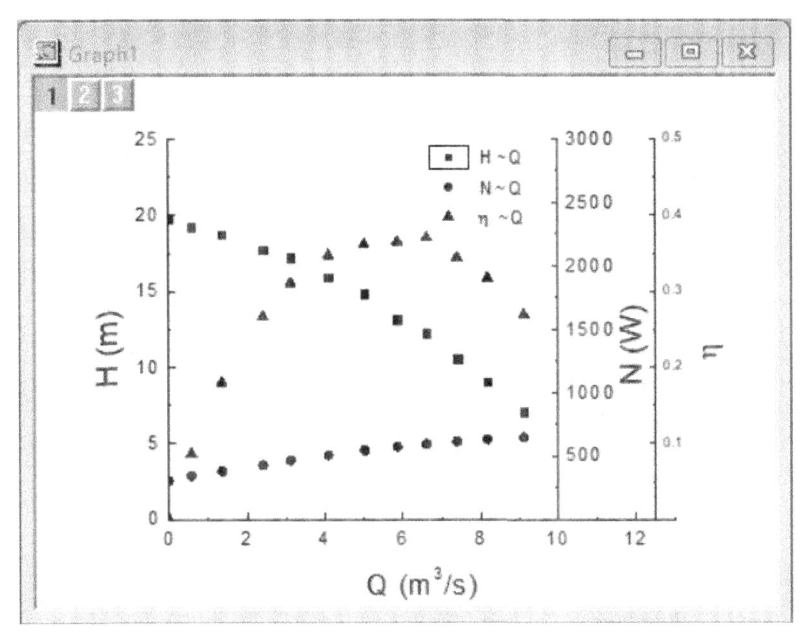

图 4-19　特性曲线数据图

在此基础上，就可以对图 4-19 中三条特性曲线进行关联回归获得扬程、功率、效率和流量的关系式，其中 N-Q、η-Q 关系式可以分别采用线性方程和多项式进行关联。对 N-Q 关系采用线性拟合，选择功率数据所在的图层 2，依次选择菜单栏“Analysis”—“Fitting”—“Fit Linear”—“Open Dialog...”打开线性拟合设置对话框（图 4-20），默认设置情况下即可进行线性拟合，拟合得到的方程及相关误差信息显示于数据表 Book1（图 4-21）中，拟合线同时会显示于图 Graph1 中，可直观看出拟合的效果。同理，选择效率数据所在图层 3，依次选择菜单栏“Analysis”—“Fitting”—“Fit Polynomial”—“Open Dialog...”打开多项式拟合设置对话框，在对话框中对“Polynomial Order”选项设置多项式项数，其他参数可根据需要设置。

扬程与流量的关系式理论上为 $H=A-BQ^2$ 的形式，在 Origin 中需要采用自定义的函数进行关联拟合。首先，激活图形窗口 Graph1 左上角图层 1，即扬程数据所在的图层；其次，选择“Analysis”—“Fitting”—“Nonlinear Curve Fit...”，打开图 4-22 所示的非线性拟合对话框，点击其中部函数编辑按钮，从而打开拟合函数管理器窗口（图 4-23），点击左侧树形菜单“User Defined”和右侧“New Category”“New Function”按钮；再次，进行函数编辑，设置自变量“Independent Variables”为 Q、因变量“Dependent Variables”为 H，参数“Parameter Names”为 A 和 B，在“Function Form”选项中选择“Equations”“Function”文本框中输入“H=A−B* Q^2”，最后，保存确定退出拟合函数管理窗口返回非线性拟合窗口，在“Category”“Function”中选择刚刚编辑的 New Category 目录下的 New Function（User）函数。而后，点击图 4-22 非线性拟合窗口中的逐次迭代按钮，逐次迭代拟合，对应拟合结果在其下方文本框中实时显示，或直接点击“Fit”按钮一次性完成迭代拟合。

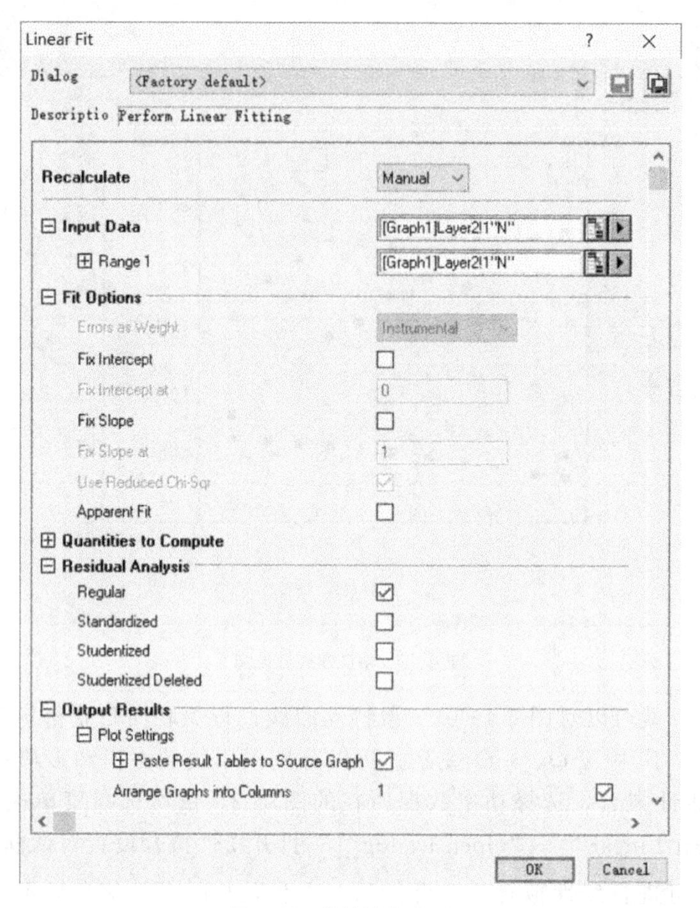

图 4-20　线性拟合对话框

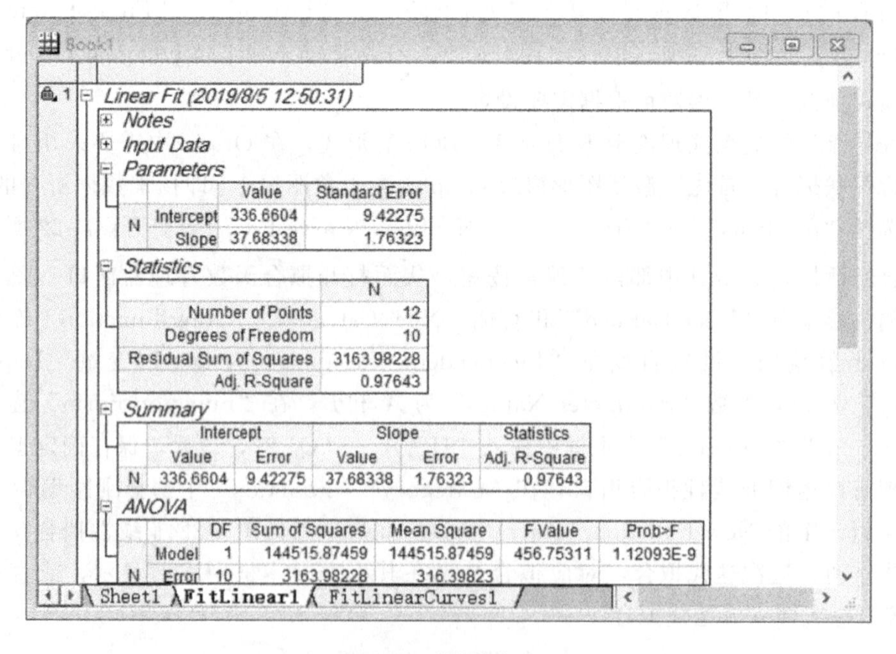

		Value	Standard Error
N	Intercept	336.6604	9.42275
	Slope	37.68338	1.76323

	N
Number of Points	12
Degrees of Freedom	10
Residual Sum of Squares	3163.98228
Adj. R-Square	0.97643

	Intercept		Slope		Statistics
	Value	Error	Value	Error	Adj. R-Square
N	336.6604	9.42275	37.68338	1.76323	0.97643

		DF	Sum of Squares	Mean Square	F Value	Prob>F
	Model	1	144515.87459	144515.87459	456.75311	1.12093E-9
N	Error	10	3163.98228	316.39823		

图 4-21　线性拟合结果

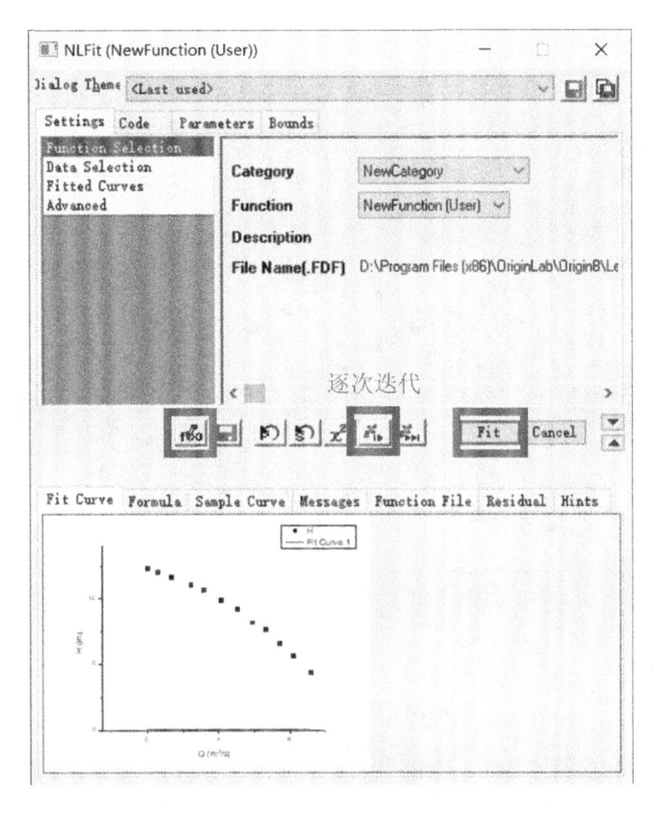

图 4-22　非线性拟合窗口

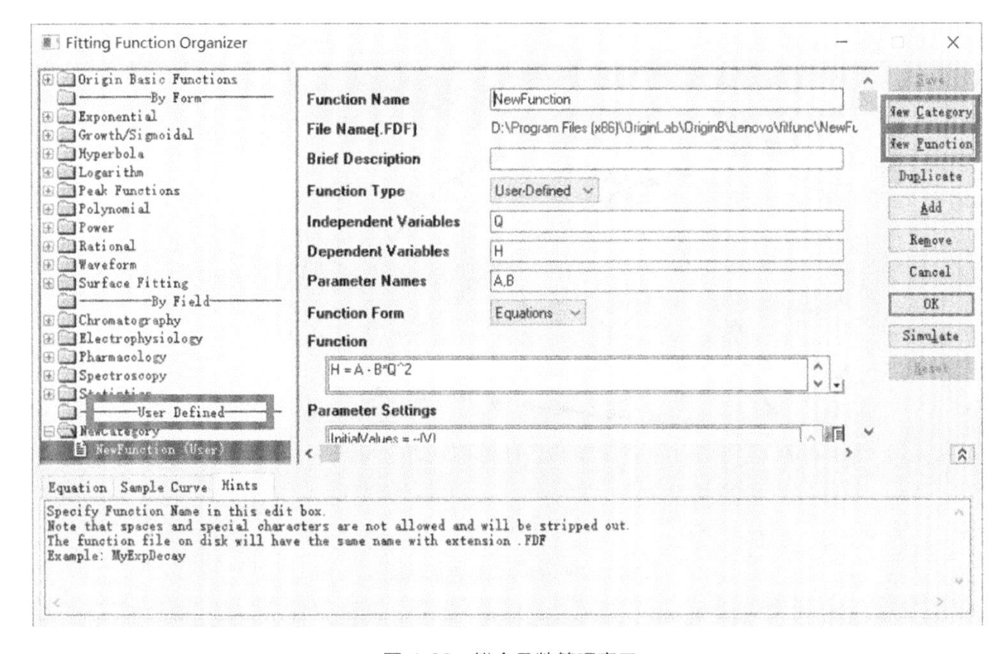

图 4-23　拟合函数管理窗口

通过以上方法最终得到图 4-24 所示的图形窗口，其中显示了离心泵特性曲线的三条线以及各线的拟合线，他们的拟合方程以及拟合误差等参数均可在 Book1 窗口中查阅。其他有关设置和操作可参阅 Origin 的相关文献。

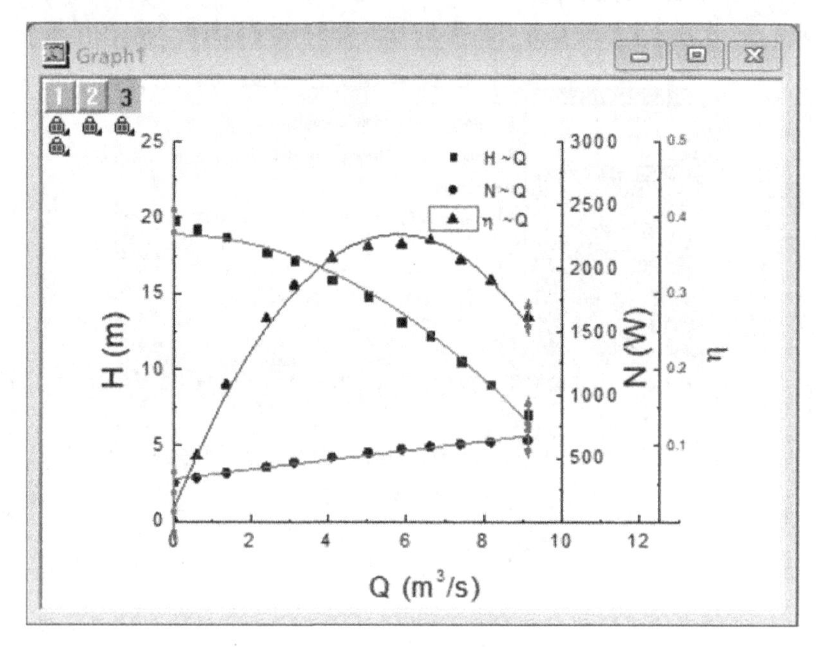

图 4-24 特性曲线拟合图窗口

在“Edit”菜单中选择“Copy Page”，即可将图 4-24 窗口中离心泵特性曲线图复制粘贴于任何文本中，具体如图 4-25 所示，当然该图还需要根据具体情况标注离心泵的型号、转速等信息。

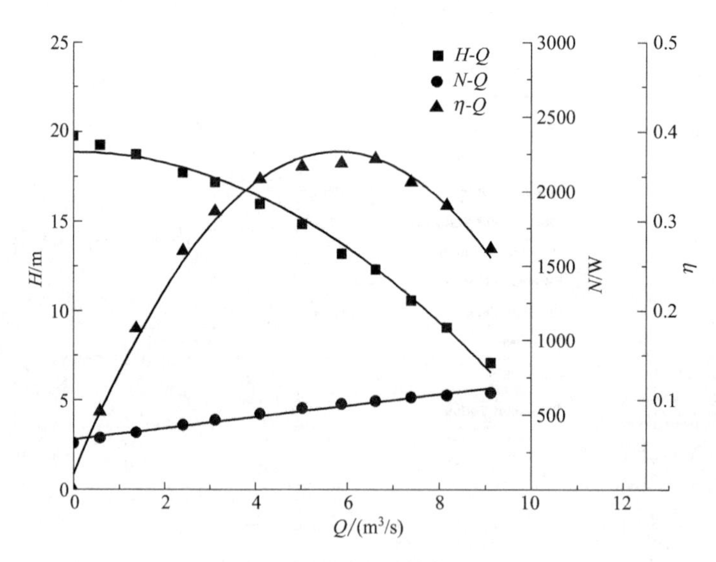

图 4-25 离心泵特性曲线图

第5章 实验室常用测量仪表

温度、压强、流量和液位是化工过程需要测量的四大参数，本章就化工原理实验室常用的测量仪表做一些简要的介绍。

5.1 温度测量

在化工生产和实验中，温度往往是测量和控制的重要参数之一，几乎每个化工原理实验设备上都装有温度测量仪表。

温度是表征物体冷热程度的物理量。温度不能直接测量，只能借助于冷热物体之间的热交换，以及物体的某些物理性质（如膨胀、电阻、热电效应等）随冷热程度不同而变化的特性进行间接测量。根据测量方式可把测温分为接触式和非接触式两种。化工原理实验中所涉及的被测温度基本上可用接触式测温仪表来测量。常用测温仪表的种类及优缺点如表5-1所列。本节重点介绍接触式测温仪表中的膨胀式温度计、热电偶温度计和热敏电阻温度计。

表5-1 常用测温仪表的种类及优缺点

测温方式	温度计种类		测温范围/℃	优点	缺点
接触式测温仪表	膨胀式	玻璃管	−50～600	结构简单，使用方便，测量准确，价格低	测量上限和精度受玻璃质量的限制，易碎，不能记录远传
		双金属	−80～600	结构紧凑，牢固可靠	精度低，量程和使用范围有限
	压力式	液体 气体 蒸汽	−30～600 −20～350 0～250	结构简单，耐震，防爆，能记录、报警，价格低	精度低，测温距离短，滞后大
	热电偶	铂铑-铂 镍铬-镍硅 镍铬-考铜	0～1600 −50～1000 −50～600	测温范围广，精度高，便于远距离、多点、集中测量和自动控制	需冷端温度补偿，在低温段测量精度较低
	热敏电阻	铂 铜	−200～600 −50～150	测量精度高，便于远距离、多点、集中测量和自动控制	不能测高温，必须注意环境温度的影响
非接触式测温仪表	辐射式	辐射式 光学式 比色式	400～2000 700～3200 900～1700	测温时不破坏被测温度场	低温段测量不准，环境条件会影响测温准确度
	红外线式	光电探测 热电探测	0～3500 200～2000	测温范围大，适于测温度分布，不破坏被测温度场，响应快	易受外界干扰，标定困难

5.1.1 膨胀式温度计

（1）玻璃管温度计

玻璃管温度计是最常用的一种测定温度的仪器，其结构简单、价格低、使用方便、测量准确，测量范围为−50～600℃。缺点是易损坏，损坏后无法修复。实验室常用的是水银温度计和有机液体（如乙醇）温度计。水银温度计测量范围广、刻度均匀、读数准确，但损坏后会造成污染。有机液体（乙醇、苯等）温度计中的液体着色后读数明显，但由于膨胀系数随温度而变化，故刻度不均匀，读数误差较大。玻璃管温度计又分为棒式、内标式和电接点式3种，如表5-2所列。

表5-2 常用玻璃管温度计

项目	棒式	内标式	电接点式
特点	实验室最常用 直径 $d=6\sim8$mm 长度 $l=250$mm，280mm，300mm，420mm，480mm	工业上常用 $d_1=18$mm，$d_2=9$mm $l_1=230$mm，$l_2=130$mm $l_3=60\sim2000$mm	用于控制、报警等，分固定接点与可调接点两种
外形图	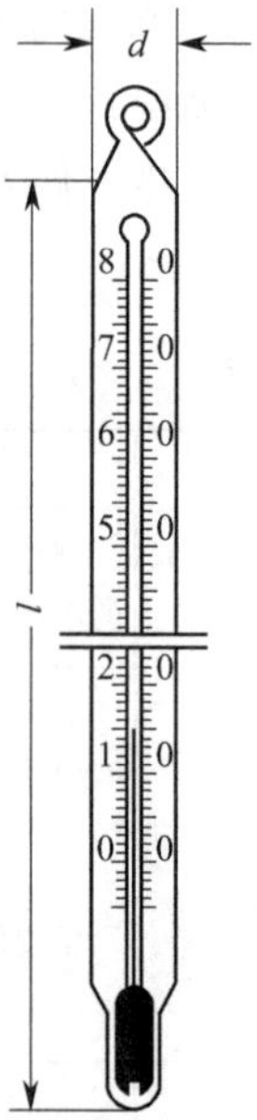 d l 8 0 7 0 6 0 5 0 2 0 1 0 0 0	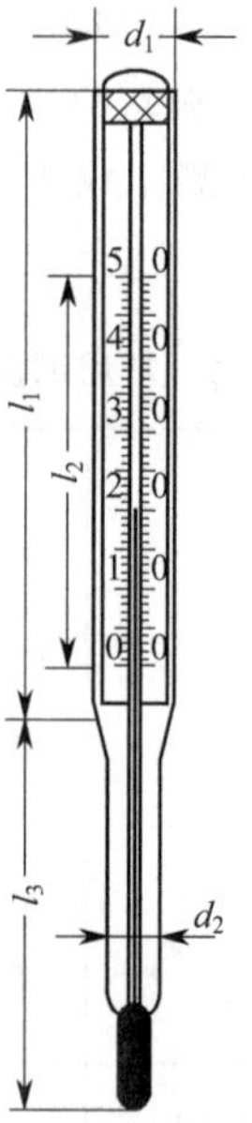 d_1 l_1 l_2 l_3 d_2 5 0 4 0 3 0 2 0 1 0 0 0	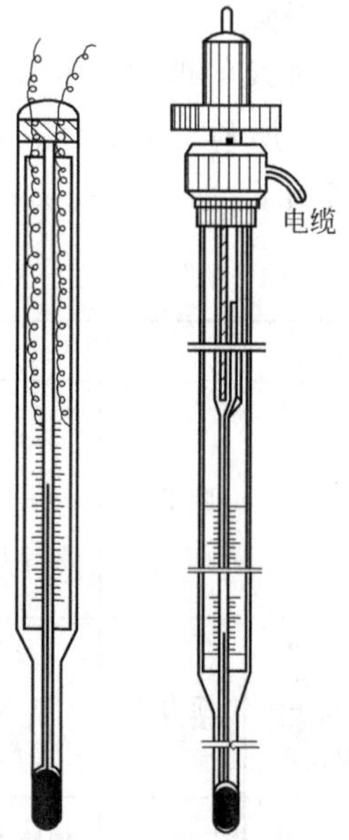 电缆

（2）玻璃管温度计的校正

用玻璃管温度计进行精确测量时需要校正，其方法有两种：一是与标准温度计在同一状况下比较；二是利用纯物质相变点校正，如冰-水、水-水蒸气系统校正。

用第一种方法进行校正时，可将被校验的玻璃管温度计与标准温度计（在市场上购买的二等标准温度计）一同插入恒温槽中，待恒温槽的温度稳定后，比较被校验温度计与标准温度计的示值。注意，在校正过程中应采用升温校验。这是因为有机液体与毛细管壁有附着力，当温度下降时，会有部分液体停留在毛细管壁上，影响准确读数。水银温度计在降温时会因摩擦发生滞后现象。

如果实验室中无标准温度计时，可用冰-水、水-水蒸气相变温度校正温度计。

1）用冰-水混合液校正 0℃　在 100mL 烧杯中，装满碎冰或冰块，然后注入蒸馏水，使液面达冰面下 2cm 为止。插入温度计，使温度计刻度便于观察或 0℃刻度露出冰面，搅拌并观察此水银柱的变化，待其所指温度恒定时，记录读数，即是校正过的 0℃。注意勿使冰块完全溶解。

2）用水-水蒸气校正 100℃　图 5-1 为校正温度计安装示意图。塞子应留缝隙，这是为了平衡试管内外的压力。向试管中加入少量沸石及 10mL 蒸馏水。调整温度计，使其水银球在液面上 3cm。以小火加热并注意蒸汽在试管壁上冷凝形成一个环，注意控制火力使该环维持在水银球上方约 2cm 处，若水银球上保持有一液滴，说明液态与气态间达到热平衡。当温度恒定时观察水银柱读数，记录读数，再经气压校正后即为校正过的 100℃。

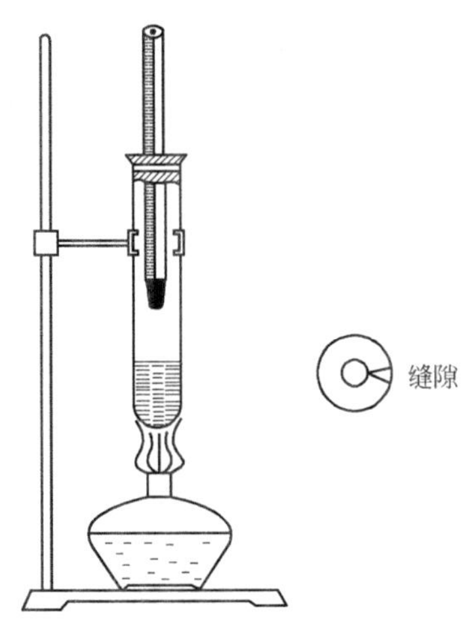

图 5-1　校正温度计安装示意

(3) 玻璃管温度计的安装和使用

① 玻璃管温度计应安装在没有大的震动、不易受碰撞的设备上，特别是有机液体玻璃管温度计，如果震动大，容易使液柱中断。

② 玻璃管温度计感温泡中心应处于温度变化最敏感处（如管道中流体流速最大处）。

③ 玻璃管温度计应安装在便于读数的位置，不能倒装，尽量不要倾斜安装。

④ 水银温度计读数时按凸面之最高点读数；有机液体玻璃管温度计则按凹面最低点读数。

⑤ 为了准确地测定温度，用玻璃管温度计测定物体温度时，应使温度计内的液体全部处于待测的物体中。

5.1.2　热电偶温度计

热电偶温度计是以热电效应为基础的测温仪表。它的测量范围大，结构简单，使用方便，测温准确可靠，便于信号的远传及自动记录和集中控制，因而在化工生产与实验中应用极为普遍。

热电偶温度计由三部分组成：热电偶（感温元件）、测量仪表（电位差计等）和连接热电偶和测量仪表的导线（补偿导线及铜导线）。热电偶温度计的测温系统示意如图 5-2 所示。

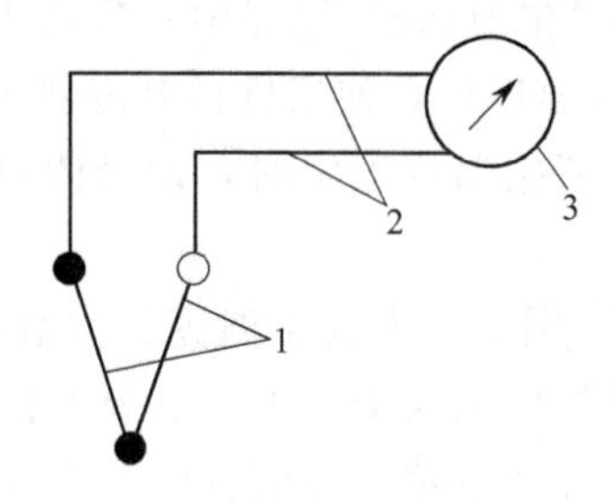

图 5-2 热电偶温度计测温系统示意

1—热电偶；2—导线；3—测量仪表

(1) 热电偶测温原理

将两种不同性质的金属丝或合金丝 A 和 B 连接成一个闭合回路。如果将它们的两个接点分别置于温度为 t_0 和 t 的热源中，则该回路中会产生电动势，如图 5-3（a）所示。

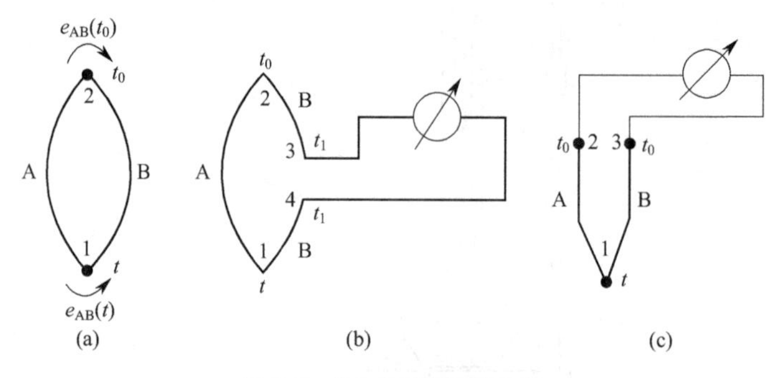

图 5-3 热电偶测量原理示意

如果在此回路中串接一只直流毫伏计（将金属 B 断开接入直流毫伏计，或者在两金属线的 t_0 接头处断开接入直流毫伏计），如图 5-3（b）、（c）所示，就可见到直流毫伏计中有电势指示，这种现象称为热电效应。这个由不同金属丝组成的闭合回路即为热电偶。闭合回路的一端插入被测介质中，感受被测温度，称为热电偶的工作端或热端，另一端与导线连接，称为自由端或冷端。导体 A、B 称为热电极。

在两种金属的接触点处，设 $t>t_0$，由于接点温度不同，就产生了两个大小不等、方向相反的热电势 $e_{AB}(t)$ 和 $e_{AB}(t_0)$，而对于同一种金属 A 或 B，由于其两端温度不同，自由电子具有的动能不同，也会产生一个相应的电动势 $e_A(t, t_0)$ 和 $e_B(t, t_0)$，这个电动势称为温差电势。热电偶回路中既有接触电势，又有温差电势，因此，回路中总电势为

$$\begin{aligned} E_{AB}(t,t_0) &= e_{AB}(t)+e_B(t,t_0)-e_{AB}(t_0)-e_A(t,t_0) \\ &= [e_{AB}(t)-e_{AB}(t_0)]-[e_A(t,t_0)-e_B(t,t_0)] \end{aligned} \tag{5-1}$$

由于温差电势比接触电势小很多，可忽略不计，故式（5-1）可简化为

$$E_{AB}(t,t_0)=e_{AB}(t)-e_{AB}(t_0) \tag{5-2}$$

当 $t=t_0$ 时，则 $E_{AB}(t, t_0)=0$；当 t_0 一定时，$e_{AB}(t_0)$ 为常数，则热电势 $E_{AB}(t, t_0)$ 就成为温度 t 的单值函数了，而和热电偶的长短及直径无关。这样，只要测出热电势的大小，就能判断测温点温度的高低，这就是利用热电效应来测温的原理。

(2) 补偿导线的选用

由热电偶测温原理知道，只有当热电偶冷端温度保持不变时，热电势才是被测温度的

单值函数。由于热电偶一般做得比较短（特别是贵重金属），这样热电偶的工作端与冷端离得很近，而且冷端又暴露在空间，容易受到周围环境的影响，因而冷端温度难以保持恒定。为了使热电偶的冷端温度保持恒定，可用一种专用导线将热电偶的冷端延伸出来，这种专用导线称为补偿导线。它也是由两种不同性质的金属材料制成，在一定温度范围内（0～100℃）与所连接的热电偶具有相同的热电特性，其材料是廉价金属。不同热电偶所用的补偿导线也不同，因此要注意型号相配。各种热电偶配用的补偿导线材料及其特点如表 5-3 所列。

表 5-3 各种热电偶配用的补偿导线材料及其特点

热电偶名称	补偿导线			
	正极		负极	
	材料	颜色	材料	颜色
铂铑-铂	铜	红	铜镍合金	绿
镍铬-镍硅 铜-康铜	铜	红	康铜	棕
镍铬-考铜	镍铬	褐绿	考铜	黄

(3) 冷端温度的补偿

与热电偶配套的仪表是根据各种热电偶的温度-热电势关系曲线在冷端温度保持为 0℃的情况下进行刻度的。采用补偿导线后，虽然把热电偶的冷端从温度较高和不稳定的地方延伸到温度较低和比较稳定的操作室内，但由于操作室内的温度往往高于 0℃，而且是不恒定的，这样热电偶所产生的热电势必然偏小，且测量值也随着冷端温度变化而变化，因此在应用热电偶测温时，只有将冷端温度保持为 0℃，或者进行一定的修正，才能得出准确的测量结果，这就称为热电偶的冷端温度补偿。

实验室采用冷端温度补偿的方法通常是把热电偶的两个冷端分别插入盛有绝缘油的试管中，然后放入装有冰水混合物的容器中，如图 5-4 所示，使冷端温度保持为 0℃。

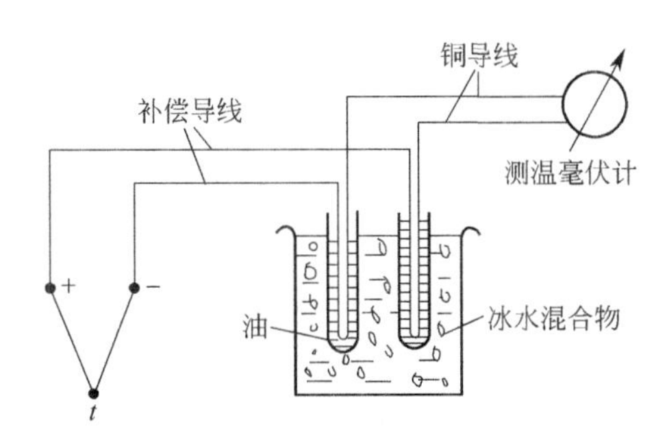

图 5-4 热电偶冷端温度保持 0℃的方法

(4) 几种常用的热电偶

目前我国广泛使用的热电偶有下列几种。

1）铂铑-铂热电偶　分度号为 S。该热电偶正极为 90%的铂和 10%的铑组合成的合金丝，负极为铂丝。此种热电偶在 1300℃以下范围内可长期使用，在良好环境中可短期测量 1600℃高温。由于容易得到高纯度的铂和铑，故该热电偶的复制精度和测量准确性较高，可用于精密温度测量和用作基准热电偶。其缺点是热电势较弱，且成本较高。

2）镍铬-镍硅热电偶　分度号为 K。该热电偶正极为镍铬，负极为镍硅。该热电偶可在氧化性或中性介质中长期测量 900℃以下的温度，短期测量可达 1200℃。该热电偶具有复制性好、产生热电势大、线性好、价格便宜等特点。缺点是测量精度偏低，但完全能满足工业测量的要求，是工业生产中最常用的一种热电偶。

3）镍铬-考铜热电偶　分度号为 EA。该热电偶正极为镍铬，负极为考铜。适用于还原性或中性介质，长期使用温度不可超过 600℃，短期测量可达 800℃。该热电偶的特点是电热灵敏度高，价格便宜。

4）铜-康铜热电偶　分度号为 CK。该热电偶正极为铜，负极为康铜。其特点是低温时精确度较高，可测量－200℃的低温，上限温度为 300℃，价格低廉。

(5) 热电偶温度计的校验

热电偶在使用过程中由于热端被氧化、腐蚀和高温下热电偶材料再结晶，其热电特性会发生变化，使测量误差变大。为了使温度的测量保证一定的精度，热电偶必须定期进行校验以得出热电势变化的情况。当热电势变化超出规定的误差范围时，可以更换热电偶丝或把热电偶的低温端剪去一段，焊接后再使用。在使用前必须重新进行校验。

根据我国的规定，各种热电偶必须在规定的温度点进行校验，且各温度点的最大误差不能超过允许的误差范围，具体如表 5-4 所列，否则不予使用。

表 5-4　常用热电偶校验允许偏差

型号	热电偶材料	校验点/℃	热电偶允许偏差			
			温度/℃	偏差/℃	温度/℃	偏差/%
S	铂铑-铂	600,800,1000,1200	0～600	±2.4	＞600	±0.4
K	镍铬-镍硅(铝)	400,600,800,1000	0～400	±4	＞400	±0.75
E	镍铬-铜镍(康铜)	300,400,600	0～300	±4	＞300	±0.1

5.1.3 热电阻温度计

除了热电偶温度计外，工业生产上经常还用到热电阻温度计。热电阻温度计是利用测温元件的电阻值会随着温度的变化而发生变化的特性进行温度测量的。热电偶温度计在 500℃以下温度的测量中输出的热电势很小，测量时容易产生误差。因此，在工业生产中，－120～500℃范围内的温度测量常采用热电阻温度计。它的主要特点是测量精度高，性能稳定。其中铂热电阻温度计的测量精确度是最高的，它不仅广泛应用于工业测温，而且被制成标准的基准仪。在特殊情况下，热电阻温度计测量的下限可达－270℃，上限可达 1000℃。

纯金属及多数合金的电阻率随温度升高而增大，在一定范围内，电阻和温度呈现线性关系。若已知金属导体在温度为 0℃时的电阻为 R_0，则在温度为 t 时的电阻为 R_t：

$$R_t = R_0(1 + \alpha t) \tag{5-3}$$

式中，α 为平均电阻温度系数。

不同金属具有不同的平均电阻温度系数，只有具有较大的平均电阻温度系数的金属才有可能作为测温用的热电阻。最佳和最常用的热电阻温度计材料是纯铂，其测量范围为

−200～500℃。工业生产上也经常用铜丝电阻温度计，它的测温范围为−150～180℃。

为了减小导线的电阻对测量的影响，常采用如图 5-5 所示的三线制连接线路来测量热电阻的阻值。注意要对通过 R_t 对电流加以限制，否则会引起较大的误差。

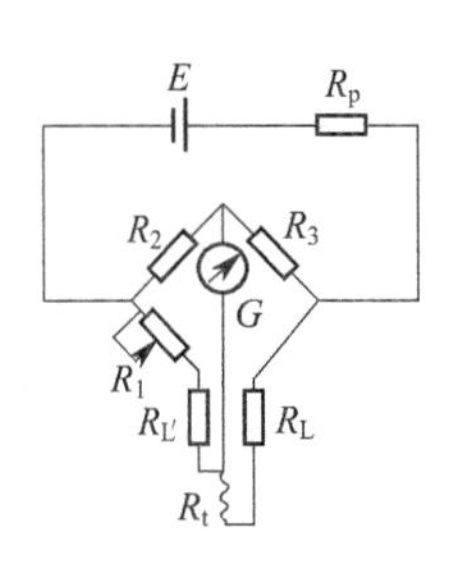

图 5-5　三线制连接线路

热电阻在使用之前要进行校验，使用一定时间后也需进行校验，以保证其准确性。工作基准或标准热电阻的校验通常在几个平衡点下进行，如 0℃冰、水平衡点等，但其要求高，方法、设备复杂，我国对于热电阻的校验有一定的规定。工业用热电阻的检验只要 R_0（0℃时的电阻值）及 R_{100}（100℃时的电阻值）的数值不超过规定的范围即可。

5.1.4　温度计的选择和使用原则

在选择和使用温度计时，应该考虑以下几点。

① 被测物体的温度是否需要指示、记录和自动控制；

② 测温范围与准确度要求；

③ 感温元件的尺寸是否会破坏被测物体的温度场；

④ 被测温度变化时，感温元件的滞后性能是否符合测温要求；

⑤ 被测物体和环境条件对感温元件有无损害；

⑥ 使用接触式温度计时，感温元件必须与被测物体接触良好，且与周围环境无热交换，否则温度计测出的温度和真实温度有差异；

⑦ 感温元件需要插入被测介质一定深度，在气体介质中，金属保护管插入深度为保护管的 10～20 倍，非金属保护管插入深度为保护管的 10～15 倍。

5.2　压力(差)测量

在化工生产和实验中，压力是重要的参数之一。例如管道阻力实验需测定流体流过管道的压降，泵性能实验需测量泵的进出口压力以便了解泵的性能和安装是否正确，精馏实验需经常观察塔顶和塔釜的压力以便了解精馏塔的操作是否正常。此外，压力测量的意义还不局限于它自身，有些其他参数的测量，如物位、流量等往往也通过测量压力或压差来换算。

测量压力的仪表很多，按照其转换原理的不同，大致可分为液柱式压力计、弹性式压力计、电气式压力计和活塞式压力计四类。

5.2.1　液柱式压力计

(1) 液柱式压力计的结构及特性

液柱式压力计是根据流体静力学原理，将被测压力转换成液柱高度进行测量的。按其结构形式的不同，有 U 形管压差计、单管压差计、斜管压差计和 U 形管双指示液压差计等。其结构及特性如表 5-5 所列。这类压力计结构简单，使用方便，但其精度受工作液的毛细作用、密度及视差等因素的影响，测量范围较窄，一般用来测量较低压力、真空度或压差。它不能进行自动指示和记录，所以应用范围受到限制。

表 5-5　液柱式压力(差)计的结构及特性

名称	示意图	测量范围	静态方程	备　注
正U形管压差计	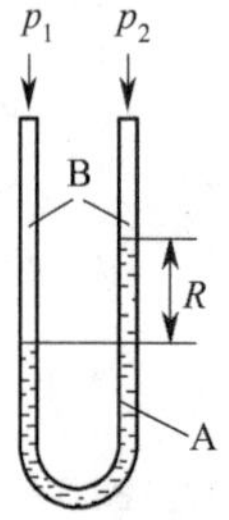 p_1 p_2 B R A	高度差 R 不超过 800mm	$\Delta p=Rg(\rho_A-\rho_B)$（液体） $\Delta p=Rg\rho$（气体）	零点在标尺中间，常用作标准压差计校正流量，适用于指示剂密度大于被测流体的情况
倒U形管压差计	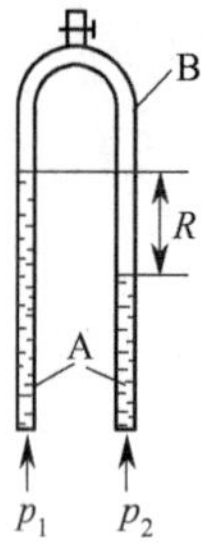 B R A p_1 p_2	高度差 R 不超过 800mm	$\Delta p=Rg(\rho_A-\rho_B)$（液体）	以待测液体为指示液，适用于较小压差、指示剂密度小于被测流体密度的测量
单管压差计	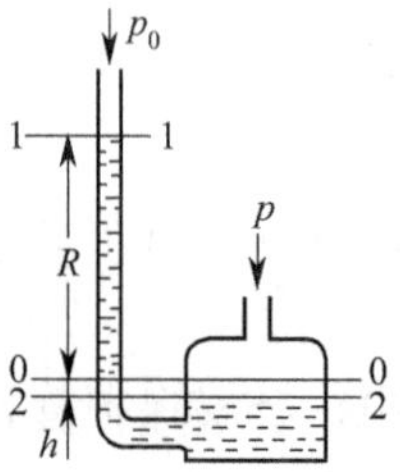 p_0 1 1 p R 0 0 2 2 h	高度差 R 不超过 1500mm	$\Delta p=R\rho(1+S_1/S_2)g$ 当 $S_1 \ll S_2$ 时 $\Delta p=Rg\rho$ S_1—垂直管截面积 S_2—扩大室截面积	零点在标尺下端，用前需调整零点，可用作标准压差计
斜管压差计	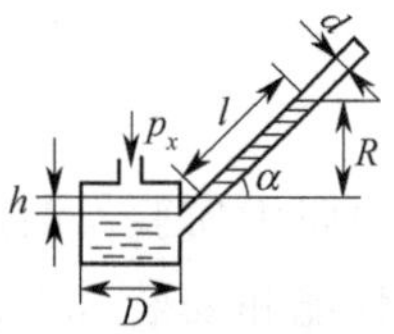 d p_x l R h α D	高度差 R 不超过 1200mm	$\Delta p=l\rho g(\sin\alpha+S_1/S_2)$ 当 $S_1 \ll S_2$ 时 $\Delta p=l\rho g\sin\alpha$ S_1—倾斜管截面积 S_2—扩大室截面积	$\alpha<15°$ 时，可改变 α 的大小来调节测量范围。零点在标尺下端，用前需调整
U形管双指示液压差计	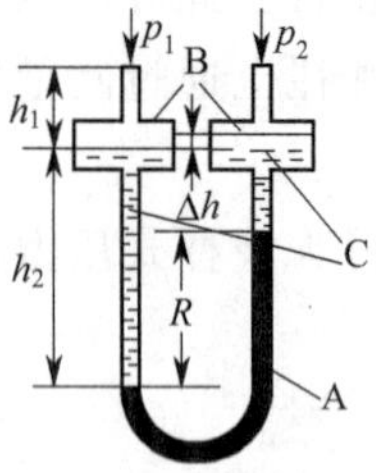 p_1 p_2 B h_1 Δh C h_2 R A	高度差 R 不超过 500mm	$\Delta p=Rg(\rho_A-\rho_C)$	U形管中装有A和C两种密度相近的指示液，且两臂上方有扩大室，旨在提高测量精度，适用于压差很小的情况

(2) 液柱式压力计使用注意事项

① 被测压力不能超过仪表测量范围。有时因被测对象突然增压或操作不当造成压力增大，会使工作液被冲走。若是水银工作液被冲走，不仅会造成损失，还会污染环境。

② 被测介质不能与工作液互溶或起化学反应。若两者互溶或起反应，则应更换工作液或采取加隔离液的方法。常用的隔离液如表 5-6 所列。

表 5-6　某些介质的隔离液

测量介质	隔离液	测量介质	隔离液
氯气	98%的浓硫酸或氟油	氨水	变压器油
氯化氢	煤油	水煤气	变压器油
硝酸	五氯乙烷	氧气	甘油

③ 液柱式压力（差）计安装位置应避开过热、过冷和有震动的地方。

④ 液柱式压力（差）计使用前应将工作液面调整到零位线上。

⑤ 在读取压力值时，视线应在液柱面上，观察水时应看凹面处，观察水银面时应看凸面处。

⑥ 工作液为水时，可在水中加入一点墨水或其他溶于水的颜料，以便于观察读数。

5.2.2　弹性式压力计

弹性式压力计是利用各种形式的弹性元件，在被测介质压力的作用下，使弹性元件受压后产生弹性变形而制成的测压仪表。这种仪表具有结构简单、使用可靠、读数清晰、价格低、测量范围宽，以及有足够的精度等优点。若增加附加装置，如记录机构、电气变换装置、控制元件等，则可以实现压力的记录、远传、信号报警、自动控制等，是一种应用最为广泛的测压仪表。弹性式压力计的结构及测量范围如表 5-7 所列。本节重点介绍弹簧管压力表。

表 5-7　弹性式压力计的结构及测量范围

类别	名称	示意	测压范围/Pa	
			最小	最大
薄膜式	平薄膜	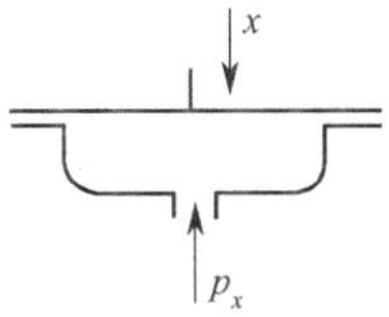 x, p_x	$0\sim10^4$	$0\sim10^8$
	波纹膜	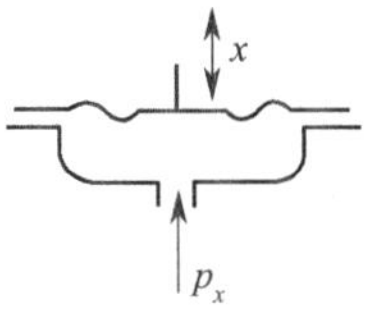 x, p_x	$0\sim1$	$0\sim10^6$
	挠性膜	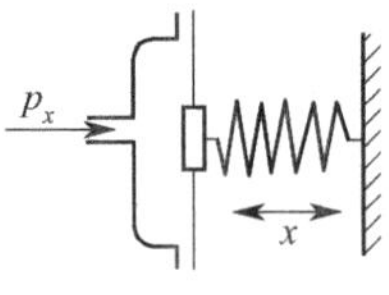 p_x, x	$0\sim10^{-2}$	$0\sim10^5$
波纹管式	波纹管	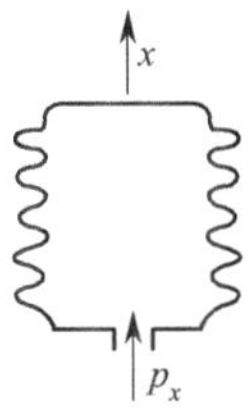 x, p_x	$0\sim1$	$0\sim10^6$

续表

类别	名称	示意	测压范围/Pa	
			最小	最大
弹簧管式	单圈弹簧管		$0\sim10^2$	$0\sim10^9$
	多圈弹簧管		$0\sim10$	$0\sim10^8$

(1) 弹簧管压力表的工作原理

弹簧管压力表的构造如图 5-6 所示。弹簧管 1 是压力表的测量元件，图中所示为单圈弹簧管，它是一根呈弧形的扁椭圆状的空心金属管。管子的自由端 B 封闭，管子的另一端固定在接头 9 上，其与测压点相接。受压后，弹簧管发生弹性变形，使自由端 B 产生位移。由于输入压力与弹簧管自由端 B 的位移成正比，所以只要测得 B 点的位移量，就能反映出压力的大小，这就是弹簧管压力表的基本测量原理。

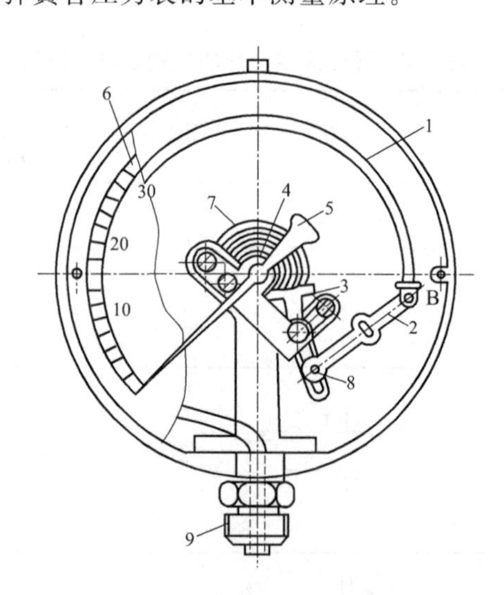

图 5-6　弹簧管压力表的构造

1—弹簧管；2—拉杆；3—扇形齿轮；4—中心齿轮；5—指针；6—面板；7—游丝；8—调整螺钉；9—接头

弹簧管的自由端 B 的位移量一般很小，直接显示有困难，所以必须通过放大机构才能指示出来。具体放大过程如下：弹簧管自由端 B 的位移通过拉杆 2，使扇形齿轮 3 作逆时针偏转，于是指针 5 通过同轴的中心齿轮 4 的带动而作顺时针偏转，在面板 6 的刻度标尺上指示出被测压力的数值。由于弹簧管自由端的位移与被测压力之间成正比关系，因此弹簧管压力表的刻度标尺是线性的。

游丝 7 用来克服因扇形齿轮和中心齿轮间的传动间隙而产生的仪表变差。改变调整螺钉 8 的位置（即改变机械传动的放大系数），可以实现压力表量程的调整。

(2) 弹簧管压力表使用安装的注意事项

正确地使用和安装压力表，是保证测量结果准确性和压力表使用寿命的重要环节。

① 应根据工艺要求正确选用仪表类型。测量爆炸性、腐蚀性、有毒流体的压力时，应使用专用的仪表。例如，普通压力表的弹簧管多采用铜合金；而氨用压力表弹簧管的材料都为碳钢，不允许采用铜合金，因为氨气对铜的腐蚀极强；又如氧用压力表禁油，因为油进入氧气系统易引起爆炸。

② 仪表应工作在允许的压力范围内。一般被测压力最大值不应超过仪表刻度的 2/3，如测量脉动压力，不应超过测量上限的 1/2，而这两种情况被测压力都不应低于仪表刻度的 1/3。

③ 仪表安装处与测定点间的距离应尽量短，以免指示迟缓。

④ 仪表必须垂直安装并无泄漏现象。

⑤ 在振动情况下使用仪表时，要装减震装置。测量蒸汽压力时，应加装凝液管，以防高温蒸汽直接与测压元件接触。测量腐蚀性介质的压力时，应加装有中性介质的隔离罐。

⑥ 当被测压力较小，而压力表与取压口又不在同一高度时，由此高度引起的测量误差应按 $\Delta p = \pm hg\rho$ 进行修正（式中，h 为高度差；ρ 为导压管中介质的密度；g 为重力加速度）。

⑦ 仪表必须定期校验。

5.2.3 电气式压力计

电气式压力计是一种能将压力转换成电信号进行传输及显示的仪表。这种仪表的测量范围广，可以远距离传送信号，可实现压力的自动控制，满足工业自动化程度不断提高的要求。

电气式压力计一般由压力传感器、测量电路和信号处理装置组成，如图 5-7 所示。常用的信号处理装置有指示仪、记录仪以及控制器、微处理器等。压力计中的压力传感器有压磁式、压电式、电容式、电感式和电阻应变式等。下面主要介绍压阻式压力传感器和电容式压力传感器。

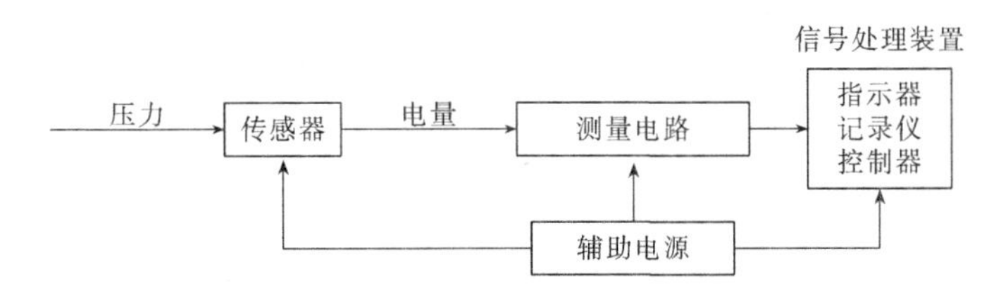

图 5-7 电气式压力计组成框图

(1) 压阻式压力传感器

压阻式压力传感器也称固态压力传感器或扩散型压阻式压力传感器。其工作原理是单晶硅的压阻效应。将单晶硅膜片和应变电阻片采用集成电路工艺结合在一起，构成硅压阻芯片，然后将此芯片封装在传感器壳内，再连接出电极线而成。典型的压阻式压力传感器的结构原理如图 5-8 所示，图中硅膜片两侧有两个腔体，通常上接管与大气或与其他参考压力源相通，下接管相连的高压腔内充有硅油并有隔离膜片与被测对象隔离。

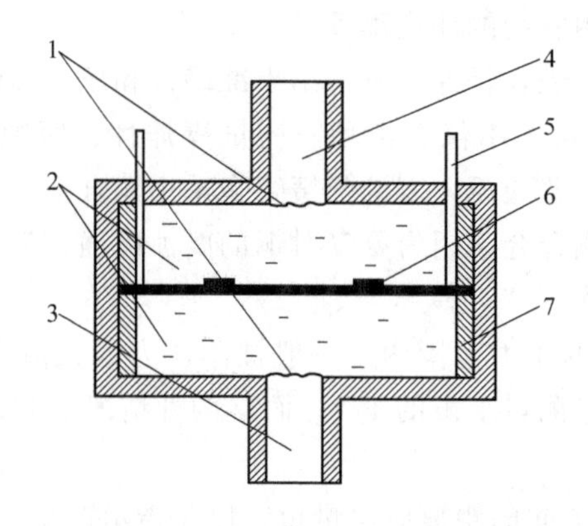

图 5-8　压阻式压力传感器的结构原理图

1—隔离膜片；2—硅油；3—高压端；4—低压端；5—引线；6—硅膜片及应变电阻；7—支架

当被测对象的压力通过下引压管线、隔离膜片及硅油，作用于硅膜片上时，硅膜片产生变形，膜片上的 4 个应变电阻片两个被压缩、两个被拉伸，使其构成的惠斯通电桥内电阻发生变化，并转换成相应的电信号输出。电桥采用恒压源或恒流源供电，减小了温度对测量结果的影响。应变电阻片的变化值与压力呈良好的线性关系，因而压阻式压力传感器的精度通常可达 0.1%。

压阻式压力传感器具有精度高、工作可靠、频率响应高、迟滞小、尺寸小、质量轻、结构简单等特点，可以在恶劣的环境下工作，便于实现显示数字化。

（2）电容式压力传感器

电容式压力传感器是利用两平行板电容测量压力的传感器，如图 5-9 所示。当压力 p 作用于膜片时，膜片产生位移，改变板间距 d，引起电容量发生变化，经测量线路的转换，可求出作用压力 p 的大小。当忽略边缘效应时，平板电容器的电容量 C 为

$$C=\frac{\varepsilon S}{d} \tag{5-4}$$

式中，ε 为介电常数；S 为极板间重叠面积；d 为极板间距离。

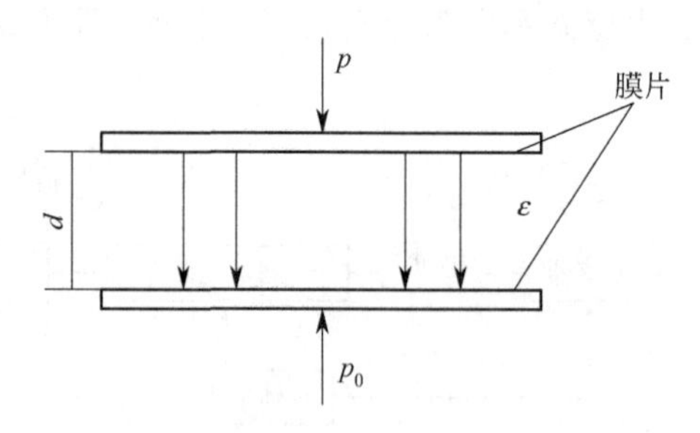

图 5-9　电容式压力传感器原理示意

由式（5-4）可知，电容量 C 的大小与 ε、S 和 d 有关，当被测压力影响三者中的任一参数，均会改变电容量。所以，电容式压力传感器可分为变面积式、变介电质式和变极间距离式 3 种。

电容式压力传感器的主要特点如下。

① 灵敏度高，特别适用于低压和微压测试。

② 内部无可动件，不消耗能量，测量误差小。

③ 膜片质量很小，因而有较高的频率，从而保证良好的动态响应能力。

④ 用气体或真空作绝缘介质，其损失小，本身不会引起温度变化。

⑤ 结构简单，多数采用玻璃、石英或陶瓷作为绝缘支架，因而可以在高温、辐射等恶劣条件下工作。

近年使用新材料、新工艺和微型集成电路，并将电容式压力传感器的信号转换电路与传感器组装在一起，有效地消除了电噪声和寄生电容的影响。电容式压力传感器的测量压力范围可从几十帕至百兆帕，使用范围得以拓展。

5.2.4 活塞式压力计

活塞式压力计是根据水压机液体传送压力原理，将被测压力转换成活塞上所加平衡砝码的质量来进行测量的。它的测量精度很高，允许误差可小到0.02%～0.05%，但结构较复杂，价格较贵，一般作为标准型压力测量仪器来检验其他类型的压力计。

5.2.5 压力(差)计的校验

新的压力（差）计在出厂使用之前要进行校验，以鉴定其技术指标是否符合规定的精度。压力（差）计使用一段时间后也要进行校验，目的是确定其是否符合原有的精度，如果确认误差超过规定值，就应对压力（差）计进行检修，检修后的压力（差）计仍需进行校验才能使用。

压力（差）计校验的方法一般有静态校验法和动态校验法两大类：a. 静态校验主要是测定静态精度，确定仪表等级，包括“标准表比较法”和“砝码校验法”；b. 动态校验主要是测定压力（差）计的动态特性，如仪表的过渡过程、时间常数和静态精度等，常用的方法是“激波管法”。

5.3 流量测量

在化工生产和实验中，经常要测量各种介质（液体、气体和蒸气等）的流量，以便为操作和控制提供依据。流量可分为瞬时流量和总量。瞬时流量指单位时间内流过管道某一截面的流体数量的大小；总量指在某一段时间内流过管道的流体流量的总和，即瞬时流量在某一段时间内的累计值。

流量和总量可以用质量表示，也可以用体积表示。单位时间内流过的流体以质量表示的称为质量流量，以体积表示的称为体积流量。测量流量的方法很多，其测量原理和所应用的仪表结构形式各不相同，大致分为速度式流量计、容积式流量计和质量流量计三类。

5.3.1 速度式流量计

速度式流量计是一种以测量流体在管道内的流速作为测量依据来计算流量的仪表。如差压式流量计、转子流量计、电磁流量计、涡轮流量计、堰式流量计等。本节介绍常用的差压式流量计、转子流量计和涡轮流量计。

(1) 差压式流量计

差压式流量计是利用流体流经节流装置或匀速管时产生的压力差来实现流量测量的。其中用节流装置和差压计所组成的差压式流量计是目前工业生产和实验装置中应用最广的一种流量测量仪表。通用的节流装置有孔板（图 5-10）、喷嘴、文丘里管等。这里重点介绍孔板流量计。

孔板流量计是通过测量流体流经孔板前后引起的压力变化来求流体的体积流量的流量计，流量计的读数与流体体积流量的关系为：

$$V_s = C_0 A_0 \sqrt{\frac{2gR(\rho_A - \rho)}{\rho}} \tag{5-5}$$

式中，C_0 为孔流系数；A_0 为孔板小孔的截面积；ρ_A 为指示液的密度；R 为指示液液面的高度差。

对于按标准规格及精度制作的孔板，用角接法取压（称标准孔板），C_0 取决于截面比 A_0/A_1（A_1 为管截面积）及管内雷诺数 Re_1。从图 5-11 中可以看出，Re_1 超过某限值之后，C_0 不再随 Re_1 而变，成为常数。显然，在孔板的设计和使用中，希望 Re_1 大于界限值。

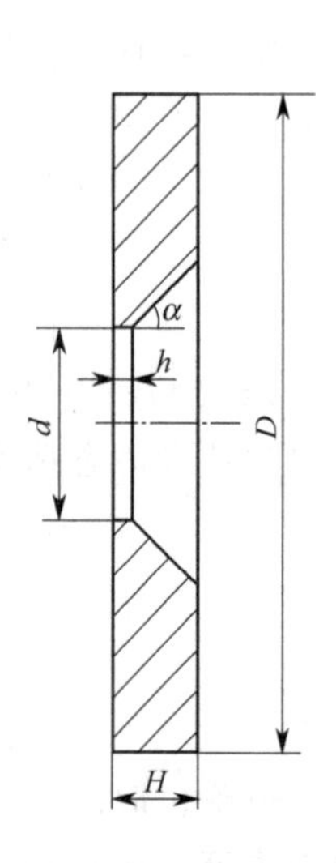

图 5-10　孔板断面示意

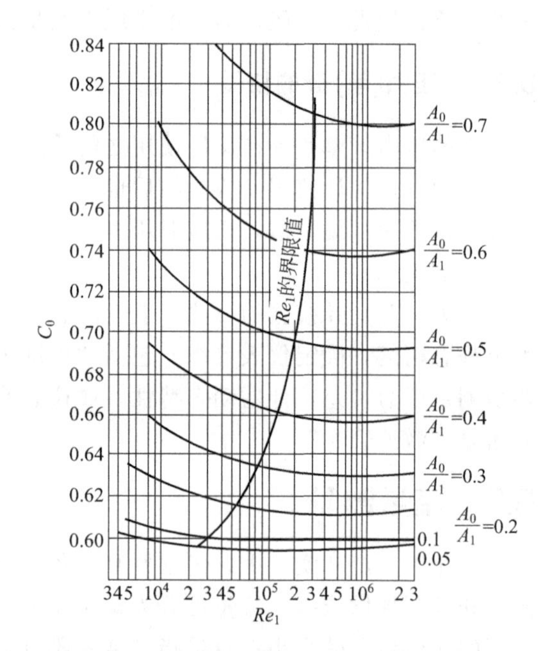

图 5-11　孔流系数 C_0 与 Re_1 与 A_0/A_1 的关系

孔板构造简单、易加工、造价低，其主要缺点是阻力损失大，使流体的最大通过能力下降颇多。

孔板流量计安装时要注意方向，不得装反；加工时要求严格且无毛刺，否则将影响测量精度。对于易使孔板变脏、磨损和变形的不洁净或腐蚀性的流体不宜使用孔板流量计。

（2）转子流量计

转子流量计与前面所述的差压式流量计在工作原理上是不同的。差压式流量计是在节流面积（如孔板流通面积）不变的条件下，以差压变化来反映流量的大小，因此又称为恒截面变压差的流量计。而转子流量计却是以压降不变，利用节流面积的变化来测量流量的大小，即转子流量计采用的是恒压差、变截面的流量测量方法。这种流量计特别适宜测量管径 50mm 以下管道的流量，测量的流量可小到每小时几升。因此，在实验室中得到广泛的应用。

1）转子流量计的工作原理　转子流量计主要由两部分组成，一个是由下往上逐渐扩大的锥形管（通常用玻璃制成，锥角为 40′～4°）；另一个是放在锥形管内可自由运动的转子（金属或其他材料制成）。当流量为零时转子沉在管下端，有流体自下而上流动时，它即被推起而悬浮在管内的流体中，随流量大小不同，转子将悬浮在不同的位置上，如图 5-12 所示。

当有流体流过锥形管时，位于锥形管中的转子受到一个向上的力，使转子浮起。当这个力正好等于浸没在流体中的转子净重力（即等于转子重力减去流体对转子的浮力）时，则作用在转子上的上下两个力达到平衡，此时转子就停浮在一定的高度上。假如被测流体的流量突然由小变大时，作用在转子上的向上的力增大，因为转子在流体中受的净重力是不变的，所以转子就上升。由于转子在锥形管中的位置升高，造成转子与锥形管的环隙增大，即流通面积增大流过此环隙的流体流速变慢，因而流体作用在转子上的向上力也就变小。当流体作用在转子上的力再次等于转子在流体中的净重力时，转子又稳定在一个新的高度上。这样，转子在锥形管中的平衡位置的高低与被测介质的流量大小相对应。这就是转子流量计测量流量的基本原理。

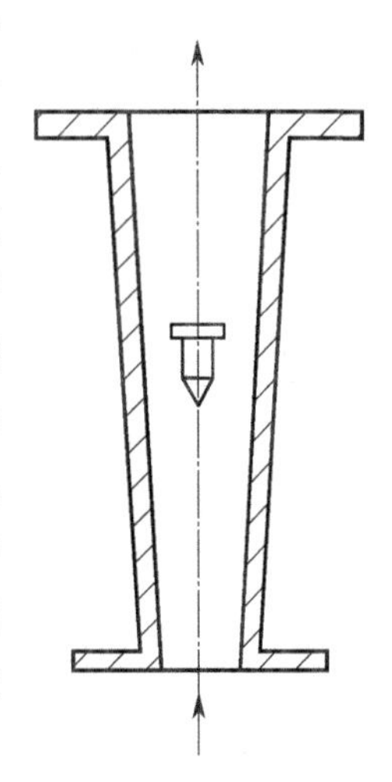

图 5-12 转子流量计工作原理图

转子流量计体积流量的计算式为：

$$V_s = C_R A_0 \sqrt{\frac{2g(\rho_f - \rho)V_f}{\rho A_f}} \tag{5-6}$$

式中，C_R 为流量系数；A_0 为环隙截面积；ρ，ρ_f 为流体与转子的密度；V_f，A_f 为转子的体积与截面积（截面最大处）。

流量系数 C_R 的值主要取决于转子的构形，也与流体通过环隙流动的 Re 有关。对于如图 5-12 中所示的转子构形，当 Re 达到 10000 以后，C_R 值便恒定等于 0.98。

由式（5-6）可知，流量与环隙面积 A_0 有关，当锥形管与转子的尺寸固定时，此 A_0 取决于转子在管内的高度，因此在锥形管外面刻上对应的流量值，那么根据转子平衡位置的高低就可以直接读出流量的大小。

读取不同形状转子的流量计刻度时，均应以转子最大截面处作为读数基准，如图 5-13 所示。

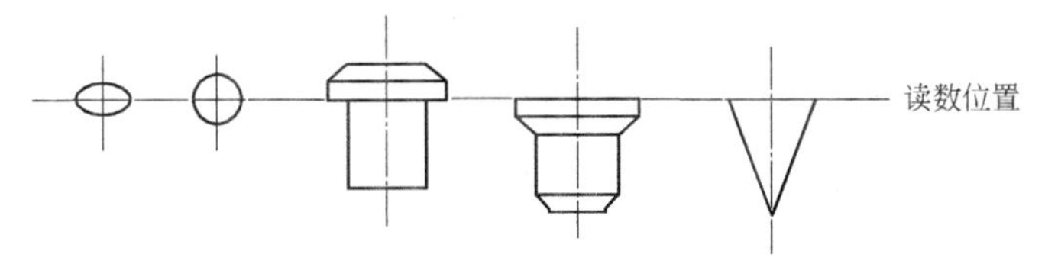

图 5-13 不同转子流量计的正确读数位置

2）转子流量计测定其他物质时流量的换算　转子流量计是一种非标准化仪表，每个转子流量计都附有出厂标定的流量数据。对用于测量液体的流量计，生产厂家是用 20℃的水标定；对用于测量气体的流量计，则是用 20℃、101.3kPa 下的空气进行标定。因此，在使用时，若不符合标定条件，则需按式（5-7）修正：

对于液体

$$V_1=V_0\sqrt{\frac{(\rho_f-\rho_1)\rho_0}{(\rho_f-\rho_0)\rho_1}} \tag{5-7}$$

式中，V_1 为工作状态下液体的实际流量；V_0 为转子流量计用水标定的读数；ρ_f 为转子的密度；ρ_1 为工作状态下液体的密度；ρ_0 为出厂标定时水的密度。

3）转子流量计量程的改变　当测量范围超出现有转子流量计的量程时，可以通过改变转子密度的方法来改变量程，方法有：改变转子的材料，将实心转子掏空，或向空心转子内加填充物。在转子形状和尺寸保持相同的情况下，流量可用式（5-8）进行换算：

$$V_0'=V_0\sqrt{\frac{\rho_f'-\rho_0}{\rho_f-\rho_0}}=V_0\sqrt{\frac{m_f'-V_f\rho_0}{m_f-V_f\rho_0}} \tag{5-8}$$

式中，V_0，ρ_0，ρ_f，m_f 分别为转子改变前的流体体积流量、流体密度、转子密度和转子质量；V_0'，ρ_f'，m_f'分别为转子改变后的流体体积流量、转子密度和转子质量。

（3）涡轮流量计

涡轮流量计是在动量矩守恒原理的基础上设计的。在流体流动的管道内，安装一个可以自由转动的叶轮。当流体通过涡轮时，涡轮的叶片因流动流体冲击而旋转，旋转速度随流量的变化而变化。在规定的流量范围和一定的流体黏度下，转速与流速成线性关系。因此，测出叶轮的转速或转数，就可确定流过管道的流体流量或总量。利用适当的装置将涡轮转速转换成脉冲电信号，通过测量脉冲频率或用适当的装置将电脉冲转换成电压或电流输出，最终测取流量。

涡轮流量计的结构示意如图 5-14 所示，它主要由下列几部分组成：涡轮 1 由高磁导率的不锈钢材料制成，叶轮芯上装有螺旋形叶片，流体作用于叶片上使之转动。导流器 2 是用以稳定流体的流向和支撑叶轮的。磁电感应转换器 4 由线圈和磁钢组成，用以将叶轮的转速转换成相应的电信号，以供给前置放大器进行放大。整个涡轮流量计安装在外壳 3 上，外壳由非导磁的不锈钢制成，两端与流体管道相连接。

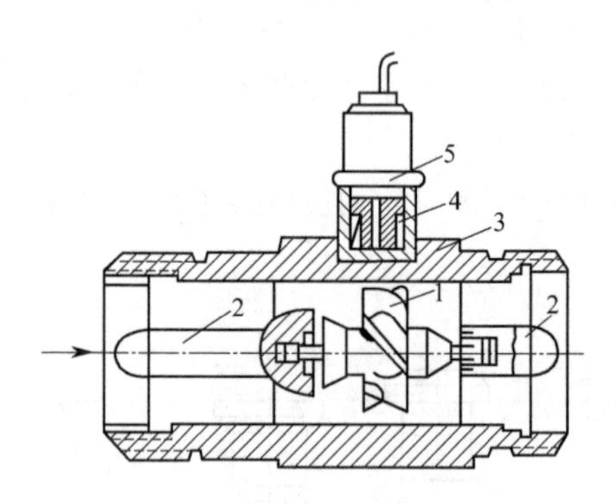

图 5-14　涡轮流量计的结构示意

1—涡轮；2—导流器；3—外壳；4—磁电感应转换器；5—前置放大器

涡轮流量计的工作过程如下：当流体通过涡轮叶片与管道之间的间隙时，由于叶片前后的压差产生的力推动叶片使涡轮旋转。在涡轮旋转的同时，高导磁的涡轮周期性地扫过磁钢，使磁钢的磁阻发生周期性变化，线圈中的磁通量也跟着发生周期性变化，从而感应产生交流电信号。交流电信号的频率与涡轮的转速成正比，也即与流量成正比。这个电信号经前置放大器放大后，送往电子计数器或电子频率计，以累积或指示流量。

涡轮流量计安装方便，磁电感应转换器与叶片间不需密封和齿轮传动机构，因而测量精确度高，可耐高压。由于基于磁感应转换原理，故反应快，可测脉动流量。输出信号为电频率信号，便于远传，不受干扰。

涡轮流量计的涡轮容易磨损，因此被测介质中不应带机械杂质，一般应加过滤器。它应水平安装，且必须保证前后有一定的直管段，以使流体流动比较稳定。一般入口直管段的长度取管道内径的 10 倍以上，出口取 5 倍以上。

5.3.2 容积式流量计

容积式流量计是一种以单位时间内所排出流体的固定容积的数目作为测量依据来计算流量的仪表。下面仅介绍实验室常用的湿式流量计。

湿式流量计主要由鼓形壳体、转鼓及传动记数机构所组成，如图 5-15 所示。转鼓是由圆筒及 4 个弯曲形状的叶片所构成，4 个叶片构成 4 个体积相等的小室。鼓的下半部浸没在水中。充水量由水位计 7 指示。气体从背部中间的进气管 9 处依次进入小室，并相继由顶部排出时，迫使转鼓转动。转鼓转动的次数通过计数机构在表盘上显示。配合秒表计时，可直接测定气体流量。湿式流量计可直接用于测量气体流量，也可作为标准仪器用于其他流量计的检定。

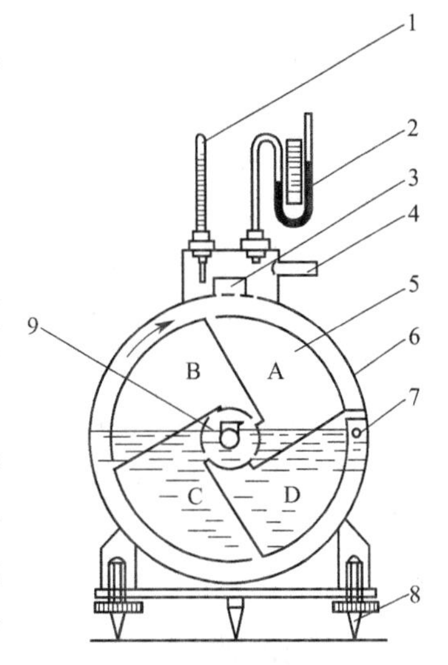

图 5-15　湿式流量计的结构示意

1—温度计；2—压差计；3—水平仪；4—排气管；5—转鼓；6—壳体；7—水位计；8—可调支脚；9—进气管

如图 5-15 所示，工作时气体由进气管进入，B 室正在进气，C 室开始进气，而 D 室排气将尽。

湿式流量计一般用标准容量瓶进行校准。标准容量瓶的体积为 V_v，湿式流量计体积示值为 V_w，则两者差值 ΔV 为 $\Delta V = V_v - V_w$。当流量计指针回转一周时，刻度盘上总体积为 5L，一般配置 1L 容量瓶进行 5 次校准，流量计总体积示值为 $\sum V_w$，则平均校正系数为：

$$C_w = \frac{\sum \Delta V}{\sum V_w} \tag{5-9}$$

因此，经校准后，湿式流量计的体积流量 V_s 与流量计示值 V'_s 之间的关系应为：

$$V_s = V'_s + C_w V'_s \tag{5-10}$$

5.3.3 质量流量计

这是一种以测量流体流过的质量为依据的流量计。质量流量计分直接式和间接式（也称推导式）两种。直接式质量流量计直接测量质量流量，有量热式、角动量式、陀螺式和科里奥利力式等。间接式质量流量计是用密度与容积流量经过运算求得质量流量的。质量流量计具有测量精度不受流体的温度、压力、黏度等变化影响的优点，是一种发展中的流量测量仪表。

流量计的种类很多，随着工业生产自动化水平的提高，出现了许多新型的流量测量仪表。超声波、激光、X 射线及核磁共振等新兴的流量测量技术正逐渐应用于工业生产中。

5.3.4 流量计的检验和标定

要想得到准确的流量值必须正确地使用流量计，应该充分了解流量计的构造和特性，采用与其相适应的方法进行测量，同时还要注意在使用中对流量计进行正确的维护和管理，每隔一定时间对其进行标定。以下几种情况均要对流量计进行标定。

① 长时间闲置的流量计；

② 要进行高精度测量时；

③ 对测量值产生怀疑时；

④ 当被测流体的特性不符合标定流量计用的流体的特征时。

液体流量计的标定有容器式、称重式、标准体积管式和标准流量计式等。

气体流量计的标定有容器式、音速喷嘴式、肥皂膜实验器式、标准流量计式、湿式流量计式等。标定气体流量计时需特别注意测量流过被标定流量计和标准器气体的温度、压力和湿度。另外在标定工作之前必须了解清楚气体的特性，如是否溶于水，其性质是否会随着温度、压力发生变化等。

5.4 液位测量

液位是表征设备或容器内液体贮量多少的量度。液位检测可为保证生产过程正常进行，如调节物料平衡、掌握物料消耗量、确定产品产量等，提供决策依据。

液位计因物系性质的变化而异，种类较多，常见液位计有：直读式液位计（玻璃管式液位计、玻璃板式液位计）、差压式液位计（压力式液位计、吹气法压力式液位计）、浮力式液位计（浮球式液位计、浮标式液位计、浮筒式液位计、磁翻板式液位计）、电气式液位计（电接点式液位计、磁致伸缩式液位计、电容式液位计）、超声波式液位计、雷达液位计、放射性液位计等。

下面介绍实验室中常用的直读式液位计、差压式液位计、浮力式液位计。

5.4.1 直读式液位计

（1）测量原理

直读式液位计的测量原理是利用仪表直接读取被测容器内的气相、液相的液位。直读式液位计测量简单，读数直观，但不便进行信号的远程传送，适于现场直读液位的测量。直读式液位计的测量原理如图 5-16 所示。

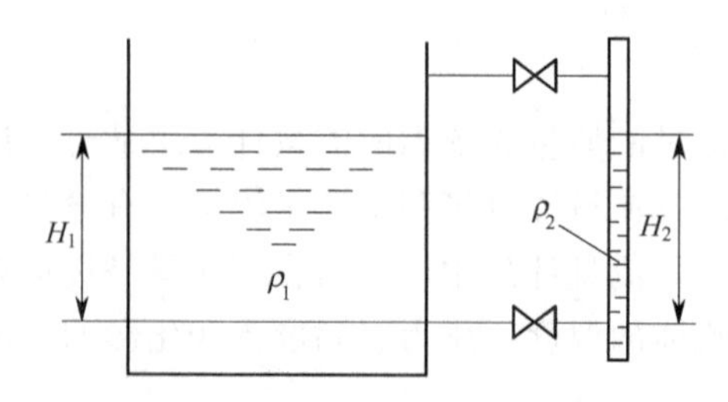

图 5-16 直读式液位计的测量原理图

利用等压面原理

$$\rho_1 g H_1 = \rho_2 g H_2 \tag{5-11}$$

当 $\rho_1=\rho_2$ 时，$H_1=H_2$。

当介质温度高时，$\rho_1\neq\rho_2$，就会出现误差。但由于其简单实用，因此得到较为广泛的应用，有时也用于自动液位计零位和最高液位的校准。

(2) 玻璃管式液位计

早期的玻璃管式液位计由于结构上的缺点，如易碎、长度有限等，只用于常压开口容器。现在由于玻璃管改成石英玻璃，同时外加了保护的金属管，克服了易碎的缺点。此外，由于石英具有耐高温高压的特点，拓宽了玻璃管式液位计的使用范围。还可以利用光线在液体与空气中折射率的不同，用滤色玻璃做成双色玻璃管式液位计，气相为红色，液相为绿色，可以方便地读取液位。

目前，常用的玻璃管式液位计如图 5-17 所示。上下两端采用法兰与设备连接并安装有阀门，上下阀内都装有钢球，当玻璃管因意外事故损坏时，钢球在容器内压力的作用下阻塞通道，这样容器便自动密封，可以防止容器内的液体继续外流。还可以采用蒸汽夹套伴热以防止易冷凝液体堵塞管道。

(3) 玻璃板式液位计

如图 5-18 所示为直读式玻璃板式液位计，其前后两侧的玻璃板交错排列，从液位计前面玻璃板可以看到其与后面的玻璃板之间的盲区，反之亦然，可以克服每段测量存在盲区的缺点。

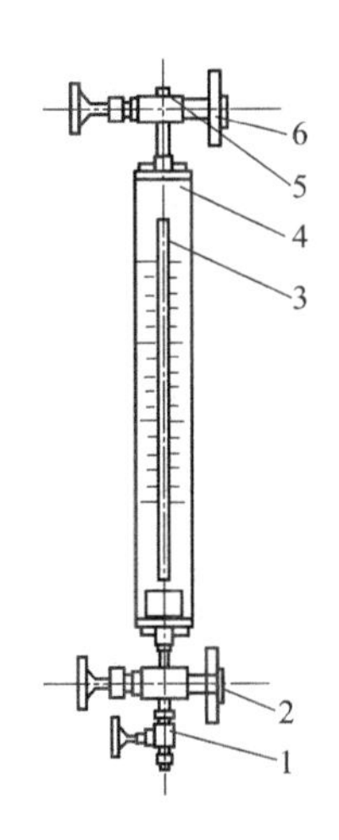

图 5-17 玻璃管式液位计

1—排污阀；2—下阀；3—石英管

4—外壳；5—堵头；6—上阀

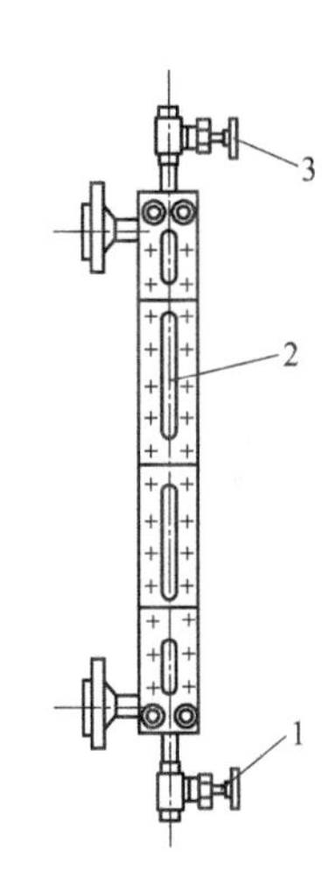

图 5-18 直读式玻璃板式液位计

1—下阀；2—玻璃板；3—上阀

5.4.2 差压式液位计

(1) 测量原理

如图 5-19 所示的液位测量系统，差压仪表 2 测得的是左右两边的压差，即

$$\Delta p=p_2-p_1=(p_0+\rho gH)-p_0=\rho gH$$

$$H=\frac{\Delta p}{\rho g} \tag{5-12}$$

式中，Δp 为压差；ρ 为被测流体密度；H 为液位高度。

由于被测液体的密度 ρ 是已知的，差压变送器测得的压差与液位高度成正比，应用式（5-12）就可以计算出液位高度。

（2）带有正负迁移的差压法液位测量原理

带有正负迁移的差压液位计测量原理图如图 5-20 所示，气相导压管中充满的不是气体而是蒸汽冷凝下来的液体。其中 ρ 为被测流体的密度，ρ_1 为介质的密度，h_1 为冷凝液的高度。当气相不断冷凝时，冷凝液会自动从气相口溢出，回到被测容器从而保持 h_1 不变。当液位在零位时，变送器的负端受到 $\rho_1 g h_1$ 的压力，这个压力在计算时必须要加以抵消，称为负迁移。若测量液位的起始点为 H_0 处，变送器的正端有 $\rho g H_0$ 的压力，也要加以抵消，称为正迁移。因此，变送器的总迁移量为 $\rho_1 h_1 g - \rho g H_0$。

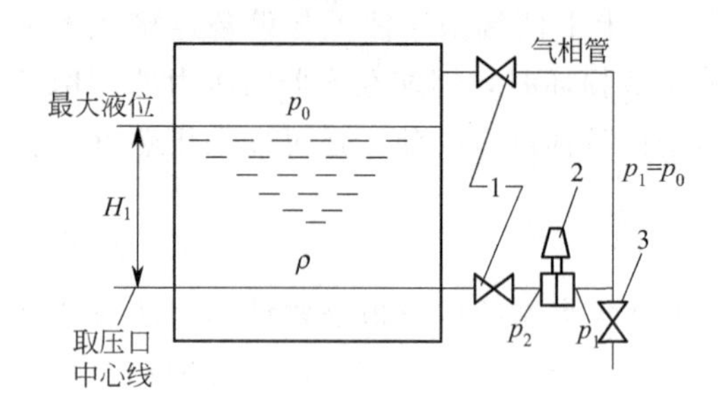

图 5-19 差压液位计测量原理图

1—切断阀；2—差压仪表；3—气相管排放阀

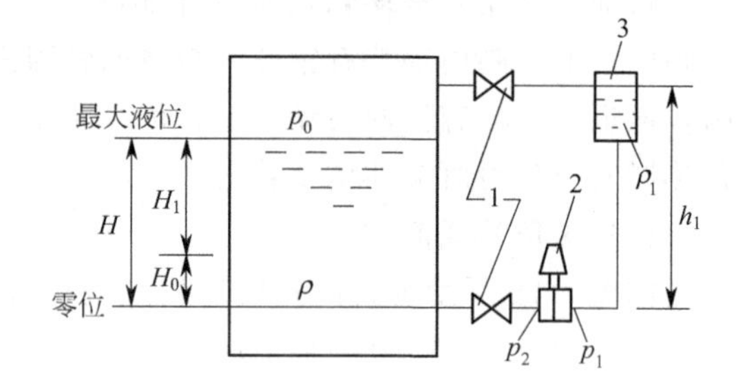

图 5-20 带有正负迁移的差压液位计测量原理图

1—切断阀；2—差压仪表；3—气相管排放阀

$$p_2 - p_1 = p_0 + \rho g (H_1 + H_0) - (p_0 + \rho_1 g h_1) = \rho g H_1 + \rho g H_0 - \rho_1 g h_1$$

$$H_1 = \frac{(p_2 - p_1) + (\rho_1 g h_1 - \rho g H_0)}{\rho g}$$

分子后面一项 $\rho_1 h_1 g - \rho g H_0$ 即为总迁移量，因此当具有迁移时，总迁移量要抵消，所以上式变为

$$H_1 = \frac{p_2 - p_1}{\rho g} \tag{5-13}$$

即仪表的量程为　$\Delta p = \rho g H_1$。

当被测流体有腐蚀性或易结晶时，可选带有隔离膜片的双法兰差压变送器，迁移量及仪表量程的计算仍然用上面的公式，但式中 ρ_1 为毛细管中所充的硅油的密度，h_1 为两个法兰中心高度差。

5.4.3 浮力式液位计

浮力式液位计是最早的一类液位仪表，这类仪表利用物体在液体中浮力的原理实现液位测量。浮力式液位计又分为浮子式、浮球式和浮筒式液位计，前面两种是恒浮力式，后面一种是变浮力式。下面分别介绍它们的测量原理。

（1）浮子式液位计

如图 5-21 所示为浮子式液位计测量原理，液位计中浮子受浮力的作用浮在液体表面上，随着液面上下而升降，通过检测浮子位置的变化进行液位测量。当液位升高时，浮子上浮，钢丝绳靠指示表中预紧发条的拉力收入表体，以保持浮子的重力、浮力与发条的拉力相平衡，此时液位值通过钢带和减速齿轮传送到指示器上，指示表指示出液位值，变送器发出正比于液位的信号。

变送器按结构可分为：

① 钢带齿轮机构、电动变送器，其可进行就地液位指示及变送输出 4～20mA 的直流信号；

② 钢带齿轮机构、带有编码孔的钢带和带有读码装置的变送器，其可进行就地液位指示，变送器输出脉冲信号到二次仪表进行指示。

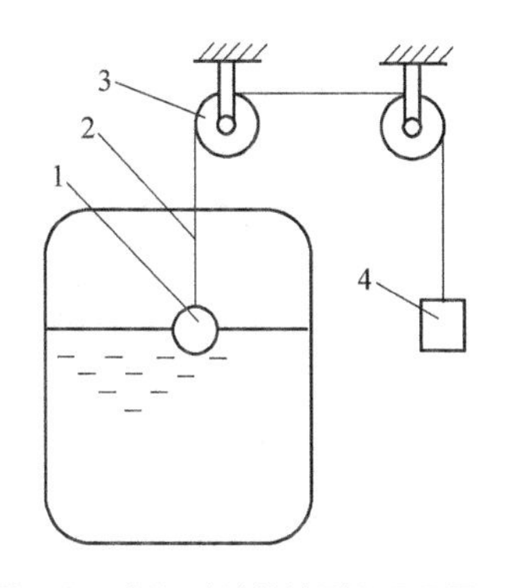

图 5-21　浮子式液位计测量原理图

1—浮子；2—钢带；3—导向滑轮装置；4—指示仪表或变送器

(2) 浮球式液位计

如图 5-22 所示为浮球式液位计测量原理，容器内液面上方的磁性浮球随着液面上下移动，通过磁性浮球的位置进行测量可得到液位信息，还可将磁性浮球的位置信号转换为电信号进行远程传送和控制。

当被测物料的密度发生改变时，还可以通过改变浮球的配重保证测量的顺利进行。

(3) 浮筒式液位计

如图 5-23 所示为浮筒式液位计测量原理。从零位到最高液位，浮筒全部浸没在液体中，浮力使浮筒有一个较小的向上位移，通过检测浮筒所受浮力的变化测量液位。当液位在零位时，扭力管受到浮筒的重力所产生的扭力矩的作用（此时扭力矩最大），处于“零”度。当液位逐渐上升到最高时，扭力管受到最大的浮力所产生的扭力矩的作用（这时的扭力矩最小），转过一个角度，变送器将这个转角换成 4～20mA 的直流信号，这个信号正比于被测量的液位。

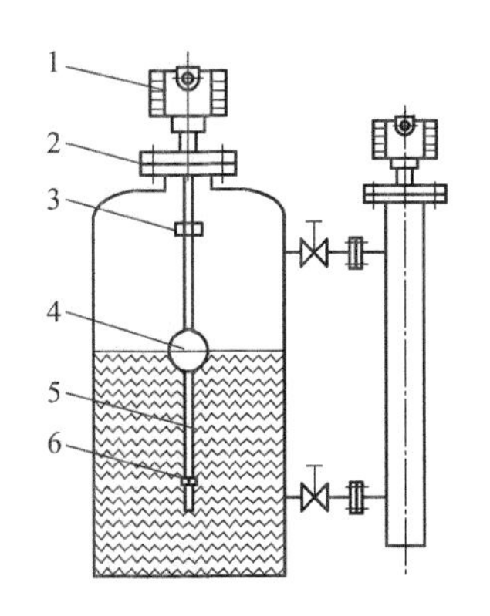

图 5-22　浮球式液位计测量原理图

1—指示仪表或变送器；2—连接法兰；3—上限位；4—浮球；5—导向连杆；6—下限位

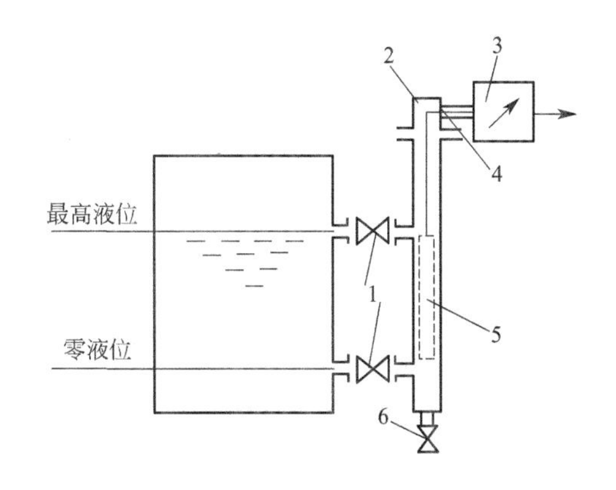

图 5-23　浮筒式液位计测量原理图

1—截止阀；2—浮筒；3—指示仪表或变送器；4—扭力管组件；5—浮筒；6—排放阀

(4) 磁翻板液位计

如图 5-24 所示为磁翻板液位计，在与容器相连的浮子室（用非导磁的不锈钢制成）内装有带磁性的钢浮子，翻板标尺贴着浮子室壁安装。当液位上升或下降时，浮子也随之升降，翻板标尺中的翻板（一半白色一半红色）受到浮子内磁钢的吸引而翻转，翻转部分显示红色，未翻转部分显示白色，红白分界处即表示液位所在。

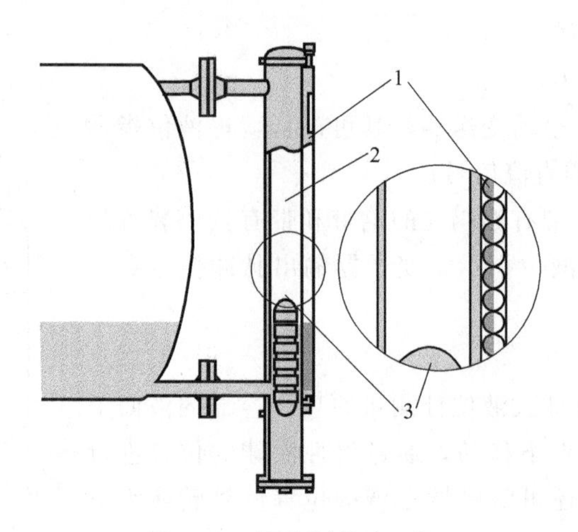

图 5-24　磁翻板液位计示意

1—翻板标尺；2—浮子室；3—磁浮子

磁翻板液位计除了配备指标标尺作就地指示外，还可以配备报警开关和远程传送装置，即可将液位转换成 4～20mA 的直流信号送到接收仪表。

第6章　化工原理虚拟仿真实验

化工原理虚拟实验室仿真软件是利用动态数学模型实时模拟真实实验现象和过程，通过3D仿真实验装置交互式操作，产生和真实实验一致的实验现象和结果。学生通过仿真实验的操作训练，可以了解实验原理，熟悉实验流程，掌握实验步骤，观察实验现象，记录实验数据，得出实验结果，达到加深对化工原理理论知识的理解和验证公式的目的。

本软件由北京欧倍尔软件技术开发有限公司开发，由9个3D虚拟仿真实验软件组成，它们分别是：雷诺演示3D虚拟仿真软件、化工流动过程综合实验3D虚拟仿真软件、离心泵综合性能测定实验3D仿真软件、恒压过滤实验3D虚拟仿真软件、传热综合实验3D虚拟仿真软件、精馏综合拓展实验3D虚拟仿真实验软件、二氧化碳吸收与解吸3D虚拟仿真软件、液液萃取塔实验3D虚拟仿真软件、洞道干燥实验3D虚拟仿真软件。

6.1　虚拟仿真实验概述

6.1.1　基本操作

以二氧化碳吸收与解吸3D虚拟仿真软件和化工流动过程综合实验3D虚拟仿真软件为例说明。

① 在仿真实验平台选择实验项目，例如二氧化碳吸收与解吸3D虚拟仿真软件的“解吸塔干填料曲线测定”，点击“启动”进入图6-1界面，若需要学习相关实验知识，点击“实验介绍”（实验介绍的内容详见每个实验）；若要进行仿真实验的学习，点击“进入系统”，出现图6-2界面，再点击“启动”出现图6-3界面。

若选择化工流动过程综合实验3D虚拟仿真软件中的“光滑管阻力测定实验”点击“启动”进入图6-4界面。

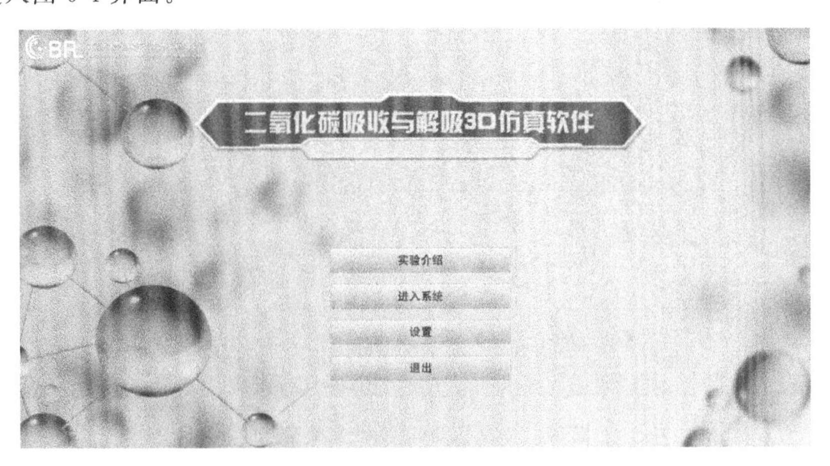

图6-1　二氧化碳吸收与解吸3D虚拟仿真软件界面

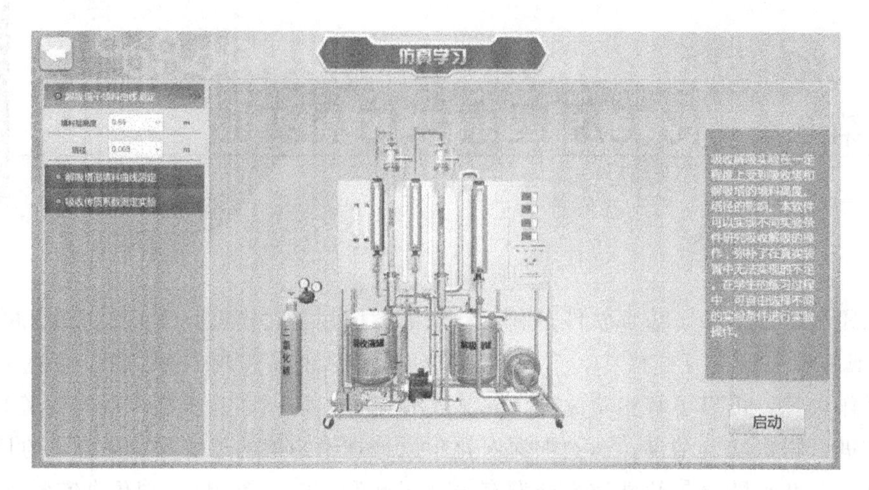

图 6-2 进入系统后的仿真学习界面

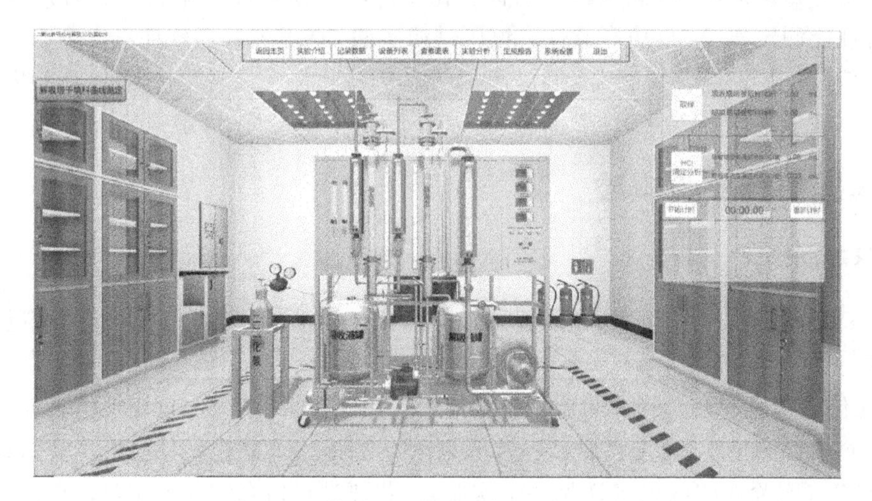

图 6-3 吸收实验室界面

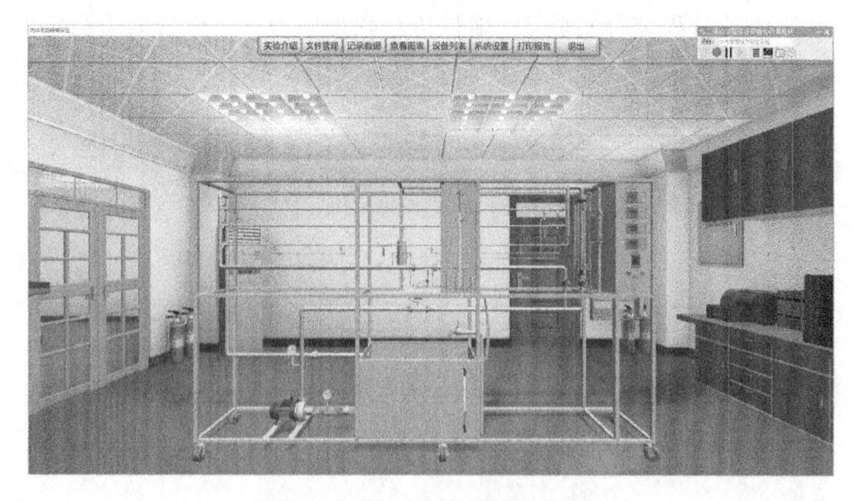

图 6-4 化工流动过程综合实验室界面

② 鼠标滑过各部件可以看到部件的名称和当前的状态。开关阀门或者其他电源键、泵开启键等，单击鼠标左键。

③ 场景控制：使用键盘的 W、S、A、D 键或上下左右键（图 6-5）可控制实验设备往前、后、左、右移动，按鼠标右键可进行各个视角的旋转。

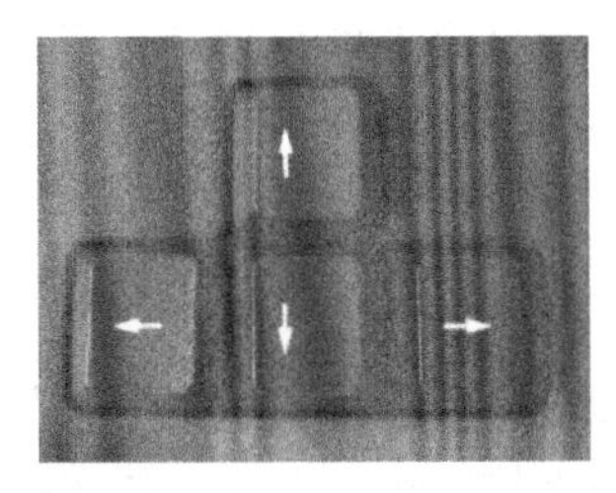

图 6-5　场景控制键

④ 拉近镜头：选择要放大的部件，双击鼠标左键即可，待放大后按住鼠标右键拖动可以上、下、左、右移动视野，按键盘任意键可复原。

⑤ 查找部件：点击菜单栏中的“设备列表”，选择要查看的部件名称，单击鼠标左键能迅速定位到放大后的目标，按住鼠标右键拖动可以上、下、左、右移动视野，按键盘任意键可复原。

⑥ 若有参数设置界面的装置或设备模型，按住鼠标左键可以旋转查看，滚动鼠标滚轮可以缩小或增大模型。

6.1.2　菜单选择项功能说明

从图 6-3 和图 6-4 可以看到在实验室界面上方出现一系列菜单选择项，如图 6-6 所示。

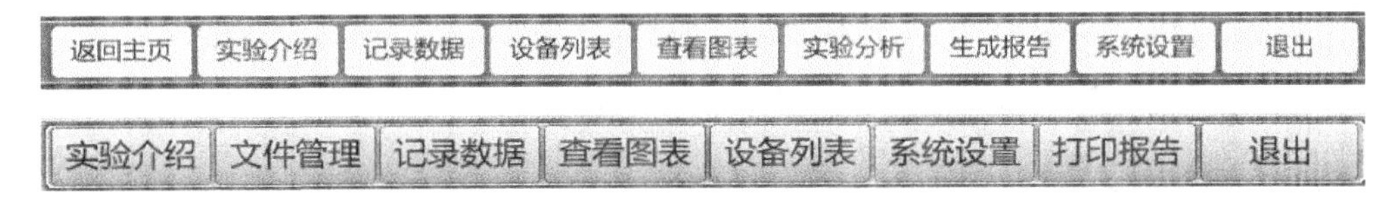

图 6-6　菜单选择项

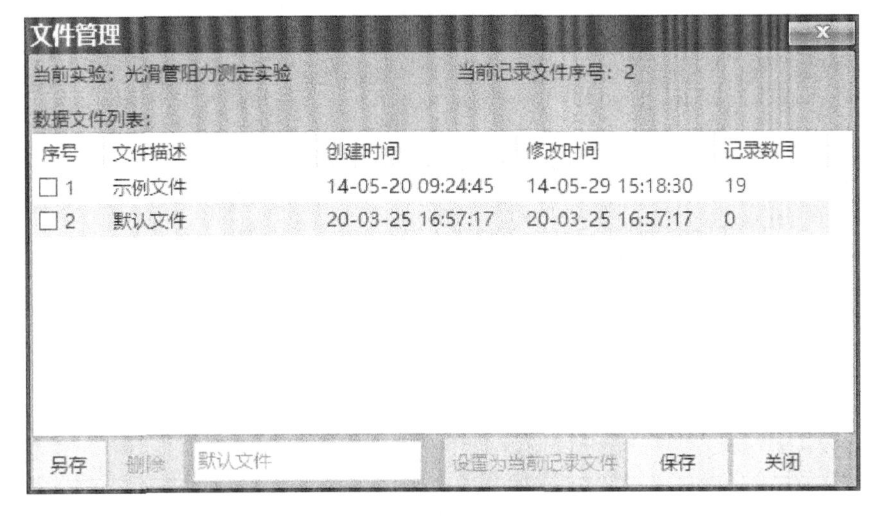

图 6-7　文件管理界面 1

菜单选择项功能说明如下：

【返回主页】点击后返回仿真学习界面，返回后会重新启动新的项目。

【实验介绍】介绍实验的基本情况，如实验目的及内容、实验原理、实验装置基本情况、实验方法、实验步骤和实验注意事项等。

【文件管理】可建立数据的存储文件名，并设置为当前记录文件。部分软件有此功能。以化工流动过程综合实验 3D 虚拟仿真软件为例说明。

操作方法：点击图 6-7 下方“另存”，出现图 6-8 界面，可以修改新建文件名称，并设置为当前记录文件，点击“保存”。若再点击“新建”又可生成新建文件。

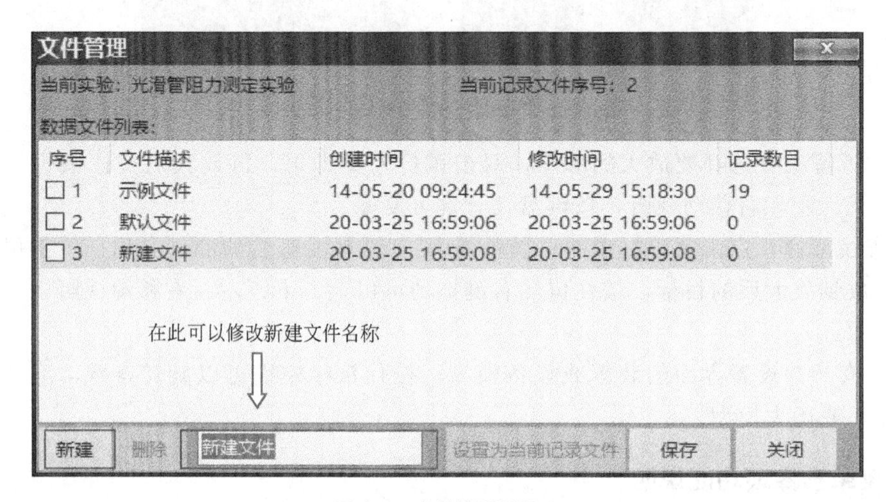

图 6-8　文件管理界面 2

【记录数据】对数据进行管理，实现数据记录和数据处理功能。

化工流动过程综合实验 3D 虚拟仿真软件的数据管理和记录数据界面如图 6-9 所示。

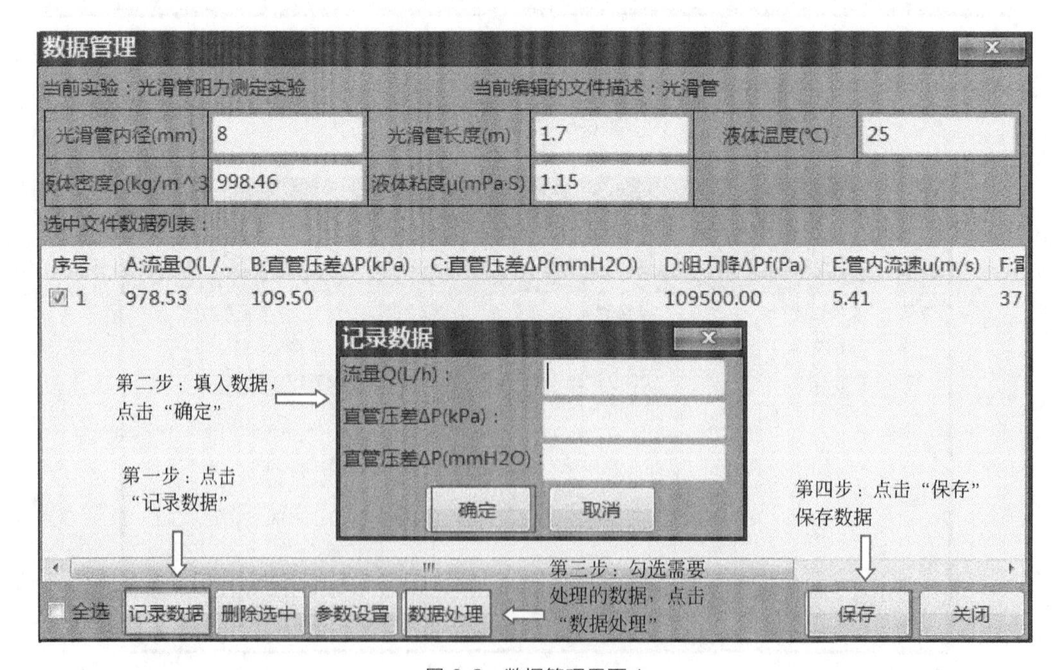

图 6-9　数据管理界面 1

操作方法：

① 点击下方“记录数据”，弹出记录数据框，在此将测得的数据填入，点击“确定”。

② 数据记录后，勾选要进行计算处理的数据（若想处理所有数据，将下方的“全选”勾选即可），选中数据后，点击“数据处理”按钮，就会将记录的数据计算出结果。

③ 如若数据记录错误，将该组数据勾选，点击“删除选中”，即可删除选中的错误数据。

④ 数据处理后，点击“保存”按钮，然后关闭窗口。

二氧化碳吸收与解吸实验 3D 虚拟仿真软件的数据管理界面如图 6-10 所示。每个序号后的栏目除了填入要记录的数据，例如空气流量和 U 形管高度差以外，其他栏目的数据要自己在软件之外进行数据处理，然后将数据处理的结果填入表中，最后点击“提交”按钮，然后关闭窗口。

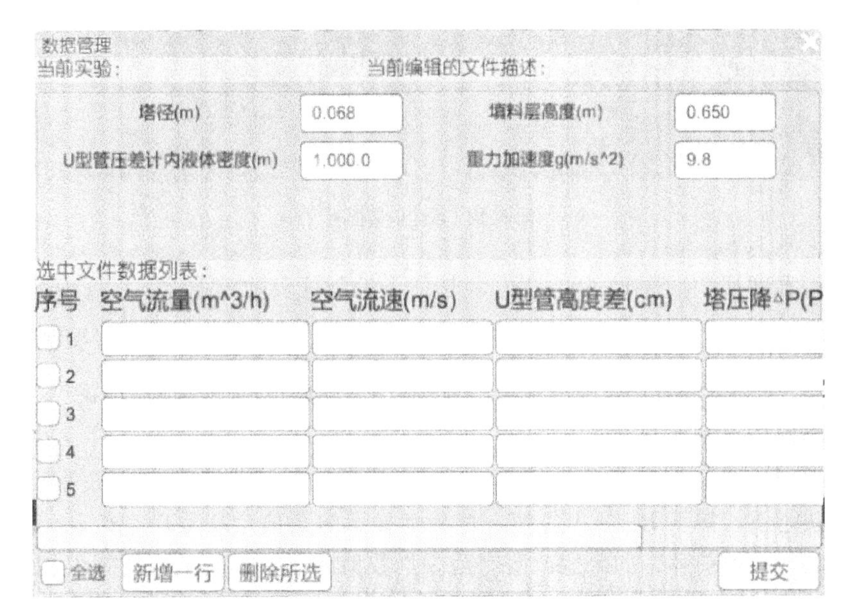

图 6-10　数据管理界面 2

【查看图表】根据记录的实验数据和数据处理的结果可以生成目标表格或显示本项目对应的关系曲线图，并可插入到实验报告中。

【设备列表】对设备进行分类，例如阀门、压力表、流量计等，单击类别能迅速定位到目标。

【实验分析】有关实验的一些选择题、判断题等。

【生成报告】虚拟仿真软件可生成报告作为预习报告。生成报告的位置可以是默认路径也可以自己选择设定。

【系统设置】可设置标签、声音和环境光等。

【退出】点击退出实验。

6.1.3　仪表和阀门调节说明

（1）数值显示表

该类表为显示表，没有任何操作，直接显示对应数值，如图 6-11 所示。

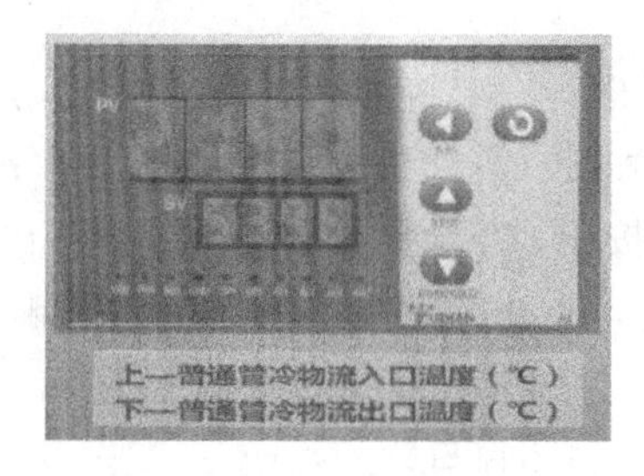

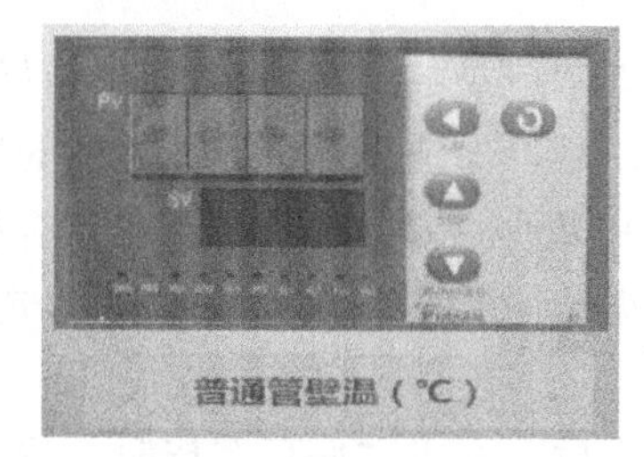

图 6-11　数值显示表面板

（2）设定仪表

仪表上行 PV 值为显示值，下行 SV 值为设定值，如图 6-12 所示。

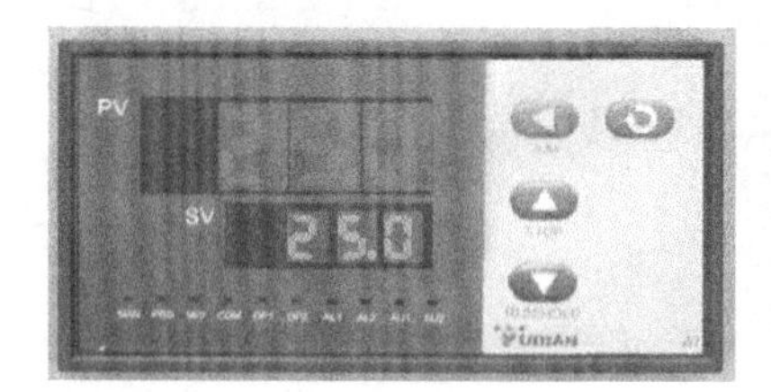

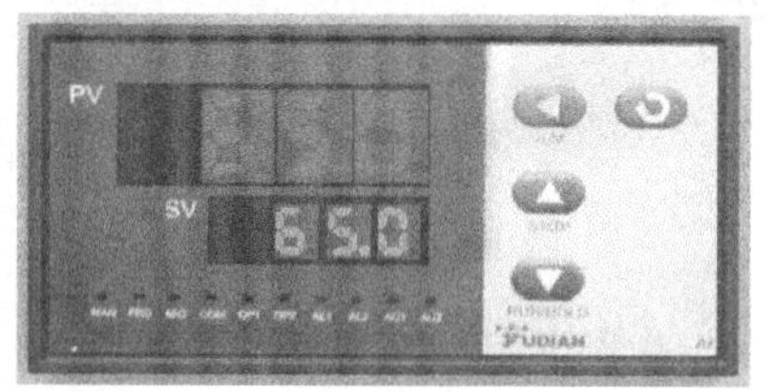

图 6-12　设定仪表面板

例如需将温度从 25.0℃改为 65.0℃，按一下控制仪表的键，在仪表的 SV 显示窗口出现一闪烁数字，每按一次键，闪烁数字便向左移动一位，哪个位置的数字闪烁就可以利用、键调节相应那个位置的数值，调好后重按确认即可。

（3）变频器使用方法

变频器面板见图 6-13。其中，RUN/STOP 为泵的启停按钮。泵的频率设定方法为：在泵启动的状态下，按下 RESET 按钮，面板上显示的数值会从最后一位开始闪烁，继续按下按钮，闪烁位数前移，如果想改变当前闪烁数值的值，按向上或向下的按钮即可改变数值大小，设定好后，按下 READ/ENTER 按钮，会自动调节至设定的数值。

图 6-13　变频器面板

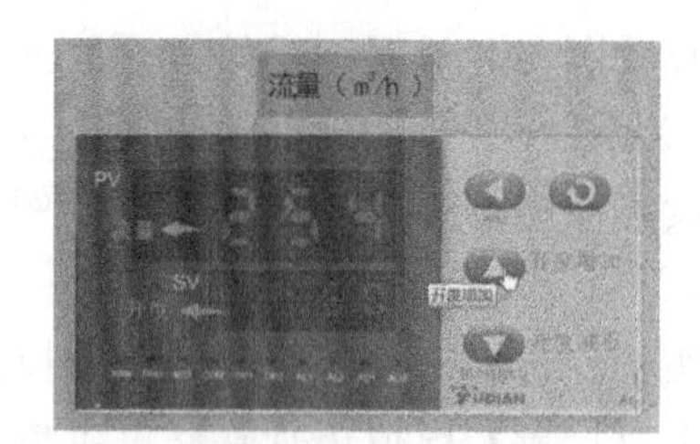

图 6-14　流量显示仪表面板

（4）离心泵实验的流量调节方法

流量显示仪表面板见图 6-14。用控制柜上的流量显示仪表来调节电动流量调节阀的阀门开度。流量显示仪表 PV 显示的是当前流量值，SV 显示的是当前电动阀的开度值，通过上、下按键调节电动流量调节阀的开度。

（5）阀门的调节

以转子流量计调节为例，点击要调节的阀门，出现调节对话框，点击“开”或拉动下方的进度条调节开度，将鼠标移动到流量计上，会显示该流量的示值，调节开度，直至所需的流量为止，如图 6-15 所示。

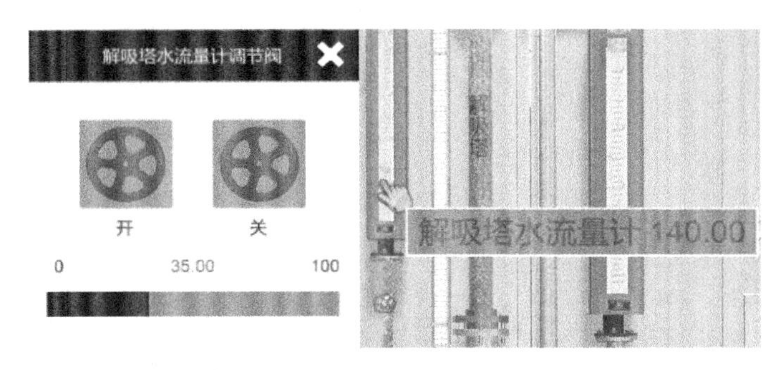

图 6-15　阀门调节示意

6.1.4　注意事项

① 部分仿真实验的数据处理需要在软件之外完成，根据菜单栏“记录数据”的提示，计算相应的内容，并将计算结果填写入菜单栏“记录数据”的数据管理系统中提交。

② 部分仿真实验需要在菜单栏“查看图表”中根据软件提供的参数，计算相关数据并填入相应栏目中。

③ 数据处理时要注意对转子流量计的读数进行校正。

④ 必须完成仿真实验菜单栏中“实验分析”的内容。

6.2　雷诺演示仿真实验

6.2.1　仿真主界面

雷诺演示仿真实验仿真主界面见图 6-16。

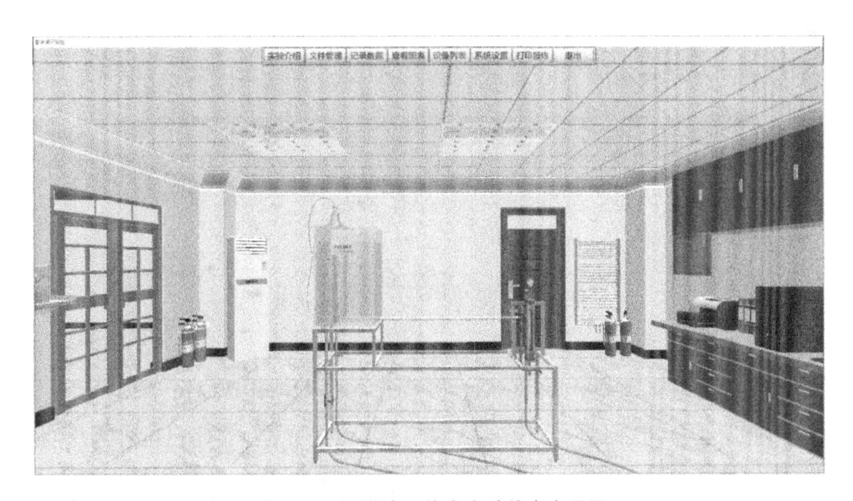

图 6-16　雷诺演示仿真实验仿真主界面

6.2.2　雷诺演示实验装置

雷诺演示实验装置示意见图 6-17。

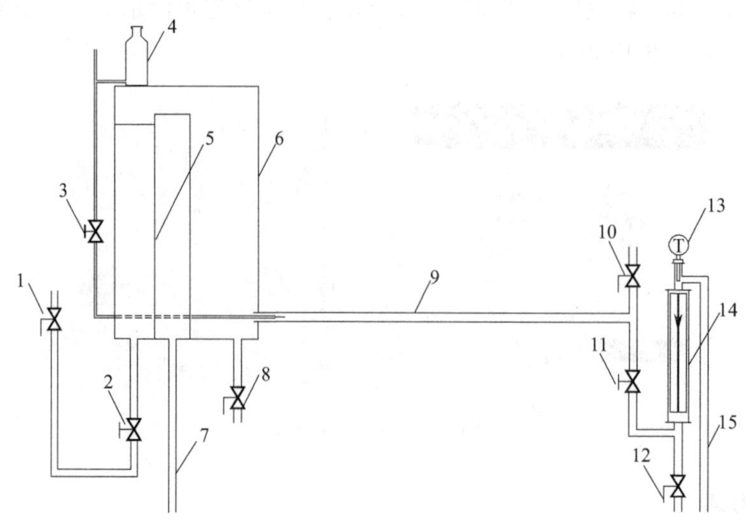

图 6-17　雷诺演示实验装置示意

1—水龙头；2—上水调节阀；3—红墨水入口阀；4—红墨水瓶；5—溢流板；6—水箱；7—溢流管；8—泄水阀 1；9—实验管；10—放空阀；11—流量调节阀；12—泄水阀 2；13—温度显示表；14—转子流量计；15—出水管

本实验装置主要由水箱、红墨水瓶、实验管和转子流量计等组成。水箱的水缓慢流过实验管道，红墨水在管中心流过。可以观察到实验管道内水的流动状况。

6.2.3　实验项目

雷诺演示仿真实验。

6.3　化工流动过程综合实验

6.3.1　仿真主界面

化工流动过程综合实验仿真主界面见图 6-18。

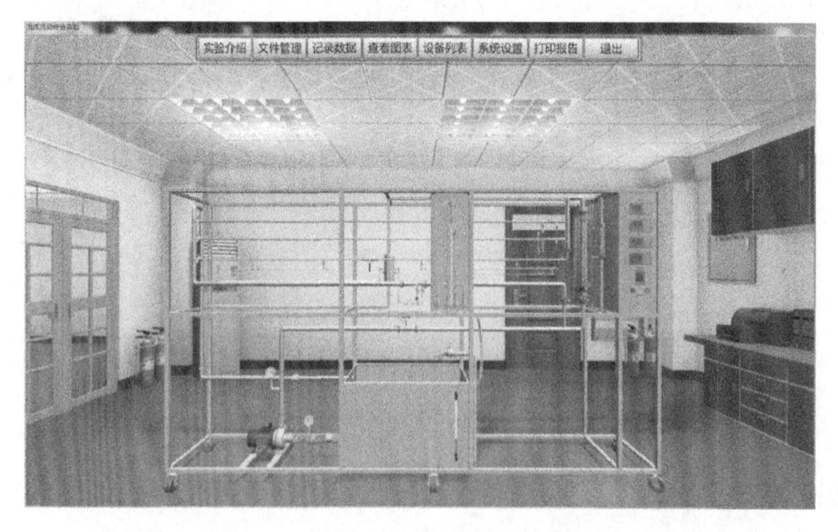

图 6-18　化工流动过程综合实验仿真主界面

6.3.2　化工流动过程综合实验装置

化工流动过程综合实验流程示意见图 6-19。

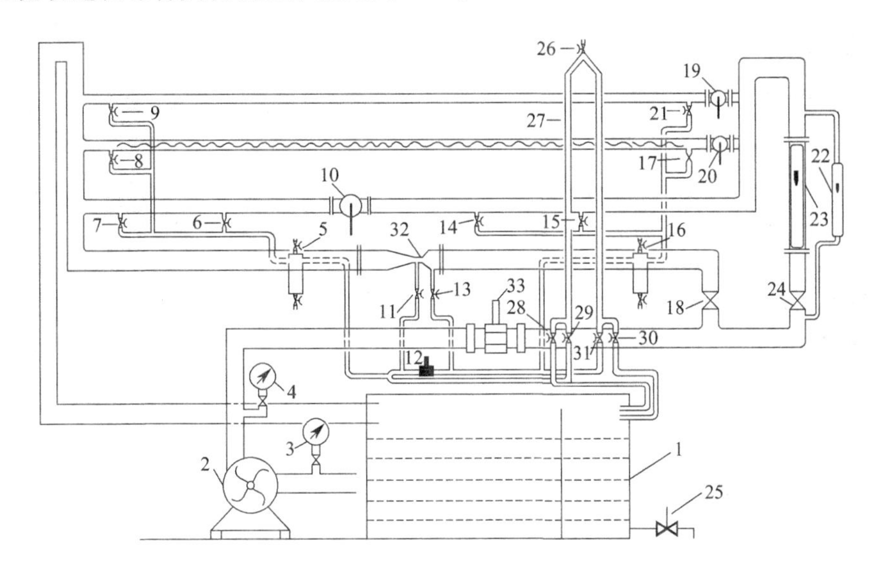

图 6-19　化工流动过程综合实验流程示意

1—水箱；2—水泵；3—入口真空表；4—出口压力表；5，16—缓冲罐顶阀；6，14—测局部阻力近端阀；7，15—测局部阻力远端阀；8，17—粗糙管测压阀；9，21—光滑管测压阀；10—局部阻力阀；11—文丘里流量计压差传感器左阀；12—压力传感器；13—文丘里流量计压差传感器右阀；18，24—阀门；19—光滑管阀；20—粗糙管阀；22—小转子流量计；23—大转子流量计；25—水箱放水阀；26—倒 U 形管放空阀；27—倒 U 形管；28，30—倒 U 形管排水阀；29，31—倒 U 形管平衡阀；32—文丘里流量计；33—涡轮流量计

本实验装置主要由水箱、离心泵、流量计、真空表、压力表、倒 U 形管和压力传感器等组成。水泵将水槽中的水抽出，送入实验系统，经流量计测量流量后送入被测管段，再经回流管流回水槽中。

6.3.3　实验项目

（1）光滑管阻力测定

（2）粗糙管阻力测定

（3）局部阻力测定

（4）离心泵特性曲线测定

（5）管路特性曲线测定

（6）流量性能测定

6.4　离心泵综合性能测定实验

6.4.1　仿真主界面

离心泵综合性能测定实验仿真主界面见图 6-20。

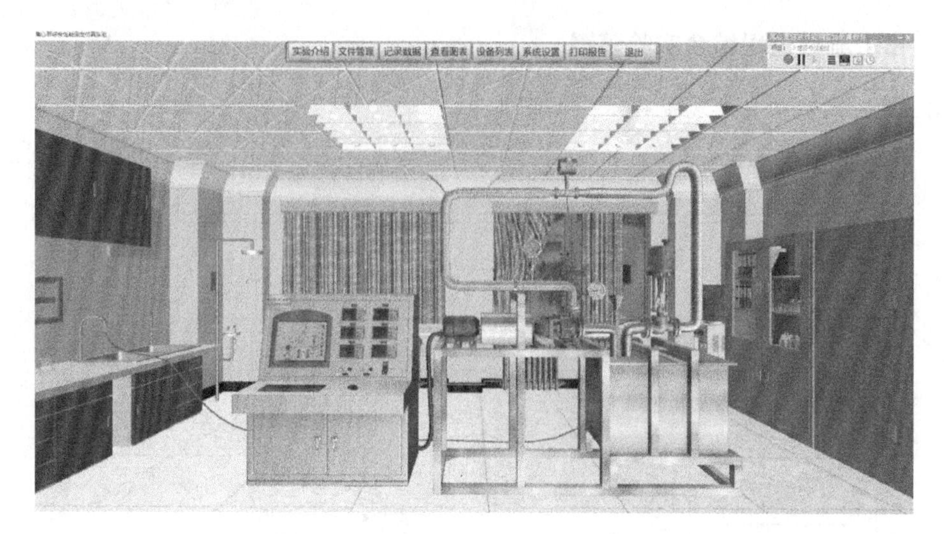

图 6-20　离心泵综合性能测定实验仿真主界面

6.4.2　离心泵实验装置

离心泵实验流程示意见图 6-21。

本实验装置主要由水槽、离心泵、涡轮流量计、真空表、压力表和温度传感器等组成。离心泵 4 将水槽 1 内的水输送到实验系统，用电动流量调节阀 8 调节流量，流体经涡轮流量计 7 计量，回到水槽。注意：离心泵安装在水槽液面之上，操作时要记得灌泵。

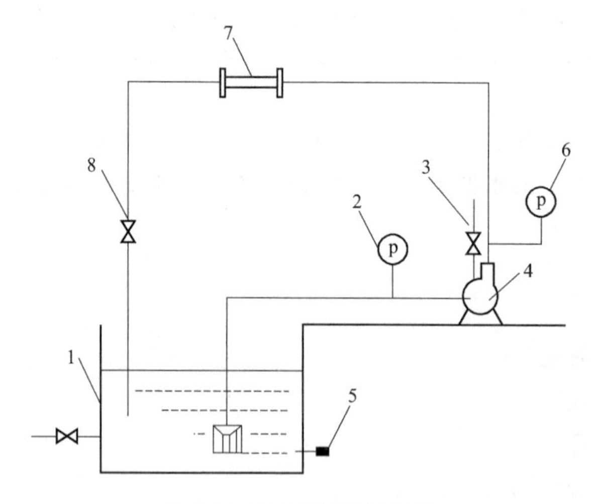

图 6-21　离心泵实验流程示意

1—水槽；2—真空表；3—灌泵阀；4—离心泵；5—温度传感器；6—压力表；7—涡轮流量计；8—电动流量调节阀

6.4.3　实验项目

（1）离心泵特性曲线测定

（2）管路特性曲线测定

6.5 恒压过滤实验

6.5.1 仿真主界面

恒压过滤实验仿真主界面见图 6-22。

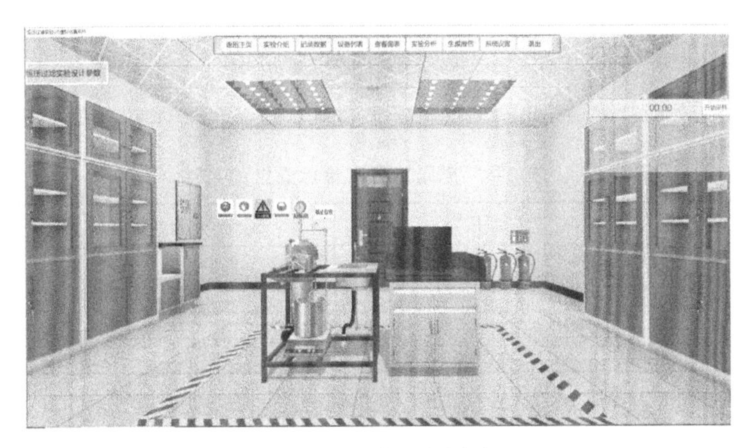

图 6-22 恒压过滤实验仿真主界面

6.5.2 恒压过滤实验装置

恒压过滤实验装置流程如图 6-23 所示。

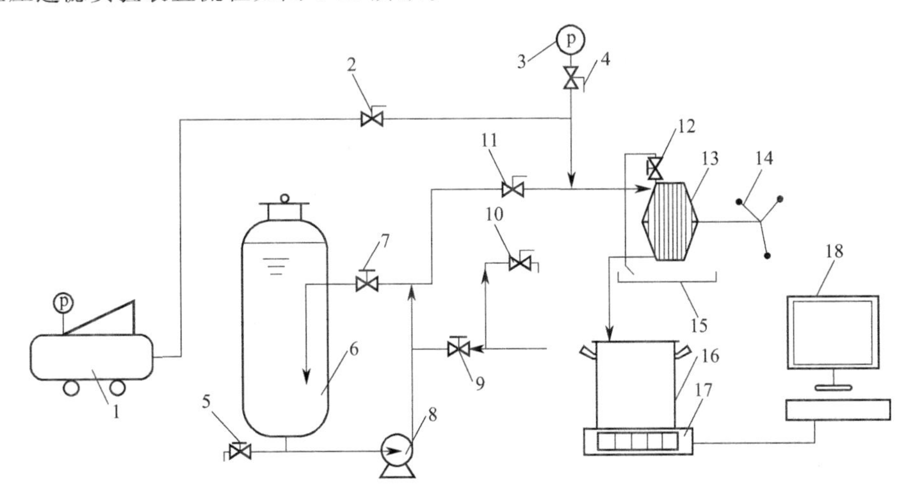

图 6-23 恒压过滤实验装置流程示意

1—空气压缩机；2—压缩空气阀；3—压力表；4—压力阀；5—排尽阀；6—悬浮液槽；7—压力调节阀；8—供料泵；9—进水阀；10—水龙头；11—进料阀；12—放气阀；13—板框过滤器；14—紧固螺杆；15—滤饼收集盆；16—清液槽；17—电子秤；18—计算机

本实验装置主要由悬浮液槽、供料泵、过滤器、清液槽、电子秤、空气压缩机和计算机等组成，板框过滤器共有 3 层滤板，过滤介质为帆布滤网。

将已配制好的碳酸钙（$CaCO_3$）悬浮液由悬浮液槽底部的供料泵 8 循环搅动，使滤浆不致沉淀；通过调节压力调节阀 7 的开度，将料液经旁路管路和进料阀 11 送入板框过滤器 13 中过滤，滤液流入清液槽 16 并由电子秤 17 称重计量；过滤完毕后，用压缩空气吹干滤饼。

6.5.3 实验项目

过滤常数的测定。

6.5.4 可变换的实验条件

实验条件可以自由选择（表 6-1），以考察不同实验条件对实验结果的影响。

表 6-1 恒压过滤实验可变换的条件

过滤直径/mm	120.0	130.0	140.0
悬浮液密度/(kg/ m^3)	1016.0	1033.0	1045.0
清液槽质量/kg	13.2	8.95	25.5
悬浮液组成	碳酸钙	淀粉	

6.5.5 数据处理注意事项

由于过滤开始时刻在过滤介质上固体颗粒尚未形成滤饼，如若实验一开始即以恒压操作，部分颗粒就可能因在过滤推动力较大时穿过过滤介质而得不到清液。因此。在实验开始后，首先在较小压力下操作片刻，待固体颗粒在过滤介质上形成滤饼后，再在预定的压力下操作至结束。因此，软件设计时考虑了在恒压过滤前的 τ_1 时间内已通过了 q_1 的滤液量，则将恒压过滤方程变换为式(6-1)。

$$\frac{\tau-\tau_1}{q-q_1}=\frac{1}{K}(q-q_1)+\frac{2}{K}(q_1+q_e)=\frac{1}{K}(q+q_1)+\frac{2}{K}q_e \tag{6-1}$$

6.6 传热综合实验

6.6.1 仿真主界面

传热综合实验仿真主界面见图 6-24。

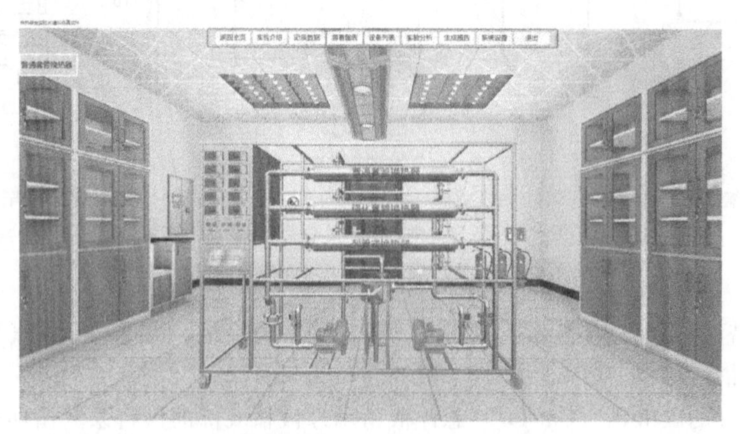

图 6-24 传热综合实验仿真主界面

6.6.2　传热综合实验装置

传热综合实验流程示意见图 6-25。

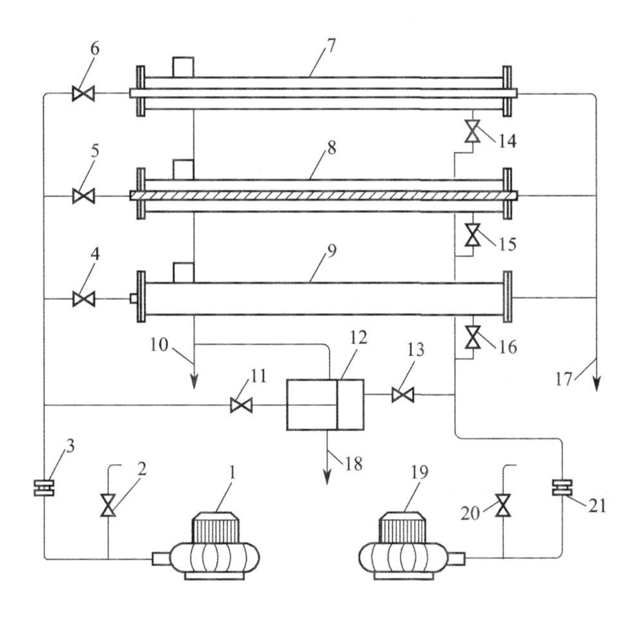

图 6-25　传热综合实验流程示意

1—冷风机；2—冷空气旁路调节阀；3—孔板流量计Ⅰ；4—列管式换热器冷物流进口阀；5—强化套管换热器冷物流进口阀；6—普通套管换热器冷物流进口阀；7—普通套管换热器；8—强化套管换热器；9—列管式换热器；10—热物流出口；11—螺旋板换热器冷物流进口阀；12—螺旋板换热器；13—螺旋板换热器热物流进口阀；14—普通套管换热器热物流进口阀；15—强化套管换热器热物流进口阀；16—列管式换热器热物流进口阀；17，18—冷物流出口；19—热风机；20—热空气旁路调节阀；21—孔板流量计Ⅱ

以介质为冷空气-热空气为例说明。冷空气由冷风机 1 吹出，由冷空气旁路调节阀 2 调节，经孔板流量计Ⅰ3，由支路控制阀选择不同的支路进入换热器。管程热空气由热风机吹出，由热空气旁路调节阀 20 调节，经孔板流量计Ⅱ21，由支路控制阀选择不同的支路进入换热器壳程，由另一端热空气出口自然喷出，达到逆流换热的效果。

6.6.3　实验项目

（1）普通套管换热器（冷空气-热空气）

（2）普通套管换热器（冷水-热水）

（3）普通套管换热器（甲苯-蒸气）

（4）强化套管换热器（冷空气-热空气）

（5）强化套管换热器（冷水-热水）

（6）强化套管换热器（甲苯-蒸气）

（7）列管式换热器（冷空气-热空气）

（8）列管式换热器（冷水-热水）

（9）列管式换热器（甲苯-蒸气）

（10）螺旋板换热器（冷空气-热空气）

（11）螺旋板换热器（冷水-热水）

（12）螺旋板换热器（甲苯-蒸气）

6.6.4　可变换的实验条件

实验条件可以自由选择（表 6-2），以考察不同实验条件对实验结果的影响。

表 6-2 传热实验可变换的条件

项目	普通套管换热器	强化套管换热器	列管式换热器	螺旋板换热器
管长/m	1.90 2.00 2.10	1.90 2.00 2.10		
内径/mm	20 25 30	20 25 30		
管数			4 5 6	
换热面积/m^2				10 20 30

6.7 二氧化碳吸收与解吸实验

6.7.1 仿真主界面

二氧化碳吸收与解吸实验仿真主界面见图 6-26。

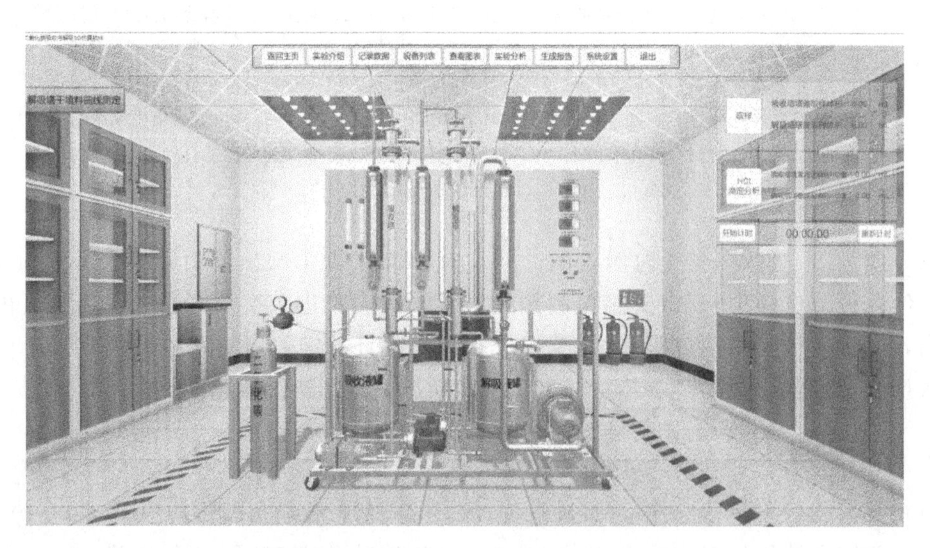

图 6-26 二氧化碳吸收与解吸实验仿真主界面

6.7.2 二氧化碳吸收与解吸实验装置

二氧化碳吸收与解吸实验流程示意见图 6-27。

本实验装置主要由 CO_2 钢瓶、风机、水泵、吸收塔、解吸塔等组成。CO_2 由 CO_2 钢瓶 1 提供，经过减压和调节流量后与来自吸收风机 4 并调节好流量的空气一同混合后进入吸收塔 9，在吸收塔中进行纯水吸收空气中 CO_2 的传质过程；由解吸风机 19 提供空气，用解吸塔空气旁路调节阀 21 调节空气的流量，对解吸塔中的吸收液进行解吸。

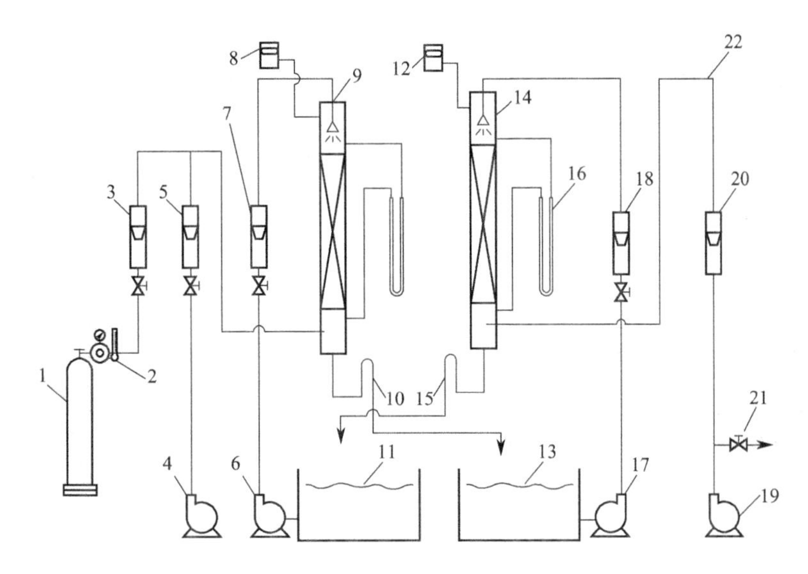

图 6-27　二氧化碳吸收与解吸实验装置流程示意

1—CO_2 钢瓶；2—减压阀；3—CO_2 流量计；4—吸收风机；5—吸收塔空气流量计；6—吸收水泵；7—吸收塔水流量计；8—吸收尾气传感器；9—吸收塔；10，15—液封；11—解吸液罐；12—解吸尾气传感器；13—吸收液罐；14—解吸塔；16—压差计；17—解吸水泵；18—解吸塔水流量计；19—解吸风机；20—解吸塔空气流量计；21—空气旁路调节阀；22—π 形管

6.7.3　实验项目

（1）解吸塔干填料曲线测定

（2）解吸塔湿填料曲线测定

（3）吸收传质系数测定

6.7.4　可变换的实验条件

实验条件可以自由选择（表 6-3），以考察不同实验条件对实验结果的影响。

表 6-3　吸收实验可变换的条件

填料层高度/m	0.65	0.78	0.90
塔径/m	0.068	0.075	0.085
HCl 浓度/(mol/L)	0.08	0.10	0.12

6.7.5　注意事项

实验时要注意吸收塔水流量和解吸塔水流量计数值要一致，两个流量计要及时调节，以保证实验时操作条件不变。

6.8　精馏综合拓展实验

6.8.1　仿真主界面

精馏综合拓展实验仿真主界面见图 6-28。

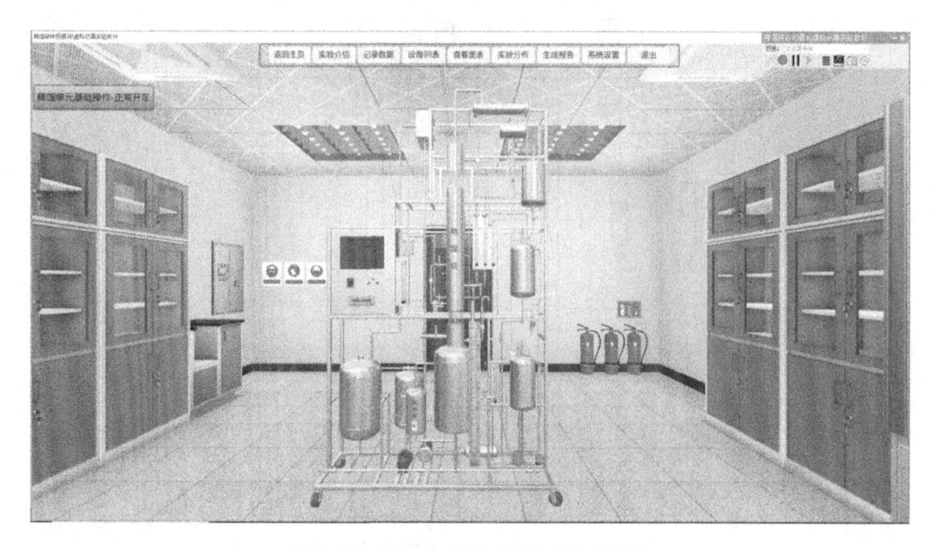

图 6-28　精馏综合拓展实验仿真主界面

6.8.2　精馏综合拓展实验装置

精馏综合拓展实验流程示意见图 6-29。主要测量点和操作控制点名称见表 6-4。

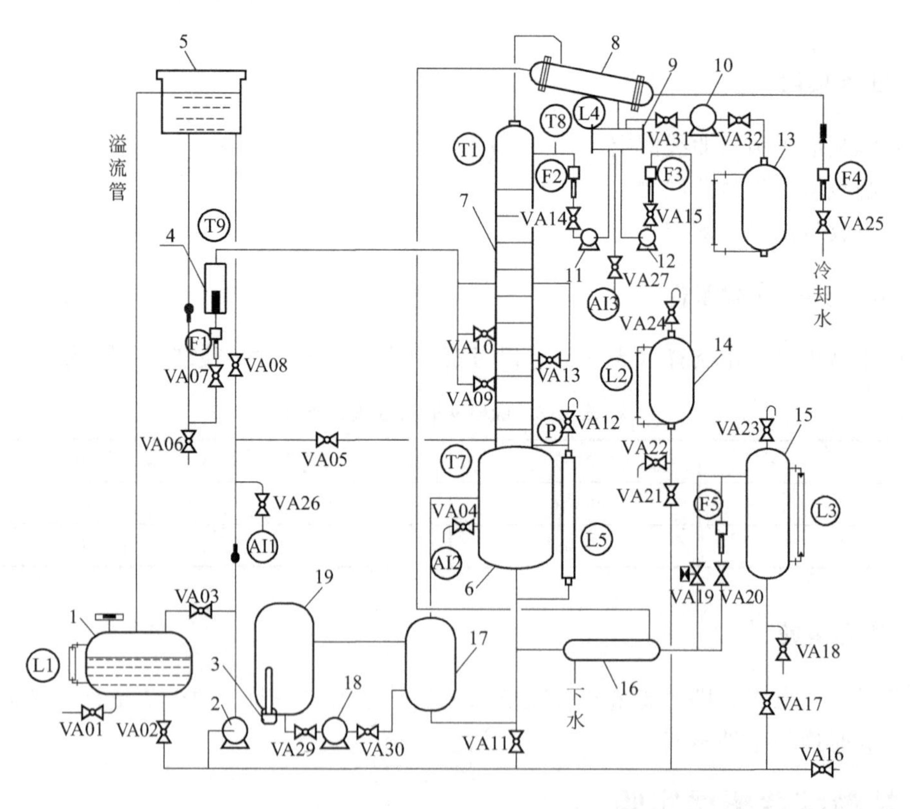

图 6-29　精馏综合拓展实验装置流程示意

1—贮料罐；2—进料泵；3—加热器；4—进料预热器；5—高位槽；6—塔釜；7—筛板精馏塔；8—冷凝器；9—回流罐；10—真空泵；11—回流泵；12—采出泵；13—贮罐；14—塔顶产品罐；15—塔釜产品罐；16—塔釜冷凝器；17—再沸器；18—导热油泵；19—导热油罐

表6-4　主要测量点和操作控制点名称

F1—原料进料流量	AI1—原料浓度	T8—回流液温度	L4—回流罐液位
F2—回流流量	AI2—塔釜浓度	T9—进料温度	L5—塔釜液位
F3—塔顶采出流量	AI3—塔顶浓度	L1—贮料罐液位	VA01～VA30—阀门
F4—冷却水流量	T1—塔顶温度	L2—塔顶产品罐液位	
F5—塔底出料流量	T7—塔釜温度	L3—塔釜产品罐液位	

精馏过程的主要设备有：精馏塔、再沸器、冷凝器、回流罐和输送设备等。一定温度和压力的料液进入精馏塔后，轻组分在精馏段逐渐浓缩，离开塔顶后全部冷凝进入回流罐，一部分作为塔顶产品（也称馏出液），另一部分被送入塔内作为回流液。重组分在提馏段中浓缩后，一部分作为塔釜产品（也称残液），另一部分则经再沸器加热后送回塔中。

精馏塔控制操作采用DCS控制系统。

6.8.3　实验项目

（1）精馏单元基础操作——指导模式

（2）精馏单元基础操作——正常开车

（3）精馏单元基础操作——正常停车

（4）异常情况及事故的紧急处理——液泛

（5）异常情况及事故的紧急处理——雾沫夹带

（6）异常情况及事故的紧急处理——严重漏液

（7）异常情况及事故的紧急处理——换热器结垢

（8）异常情况及事故的紧急处理——汽蚀

（9）常压单元操作参数变化对精馏过程的影响——回流比

（10）常压单元操作参数变化对精馏过程的影响——进料温度

（11）常压单元操作参数变化对精馏过程的影响——导热油加热功率

（12）不同压力对精馏过程的影响——常压

（13）不同压力对精馏过程的影响——加压

（14）不同压力对精馏过程的影响——减压

（15）设备参数对精馏过程的影响——设备参数

（16）实验物系的变化对精馏过程的影响——实验物系

6.8.4　注意事项

① 要先通冷却水后再开导热油加热开关，实验结束后先关闭加热开关，待塔顶温度降至70℃以下，方可闭关冷却水。

② 开启加热后每隔一定的时间记录一次实验数据（温度、压力、液位、加热功率），实时观测塔内现象。

③ 操作过程中要随时观察加热功率、塔内温度、塔釜压力、回流量、进料量和采出量等并始终处于稳定状态。每隔一定的时间记录一次数据（温度、压力、液位、加热功率、流量）。每隔10min取样分析浓度（塔顶、塔釜、进料）。

④ 要维持回流罐内液位恒定，记录液位恒定后的回流转子流量计读数。

6.9 液液萃取实验

6.9.1 仿真主界面

液液萃取实验仿真主界面见图 6-30。

图 6-30 液液萃取实验仿真主界面

6.9.2 液液萃取实验装置

液液萃取实验流程示意见图 6-31。

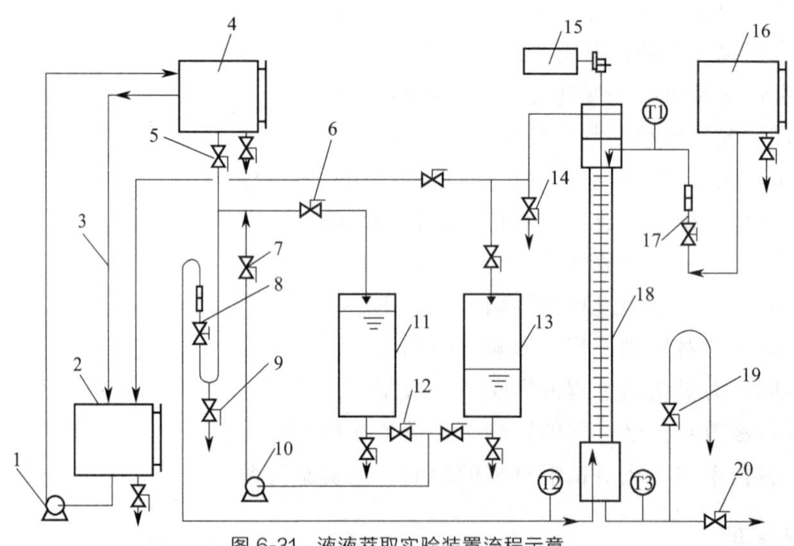

图 6-31 液液萃取实验装置流程示意

1—原料输送泵Ⅰ；2—配料槽；3—溢流管；4—原料高位槽；5—原料进口阀；6—旁路阀；7—泵出口阀；8—原料（油）转子流量计及阀门；9—原料（油进口）取样阀；10—原料输送泵Ⅱ；11—原料槽；12—泵进口阀；13—萃余相槽；14—萃取相（油出口）取样阀；15—直流电机；16—水相高位槽；17—水转子流量计及阀门；18—萃取塔；19—π形管调节阀；20—排尽阀；T1—水相进口温度表；T2—原料进口温度表；T3—萃取相进口温度表

液液萃取实验装置主要由萃取塔、原料高位槽、水相高位槽、萃余相槽、原料输送泵、直流电机等设备和部件构成。本装置操作时先向萃取塔中注入连续相（水）至萃取塔上半部分的颈部处，然后按相比 1∶1（校正后的萃取剂与原料液质量之比）的要求注入分散相（煤油），分散相在塔内向上运动，连续相向下流动，两相逆流接触传质，分散相在塔顶汇聚并形成一定厚度的轻液层，在萃取塔上部两相之间会形成明显的界面，通过连续相出口 π 形管上的调节阀，将两相界面调节至连续相进口与水相出口中间的位置，煤油由塔顶采出进入萃余相槽，水由塔底部通过 π 形管导出。

6.9.3　实验项目

液液萃取塔的操作和萃取传质单元高度的测定。

6.9.4　可变换的实验条件

实验条件可以自由选择（表 6-5），以考察不同实验条件对实验结果的影响。

表 6-5　萃取实验可变换的条件

塔高/m	0.90	1.00	1.10
塔径/m	0.025	0.030	0.035
原料油浓度/(kg/kg)	0.00400	0.00425	0.00450
NaOH 浓度/(mol/L)	0.010	0.015	0.020
筛板振幅/mm	20.0	23.0	26.0
分散相	煤油	水	
连续相	水	煤油	

6.9.5　注意事项

① 通过 π 形管上的调节阀调节两相界面的位置过程中，不允许水相从萃取塔顶部流出进入萃余相槽。

② 一定要系统稳定之后方可取样分析。

6.10　洞道干燥实验

6.10.1　仿真主界面

洞道干燥实验仿真主界面见图 6-32。

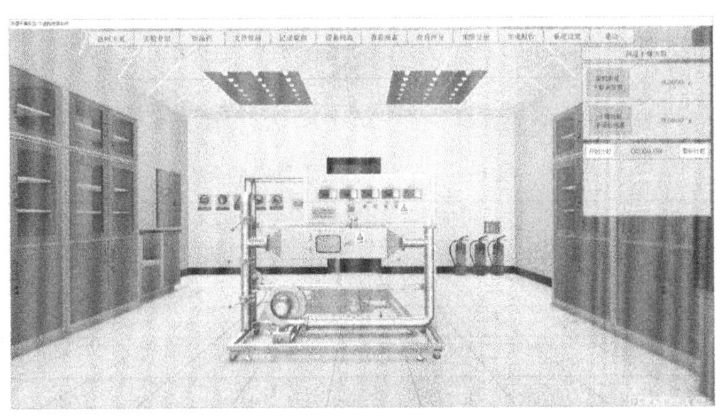

图 6-32　洞道干燥实验仿真主界面

6.10.2　洞道干燥实验装置

洞道干燥实验装置流程示意见图 6-33。

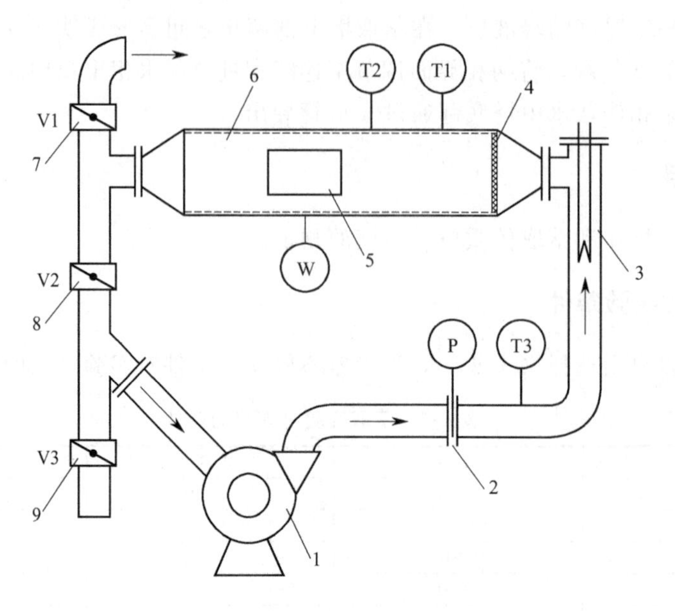

图 6-33 洞道干燥实验装置流程示意

1—离心风机；2—孔板流量计；3—预热器；4—气流分布器；5—干燥器舱门；
6—洞道干燥器；7—风机排出口阀门；8—循环阀门；9—风机吸入口阀门；
T1—干球温度；T2—湿球温度；T3—空气入口温度；P—孔板压差；
W—重量传感器显示仪表；V1，V2，V3—蝶阀

空气由离心风机送出经过孔板流量计计量后进入预热器加热，加热后的空气进入洞道干燥器中，湿物料放在干燥器内重量传感器的支架上与热空气接触进行干燥，热空气将湿物料表面的水分传递至空气中带出干燥器，一部分废气通过阀门 V1 放空，另一部分废气经由阀门 V2 返回循环使用。

6.10.3 实验项目

洞道干燥实验。

6.10.4 可变换的实验条件

实验条件可以自由选择（表 6-6），以考察不同实验条件对实验结果的影响。

表 6-6 干燥实验可变换的条件

干燥框架长度/m	0.150	0.160	0.180
干燥框架宽度/m	0.070	0.080	0.100
干燥框架质量/g	79.5	88.6	93.8

6.10.5 注意事项

① 干燥器内必须确认有空气流过且空气流量稳定后才能开启加热开关对空气进行加热。
② 实验结束关闭加热开关后要待干球温度降低到 45℃后再关闭风机。
③ 要记得在实验结束取出物料关上舱门后，点击“干燥物料干燥后称重”。

第 7 章 化工原理演示实验

7.1 雷诺实验

7.1.1 实验目的

① 了解圆形直管内流体质点的运动方式，认识不同流动型态的特点，掌握判别管内流体流动型态的准则。

② 观察圆形直管内流体作层流、湍流以及过渡区的流动型态，测定临界雷诺数。

③ 观察流体层流流动的速度分布。

7.1.2 实验原理

1883 年英国的雷诺（O. Reynolds）通过实验观察了流体在圆管内的流动有两种截然不同的流动型态：层流（滞流）和湍流（紊流），同时研究发现流体的流动型态与流体的流速 u、黏度 μ、密度 ρ 及流体流经的管道直径 d 有关，并将这四个因素用雷诺数表示为

$$Re=\frac{du\rho}{\mu} \tag{7-1}$$

① $Re\leqslant 2000$，层流流动，流体质点运动为直线，且相互平行。层流流动时，圆管内流通截面上速度成抛物线分布。

② $Re\geqslant 4000$，湍流流动，流体质点除了沿主体流动方向外，还有其他方向的不规则的脉动现象。

③ $2000<Re<4000$，处于层流流动和湍流流动之间，又称为过渡区。在过渡区内，有可能为层流也有可能为湍流，与外界环境有关，流动型态不稳定。

7.1.3 实验装置

（1）实验流程

雷诺实验装置流程图如图 7-1 所示。

（2）设备主要技术数据

实验管道有效长度 $L=1100$ mm，外径 $D_o=30$ mm，内径 $D_i=24.2$mm。

7.1.4 实验步骤

（1）实验前的准备工作

①实验前应观察并适当调整细管位置，使其处于实验观察管道的中心线上。

② 向示踪剂瓶中加入适量用水稀释过的红墨水作为实验用的示踪剂。

③ 关闭水流量调节阀、排气阀，打开上水阀、排水阀，向高位水箱注水，使水充满

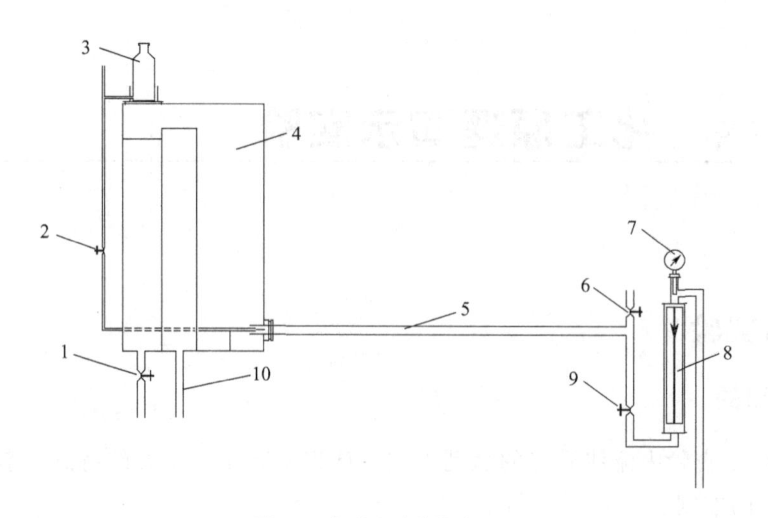

图 7-1　雷诺实验装置流程图

1—上水阀；2—调节阀；3—示踪剂瓶；4—高位水箱；5—观察管；
6—排气阀；7—温度计；8—流量计；9—流量调节阀；10—溢流管

水箱产生溢流，并保持一定的溢流量，以保证高位槽的液位恒定。

④ 轻轻打开水流量调节阀，让水缓慢流过实验管道，排出红墨水注入管中的气泡，使红墨水充满细管道中。

（2）实验过程

① 在做好以上准备的基础上，调节上水阀，维持尽可能小的溢流量。轻轻打开流量调节阀，让水缓慢流过实验管道。

② 缓慢有控制地打开红水流量调节阀，红水流束即呈现不同流动状态，红水流束所表现的就是当前水流量下实验管内水的流动状况（如图 7-2 所示层流流动）。读取流量数值并计算出对应的雷诺数。

③ 进水和溢流造成的震动，有时会使实验管道中的红水流束偏离管内中心线或发生不同程度的左右摆动，此时可立即关闭上水阀，稳定一段时间，即可看到实验管道中出现的与管中心线重合的红色直线。

④ 逐步加大上水阀开度和流量调节阀的开度，在维持尽可能小的溢流量情况下增大实验管道中水的流量，观察实验管道内水的流动状况。过渡流、湍流流动如图 7-3 所示，同时记录流量数值并计算对应的雷诺准数。

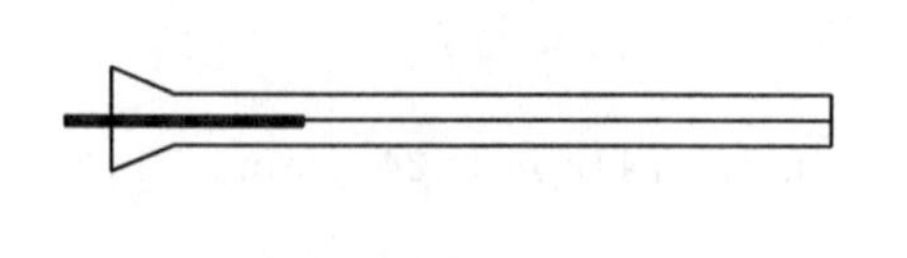

图 7-2　层流流动示意

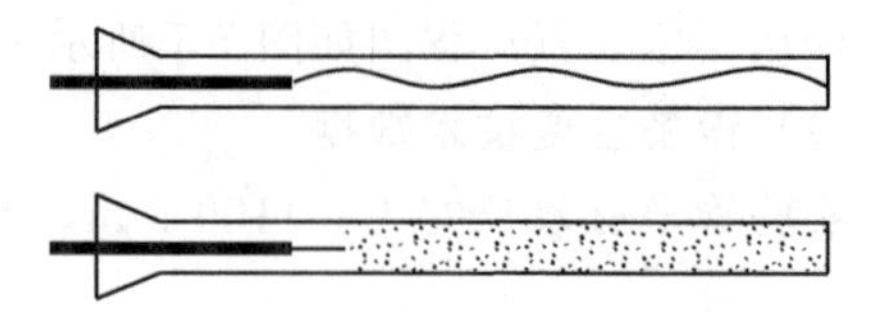

图 7-3　过渡流、湍流流动示意

（3）流体在圆管内流体速度分布演示实验

① 将上水阀打开，关闭水流量调节阀。

② 将红水流量调节阀打开，使少量红水流入实验管道入口端。

③ 突然打开水流量调节阀，在实验管路中可以清晰看到红水线流动所形成如图 7-4 所

示的速度分布。

(4) 实验结束操作

① 关闭红水流量调节阀，使红墨水停止流动。

② 关闭上水阀，使水停止流入高位槽。

③ 待实验管道冲洗干净红色消失时，关闭水流量调节阀。

④ 如果日后较长时间不再使用该套装置，应将设备内各处存水放净。

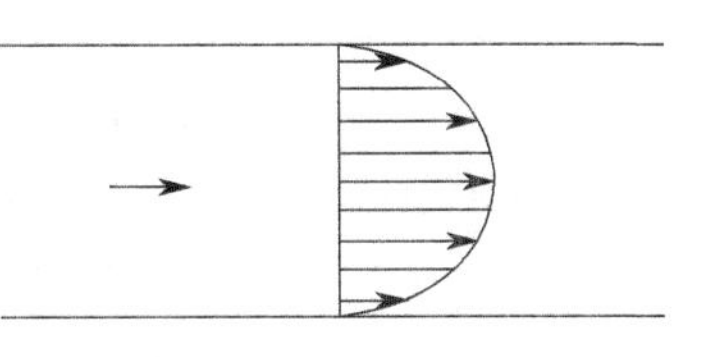

图 7-4　流速分布示意

7.1.5　注意事项

演示层流流动时，为了使层流状态较快形成并保持稳定，请注意以下几点。

① 水槽溢流量尽可能较小，因为溢流过大，上水流量也大，上水和溢流两者造成的震动都比较大，会影响实验结果。

② 尽量不要人为地使实验架产生震动，为减小震动，保证实验效果，可对实验架底面进行固定。

7.1.6　思考题

① 若红墨水注入管不设在实验管道中心，能得到实验预期的结果吗？

② 如何计算某一流量下的雷诺数？用雷诺数判别流型的标准是什么？

③ 根据实验现象说明流体作层流和湍流流动时红墨水的流动状态。

④ 如果流体为理想流体，所看到的现象应该是怎样的？

7.2　能量转化(流体机械能转化)演示实验

7.2.1　实验目的

① 观察和测试不同流量下流体流过不同管径和位置时动能、位能、静压能的变化。

② 掌握流体流动时各能量间的相互转化关系，在此基础上理解伯努利方程。

③ 了解流体在管内流动时能量损失现象。

7.2.2　实验原理

流体在流动时具有三种机械能：动能、位能、静压能，这三种能量可以相互转化。对实际流体，因为存在内摩擦，流动过程中会有一部分机械能因摩擦和碰撞转化为热能，这部分机械能在流动过程中不能恢复，因此对于实际流体来说，上、下游两个截面上的机械能总和是不相等的，两者的差别即为能量（机械能）损失。

动能、位能、静压能三种机械能都可以用液柱高度来表示，分别称为动压头、位压头和静压头；任意两个截面上位压头、动压头和静压头三者总和之差即为压头损失。

7.2.3　实验装置

实验装置如图 7-5 所示，测试管由不同直径、不同高度的管连接而成。在测试管的不同位置选择若干个测量点，每个测量点连接有两个垂直测量管，其中一个测量管直接连接在管壁处，其液位高度反映测量点处静压头的大小，为静压头测量管；另一个测压管开口在管中心处，正对水流方向，其液位高度为静压头和动压头之和，称为冲压头测量管。测量管液位高度可由装置上的刻度读出。水由高位槽经测试管回到水箱，水箱中的水用泵打到高位槽，以保证高位槽始终保持溢流状态。实验测试导管管路图如图 7-6 所示。

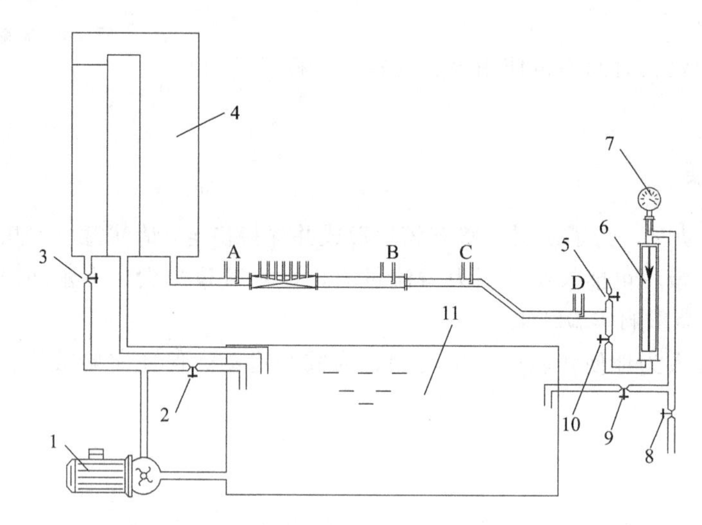

图 7-5　能量转换实验流程示意

1—离心泵；2—循环水阀；3—上水阀；4—高位水箱；5—排气阀；
6—流量计；7—温度计；8—排水阀；9—回水阀；10—流量调节阀；11—水箱

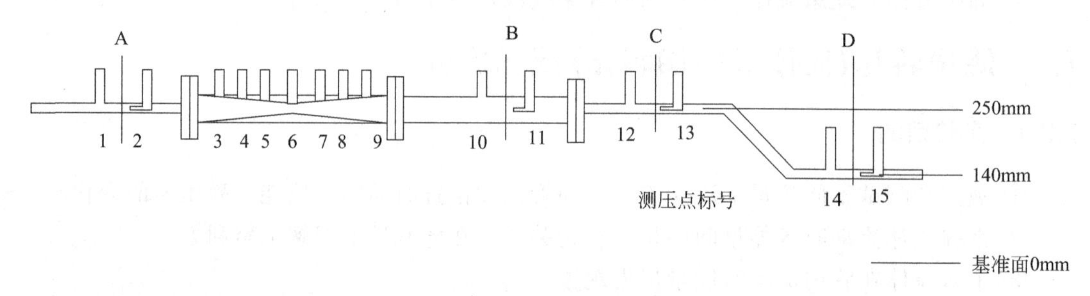

图 7-6　实验测试导管管路图

7.2.4　实验步骤

① 往水箱加入 3/4 体积的蒸馏水，关闭离心泵出口上水阀、循环水阀及实验测试导管出口流量调节阀、排气阀、排水阀，打开回水阀，启动离心泵。

② 打开循环水阀，全开管路出口流量调节阀，逐步开大离心泵出口上水阀至高位槽溢流管中有水溢流，待流动稳定后观察并读取各测量管的液位高度。

③ 逐渐关小流量调节阀，改变流量，观察同一测量点及不同测量点各测量管液位的变化。

④ 关闭离心泵上水阀、循环水阀、出口流量调节阀和回流阀后，关闭离心泵，实验结束。

7.2.5 注意事项

① 不要将离心泵出口上水阀开得过大，以免使水流冲出或导致高位槽液面不稳定。

② 水流量增大时，应检查一下高位槽内水面是否稳定，当水面下降时要适当开大上水阀补充水量。

③ 流量调节阀关小时要缓慢，以免造成流量突然下降使测量管中的水溢出管外。

④ 若系统内含有气泡，必须排出实验管路和测量管内的气泡。

7.2.6 思考题

① 流体在管道中流动时涉及哪些能量?

② 如何测量某截面上的静压头和总压头? 如何测得某截面上的动压头?

③ 若两测压面距基准面的高度不同，两截面间的静压差仅是流动阻力造成的吗?

④ 观察和比较各测压点各项机械能数值的相对大小，得出结论。

7.3 流线演示实验

7.3.1 实验目的

① 通过演示实验，学生进一步理解流体流动的轨迹及流线的基本特征。

② 观察液体流经不同固体边界时的流动现象以及漩涡发生的区域和形态等，增强对流体流动特性的感性认识。

7.3.2 实验原理

实际流体沿着固体壁面流过时，由于黏性作用，黏附在固体壁面上，静止的流体层与其相邻的流体层之间产生摩擦力，使相邻流体层的流动速度减慢。因此，在垂直于流体流动的方向上便产生速度梯度 du/dy，有速度梯度存在的流体层称为边界层。

在化学工程学科中非常重视对边界层的研究。边界层概念的意义在于研究真实流体沿着固体壁面流动时，要集中注意流动边界层内的变化，它的变化将直接影响到动量传递、能量传递和质量传递。

当流体流过曲面，或者流体的流道截面大小或流体流动方向发生改变时，若此时流体出现逆压强梯度 dp/dx（沿着流动方向的流体压强变化率），那么流体边界层将会与壁面脱离而形成漩涡（或称涡流），加剧了流体质点间的互相碰撞，造成流体能量的损耗。边界层从固体壁面脱离的现象称为边界层的分离或脱体。由此，我们可寻找到流体在流动过程中能量消耗的原因。同时，这种漩涡造成的流体微团的杂乱运动并相互碰撞混合也会使传递过程大大强化。因此，流体流线研究的现实意义就在于，可对现有流动过程及设备进行分析研究，进而强化传递过程，为开发新型高效设备提供理论依据，并在选择适宜的操作控制条件方面作出指导。

本演示实验采用气泡示踪法，可以把流体流过不同几何形状的固体时，流体的流线、边界层分离现象以及漩涡发生的区域和强弱等流动图像清晰地显示出来。

7.3.3 实验装置

流线演示实验装置流程、演示板外形及绕流演示板分别如图 7-7～图 7-9 所示。水箱

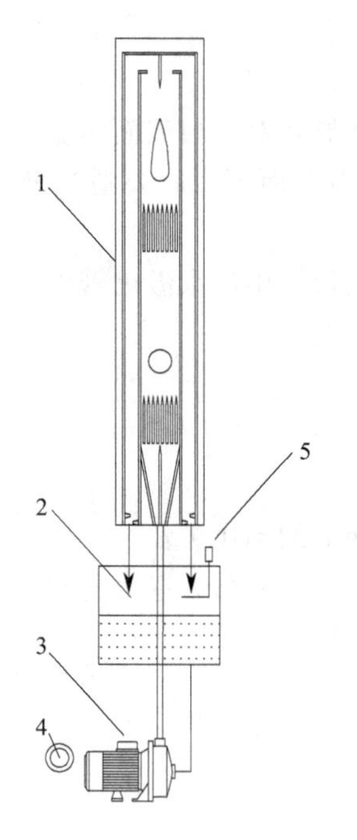

图 7-7 流线演示实验装置流程示意

1—实验面板；2—实验水箱；3—水泵；4—调压旋钮；5—掺气旋钮

中的水被离心泵送入演示仪中，再通过演示仪的溢流装置返回水箱。在每个演示仪中，水从狭缝式流道流过，通过在水流中掺入气泡的方法演示出不同形状边界下的多种水流现象，并显示相应的流线。装置中的每个演示仪均可作为独立的单元使用，也可以同时使用。为便于观察，演示仪用有机玻璃制成。

几种流动演示仪的说明如下。

(a) 型：带有气泡的流体经过逐渐扩大、稳流、单圆柱绕流、稳流、流线体绕流、直角弯道后流入循环水箱。

(b) 型：带有气泡的流体经过逐渐缩小、稳流、转子流量计、直角弯道后流入循环水箱。

(c) 型：带有气泡的流体经过逐渐扩大、稳流、孔板流量计、稳流、喷嘴流量计、直角弯道后流入循环水箱。

(d) 型：带有气泡的流体经过逐渐扩大、稳流、多圆柱绕流（顺排)、稳流、多圆柱绕流（错排)、直角弯道后流入循环水箱。

(e) 型：带有气泡的流体经过 45°角弯道、圆弧形弯道、直角弯道、突然扩大、稳流、突然缩小后流入循环水箱。

(f) 型：带有气泡的流体经过阀门、突然扩大、直角弯道后流入循环水箱。

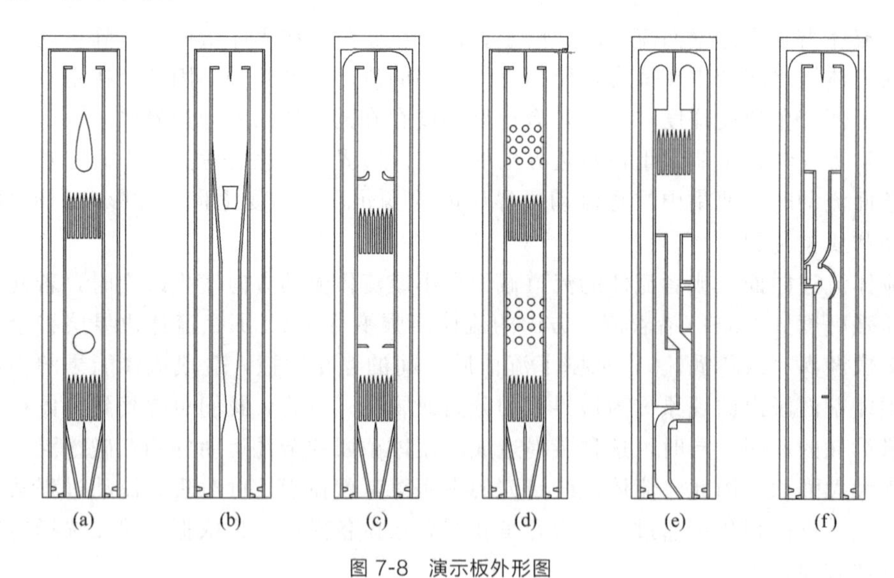

图 7-8 演示板外形图

7.3.4 实验步骤

① 实验前将加水开关打开，将水箱灌水至 2/3 处。

② 打开调速旋钮，在最大流速下使显示面两侧下水道充满水。

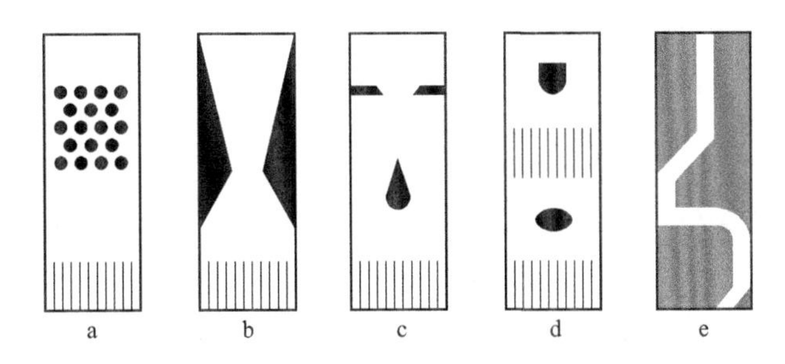

图7-9　绕流演示板

a—列管换热模拟；b—文丘里模拟；c—流线体及孔板模拟；d—圆形体及直线收尾体模拟；e—转角模拟

③ 调节掺气量到最佳状态（现象最清晰为最佳），观察实验现象。

④ 其他几个实验装置的流线演示，均按上述要求进行操作，仔细观察不同流线。

⑤ 实验结束时应将调速旋钮关闭后切断总电源。

注：为了达到更好的实验效果，可往水中添加颜料；实验中应注意调节进气阀的进气量，使气泡大小适中，流动演示更清晰。

7.3.5　思考题

① 在输送流体时，为什么要避免漩涡的形成？

② 为什么在传热、传质过程中要形成适当的漩涡？

③ 流体绕圆柱流动时，边界层分离发生在什么地方？流速不同，分离点是否相同？边界层分离后流体的流动状态时怎样的？

④ 流体经过突然扩大、突然缩小与渐缩渐扩管的流动状态有何区别？

⑤ 多圆柱顺排绕流和多圆柱错排绕流的流动状态有何区别？对列管换热器的管束排列有何启示？

7.4　非均相气固分离演示实验

7.4.1　实验目的

① 了解沉降室、旋风分离器及布袋除尘器的结构、特点和工作原理。

② 测定旋风分离器内的静压强分布，认识出灰口和集尘室良好密封的必要性。

③ 测定进口气速对旋风分离器分离性能的影响，理解适宜操作气速的计算方法。

7.4.2　实验原理

(1) 重力沉降的除尘原理

重力沉降过程是含尘气体沿水平方向进入重力沉降设备，在重力的作用下，粉尘粒子逐渐沉降到设备底部，而气体沿水平方向继续前进流出沉降设备，从而达到除尘的目的，属于粗除尘。常见的用于分离气固混合物的重力沉降设备有沉降槽。

在重力除尘设备中，气体流动的速度越小，越有利于粒径小的粉尘的沉降分离，提高除尘效率。因此，一般控制气体流动的速度为1～2m/s，除尘效率为40%～60%。但气体速度太小，设备相对庞大，投资费用较高。在气体流速基本固定的情况下，重力沉降设备设计得越长，越有利于提高除尘效率。

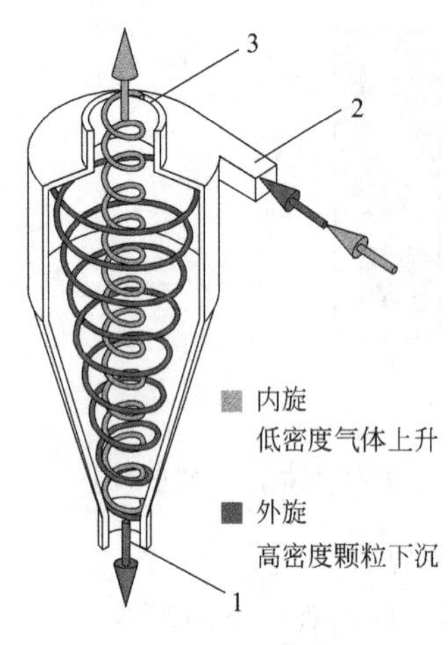

图 7-10　旋风分离器示意

1—出灰口；2—进气口；3—排气口

(2) 旋风分离器的工作原理

旋风分离器如图 7-10 所示。含尘气体由分离器圆筒部分上的进气口，沿切线方向进入，在器壁的约束下作向下的螺旋运动。在旋风分离器内气体和尘粒同时受到离心力作用，因尘粒密度远大于气体密度，所以尘粒所受到的离心力远大于气体。在离心力作用下，尘粒在向下旋转运动的同时还作向外的径向运动，其结果是尘粒产生径向且朝器壁的离心沉降运动，当运动到器壁后失去动能下落而与气体分离，然后在气流摩擦力和重力的作用下，再沿器壁表面作向下的螺旋运动，最后落入锥底的出灰口内。含尘气体在向下螺旋运动中逐渐被净化。在到达分离器的圆锥部分时，被净化了的气流由以靠近器壁的空间为范围的下行螺旋运动改为以中心轴附近空间为范围的上行螺旋运动，最后由分离器顶部的排气管排出。下行螺旋在外，上行螺旋在内，但两者的旋转方向是相同的。下行螺旋流的上部是主要的除尘区。在演示实验中所看到的螺旋状轨迹，是已经被甩到器壁上的粉粒被下行螺旋气流吹扫着器壁表面向下螺旋运动的情况。

旋风分离器内的压力从器壁附近静压强最高到中心静压强逐渐降低。这是由于下行和上行螺旋以相同的方向旋转，气体受惯性离心力作用被推向外的结果。这种静压强分布的一个后果是：在压强差的驱使下，不断有一部分气体由压强较高的下行旋流，沿径向窜入压强较低的上行旋流。因此在器内任何位置上，气体都有三个方向的速度，即切向速度、径向速度和轴向速度。

从器壁到中心，气体的切向速度先增大后减小。在切向速度最大的圆周以内的气流，称为“气芯”，具有以下几个特点：a. 上升的轴向速度颇大；b. 气芯内的静压强可小至排气管出口压强和当地大气压强以下；c. 低压的气芯通常由排气管的下端一直延伸到锥底的出灰口。因此，出灰口及其下方的集尘室均应密封良好，否则易漏入空气，把已收集在锥底的尘粒重新吹起，严重降低分离效果。

(3) 布袋除尘器的工作原理

如图 7-11 所示，过滤时含尘气体从入口 2 进入除尘器后，气体折转向上进入箱体 6，通过内部装有金属骨架的布袋时，粉尘被阻留在布袋的外表面。净化后的气体进入布袋上部的净气室 3 汇集后由净气出口 5 排出。过滤结束后进行清灰，空气经过喷吹管 4 通入布袋内部将布袋外部的尘粒吹下。

7.4.3 实验装置

非均相气固分离装置流程如图 7-12 所示。实验装置由重力沉降室、旋风分离器、布袋除尘器及风机等设备组成。含尘气体经过重力沉降室、旋风分离器、布袋除尘器后颗粒与气体分离，被净化的气体排入大气。

7.4.4 实验步骤

① 让风机旁路阀 10 处于全开状态。接通鼓风机的电源开关，开启鼓风机。

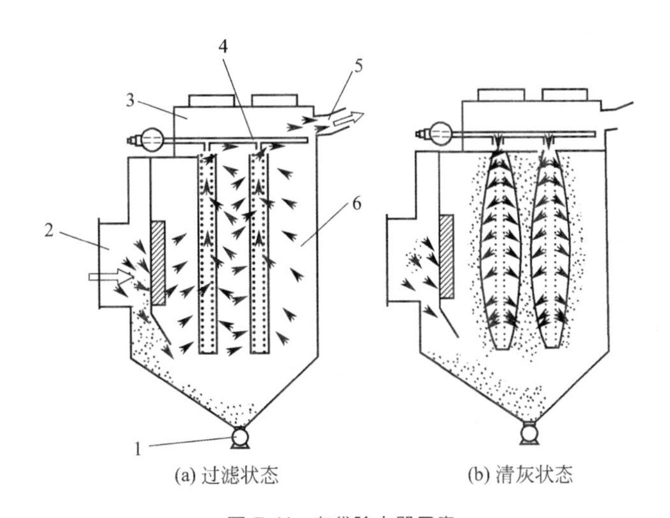

图7-11 布袋除尘器示意

1—出灰口；2—含尘气体入口；3—净气室；4—喷吹管；5—净气出口；6—箱体

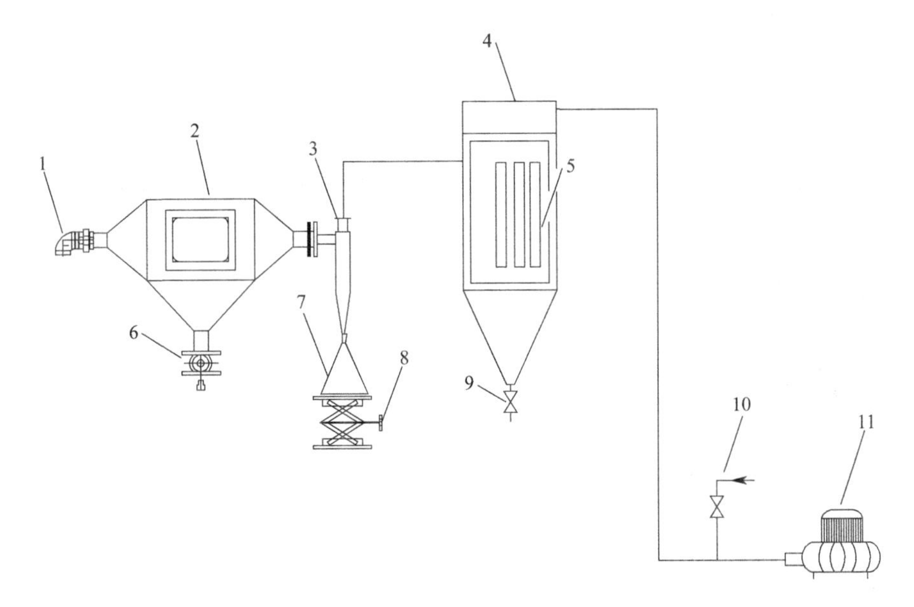

图7-12 非均相气固分离装置流程示意

1—进料口；2—沉降室；3—旋风分离器；4—布袋除尘器；5—布袋；
6—放料阀；7—收集瓶；8—升降台；9—放料阀；10—旁路阀；11—风机

② 逐渐关小旁路阀10，增大通过沉降室、旋风分离器的吸风量，了解气体流量的变化趋势。

③ 将空气流量调节至阀门全部关闭状态。将实验用的固体物料（玉米面、洗衣粉等）倒入进料容器中靠近物料进口，观察、分析含尘气体及其中的尘粒和气体在分离器中的运动情况。为了能够在较长的时间内连续观察到上述情况，可用手轻轻地移动容器，推动尘粒连续加入。虽然观察者实际上所看到的是尘粒的运动轨迹，但因尘粒沿器壁的向下螺旋运动是由于气流带动所致，故完全可以由此推断出含尘气流和气体的流动路线。

④ 结束实验时，先将旁路阀10全开，然后切断鼓风机的电源开关。若今后一段时间

长期不用，停车后从收集瓶内取出固体粉粒。

7.4.5 注意事项

① 开车和停车时，均应先让旁路阀处于全开状态，后接通或切断鼓风机的电源开关。

② 旋风分离器的排灰管与沉降室的连接应比较严密，以免因内部呈负压漏入空气将已分离下来的尘粒重新被吹起带走。

③ 实验时，若气体流量足够小，且固体粉粒比较潮湿，则固体粉粒会出现沿着向下螺旋运动的轨迹贴附在器壁上的现象。若想去掉贴附在器壁上的粉粒，可在大流量下，利用从含尘气体中分离出来的高速旋转的新粉粒，将原来贴附在器壁上的粉粒冲刷掉。

7.4.6 思考题

① 颗粒在旋风分离器内径向沉降的过程中，沉降速度是否为常数？

② 离心沉降与重力沉降有何异同？

③ 评价旋风分离器的主要指标是什么？影响其性能的因素有哪些？

④ 沉降室、旋风分离器及布袋除尘器的分离原理有何不同？

⑤ 该实验装置三种分离设备的先后顺序是沉降室、旋风分离器、布袋除尘器，其原因是什么？

⑥ 风量提高之后，沉降室、旋风分离器及布袋除尘器截留的固体颗粒在大小和量上有何区别？

7.5 板式塔流体力学性能演示实验

7.5.1 实验目的

① 通过实验了解塔设备和塔板（筛孔、浮阀、泡罩、舌形）的基本结构。

② 观察气、液两相在不同类型塔板上的流动与接触状况，观察实验塔内正常与不正常操作下的实验现象，掌握塔板压降的测量方法，比较不同塔板的流体力学性能。

7.5.2 实验原理

板式塔是一种重要的气液传质设备，广泛应用于精馏和吸收操作中。塔板是板式塔的核心部件，决定了塔的基本性能。为了有效地实现气、液两相间的物质传递和热量传递，要求塔板具有以下两个条件：其一，必须创造良好的气液接触条件，形成较大的接触面积，而且接触面应不断更新以增大传质、传热的推动力；其二，全塔总体上应保证气液逆流流动，避免返混合气液短路。

塔是靠自下而上的气体和自上而下的液体在塔板上流动时进行接触而达到传质和传热目的的，因此，塔板传质、传热性能的好坏主要取决于板上气、液两相的流体力学状态。

（1）塔板上气、液两相的接触状态

当气体速度较低时，气、液两相呈鼓泡接触状态。塔板上存在明显的清液层，气体以气泡形态分散在清液层中，气、液两相在气泡表面进行传质。当气体速度较高时，气、液两相呈泡沫接触状态。此时塔板上的清液层明显变薄，只有在塔板表面处才能看到清液，清液层随气速增大而变薄，塔板上存在大量泡沫，液体主要以不断更新的液膜形态存在于十分密集的泡沫之间，气、液两相在液膜表面进行传质。当气体速度很高时，气、液两相呈喷射接触状态，液体以不断更新的液滴形态分散在气相中，气、液两相在液滴表面进行传质。

(2) 塔板上不正常的流动现象

在板式塔的操作过程中，塔内要维持正常的气液负荷，避免以下不正常的状况。

① 严重漏液：当上升的气体速度很小时，气体通过塔板升气孔动压不足以阻止塔板上的液层下降，大量液体将从塔板的开孔处往下漏，出现严重漏液现象。

② 雾沫夹带：上升的气体穿过塔板的液层时，将板上的液滴裹挟到上一层塔板，引起液相返混的现象称为雾沫夹带。

③ 液泛：塔内气液两相之一的流量增大，使降液管内的液体不能顺利流下，降液管内液体积累，当管内液体越过溢流堰顶部时，两板间液体相连，并依次上升，这种现象称为液泛，也称淹塔。此时，塔板压降上升，全塔操作被破坏。

因此，塔板的设计应力求结构简单、传质效果好、气液通过能力强、压降低、操作弹性大。

7.5.3 实验装置

板式塔实验流程如图 7-13 所示，该流程包括 4 个塔，分别是泡罩塔、筛板塔、舌型塔和浮阀塔，4 个塔并联。空气由旋涡气泵经过孔板流量计计量后输送到每个板式塔塔

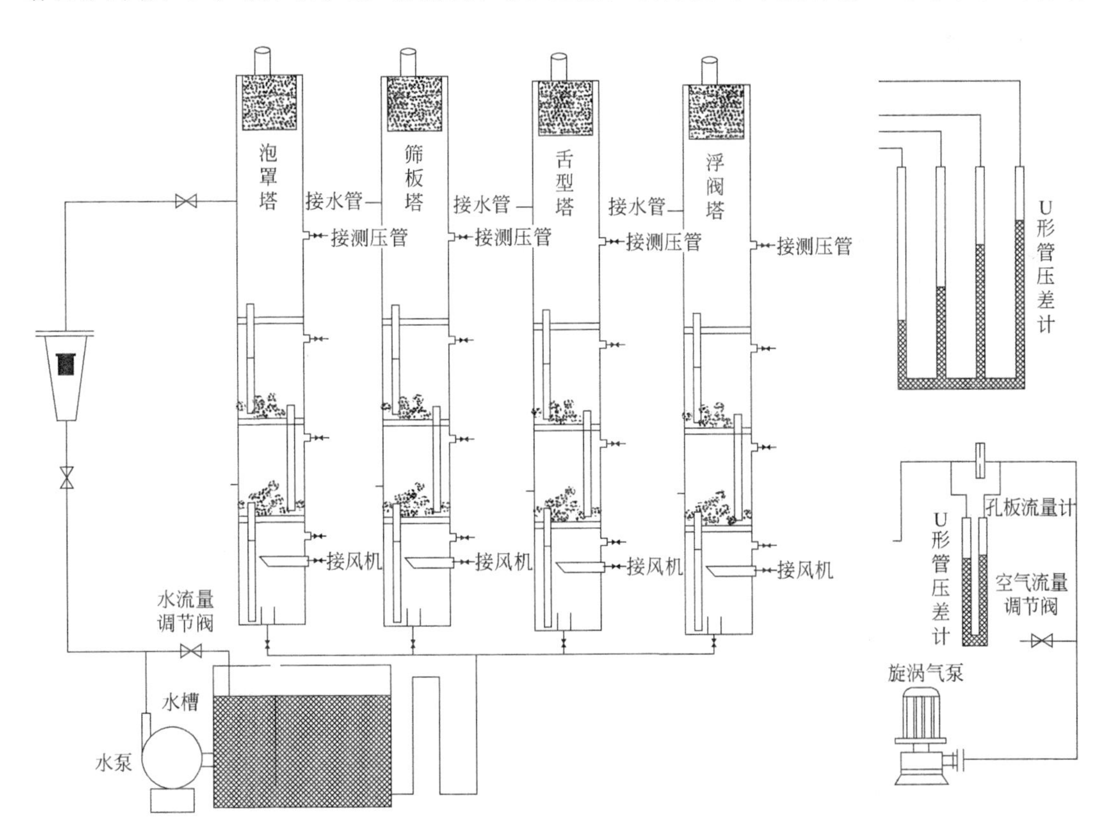

图 7-13　板式塔实验流程示意

底，向上经过塔板，从塔顶流出；液体则由离心泵经过转子流量计计量后由塔顶进入塔内并与空气进行接触，由塔底流回水槽。

7.5.4 实验步骤

① 向水箱内灌满蒸馏水，将空气流量调节阀置于全开的位置，关闭离心泵流量调节阀。

② 启动旋涡气泵，向塔内通入空气，同时打开离心泵向该塔输送液体，改变气液流量，观察塔板上气液流动与接触状况，并记录塔压降、空气流量、液体流量。

③ 用同样的方法测定并观察其他塔的压降和气液流动与接触状况。

④ 实验结束后先关闭调节阀和离心泵，待塔内液体大部分流到塔底时再关闭旋涡气泵，阻止设备和管道内进水。

7.5.5 注意事项

① 为保护有机玻璃塔的透明度，实验用水必须采用蒸馏水。

② 开车时先开旋涡气泵后开离心泵，停车反之，这样避免板式塔内的液体灌入风机中。

③ 实验过程中改变空气流量或水流量时，必须待其稳定后再观察现象和记录数据。

④ 若 U 形管压差计指示液面过高，将导压管取下，用洗耳球吸出指示液。

⑤ 水槽必须充满水，否则空气压力过大时易走短路而从水箱溢出。

7.5.6 思考题

① 在板式塔中气、液两相的传质面积是固定不变的吗？

② 评价塔板性能的指标是什么？讨论筛板、浮阀、泡罩、舌形塔板 4 种塔板的优缺点。

③ 由传质理论可知，流动过程中接触的两相湍动程度越大，传质阻力越小，如何提高两相的湍动程度？湍动程度的提高受不受限制？

④ 定性分析影响液泛的因素。

第8章　化工原理“三型”实验

8.1　流体流动阻力实验

8.1.1　实验目的

① 了解测定摩擦系数、局部阻力系数的工程意义。

② 掌握圆形直管管路流动阻力损失 Δp_f、摩擦系数 λ 和局部阻力系数 ζ 的测定方法，以及它们与 Re 的关系。

③ 学习并掌握对数坐标的使用方法，掌握倒U形管压差计和转子流量计的使用方法。

④ 了解各类管件、阀件在管路中的作用。

8.1.2　实验原理

由于流体存在黏性，流体在管道中流动会产生阻力损失而消耗一定的机械能。管路是由直管和管件（如三通、弯头、阀门）等组成，流体在直管中流动造成的机械能损失称为直管阻力损失；流体流经管件、阀件等局部地方时由于流道突然变化而引起的机械能损失称为局部阻力损失。

(1) 圆形直管摩擦阻力损失 Δp_f 和摩擦系数 λ 的测定

根据流体力学的基本理论，无论是层流还是湍流，流体在直管中流过时，摩擦系数与阻力损失之间的关系符合范宁公式：

$$\Delta p_f = \lambda \frac{l}{d}\frac{\rho u^2}{2} \tag{8-1}$$

在一根等径的水平放置的圆形直管上，如果没有流体输送机械做功，流体流经一定长度直管引起的阻力损失 Δp_f 等于此段管路的压力降 $p_1 - p_2$，即

$$\Delta p_f = -\Delta p = p_1 - p_2 \tag{8-2}$$

因此，通过测定两截面的压差可得到阻力损失。在一已知长度和管径的等径水平管段上，通过改变流体的流速，即可测量出不同 Re 下的直管阻力损失 Δp_f，按式(8-1) 求出摩擦系数 λ，即可得到 λ-Re 的关系。

层流时，摩擦阻力损失的计算式可由理论推导得到，即哈根-泊谡叶公式：

$$\Delta p_f = \frac{32\mu l u}{d^2} \tag{8-3}$$

式中，Δp_f 为摩擦阻力损失，Pa；μ 为流体的黏度，Pa·s；l 为管段长度，m；u 为流速，m/s；d 为管径，m。

由式(8-3) 可知层流时的压力损失与速度的一次方成正比，对比式(8-1) 和式(8-3) 可知层流时的摩擦系数为

$$\lambda=\frac{64}{Re} \tag{8-4}$$

湍流时，由于流动情况复杂得多，未能获得 λ 的理论计算公式，但可以应用因次分析方法确定它们之间的关系。通过因次分析方法可知直管的摩擦系数是雷诺数 Re 和管壁相对粗糙度 ε/d 的函数，即

$$\lambda=\phi(Re,\varepsilon/d) \tag{8-5}$$

若相对粗糙度一定，λ 是 Re 的函数，即 $\lambda=\phi(Re)$。

依据上述方法，通过实验可测出层流或湍流时摩擦系数 λ 与 Re 的关系并绘制出 λ 与 Re 的关系曲线，还可将实验结果与已知的式(8-4)、式(8-5) 进行对比。

(2) 局部阻力损失 $\Delta p_f'$ 和局部阻力系数 ζ 的测定

流体流经管件、阀门等局部地方引起的局部阻力损失可以表示为动能的一个倍数

$$\Delta p_f'=\zeta\frac{\rho u^2}{2} \tag{8-6}$$

式中，ζ 为局部阻力系数。

阻力系数测定方法与摩擦系数一样，只要测量出流体经过管件时的阻力损失 $\Delta p'_f$ 以及流体通过管路的流速 u，即可通过式(8-6) 计算出局部阻力系数 ζ。但是，在测定局部阻力时，由于管件前后流体的流动属于不稳定流动，测压口不能设置在紧靠管件处，而是取在距管件一定距离的管子上，以避免管件前后流动的不稳定性对测量结果的影响。如此，两测压口之间的流体流动阻力不仅包含管件的局部阻力，还包括管件前后直管段的流动阻力。因此，局部阻力测量的取压口需采用如图 8-1 所示的方式设置，以便准确测量管件的阻力。

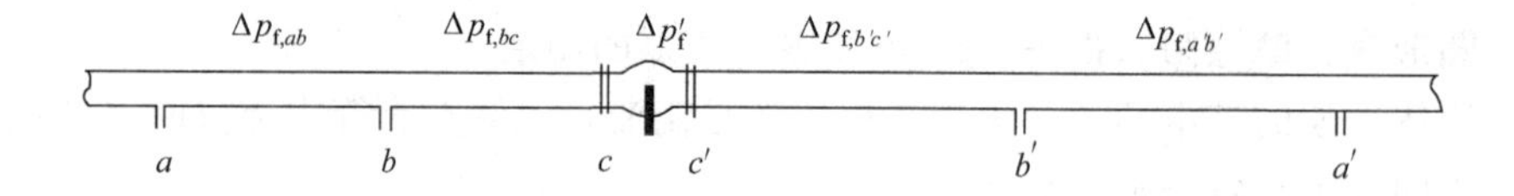

图 8-1 局部阻力测量取压口布置图

如图 8-1 所示，在一条等径的直管段上，安装待测局部阻力的阀门，在其上、下游各开两个测压口 a—a' 和 b—b'，并使

$$ab=bc;a'b'=b'c'$$

则

$$\Delta p_{f,ab}=\Delta p_{f,bc};\Delta p_{f,a'b'}=\Delta p_{f,b'c'}$$

在 a—a' 之间列伯努利方程式：

$$p_a-p_{a'}=2\Delta p_{f,ab}+2\Delta p_{f,a'b'}+\Delta p_f' \tag{8-7}$$

在 b—b' 之间列伯努利方程式：

$$p_b-p_{b'}=\Delta p_{f,bc}+\Delta p_{f,b'c'}+\Delta p_f'=\Delta p_{f,ab}+\Delta p_{f,a'b'}+\Delta p_f' \tag{8-8}$$

联立式(8-7) 和式(8-8)，则：

$$\Delta p_f'=2(p_b-p_{b'})-(p_a-p_{a'}) \tag{8-9}$$

式中，$(p_b-p_{b'})$ 称为近端压差；$(p_a-p_{a'})$ 为远端压差。它们可采用差压传感器或倒 U 形管压差计测量。

8.1.3　实验装置与流程

流体流动阻力实验装置流程如图 8-2 所示。

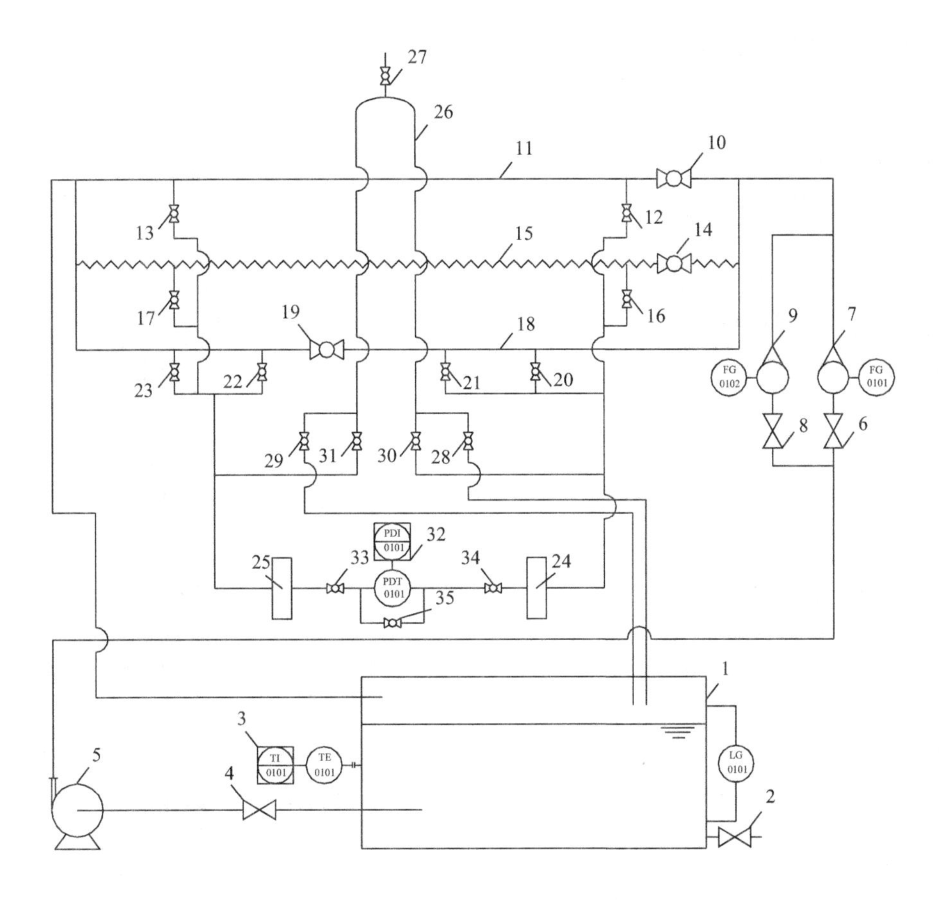

图 8-2　流体流动阻力实验装置流程图

1—水箱；2—水箱放水阀；3—测温点；4—离心泵入口阀；5—离心泵；6，8—流量调节阀；7，9—转子流量计；10—光滑管阀；11—光滑管；12，13—光滑管测压阀；14—粗糙管阀；15—粗糙管；16，17—粗糙管测压阀；18—局部阻力测量管；19—球阀；20，23—局部阻力远端测压阀；21，22—局部阻力近端测压阀；24，25—缓冲罐；26—倒 U 形管；27—倒 U 形管放空阀；28，29—倒 U 形管排水阀；30，31—倒 U 形管进水阀；32—差压传感器；33，34—差压传感器阀门；35—平衡阀

设备主要技术数据如下。

① 被测直管段

光滑管：管径　ϕ10mm×1mm，管长　l=1.70m，材料　不锈钢；

粗糙管：管径　ϕ12mm×1mm，管长　l=1.70m，材料　不锈钢；

局部阻力管：管径 ϕ25mm×1.5mm，材料　不锈钢球阀。

② 玻璃转子流量计

型号　LZB-25，测量范围　100～1000L/h；

型号　VA10-15F，测量范围　10～100L/h。

③ 差压传感器：型号　SM9320DP，差压范围　0～200kPa。

④ 倒 U 形管压差计：差压范围 0～900mmH_2O。

⑤ 离心泵：型号 WB70/0.55，额定流量 1.2～7.2m^3/h，额定扬程 15～20.5m；供电 220V/380V，0.55kW。

8.1.4 实验步骤及注意事项

8.1.4.1 实验步骤

① 检查水箱水位是否符合实验要求并采取相应措施。

② 打开电源总开关。

③ 将阀门 4、10、12、13、14、16、17、19、20、21、22、23、30、31、33、34、35 打开，其余阀门关闭，启动离心泵。

④ 将阀门 6、8 缓慢打开，在大流量状态下把实验管路中的气泡赶出。

⑤ 分别缓慢打开两个缓冲罐上方的阀门，直到有水微微溢出时即关闭，以放净罐中的空气。

⑥ 关闭阀门 6、8，将流量调为零，关闭阀门 30、31，打开阀门 27，分别用阀门 28、29 缓慢地将倒 U 形管内两液面调到管中心位置。再关闭阀门 27 和 35，而后打开阀门 30、31，若此时倒 U 形管压差计两端的液面相平，则说明管路完成排气，可以开始实验，若倒 U 形管压差计两端液面不平则需要重复步骤④和⑤进行排气，直至倒 U 形管压差计两端的液面相平。

⑦ 待管路排气完成后可开始实验，将被测管路的阀门全部打开，其他无关管路的阀门关闭，阀门 28、29 也关闭。

⑧ 在流量稳定的情况下，测取直管阻力压差。测量顺序从大流量至小流量，测 15～20 组数据。流量大于 100L/h 时的阻力压差值通过差压传感器测量并从仪器面板上读取，当流量读数低于 100L/h 时，打开阀门 30、31，用倒 U 形管压差计测取压差。

⑨ 测完一套管路的数据后，关闭流量调节阀，再次检查倒 U 形管压差计的液面是否相平，然后重复以上步骤，测取其他管路的数据。

测量球阀局部阻力系数时，需测近端压差 $p_b-p_{b'}$ 和远端压差 $p_a-p_{a'}$，局部阻力只在较大流量下测 3 组数据；球阀全开时采用倒 U 形管压差计测量近端压差和远端压差，球阀半开时采用差压传感器测量近端压差和远端压差。

⑩ 待数据测量完毕，关闭流量调节手阀，停泵，关闭仪表电源和设备总电源。

8.1.4.2 注意事项

① 启动离心泵之前，以及从光滑管阻力测量过渡到其他测量之前，都必须检查所有流量调节阀是否关闭。

② 系统要先排净气体，以使流体能够连续流动。

③ 启动离心泵前，必须关闭压力表和真空表的开关，以免损坏仪表；并全开离心泵入口阀门 4，避免发生汽蚀现象。

④ 利用差压传感器测量大流量下的压力差时，应关闭阀门 30、31，防止形成并联管路而影响测量结果。

⑤ 在实验过程中每调节一个流量之后应待流量和直管压差的数据稳定以后方可记录数据。

⑥ 不论采用倒 U 形管压差计还是差压传感器测量压差，除了所测量测压点的测压阀打开之外，其他测压阀均需关闭。

8.1.4.3　水的物性参数

本实验水的物性参数包括密度 ρ 和黏度 μ，可以通过手册查得，也可以用以下公式计算得到。

① 密度

$$\rho=-0.003589285t^2-0.0872501t+1001.44(\mathrm{kg/m^3}) \tag{8-10}$$

式中，t 为水的平均温度，℃。

② 黏度

$$\mu=0.000001198\exp\left(\frac{1972.53}{273.15+t}\right)(\mathrm{Pa\cdot s}) \tag{8-11}$$

8.1.5　思考题

① 流体在直管内稳定流动时，产生直管阻力损失的原因是什么？阻力损失是如何测量的？

② 流体流动时，产生局部阻力损失的原因是什么？局部阻力损失是如何测量的？局部阻力系数又是如何确定的？通过局部阻力系数如何确定当量长度？

③ U 形管压差计的测压原理是什么？用它能直接测定绝对压力吗？

④ 如何选择 U 形管压差计中的指示剂？

⑤ 简述孔板流量计和转子流量计的结构、工作原理、特点、安装注意事项及使用方法。

⑥ 在测量前为什么要将设备中的空气排尽？怎样才能迅速排尽空气？

⑦ 在本实验中，若离心泵在启动后，管路中没有水流出，可能的原因有哪些？

⑧ 本实验测量直管阻力的管路采用的是等径直管管段，且水平放置，该管段上、下游的压差与该管段的流动阻力损失相等，即 $\Delta p_f=p_1-p_2$；若该管段倾斜放置，则 $\Delta p_f=p_1-p_2$ 是否成立？此时压差计的读数大小能否反映两测压点之间流动阻力的大小？为什么？

⑨ 在粗糙管的测量过程，若光滑管的阀门没有完全关闭，则对粗糙管两端压差的测量结果会产生什么影响？在摩擦系数图上，测量点会产生正偏差还是负偏差？

⑩ 在实验过程中，除了被测管路外，其他管路上的上、下游的两个测压阀若也处于开启状态，对测量结果有何影响？

⑪ 在不同设备上（包括相对粗糙度相同而管径不同），不同温度下，应用不同流体（均为牛顿型流体）测定的 λ-Re 数据是否能关联在一条曲线上？为什么？

⑫ 通过对管路中各种阻力的测定，你认为减少流体在管路中的流动阻力可采取哪些措施？

⑬ 试设计一套适宜的实验方案，测量 90°弯头的局部阻力系数 ζ，以水为实验流体，要求画出实验装置流程图、指明需测量的数据及所用的仪器和仪表、并说明数据处理和计算方法。

8.2 离心泵实验

8.2.1 实验目的

① 熟悉离心泵的基本构造，掌握离心泵开、停车等操作方法。

② 掌握离心泵特性曲线和管路特性曲线的测定方法。

③ 掌握双泵联合操作方法以及双泵串联、双泵并联特性。

8.2.2 实验原理

离心泵是借助叶轮的高速旋转使充满在泵体内的液体在离心力的作用下，从叶轮中心甩向叶轮边缘的过程中获得机械能，提高静压能和动能，液体离开叶轮进入泵壳（蜗壳），蜗壳的特殊结构使部分动能转化成静压能，最后以高压液体进入排出管路。离心泵的理论压头是在理想情况下从理论上对离心泵中液体质点的运动情况进行分析研究后，获得离心泵压头与流量的关系。由于离心泵的性能受到泵的内部结构、叶轮形式和转速等因素的影响。在实际工作中，流体在泵内流动过程中会产生各种各样的阻力损失，实际压头要小于理论压头，且泵内部液体流动的情况比较复杂。因此，离心泵的扬程、效率等特性参数尚不能从理论上进行精确计算，只能通过实验测定。

(1) 离心泵特性曲线

在一定转速下，离心泵的扬程、功率、效率与其流量之间的关系称为离心泵的特性曲线。

1）流量 Q　离心泵的流量采用涡轮流量计测定。

2）扬程（压头）H　泵的扬程可由泵进、出口间的机械能衡算求得。在泵的进口（吸入口）和出口（排出口）之间列伯努利方程可得：

$$z_{入}+\frac{p_{入}}{\rho g}+\frac{u_{入}^{2}}{2g}+H=z_{出}+\frac{p_{出}}{\rho g}+\frac{u_{出}^{2}}{2g}+h_{f,入-出} \tag{8-12}$$

$$H=(z_{出}-z_{入})+\frac{p_{出}-p_{入}}{\rho g}+\frac{u_{出}^{2}-u_{入}^{2}}{2g}+h_{f,入-出} \tag{8-13}$$

上式中 $h_{f,入-出}$ 是泵的进口和出口之间的流体流动阻力，它在离心泵的效率中给予考虑，故上述伯努利方程中的 $h_{f,入-出}$ 可不予考虑，于是式（8-13）为：

$$H=(z_{出}-z_{入})+\frac{p_{出}-p_{入}}{\rho g}+\frac{u_{出}^{2}-u_{入}^{2}}{2g}=h_{0}+\frac{p_{出}-p_{入}}{\rho g}+\frac{u_{出}^{2}-u_{入}^{2}}{2g} \tag{8-14}$$

式中，H 为离心泵扬程，m；h_0 为离心泵出口压力表与进口真空表测压口之间的垂直距离，m；$p_{入}$、$p_{出}$ 分别为泵进、出口的压力，Pa；$u_{入}$、$u_{出}$ 分别为泵进、出口的流速，m/s；ρ 为流体密度，kg/m^3；g 为重力加速度，m/s^2。

由上式可知，只要测量出泵进口和出口的压力，以及两表测压点的高度差，并通过流量计算出流速即可求得泵的扬程。

3）轴功率 N

$$N=N_{电}\,k \tag{8-15}$$

式中，$N_{电}$ 为电功率表显示值，W；k 为电机传动效率，本实验装置可取 $k=0.95$。

4）效率 η　泵的效率 η 是泵的有效功率 N_e 与轴功率 N 的比值。有效功率 N_e 是单位时间内流体经过泵时所获得的实际功，轴功率 N 是单位时间内泵轴从电机得到的功，两者差异反映了水力损失、容积损失和机械损失的大小。泵的有效功率 N_e 可用下式计算：

$$N_e = HQ\rho g \tag{8-16}$$

故泵效率为

$$\eta = \frac{HQ\rho g}{N} \times 100\% \tag{8-17}$$

综上所述，在一定转速下，离心泵的扬程、功率、效率与流量之间的关系线即特性曲线，可以通过调节管路阀门开度，改变管路流量测得。

(2) 管路特性曲线

离心泵的特性曲线只是泵本身的特性，与管路状况无关。而离心泵使用时总是安装于某一特定的管路中，它提供了液体在管路中流动所需的机械能。因此，离心泵的实际工作情况是由泵的特性和管路本身的特性共同决定的。管路特性曲线方程可由伯努利方程整理得到：

$$H = \frac{\Delta p}{\rho g} + \Delta z + \frac{\Delta u^2}{2g} + \sum h_f = A + BQ^2 \tag{8-18}$$

式中，Δp 为管路两端压差，Pa；Δz 为管路两端的垂直距离，m；u 为流速，m/s；$\sum h_f$ 为流动阻力，m；A 为管路两端的总势能差，m；B 为管路特性曲线系数，h^2/m^5；Q 为管路流量，m^3/h。

其中扬程 H 的计算与流量 Q 的测定与离心泵特性曲线中 H 的计算与 Q 的测定方法相同。A 由管路两端实际条件决定，B 由管路状况决定，当液体处于高度湍动时，B 为常数。需要注意的是，管路特性曲线的测定不能通过管路流量调节阀来改变流量，因为流量调节阀开度变化即导致管路特性曲线变化，所以测量管路特性曲线时，只能采用变频器改变离心泵转速的方式调节流量。

若把离心泵的特性曲线与管路特性曲线标绘于同一坐标中，则两曲线的交点即为离心泵在管路中的工作点。

8.2.3 实验装置与流程

离心泵实验装置流程见图8-3。

设备主要技术数据如下所示。

① 涡轮流量计：型号　LWGY-40A05WSN，测量范围　1～20m^3/h。

② 离心泵：型号　MS100/0.55，额定流量　6m^3/h，额定扬程　14m；供电　三相AC380V，0.55kW。

③ 功率表：型号　GPW201-V3-A3-F1-P2-O3，精度　0.5%。

④ 真空表与压力表测压口之间的垂直距离 h_0=0.29m。

⑤ 真空表测压位置管内径 d_1=0.035m。

⑥ 压力表测压位置管内径 d_2=0.042m。

⑦ 真空表：型号　Y-100，测量范围　−0.1～0MPa，精度　1.5级。

⑧ 压力表：型号　Y-100，测量范围　0～0.6MPa，精度　1.5级。

⑨ 变频器：三菱变频器，型号　FR-D740-0.75K-CHT，功率　0.75kW/380V。

⑩ 温度表：铂电阻，型号　WZP-270501BX，测量范围　0～100℃，精度　B级。

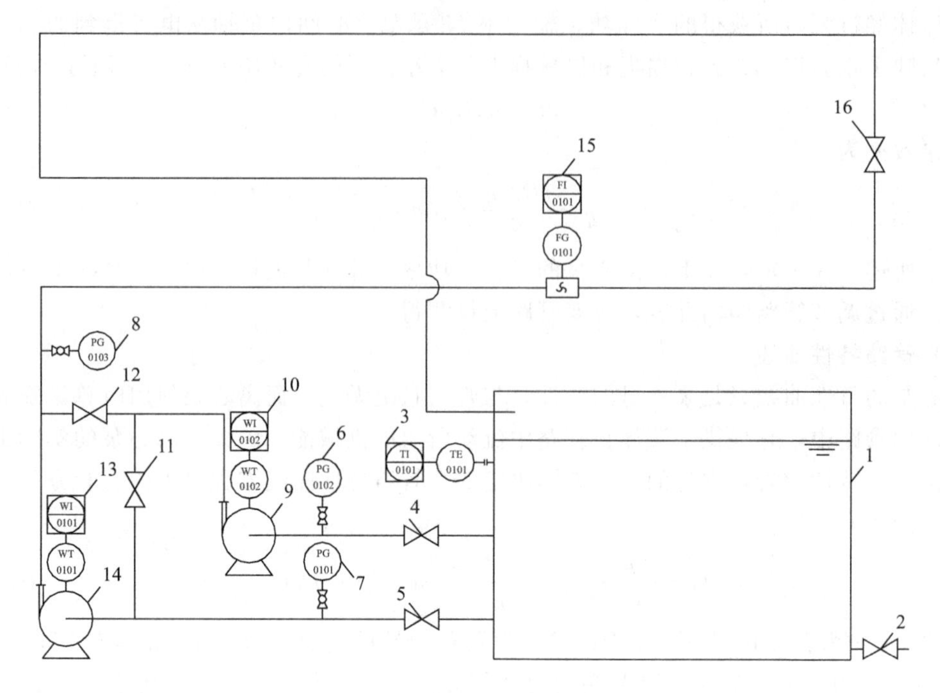

图 8-3　离心泵实验装置流程图

1—水箱；2—水箱放水阀；3—测温点；4—2＃水泵入口阀；5—1＃水泵入口阀；6—2＃水泵入口真空表及开关；7—1＃水泵入口真空表及开关；8—水泵总出口压力表及开关；9—2＃水泵；10，13—功率表；11，12—串并联调节手阀；14—1＃水泵；15—涡轮流量计；16—流量调节阀

8.2.4　实验步骤及注意事项

8.2.4.1　实验步骤

(1) 离心泵特性曲线的测定

① 检查水箱的水位是否符合实验要求。

② 打开电源总开关及仪表电源开关。

③ 全开 1＃水泵的入口阀 5，其他阀门全部关闭，启动 1＃水泵。

④ 缓慢打开流量调节阀 16 至全开，待系统内流体稳定，即系统内已没有气体，打开离心泵入口真空表的开关 7 和出口压力表的开关 8。

⑤ 将流量调节阀开至最大，确定流量测量范围，测取数据的顺序可从最大流量至零或反之，一般测 12 组数据。

⑥ 每次测量应同时记录涡轮流量计流量、泵入口真空度、泵出口压力、功率表读数及流体温度。

⑦ 测试结束，关闭离心泵出口流量调节阀 16，停 1＃水泵。

(2) 管路特性曲线的测定

① 打开电源总开关及仪表电源开关。

② 全开 1＃水泵的入口阀 5，其他阀门全部关闭，启动 1＃水泵；将流量调节阀 16 调至某一状态使该管路具备高阻（流量调节阀 16 调节至流量约为 $6m^3/h$）或低阻（流量调节阀 16 全开）特性。

③ 通过离心泵电机频率调节管路流量测定管路特性曲线，调节范围为电机全速的50%～100%。

④ 每改变电机频率一次，记录一组数据，包括流量、泵入口真空度、泵出口压力，共记录6组数据。

⑤ 测试结束，将电机频率恢复到全速的100%，关闭出口流量调节阀16，停1♯水泵。

(3) 双泵并联和双泵串联特性曲线的测定

① 打开电源总开关及仪表电源开关。

② 并联实验时，全开1♯水泵的入口阀5和2♯水泵的入口阀4以及阀门12，其他阀门全部关闭，启动1♯水泵及2♯水泵，打开入口真空表的开关6和7以及泵出口压力表的开关8；通过调节阀门16改变流量，记录泵入口真空度，泵出口压力、两功率表读数及流体温度。

③ 全开2♯水泵的入口阀4和阀门11，其他阀门全部关闭，启动1♯水泵及2♯水泵，打开入口真空表的开关6和7以及泵出口压力表的开关8；通过调节阀门16改变流量，记录泵入口真空度，泵出口压力、两功率表读数及流体温度。

④ 测试结束，关闭出口流量调节阀16，停泵，切断电源。

8.2.4.2　注意事项

① 出口流量调节阀关闭条件下，才能启动或关闭离心泵。

② 实验前管路需排气，确保流体连续流动。

③ 启动离心泵前关闭压力表、真空表的开关，避免损坏。

④ 流量调节阀需缓慢调节，避免传感器进水。

⑤ 离心泵入口阀门必须全开，避免汽蚀。

8.2.5　思考题

① 选择离心泵的原则是什么？

② 离心泵的 H-Q 特性曲线与管路的特性曲线有何不同？

③ 为什么本实验所测出的管路特性曲线是一条无明显截距且近似通过原点的曲线？

④ 根据你所测定的特性曲线，分析如果要增加该管路的流量范围，可采取哪些措施？

⑤ 双泵串联、双泵并联操作具有什么特点？

⑥ 实验过程中涉及哪些流量调节方式？说明它们的特点与区别。

⑦ 当离心泵出口流量调节阀关小时，管路中的流量、离心泵入口真空表和出口压力表的读数如何变化？为什么？

⑧ 管路特性曲线测定的实验中，为什么采用改变电机频率调节流量，而不是采用阀门调节？

⑨ 根据所绘出的双泵并联、双泵串联操作的 H-Q 特性曲线与管路的特性曲线，试解释什么情况下适宜采用双泵并联操作或双泵串联操作。

⑩ 在离心泵特性曲线的测定实验中，随着流量的减小，真空表的读数逐渐减小，直至为零，原因何在？

⑪ 结合实验测定结果说明：a. 为什么离心泵在启动时，应关闭出口阀门；b. 离心泵

适宜的工作区域如何确定，为什么？

⑫ 本实验装置中离心泵发生汽蚀最可能的原因是什么？当离心泵发生汽蚀现象时，其扬程线发生什么变化？

⑬ 若将本次实验装置的离心泵改装在水箱液面之上，请设计出测定离心泵特性曲线的实验方案。要求：画出实验装置流程图、写出实验操作步骤和设计出原始数据记录表。

8.3 过滤实验

8.3.1 实验目的

① 熟悉板框过滤基本流程、板框结构，熟练掌握板框过滤机的操作方法。

② 掌握恒压过滤操作条件下，过滤常数、压缩性指数等过滤参数的测定方法。

③ 掌握过滤问题的工程简化处理方法和实验研究方法。

8.3.2 实验原理

过滤是利用多孔介质使悬浮液（固液混合物，或称为滤浆）中固体颗粒被过滤介质截留形成滤饼（滤渣），而液体通过滤饼层和过滤介质，实现固、液分离的单元操作。无论是过滤操作还是过滤机的设计，过滤常数都是一个非常重要的基础数据。不同的悬浮液，其过滤常数差别很大，即使是同一种悬浮液，固体颗粒浓度不同、滤浆温度不同、过滤推动力不同，其过滤常数也不尽相同，故需要通过实验测定准确可靠的过滤常数等数据。

恒压过滤方程为

$$q^2+2qq_e=K\theta \tag{8-19}$$

式中，q 为单位过滤面积获得的滤液体积，m^3/m^2；q_e 为单位过滤面积的当量滤液体积，m^3/m^2；K 为过滤常数，m^2/s；θ 为过滤时间，s。

过滤常数的实验测定方法主要有微分法与积分法两种，其原理分别叙述如下。

(1) 微分法测定过滤常数

将式(8-19) 微分得

$$\frac{d\theta}{dq}=\frac{2}{K}q+\frac{2}{K}q_e \tag{8-20}$$

当各数据点的时间间隔不大时，$\frac{d\theta}{dq}$可以用增量之比$\frac{\Delta\theta}{\Delta q}$来代替，即

$$\frac{\Delta\theta}{\Delta q}=\frac{2}{K}\bar{q}+\frac{2}{K}q_e \tag{8-21}$$

上式为一直线方程。在过滤推动力、过滤温度恒定的条件下，应用一定的过滤介质对一定浓度的悬浮液进行恒压过滤，测出过滤时间 θ 及滤液累积量 q 的数据，在直角坐标纸上标绘$\frac{\Delta\theta}{\Delta q}$对 $\bar{q}$ 的关系（$\bar{q}$ 为 $\Delta\theta$ 时间内 q 的平均值），所得直线斜率为$\frac{2}{K}$，截距为$\frac{2}{K}q_e$，由此直线的斜率和截距即可求得过滤常数 K 和 q_e。

(2) 积分法测定过滤常数

式(8-19) 为过滤基本方程积分得到的恒压过滤方程，整理式(8-19) 可得：

$$\frac{\theta}{q}=\frac{1}{K}q+\frac{2}{K}q_e \tag{8-22}$$

同样，上式亦为直线方程。在恒压条件下过滤待测定的悬浮液，测出过滤时间 θ 及滤液累积量 q 的数据，在直角坐标纸上标绘 $\frac{\theta}{q}$ 对 q 的关系，所得直线斜率为 $\frac{1}{K}$，截距为 $\frac{2}{K}q_e$，由此直线的斜率和截距可求得过滤常数 K 和 q_e。

(3) 压缩性指数的测定

压缩性指数反映了滤饼的压缩性能，压缩性指数可通过过滤常数与过滤推动力的关系来获得，过滤常数的定义式为

$$K=\frac{2\Delta p}{\mu rc}=\frac{2\Delta p^{1-s}}{\mu r_0 c}=2k\Delta p^{1-s} \tag{8-23}$$

式(8-23) 中滤饼比阻 r 与过滤推动力 Δp 的经验公式为

$$r=r_0\Delta p^s \tag{8-24}$$

以上两式中，Δp 为过滤推动力，Pa；μ 为滤液的黏度，Pa·s；r_0 为 Δp 为 1Pa 时的滤饼比阻，$1/m^2$；r 为滤饼比阻，$1/m^2$；c 为过滤得到单位体积滤液时所形成的滤饼体积，m^3/m^3；s 为压缩性指数。

由式(8-23) 可得

$$r=\frac{2\Delta p}{K\mu c} \tag{8-25}$$

根据不同过滤推动力 Δp 下测得过滤常数 K 的实验数据，由式(8-25) 可求得滤饼比阻 r。而后将不同过滤推动力条件下，r 与 Δp 的数据绘制在双对数坐标上，由式(8-24) 可知 r 与 Δp 的数据应为一条直线。通过该直线的斜率和截距即可求出压缩性指数 s 和 r_0。或将 K 与 Δp 的数据绘制在双对数坐标上，由式(8-23) 可知 K 与 Δp 的数据为一条直线，其斜率和截距分别为 s 和 $2k$，而后根据式(8-23) 中 k 的定义可求得 r_0。

8.3.3 实验装置与流程

板框过滤实验装置流程如图 8-4 所示，主要由滤浆桶、搅拌器、板框过滤机、离心泵等组成。该板框过滤机是一种小型的工业过滤机，由非洗板、滤框、洗板按照一定的顺序组装而成，图 8-5 为板框结构示意。

设备主要技术数据如下所述。

① 轻型不锈钢卧式单级离心泵：型号　MS60/0.75，额定流量　60L/min，额定功率　0.75kW。

② 电动搅拌器：额定功率　400W，转速　1400r/min。

③ 过滤板框：不锈钢材质，框直径　0.09m，框厚度　0.02m。

④ 过滤介质：工业滤布。

⑤ 电子天平：型号　LNW-15，最大称量　15kg，最小称量　100g，工作环境温度　0～40℃。

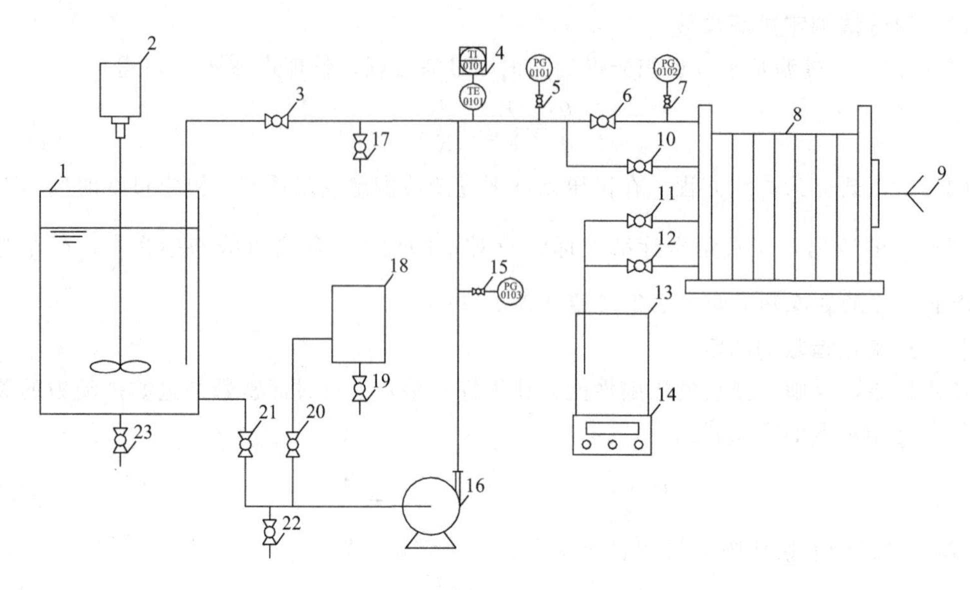

图 8-4　板框过滤实验装置流程图

1—滤浆桶；2—电动搅拌器；3—压力调节阀；4—测温点；5—阀前压力表；6—滤浆入口控制阀；7—阀后压力表；8—板框过滤机；9—压紧装置；10—洗涤液入口控制阀；11—滤液出口控制阀；12—洗涤液出口控制阀；13—滤液接收桶；14—电子天平；15—离心泵出口压力表；16—离心泵；17—管道排净阀；18—清水桶；19—清水桶放净阀；20—清水桶出口阀；21—滤浆桶出口阀；22—放净阀；23—滤浆桶放净阀

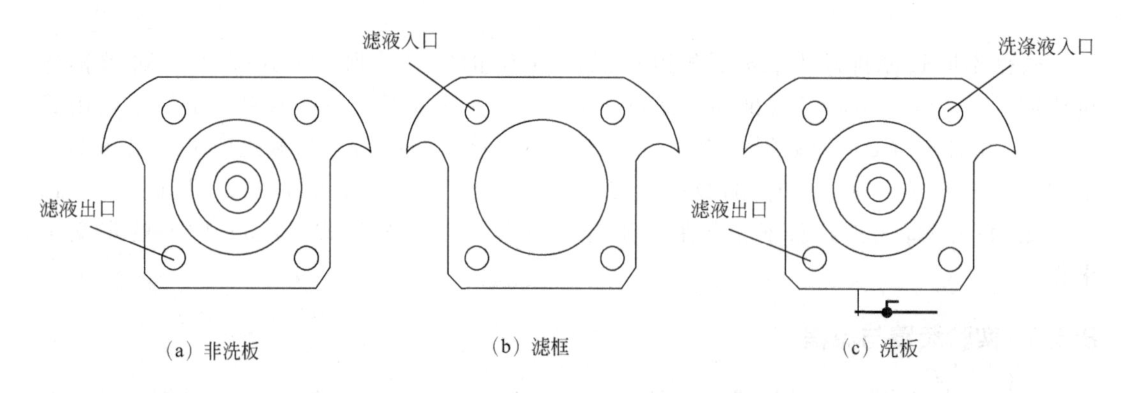

图 8-5　板框结构示意

8.3.4　实验步骤及注意事项

8.3.4.1　实验步骤

(1) 过滤

① 配料与搅拌：在滤浆槽内配制一定浓度的 $CaCO_3$ 水悬浮液。在仪表控制柜上，依次打开电源开关、仪表开关和搅拌器开关，使 $CaCO_3$ 悬浮液搅拌均匀。搅拌时，应将配料罐的顶盖合上。

② 板框的安装：板、框按照固定头—非洗涤板—框—洗涤板—框—非洗涤板—可动头顺序安装，滤布使用前用水浸湿，滤布要绷紧，不能起皱。滤布紧贴滤板，密封垫贴紧

滤布，而后板框由压紧装置压紧后待用。注意过滤实验时洗涤板下方的通孔切换阀必须是打开的。

③ 启动离心泵：打开阀门21和11，其余阀门全部关闭。打开仪表控制柜上的离心泵开关，启动离心泵。将智能仪表调节到显示频率的页面，用增加键将离心泵的电机频率调至100%，将压力调节阀3全开。

④ 设定压力：用压力调节阀3将过滤压力调节到所需要的数值。

⑤ 过滤：打开滤浆入口控制阀6，看到板框滤液出口的汇集管流出滤液时开始计时，建议每收集滤液0.5kg时记录相应的过滤时间θ或$\Delta\theta$，直至滤饼接近满框为止。

⑥ 一个压力的过滤实验完成后，关闭滤浆入口控制阀6，全开压力调节阀3，将压紧装置松开，卸下滤框、滤板、滤布并进行清洗。注意，为使滤浆桶的浓度保持一致，应将滤饼放回滤浆桶，用所收集的滤液清洗滤框、滤板、滤布，并将清洗液和板框下收集槽内的料液倒回滤浆桶。

⑦ 重复④、⑤、⑥步骤，测定三个不同压力下的过滤常数后即可结束过滤实验。

(2) 洗涤过程

① 一个压力的过滤实验完成后，关闭板框过滤机滤液进出口阀门6和11。将洗涤板下方的通孔切换阀关闭。

② 关闭压力调节阀3，停止进料泵，关闭滤浆桶出口阀21，然后打开清水桶出口阀20，在仪表控制柜上启动泵，打开管道排净阀17，清洗管道中残存的料液到地沟。

③ 关闭管道排净阀17，打开洗涤液出口控制阀12，打开洗涤液入口控制阀10，此时，压力表指示清洗压力，汇集管流出清洗液。

④ 清洗液流动约2min，可根据清洗液混浊变化情况来判断洗涤实验是否可以结束。一般物料可不进行清洗过程。

(3) 实验结束

① 关闭离心泵出口阀门，停泵。

② 关闭滤浆桶出口阀21，关闭搅拌电机。

③ 卸下滤框、滤板、滤布进行清洗。

④ 对料液泵及管道进行清洗，全开压力调节阀3，打开管道底部排净阀22，将管道内料液排尽。关闭压力调节阀3，打开清水桶18的出口阀20和管道排净阀17，启动料液泵，对料液泵及管道进行清洗，直至清洗液清澈为止。注意若清水桶水量不足，可补充一定量清水。

8.3.4.2 注意事项

① 清洗滤布时尽量不要折，要平整；过滤板与框之间的密封垫应注意放正；用摇柄把过滤设备压紧，以免漏液。

② 板、框的排列顺序和方向要正确。

③ 洗涤板下方的通孔切换阀阀门手柄与滤板平行为过滤状态，垂直为清洗状态。

④ 滤饼和清洗液以及板框下收集槽内的料液必须倒回滤浆桶。

⑤ 恒压过滤实验过程中，不可调节压力调节阀。

8.3.5 思考题

① 什么是滤浆、滤饼、滤液、过滤介质、助滤剂？

② 板框过滤机的优缺点是什么？适用于什么场合？

③ 简述恒压过滤的特点。

④ 过滤常数与哪些因素有关？

⑤ 为什么过滤开始时，滤液常常有点浑浊，而过段时间后才变得澄清？过滤推动力不同时对此现象有何影响？

⑥ 恒压过滤中，不同过滤压力得到的滤饼结构是否相同？其空隙率随压力如何变化？

⑦ 不同过滤推动力条件下，恒压过滤至满框时，得到的滤液量是否相同？有何规律？为什么？

⑧ 恒压过滤实验中，悬浮液贮槽为什么要用搅拌器对悬浮液进行搅拌？悬浮液贮槽内侧的挡板有何作用？

⑨ 过滤实验接近满框的最后一点数据与其他数据相比是偏低还是偏高？为什么？

⑩ 若过滤过程中，发现在同样过滤压力下过滤速率明显下降，可能是何原因？

⑪ 加快过滤速率的途径有哪些？

⑫ 滤饼中不同位置的空隙率有何不同？为什么？

⑬ 对恒压过滤，通过延长过滤时间来提高板框过滤机的生产能力是否可行？为什么？

⑭ 简述影响间歇过滤机生产能力的主要因素及提高间歇过滤机生产能力的途径。

⑮ 影响过滤速率的主要因素有哪些？当你在某一恒压条件下测得 K 和 q_e 值后，若将过滤压差提高 1 倍，则 K 和 q_e 将有何变化？

8.4 传热实验

8.4.1 实验目的

① 了解传热设备的主要结构和传热装置流程，掌握传热设备的操作方法。

② 掌握传热系数 K、圆管内对流给热系数 α_1 的测定方法，加深对其概念和影响因素的理解。

③ 掌握通过作图法或最小二乘法确定经验关联式 $Nu=ARe^m Pr^n$ 中常数的方法。

④ 通过对普通套管换热器和强化套管换热器的比较，了解工程上强化传热的措施。

8.4.2 实验原理

流体在圆形直管管内作强制湍流时，对流给热系数的准数关联式为：

$$Nu=ARe^m Pr^n \tag{8-26}$$

系数 A 与指数 m 和 n 则需由实验加以确定。本实验装置采用水蒸气加热空气，对管内冷流体气体而言，$n=0.4$。因此，式(8-26) 可写为：

$$\frac{Nu}{Pr^{0.4}}=ARe^m \tag{8-27}$$

式中：
$$Nu=\frac{\alpha_1 d_1}{\lambda} \quad Re=\frac{d_1 u_1 \rho}{\mu} \quad Pr=\frac{c_p \mu}{\lambda}$$

Re 中流速 u_1 通过流量计测量流量后计算获得，空气的密度 ρ、黏度 μ、热导率 λ、比热容 c_p 等物性参数根据流体的定性温度查物性数据或由如下关联公式（式中的温度 t 为定性温度，各式使用温度范围为 0℃≤t≤100℃）计算得到。

空气密度（kg/m^3）

$$\rho=1.2916-0.0045t+1.05828\times10^{-5}t^2 \tag{8-28}$$

空气比热容[kJ/(kg·℃)]

$$c_p=1.00492-2.88378\times10^{-5}t+8.88638\times10^{-7}t^2-1.36051\times10^{-9}t^3 +9.38989\times10^{-13}t^4-2.57422\times10^{-16}t^5 \tag{8-29}$$

空气黏度（Pa·s）

$$\mu=1.71692\times10^{-5}+4.96573\times10^{-8}t-1.74825\times10^{-11}t^2 \tag{8-30}$$

空气热导率[W/(m·℃)]

$$\lambda=0.02437+7.83333\times10^{-5}t-1.51515\times10^{-8}t^2 \tag{8-31}$$

Nu 通过 α_1 求得。对于一侧为饱和蒸汽加热另一侧空气的情况，由于蒸汽侧对流给热系数 $\alpha_2\gg\alpha_1$，且换热器传热管为金属管，其热导率很大、管壁很薄，所以

$$K\approx\frac{\alpha_1 d_1}{d_2} \tag{8-32}$$

又

$$Q=m_{s2}c_{p2}(t_2-t_1)=KA\Delta t_m\approx\alpha_1\frac{d_1}{d_2}A\Delta t_m \tag{8-33}$$

式（8-33）可通过空气的质量流量、空气的进口和出口温度、蒸汽温度求出 α_1，从而获得不同流量下的 Nu 和 Re，然后采用作图法或线性回归方法（最小二乘法）确定关联式（8-27）中的常数 A 和指数 m 值，确定对流给热系数经验关联式。

8.4.3 实验装置与流程

(1) 装置一实验流程

传热实验装置流程图（装置一）如图 8-6 所示，装置主要由套管换热器和列管换热器构成，其中套管换热器中装有相互平行的普通管和螺旋线圈强化管各一根。蒸汽发生器为电加热釜，产生的蒸汽压力为 0.4MPa，通过蒸汽控制阀调节蒸汽流量和压力，空气则由旋涡风机提供，使用旁路调节阀调节空气流量，蒸汽和空气分别通过管路输入各传热设备。

(2) 装置一设备主要技术数据

① 套管换热器

内管（普通管）规格：内径 19mm，外径 21 mm，换热管长度 $L=960$mm，紫铜。

内管（强化管）规格：内径 19mm，外径 21 mm，换热管长度 $L=960$mm，紫铜；管内添加螺旋线圈。

外管规格：ϕ100mm×3mm，长度 $L=1000$mm，不锈钢。

② 列管换热器

管程规格：ϕ20mm×1.5mm，长度 L＝800mm，数量 7 根，不锈钢。

壳程规格：ϕ139mm×2mm，不锈钢。

控制面板及仪表显示屏如图 8-7 及图 8-8 所示。

(3) 装置二实验流程

传热实验装置流程图（装置二）如图 8-9 所示，装置主要由套管换热器和列管换热器构成，其中套管换热器可通过内部拆装螺旋线圈实现普通套管和强化套管的转换，列管换热器可利用管堵改变传热面积。蒸汽发生器为电加热釜，空气由旋涡风机提供，使用旁路调节阀调节空气流量，蒸汽和空气分别通过管路输入各传热设备。

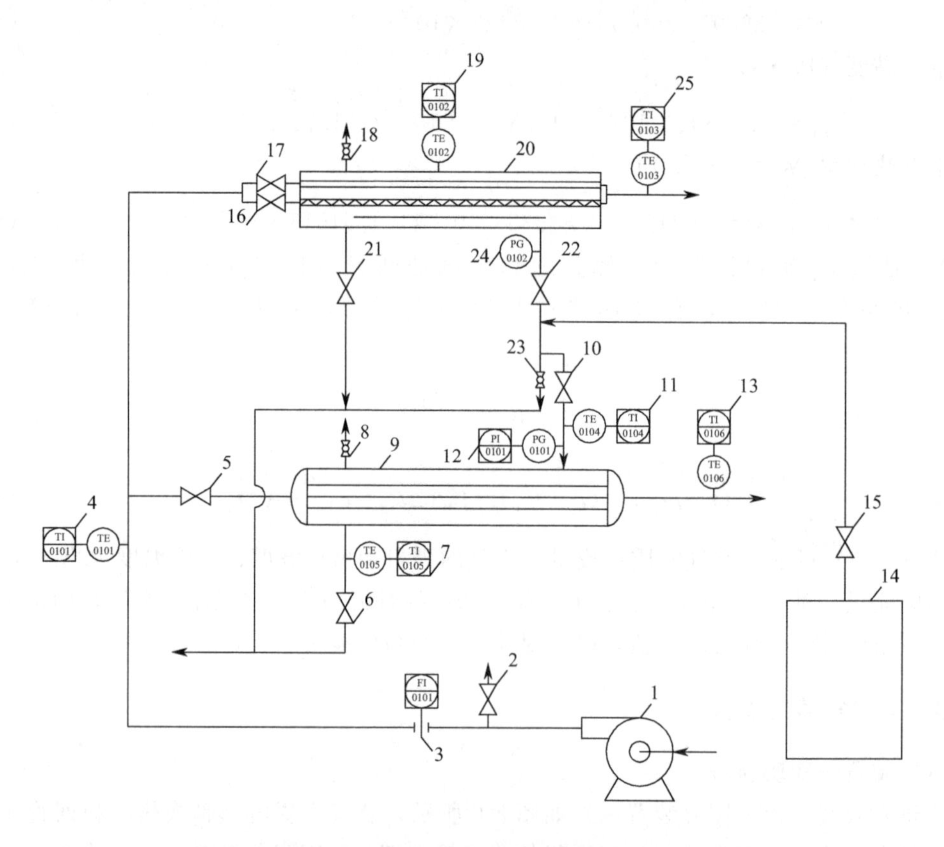

图 8-6 传热实验装置流程图（装置一）

1—旋涡风机；2—空气旁路调节阀；3—孔板流量计；4—空气进口测温点；
5—列管换热器空气进口阀；6—列管换热器冷凝水出口阀；7—列管换热器蒸汽出口测温点；
8—列管换热器壳程不凝性气体放空阀；9—列管换热器；10—列管换热器蒸汽进口阀；
11—列管换热器蒸汽进口测温点；12—列管换热器蒸汽进口压力表；
13—列管换热器空气出口测温点；14—蒸汽发生器；15—蒸汽控制阀；
16—套管换热器（强化管）空气进口阀；17—套管换热器（普通管）空气进口阀；
18—套管换热器壳程不凝性气体放空阀；19—套管换热器壳程测温点；
20—套管换热器；21—套管换热器冷凝水出口阀；22—套管换热器蒸汽进口阀；
23—套管换热器蒸汽管路排水阀；24—套管换热器壳程蒸汽压力表；
25—套管换热器空气出口测温点

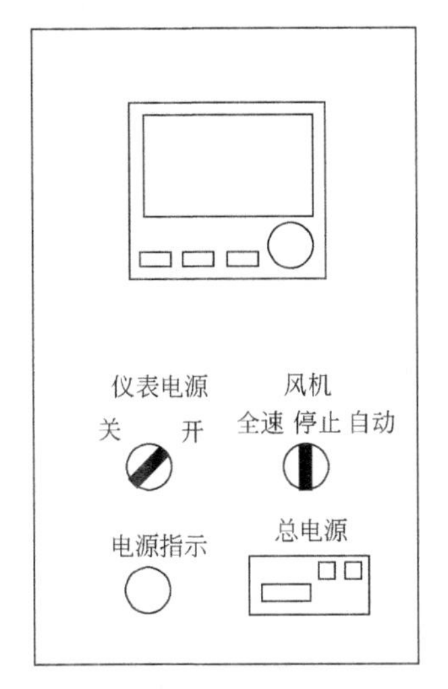

图 8-7　控制面板

TI101　℃	TI104　℃	TI105　℃
TI106　℃	TI102　℃	TI103　℃
PI103　kPa	FIC101　m^3/h	FIC101　Pa

图 8-8　仪表显示屏

TI101—冷流体进口温度（套管、列管）；TI102—套管换热器蒸汽温度；TI103—套管换热器冷流体出口温度；TI104—列管换热器进口蒸汽温度；TI105—列管换热器出口蒸汽温度；TI106—列管换热器冷流体出口温度；PI103—蒸汽压力；FIC101—冷流体流量

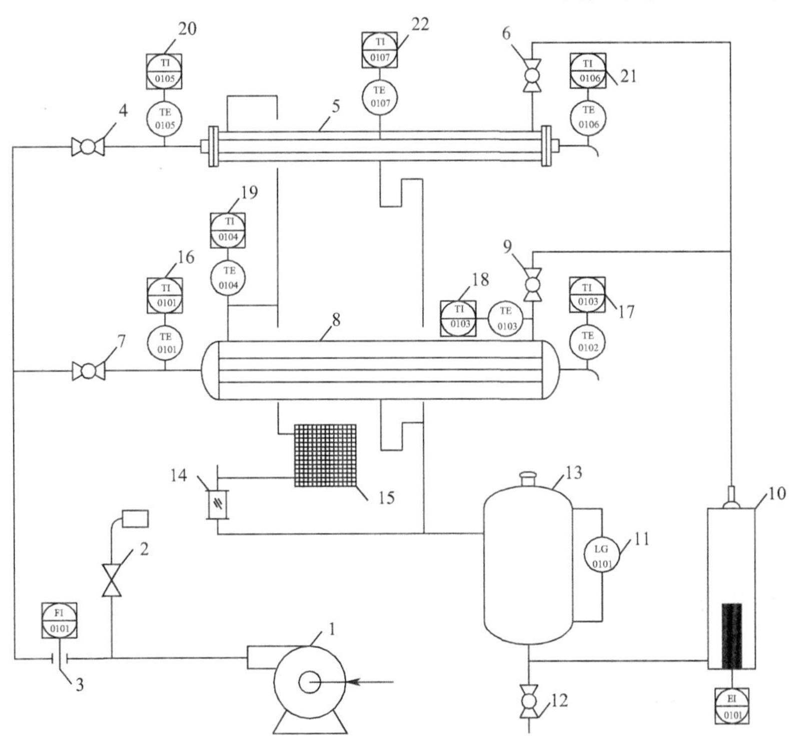

图 8-9　传热实验装置流程图（装置二）

1—旋涡风机；2—空气旁路调节阀；3—孔板流量计；4—套管换热器空气进口阀；5—套管换热器；6—套管换热器蒸汽进口阀；7—列管换热器空气进口阀；8—列管换热器；9—列管换热器蒸汽进口阀；10—蒸汽发生器；11—液位计；12—排净阀；13—贮水罐；14—视盅；15—散热器；16—列管换热器空气进口测温点；17—列管换热器空气出口测温点；18—列管换热器蒸汽进口测温点；19—列管换热器蒸汽出口测温点；20—套管换热器空气进口测温点；21—套管换热器空气出口测温点；22—套管换热器壁面测温点

(4) 装置二设备主要技术数据

① 套管换热器

内管规格：内径 20mm，外径 22mm，换热管长度 $L=1200$mm，紫铜。管内可添加螺旋线圈（线圈丝径为 1mm，节距为 40mm）。

外管规格：内径 50mm，外径 57mm，管子数 6 根。

② 列管换热器

内管规格：内径 16mm，外径 19 mm，换热管长度 $L=1200$mm，管子数 6 根。

外管规格：内径 82mm，外径 89mm。

③ 孔板流量计

孔流系数 0.65，孔径 0.017m。

控制面板如图 8-10 所示。

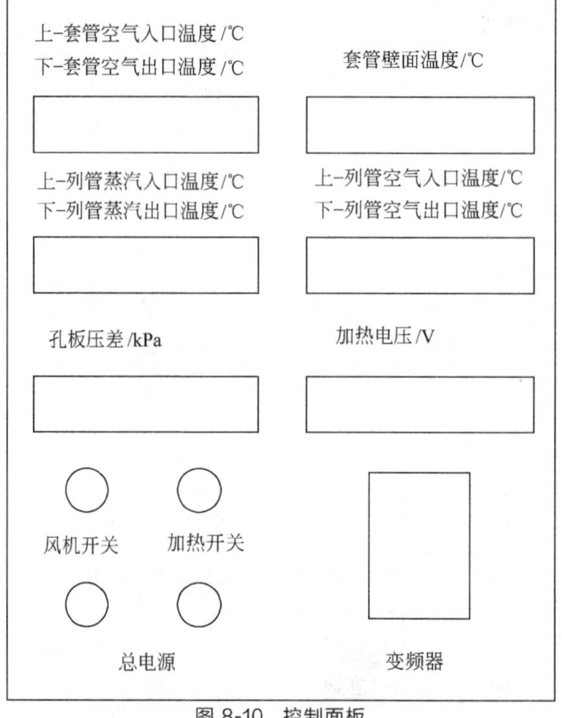

图 8-10 控制面板

8.4.4 实验步骤与注意事项

8.4.4.1 实验步骤

(1) 实验步骤（适用于装置一）

① 打开控制台上的总电源开关，打开仪表电源开关。

② 根据实验使用的换热器，打开换热器蒸汽入口阀（阀 22 或阀 10），打开蒸汽发生器的进水阀，确保蒸汽发生器能自动补水；确认蒸汽发生器已加满水后，打开水蒸气发生器电源开关开始加热，同时微微打开蒸汽控制阀 15。

③ 根据实验使用的换热器，全开相应换热器的空气进口阀（阀 16、阀 17、阀 5

之一)。

④ 将空气旁路调节阀 2 全开，打开控制台上的风机电源开关（置于全速挡），启动风机。

⑤ 打开换热器冷凝水出口阀 21 或阀 6，让管路中的冷凝水排净后再关闭（实验过程中由于蒸汽会在管路中冷凝累积，所以实验过程中可间断性地开启阀 21 或阀 6 排放冷凝水）。

⑥ 根据实验使用的换热器，打开不凝性气体放空阀（阀 18 或阀 8），注意阀门开度不宜太大；同时适当开启冷凝水排放阀（阀 21 或阀 6）。

⑦ 调节实验操作换热器的蒸汽进口阀（阀 22 或阀 10）的开度（开度不宜太大），让蒸汽缓慢流入换热器中，逐渐加热换热器，使换热器由“冷态”转变为“热态”，暖管时间不得少于 10min。

⑧ 当换热器预热结束后，用蒸汽进口阀（可用冷凝水出口阀辅助调节）将蒸汽压力调节至 0.01MPa 左右（温度略高于 100℃），维持温度稳定即可。

⑨ 通过手动调节空气旁路调节阀（阀 2），改变空气流量（也可通过风机变频器的频率来调节流量），在每个流量条件下，待热交换过程稳定后方可记录实验数值，一般每个流量下至少应使热交换过程保持 3～5min 方为稳定；改变不同流量，记录不同流量下的各个实验参数值。

⑩ 空气流量按照从小到大的顺序变化，记录 7 组实验数据，完成实验。关闭蒸汽发生器，待换热器温度降至常温后，关闭风机电源，关闭换热器蒸汽进口阀与换热器空气进口阀，排放系统冷凝水，关闭蒸汽发生器进水阀，关闭仪表电源和设备总电源。

(2) 实验步骤（适用于装置二）

① 向贮水罐 13 中加入蒸馏水至液位计高度 2/3 处。

② 根据实验使用的换热器，全开换热器蒸汽入口阀门 6 或阀门 9。打开总电源和加热开关，对蒸汽发生器内的水进行加热。

③ 根据实验使用的换热器，全开换热器空气入口阀门 4 或阀门 7，全开空气旁路调节阀 2，当套管换热器壁温或列管换热器入口蒸汽温度接近 90℃时，启动风机。

④ 通过手动调节空气旁路调节阀 2，改变空气流量，在每个流量条件下，待热交换过程稳定后方可记录实验数值，一般每个流量下至少应使热交换过程保持 3～5min 方为稳定；改变不同流量，记录不同流量下的各个实验参数值。

⑤ 空气流量按照从小到大的顺序变化，记录 7 组实验数据，完成实验。关闭蒸汽发生器加热电源，待换热器温度降至常温后，关闭风机电源，关闭换热器蒸汽进口阀与换热器空气进口阀，关闭总电源。

8.4.4.2 注意事项

(1) 装置一注意事项

① 一定要在换热器内管输入一定量的空气后，方可开启蒸汽阀门，且必须在排除蒸汽管线上原先积存的冷凝水后，方可把蒸汽通入套管换热器中。

② 操作过程中，蒸汽压力控制在 0.04MPa（表压）以下。

③ 随时注意惰性气的排空和蒸汽压力或蒸汽温度的调整。

④ 实验过程中或换热器间需要进行切换操作时，套管换热器与列管换热器蒸汽入口阀不可同时处于全关状态，套管换热器与列管换热器空气入口阀也不可同时处于全关

状态。

（2）装置二注意事项

① 检查蒸汽加热釜中的水位是否在正常范围内，否则应及时补充水量。

② 实验过程中或换热器间需要进行切换操作时，套管换热器与列管换热器蒸汽入口阀不可同时处于全关状态，套管换热器与列管换热器空气入口阀也不可同时处于全关状态。

③ 实验过程中保持上升蒸汽量的稳定，不应改变加热电压。

④ 需要在套管换热器中添加螺旋线圈或列管换热器需要安装堵头时，需要对正在操作的换热器停止供汽（蒸汽）并适当冷却，而后停止供气（空气），方可进行拆装。

8.4.5 思考题

① 根据实验测定的 *Nu-Re* 经验关联式，讨论影响圆管内对流给热系数的主要因素有哪些？这些因素是如何对 α 产生影响的？

② 比较普通管与强化管的 *Nu-Re* 关联式，可得到什么结论？强化管强化传热的机理是什么？螺旋线圈传热强化的代价是什么？

③ 在普通管内添加螺旋线圈，可强化传热，除此以外，还有哪些强化传热措施？

④ 在传热实验过程中，理论上空气出口温度随空气流量的增大应该如何变化？而实际情况是否与此结论相符，若不相符，试分析原因。

⑤ 在本实验装置中采用饱和蒸汽加热空气，则传热管壁温与蒸汽温度相近还是与空气温度相近？原因何在？

⑥ 以空气为介质的传热实验，其雷诺数 *Re* 如何计算比较方便？

⑦ 在本传热实验中，蒸汽侧需要排放什么流体？原因何在？

⑧ 当空气流量增大时，蒸汽的冷凝量和传热量如何变化？

⑨ 普通管和强化管的尺寸相同，管路相同，使用同一台风机，为什么强化管内空气的流量在同样条件下，要比普通管内空气的流量小？

⑩ 排放不凝性气体的作用是什么？

⑪ 实验测定的 *Nu-Re* 经验关联式，能否用于水对流传热系数的计算？

⑫ 若在进行套管换热器传热实验时，列管换热器入口空气阀门没有完全关闭，则对测量结果 *Nu-Re* 关联式有何影响？为什么？

⑬ 列管换热器和普通管套管换热器测量得到的 *Nu-Re* 经验关联式是否相同？

⑭ 用恒定温度的水蒸气加热空气，若换热器的面积无限大，则空气出口温度为多少？

⑮ 本实验装置中，蒸汽发生器工作正常，若发现换热器入口蒸汽压力增大，可能的原因有哪些？针对不同的原因分别应采取什么措施使换热器入口蒸汽压力恢复正常？

8.5 吸收实验

8.5.1 实验目的

① 了解填料吸收塔基本结构及吸收装置流程。

② 掌握填料吸收塔流体力学性能的测定方法，了解填料吸收塔在干塔、湿塔条件下，气、液流量对全塔压降的影响规律。

③ 熟悉填料吸收塔的操作，塔内气、液正常流动状态及液泛现象。

④ 掌握填料吸收塔体积总传质系数的测定方法。

8.5.2 实验原理

(1) 流体力学性能

填料塔是一种气液传质设备，气液两相通常在塔内逆流流动，填料在为气液两相的传质提供接触面积的同时，也对气液的流动形成了阻力。因此，了解填料层压降、液泛等流体力学性能及其影响因素和变化规律，对确定填料塔的适宜操作范围、实现填料塔的操作和优化具有重要意义。

填料层的压降除了和填料的类型、结构、大小、材质等因素有关，在操作过程中主要与填料塔中气液负荷有关，由于填料塔中气液两相逆流流动，因此，在一定的喷淋量下，气速越大，则填料层压降越大；在一定的气速下，喷淋量越大，填料层压降也越大。将单位填料层高度的压降和气、液流量的关系绘图即得到填料层压降曲线图（图 8-11）。

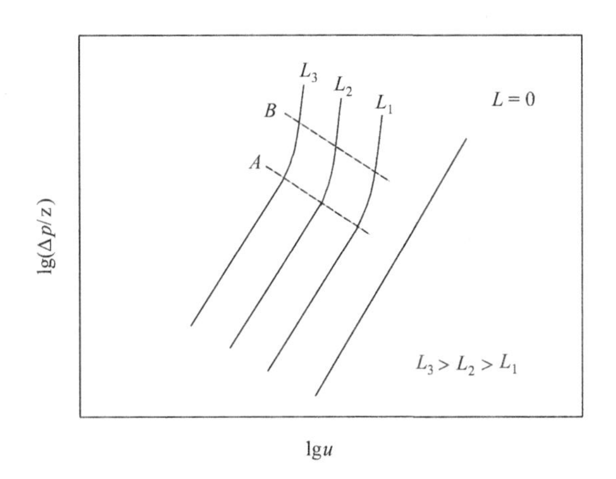

图 8-11 填料层压降曲线

如图 8-11 所示，气体在填料层中的流动状态一般为湍流状态，当液体喷淋量 $L=0$ 时，提高气体流速，填料层压降成比例增大，压降曲线为直线，其斜率约为 1.8～2.0，称为干填料层压降曲线。而在一定的液体喷淋量下的压降曲线可分为三段，当气速较小时，气体对填料表面液膜的曳力较小，提高气速对液体流动影响不大，填料表面液膜厚度变化不大，所以填料层中液体的持液量基本不变，压降与气速的关系为直线，且与干填料层压降线基本平行，故在 A 点以下区域为恒持液区。当气速超过 A 点时，气体对液体产生的曳力较大，气速的提高将阻碍液体的流动，从而使液膜厚度增大，持液量增大，而持液量增大，缩小了气体的通道，使气体流动阻力明显增大，此时填料层压降曲线斜率变大，随着气速的提高，填料层持液量持续提高，故 A 点称为载点，A 点的气速为载点气速，A、B 之间的区域为载液区。当气速达到 B 点时，液体已无法顺利向下流动，塔内持液量急剧增大，液体几乎充满整个填料层，液体由分散相转化为连续相，气体则以气泡形式通过填料层，由连续相转化为分散相，此时气速的微小提高，都会导致填料层压降的显著提升，正常的流动和传质都受到严重破坏，大量液体被气体夹带出塔顶，吸收塔操作极不稳定。故 B 点以上区域称为液泛区，B 点称为液泛点。

（2）传质性能

传质系数是反映填料塔传质性能的重要参数之一，体系性质、操作条件、气液流动状态、填料类型、填料结构和大小等均会影响传质系数。由于传质系数影响因素多而复杂，所以传质系数通常通过实验测定，为工程设计与操作提供准确、可靠的基础数据。本实验采用水吸收空气中的CO_2组分，CO_2气体的吸收过程属于液膜控制，由于CO_2在水中的溶解度很小，所以该吸收过程可视为低浓度气体吸收且吸收过程温度恒定，相平衡常数亦为常数。在填料层高度一定的实验装置中，可通过测量吸收塔进出口气液相组成计算传质单元数，而后由下式计算传质单元高度，从而获得传质系数。

$$h_0=\int_0^{h_0}\mathrm{d}h=\frac{L}{K_\mathrm{x}a}\int_{x_a}^{x_b}\frac{\mathrm{d}x}{x^*-x}=H_\mathrm{OL}N_\mathrm{OL}=\frac{L}{K_\mathrm{y}a}\int_{y_a}^{y_b}\frac{\mathrm{d}y}{y-y^*}=H_\mathrm{OG}N_\mathrm{OG} \quad (8\text{-}34)$$

式中，h_0为填料层高度，m;；L为液相摩尔流率，kmol/（m^2·h）；G为气相摩尔流率，kmol/（m^2·h）；x_a、x_b为液相进口、出口摩尔分数；y_a、y_b为气相出口、进口摩尔分数；H_OL为液相总传质单元高度，m；N_OL为液相总传质单元数；H_OG为气相总传质单元高度，m；N_OG为气相总传质单元数。

其中

$$N_\mathrm{OG}=\frac{1}{1-S}\ln\left[(1-S)\frac{y_b-mx_a}{y_a-mx_a}+S\right]\quad H_\mathrm{OG}=\frac{G}{K_\mathrm{y}a}\quad S=\frac{mG}{L}$$

$$N_\mathrm{OL}=\frac{1}{1-A}\ln\left[(1-A)\frac{y_b-mx_a}{y_b-mx_b}+A\right]\quad H_\mathrm{OL}=\frac{L}{K_\mathrm{x}a}\quad A=\frac{L}{mG}$$

式中，$K_\mathrm{y}a$为气相体积总传质系数，kmol/(m^3·h)；$K_\mathrm{x}a$为液相体积总传质系数，kmol/(m^3·h)；S为解吸因子；A为吸收因子。

低浓度等温物理吸收相平衡关系符合亨利定律，即：

$$y^*=mx \quad (8\text{-}35)$$

式中，m为相平衡常数。

在此特定的吸收塔（塔径、填料层高度、填料类型、填料大小、填料材质等）中，一定操作条件下（如温度、压力、气液比等），可通过吸收率评价其吸收效果。

$$\eta=\frac{y_b-y_a}{y_b} \quad (8\text{-}36)$$

式中，η为吸收率。

8.5.3 实验装置与流程

实验装置包括吸收塔和解吸塔，空气和来自钢瓶的二氧化碳混合后从塔底进入吸收塔，吸收剂水来自吸收液贮槽，经吸收液泵打入，水由塔顶喷淋而下与二氧化碳-空气混合气在填料层中接触传质。经过吸收后的混合气从吸收塔塔顶排出；吸收了二氧化碳的液体从吸收塔塔底排出，进入解吸液贮槽，经解吸液泵打入解吸塔塔顶；环境空气由风机输送从解吸塔塔底输入，作为解吸剂与液体同样在填料层中接触传质，液体在解吸塔中得到解吸再生从解吸塔塔底排出，进入吸收液贮槽作为吸收剂循环使用。吸收实验装置流程如图 8-12 所示。

设备主要技术数据如下所述。

① 吸收塔、解吸塔：塔径 ϕ76mm×3.5mm，填料层高度 850mm。

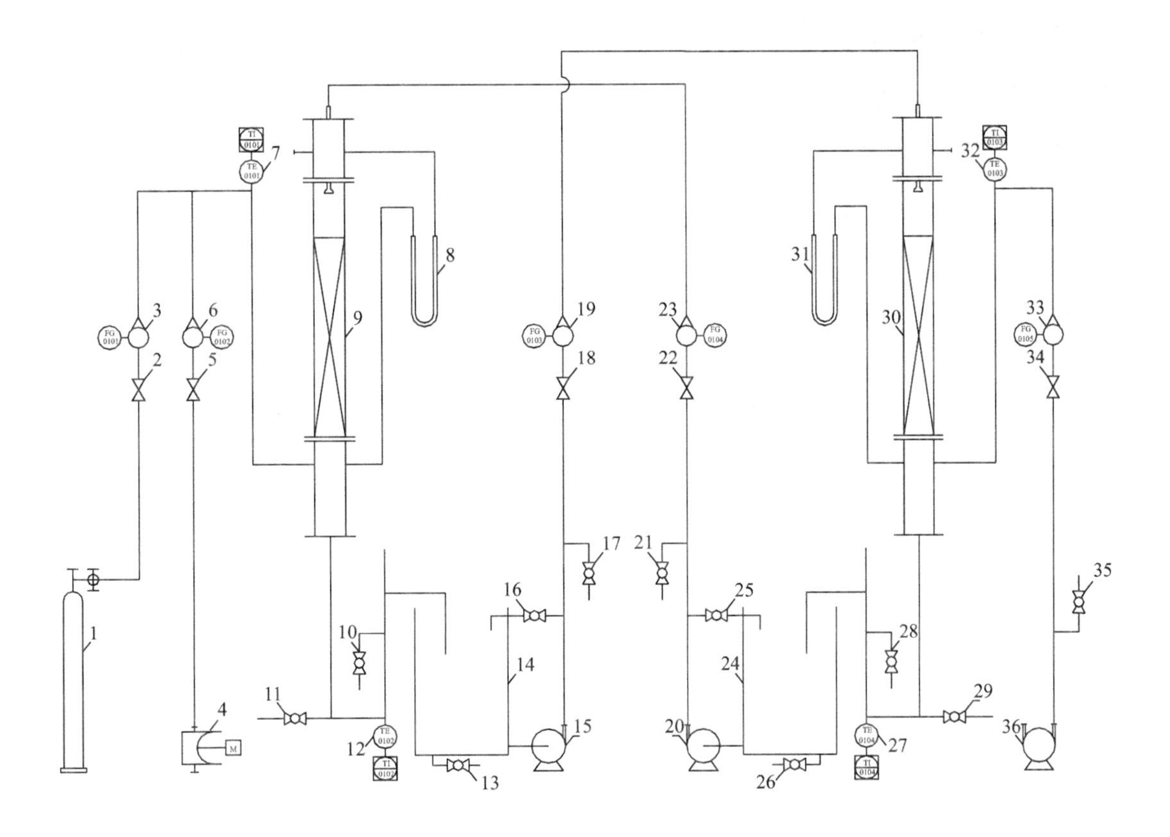

图 8-12　吸收实验装置流程图

1—二氧化碳钢瓶；2—二氧化碳流量调节阀；3—二氧化碳流量计；4—吸收塔空气泵；5—吸收塔空气流量调节阀；6—吸收塔空气流量计；7—吸收塔混合气温度计；8—吸收塔 U 形管压差计；9—吸收塔；10—吸收塔液体出口取样阀；11—吸收塔液体出口管排净阀；12—吸收塔液体温度计；13—解吸液（吸收后液体）贮槽放净阀；14—解吸液贮槽；15—解吸液泵；16—解吸液泵循环阀；17—解吸塔液体入口取样阀；18—解吸液流量调节阀；19—解吸液流量计；20—吸收液泵；21—吸收塔液体入口取样阀；22—吸收液流量调节阀；23—吸收液流量计；24—吸收液贮槽；25—吸收液泵循环阀；26—吸收液贮槽放净阀；27—解吸塔液体温度计；28—解吸塔液体出口取样阀；29—解吸塔液体出口管排净阀；30—解吸塔；31—解吸塔 U 形管压差计；32—解吸塔解吸气温度计；33—解吸塔解吸气流量计；34—解吸塔解吸气流量调节阀；35—解吸塔解吸气风机旁路阀；36—解吸塔解吸气风机

② 填料：陶瓷拉西环填料，比表面积 $440m^2/m^3$；不锈钢鲍尔环填料，比表面积 $480m^2/m^3$。

③ 流量计：CO_2 转子流量计，LZB-6（0.06～$0.6m^3/h$）；吸收塔空气转子流量计，LZB-10（0.25～$2.5m^3/h$）；水转子流量计，LZB-15（40～400L/h）；解吸塔解吸气转子流量计，LZB-40（4～$40m^3/h$）。

④ 风机：HG-250-C 旋涡式气泵。

⑤ 离心泵：WB50/025。

⑥ 温度计：PT100、AI501 数显仪表

8.5.4 实验步骤与注意事项

8.5.4.1 实验步骤

(1) 流体力学性能实验

流体力学性能实验在解吸塔中进行，主要步骤如下：

① 检查设备、阀门、仪表等是否正常，水箱水位是否合适。

② 打开总电源开关和仪表电源开关，仪表控制面板图 8-13 所示。

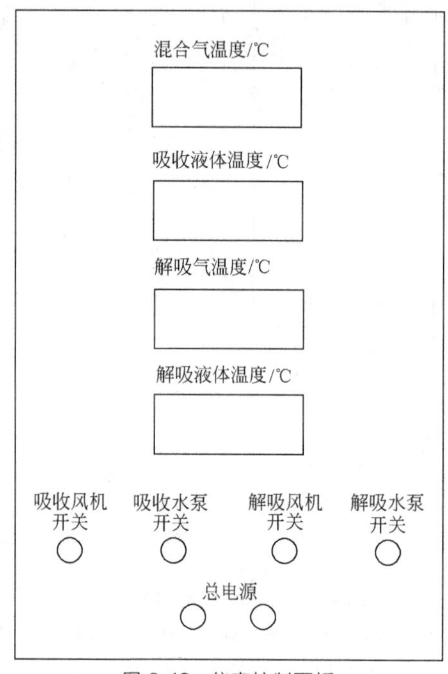

图 8-13 仪表控制面板

③ 开启解吸液泵 15（关闭泵所有出口阀 16、17、18，开启泵 15 电源，而后开启循环阀 16，稳定后开启流量调节阀 18；后续涉及的离心泵启动操作不再赘述），对解吸塔填料进行充分润湿 10～20min（若实验前填料已润湿，可不执行该步骤）。

④ 调节流量调节阀 18 至一定流量（流量为 0 时，测量干填料层压降；流量为一定值时，测量湿填料层压降）。

⑤ 全开解吸塔解吸气风机旁路阀 35，开启风机电源，通过调节阀 34（可以利用旁路阀 35 辅助调节）由小到大调节解吸气流量，测量填料塔压降，并观察解吸塔内气液流动状态，直至液泛。

⑥ 关闭解吸气流量调节阀 34，全开旁路阀 35，关闭解吸气风机（后续涉及的风机关闭操作不再赘述）；关闭解吸液泵流量调节阀 18，关闭循环阀 16，关闭解吸液泵 15（后续涉及的离心泵关闭操作不再赘述）。

(2) 传质性能实验

① 开启吸收液泵 20，调节吸收液流量调节阀 22 至一定流量。

② 开启空气泵 4，通过调节阀 5 调节空气流量至一定值；减压阀关闭的情况下开启二氧化碳钢瓶总阀，而后缓慢打开减压阀，调节二氧化碳流量调节阀 2 至一定流量。

③ 全开解吸气风机旁路阀35，开启风机电源，通过流量调节阀34和解吸气风机旁路阀35将解吸气流量调节至较大值（但不可出现液泛），保证解吸液得到充分的解吸。

④ 开启解吸液泵15，调节解吸液流量调节阀18，使其流量与吸收液的流量保持一致。

⑤ 维持系统温度、流量、液位等参数恒定，待系统稳定后，对吸收塔的进出口液相取样分析。

⑥ 实验完毕，先关闭二氧化碳钢瓶总阀，待二氧化碳流量降为零时，关闭二氧化碳转子流量计和钢瓶减压阀；关闭空气泵4。

⑦ 关闭吸收液流量调节阀22，关闭吸收液泵20；关闭解吸液流量调节阀18，关闭解吸液泵15。

⑧ 全开解吸气风机旁路阀35，关闭解吸气流量调节阀34，关闭解吸气风机36。

⑨ 关闭仪表电源开关和总电源开关，清理实验仪器和实验场地。

(3) 分析操作

① 各液体样品中CO_2的含量采用酸碱滴定法分析。取液体样品20mL，加入0.1mol/L左右的Ba（OH）$_2$溶液5mL，充分反应后，加入酸碱指示液（酚酞）若干滴，利用0.1mol/L左右的盐酸溶液进行滴定分析。

② 按下式计算得出溶液中CO_2的浓度。

$$c_{CO_2}=\frac{2c_{Ba(OH)_2}V_{Ba(OH)_2}-c_{HCl}V_{HCl}}{2V_{样品}}$$

③ 滴定空白样对实验样品分析结果进行校正。

④ 分析测量需测试平行样。

8.5.4.2 注意事项

① 流体力学性能实验达到液泛状态时，注意控制操作，避免大量液体从塔顶夹带出塔，甚至液体溢出塔顶。

② 即将进入液泛时，快速调节流量并测量压力等相关数据。

③ 开启钢瓶总阀前，先确保钢瓶的减压阀处于关闭状态、CO_2气体流量计的调节阀处于关闭状态；开启总阀后，缓慢打开减压阀并调节二氧化碳流量调节阀。

④ 传质性能实验中，保证两水箱水位稳定，避免水箱内的水被抽干；流体力学性能实验也需注意水箱水位。

⑤ 样品取样前要置换管路内的液体，过程迅速并及时将取样瓶盖好，尽量减少与大气的接触，尽快分析。

8.5.5 思考题

① 测定填料塔的$\Delta p/z\text{-}u$曲线有何实际意义？

② 通过水力学性能实验，可知填料塔的液泛与哪些因素有关？

③ 测定总体积传质系数有何工程意义？

④ 为什么二氧化碳吸收过程属于液膜控制？

⑤ 当气体温度和液体温度不同时，应用什么温度计算亨利系数？

⑥ 塔釜液体出口管路上为什么要设置放空管？

⑦ 本实验装置中，吸收塔和解吸塔的液相出口管路都是U形管，它的作用是什么，它的高度如何设计？

⑧ 其他条件不变，液体流量提高，填料层压降如何变化？液体流量越大，泛点气速如何变化？

⑨ 其他条件不变，吸收塔液体流量提高，液体出口组成如何变化？吸收率如何变化？

⑩ 若解吸塔解吸气流量减小，对吸收过程有何影响？

⑪在其他条件相同的情况下，该吸收实验在冬季操作和在夏季操作有何区别？

⑫ 当气体温度和吸收剂温度不同时，应采用哪个温度计算相平衡常数？

⑬ 吸收塔开车时，为什么要先输入吸收剂，而后送入混合气？

⑭ 本实验解吸塔采用的填料是什么填料，通过观察请判断对该塔而言选择该尺寸的填料是否合适？原因是什么？

⑮ 对散装填料而言，一般要求填料润湿率不低于 $0.08m^3/(m \cdot h)$，在实验条件下，请计算说明吸收塔填料的润湿率是否能够满足上述要求？已知润湿率＝液体喷淋密度÷填料比表面积。

⑯ 对吸收传质过程，若已知混合气入塔组成和混合气流量以及操作条件（如温度、压力），规定吸收率要达到10％，那么如何确定吸收剂的流量？

⑰ 散堆填料上方有一段规整填料，它的作用是什么？

⑱ 本实验中两离心泵与水槽间均有一循环管路，离心泵为什么要设置循环管路？

⑲ 填料塔发生液泛时，典型的特征是什么？

⑳ 吸收塔采用陶瓷拉西环，解吸塔使用不锈钢鲍尔环，这两种填料哪种填料的压降大？实验条件下解吸塔的压降比吸收塔的大许多，主要原因是什么？

㉑ 在同样的实验条件下，如果你的实验装置上解吸气的温度明显高于其他实验装置上的解吸气温度，可能的原因是什么（温度测量仪表等正常）？

㉒ 吸收解吸传质实验中，若解吸塔的液体流量低于吸收塔的液体流量，对吸收过程有何影响？

8.6 精馏实验

8.6.1 实验目的

① 了解精馏塔的结构、熟悉精馏单元操作的工艺流程。

② 掌握精馏塔的开车、停车及稳定运行操作方法。

③ 掌握全塔效率和单板效率的测定方法。

④ 了解各操作参数对精馏塔传质分离性能和流体力学性能的影响，掌握精馏塔稳定运行的控制与调节方法。

8.6.2 实验原理

在板式精馏塔中，由塔釜产生的蒸汽沿塔高方向逐板上升，与来自塔顶逐板下降的回流液在塔板上实现多次接触传热与传质，使混合液达到一定程度的分离。塔顶液相回流和塔釜汽相回流是精馏操作得以实现的基础，回流比是精馏操作的重要参数之一，它的大小影响着精馏操作的分离效果和能耗。此外，进料位置、进料浓度、进料量等同样影响精馏操作的分离效果。在塔设备的实际操作中，由于受到传质时间和传质面积的限制，以及其他一些因素的影响，塔板上汽液两相不可能达到平衡状态，因此，实际塔板的分离作用低于理论塔板。故通常采用全塔效率（总板效率）和单板效率分别评价全塔和单个塔板的分离效果。

(1) 全塔效率（总板效率）E

$$E=\frac{N}{N_e} \tag{8-37}$$

式中，E 为全塔效率；N 为理论塔板数（不包括塔釜）；N_e 为实际塔板数。

(2) 单板效率 E_m

单板效率又称为默弗里（Murphree）效率，分为液相单板效率 E_{ml} 和汽相单板效率 E_{mv}。

$$E_{ml,n}=\frac{x_{n-1}-x_n}{x_{n-1}-x_n^*} \tag{8-38}$$

式中，$E_{ml,n}$ 为第 n 块实际板的液相单板效率；x_n，x_{n-1} 分别为第 n 块实际板和第（$n-1$）块实际板的液相组成，摩尔分率；x_n^* 为与第 n 块实际板汽相浓度相平衡的液相组成，摩尔分率。

$$E_{mv,n}=\frac{y_n-y_{n+1}}{y_n^*-y_{n+1}} \tag{8-39}$$

式中，$E_{mv,n}$ 为第 n 块实际板的汽相单板效率；y_n，y_{n+1} 分别为第 n 块实际板和第（$n+1$）块实际板的汽相组成，摩尔分率；y_n^* 为与第 n 块实际板液相浓度相平衡的汽相组成，摩尔分率。

其中，在全回流条件下，由于操作线为 $y_{n+1}=x_n$，所以单板效率的测量与计算可得到简化：

$$E_{ml,n}=\frac{x_{n-1}-x_n}{x_{n-1}-x_n^*} \tag{8-40}$$

$$E_{mv,n}=\frac{y_n-y_{n+1}}{y_n^*-y_{n+1}}=\frac{x_{n-1}-x_n}{y_n^*-x_n} \tag{8-41}$$

式中 $x_n^*=f(y_n)=f(x_{n-1})$，$y_n^*=f(x_n)$，因此，在全回流条件下，通过上下两块塔板的液相组成即可测量塔板的液相与汽相默弗里板效率。

(3) 理论塔板数 N

理论塔板数可采用逐板计算法或图解法计算，以乙醇-水体系为例，由于乙醇-水体系为高度非理想体系，本文采用图解法计算理论塔板数，其基本原理与逐板计算法类似，全回流和部分回流条件下图解法分别如图 8-14 和图 8-15 所示。

1）全回流条件下理论塔板数的计算　全回流条件下，操作线与 x-y 相图上对角线重合，故只需确定塔顶组成 x_D 与塔釜组成 x_W 即可在 x-y 相图上平衡线与对角线之间进行逐板计算即可得到理论塔板数。

2）部分回流条件下理论塔板数的计算　部分回流条件下，需要在平衡线与操作线之间进行逐板计算，因此，首先需确定操作线。如图 8-15 所示，在确定塔顶与塔釜组成之后，可通过精馏段操作线方程和进料方程作图。

$$y=\frac{R}{R+1}x+\frac{x_D}{R+1} \tag{8-42}$$

$$y=\frac{q}{q-1}x-\frac{x_F}{q-1} \tag{8-43}$$

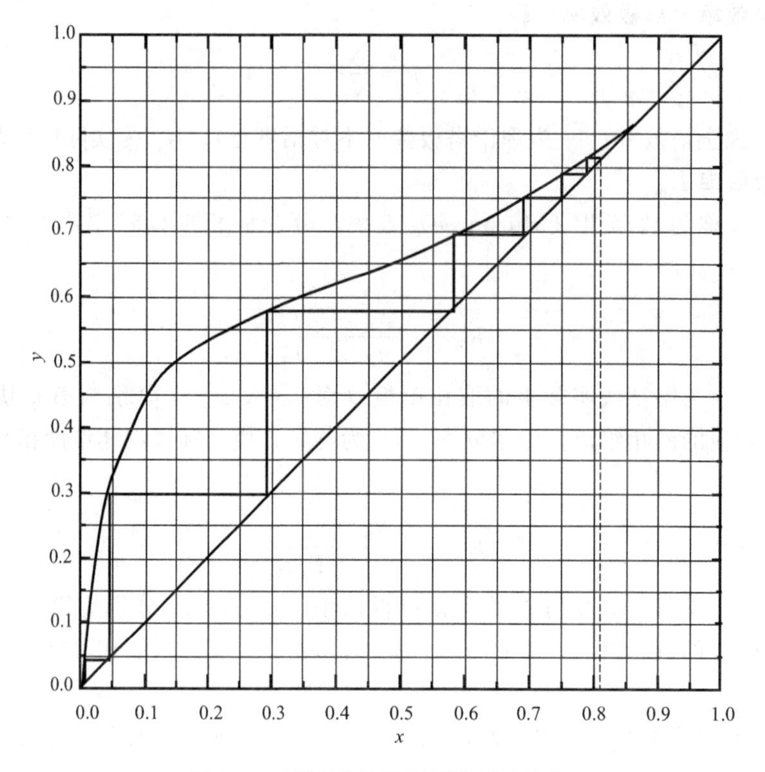

图 8-14　全回流条件下理论塔板数的计算

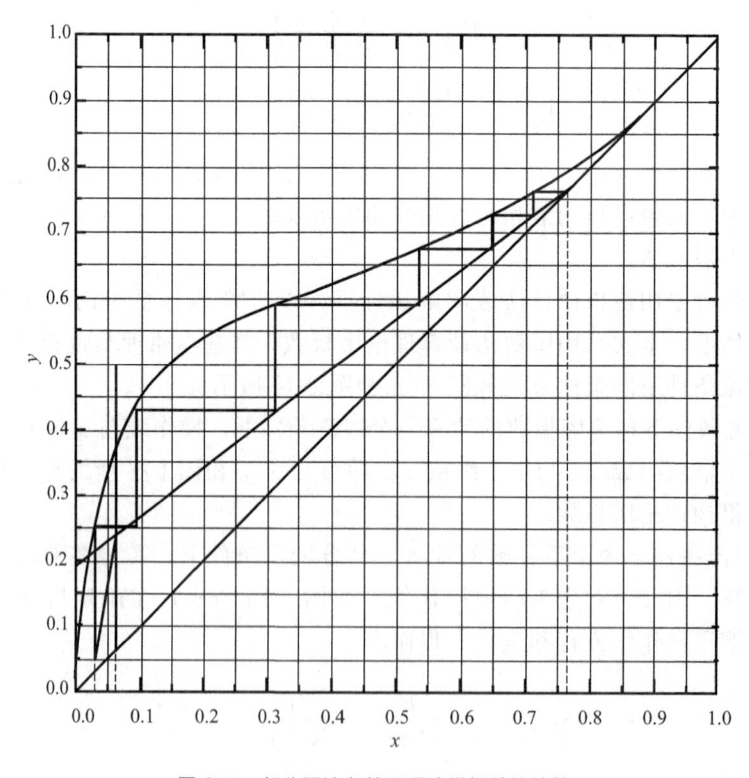

图 8-15　部分回流条件下理论塔板数的计算

在相图上确定精馏段操作线和 q 线，而后连接这两条操作线交点和对角线上塔釜组成 x_W 对应的状态点即可确定提馏段操作线。

可见上述操作线的确定关键在于回流比 R 和 q 值的计算。由于进料液体状态为冷液，根据进料口的热量衡算可得

$$q=1+\frac{c_{pF}(t_S-t_F)}{r_F} \tag{8-44}$$

式中，q 为进料热状况参数；r_F 为进料液组成下的汽化潜热，kJ/kmol；c_{pF} 为进料液在平均温度（t_S+t_F）/2 下的比热容，kJ/（kmol·℃）；t_S 为进料液的泡点温度,℃；t_F 为进料液温度,℃。

同理，当冷液回流时，过冷液体回流入塔后会将塔内上升的蒸汽部分冷凝，从而使实际向下流动的液体量增大，从而提高了塔内的汽液两相摩尔流量之比（塔内汽液两相的流量分别为 $\bar{L}$、$\bar{V}$）

$$\frac{\bar{L}}{\bar{V}}=\frac{R'}{R'+1}$$

即相当于改变了回流比（此时用 R'表示，数值上大于 $R=L/D$）。因此，式（8-42）精馏段操作线中的回流比 R 应更换为实际回流比 R'，塔内实际回流比为

$$R'=\frac{L'}{D}=\frac{L}{D}\left[1+\frac{c_p(t_1-t_R)}{r}\right]=R\left[1+\frac{c_p(t_1-t_R)}{r}\right] \tag{8-45}$$

式中，R'为实际回流比；R 为泡点回流回流比，即回流液与采出液流量之比 L/D；c_p 为回流液在平均温度（t_1+t_R）/2 下的平均比热容，kJ/（kmol·℃）；t_1 为塔顶饱和蒸汽温度,℃；t_R 为回流液温度,℃；r 为塔顶饱和蒸汽冷凝潜热，kJ/mol。

其中，精馏体系主要物性如下。

乙醇水溶液比热容：

$c_p=$（$1.0365-1.3485\times10^{-3}w-9.3326\times10^{-4}t-4.3944\times10^{-5}w^2+3.863\times10^{-5}wt+9.962\times10^{-6}t^2$）$\times4.187$　[c_p，kJ/（kg·℃）；w，乙醇质量分数,%；t,℃]

乙醇水溶液汽化潜热：

$r=(4.745\times10^{-4}w^2-3.315w+5.3797\times10^{-2})\times4.187$　[r,kJ/kg;w,乙醇质量分数,%]

乙醇汽化潜热：$r=-0.0042t^2-1.5074t+985.14$（$r$，kJ/kg；$t$,℃）

正丙醇汽化潜热：$r=-0.0031t^2-1.1843t+839.79$（$r$，kJ/kg；$t$,℃）

乙醇比热容：$c_p=4.3357\times10^{-5}t^2+0.00621t+2.2332$ [c_p，kJ/（kg·℃）；t,℃]

正丙醇比热容：$c_p=-8.3528\times10^{-7}t^3+1.2144\times10^{-5}t^2+0.00365t+2.222$ [c_p，kJ/（kg·℃）；t,℃]

8.6.3 实验装置与流程

(1) 装置一实验流程

原料罐中乙醇水溶液由进料泵加压后进入釜液冷却器，与塔釜采出的釜液进行热交换，经预热升温的原料液进入精馏塔，精馏塔塔釜内的液体由电加热器加热汽化后逐板上升，与各塔板上的液体传热传质，最后进入精馏塔塔顶经盘管式冷凝器冷凝，冷凝液从集

液器流出，一部分作为塔顶产品馏出，进入产品罐，另一部分作为回流液从塔顶流入塔内，逐板自上而下流动，最终进入塔釜进行部分汽化，产生的汽体沿塔高方向向上流动，塔底残液经釜液冷却器冷却后进入釜液罐。具体实验流程如图 8-16 所示。

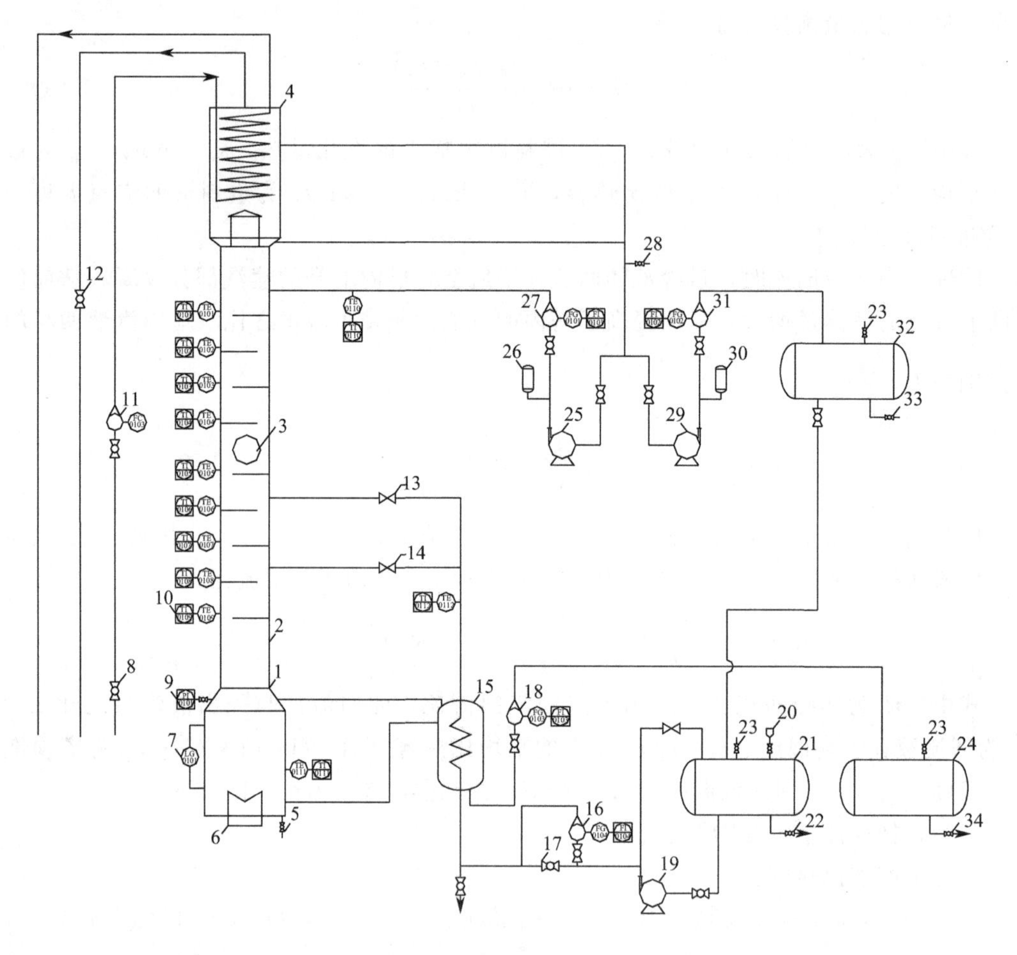

图 8-16 精馏实验装置流程图（装置一）

1—塔釜；2—精馏塔；3—精馏塔视盅；4—塔顶冷凝器；5—塔釜取样阀；6—塔釜加热管；7—塔釜液位计；8—冷却水进水阀；9—塔釜压力表；10—温度表；11—冷却水流量计；12—塔顶放空阀；13—上进料阀；14—下进料阀；15—釜液冷却器；16—进料流量计；17—快速进料阀；18—釜液流量计；19—进料泵；20—原料罐加料口；21—原料罐；22—原料罐排净口及取样阀；23—放空阀；24—釜液罐；25—回流泵；26—回流泵缓冲罐；27—回流液流量计；28—塔顶取样阀；29—产品泵；30—产品泵缓冲罐；31—产品流量计；32—产品罐；33—产品罐排净口；34—釜液罐排净口

(2) 装置一设备主要技术数据

筛板塔主要结构参数：塔内径 $D=68\text{mm}$，厚度$\delta=2.5\text{mm}$，塔节 $\phi 73\text{mm}\times 2.5\text{mm}$，塔板数 $N=10$ 块，板间距 $H_T=100\text{mm}$。加料位置由上向下数第 6 块和第 8 块。降液管为弓形，降液管底隙高度 4.5mm。溢流堰为齿形堰，堰长 56mm、堰高 7.3mm、齿深 4.6mm，齿数 9 个。筛孔直径 $d_0=1.5\text{mm}$，正三角形排列，孔间距 $t=5\text{mm}$，开孔数为 54 个。塔釜为内电加热式，加热功率 2.5kW，有效容积为 10L。塔顶冷凝器、釜液冷却器均为盘管式。单板取样为自上而下第 1 块、第 2 块、第 9 块和第 10 块。

原料罐：ϕ250mm×540mm，卧式；产品罐：ϕ200×520mm，卧式；釜液罐：ϕ200×520mm，卧式。

(3) 装置二实验流程

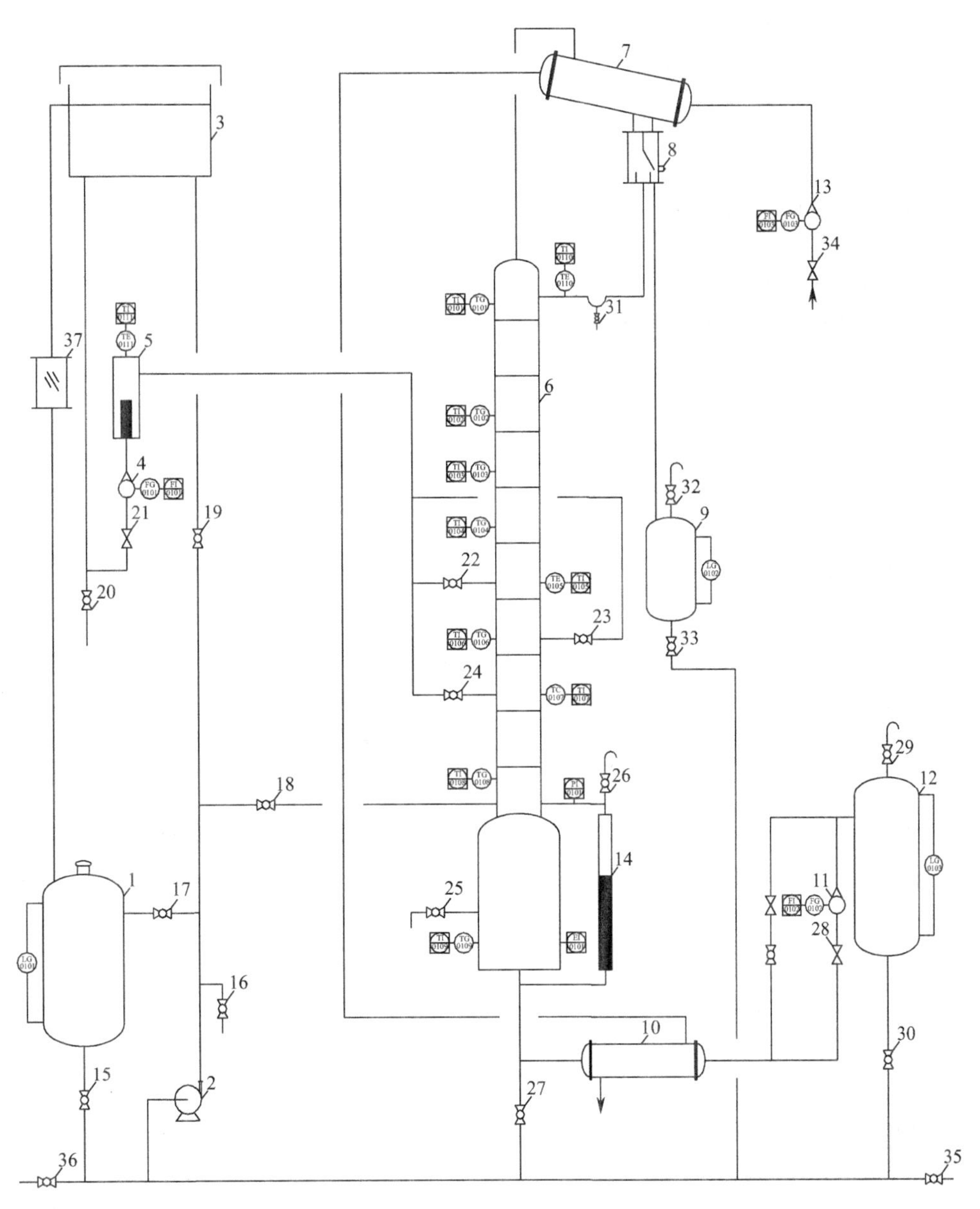

图 8-17　精馏实验装置流程图（装置二）

1—原料罐；2—进料泵；3—原料高位槽；4—进料流量计；5—进料预热器；6—精馏塔；7—塔顶冷凝器；8—回流比控制器；9—塔顶产品罐；10—塔釜残液冷却器；11—塔釜残液流量计；12—塔釜残液罐；13—冷却水流量计；14—磁翻板液位计；15—原料罐底阀；16—阀门；17—原料泵循环阀；18—塔釜快速进料阀；19—高位槽进料阀；20—原料取样阀；21—进料总阀；22—进料阀1；23—进料阀2；24—进料阀3；25—塔釜取样阀；26—塔釜放空阀；27—塔釜底阀；28—塔釜残液出料阀；29—残液罐放空阀；30—残液罐底阀；31—塔顶产品取样阀；32—塔顶产品罐放空阀；33—塔顶产品罐底阀；34—冷却水阀；35，36—放净阀；37—玻璃观测罐

（4）装置二设备主要技术数据

筛板塔主要结构参数：塔内径 $D=50$mm，塔板数 $N=9$ 块，板间距 $H_T=120$mm。降液管为圆形，降液管底隙高度 4.5mm。塔釜为电加热式，加热功率 2.5kW。塔顶冷凝器、釜液冷却器均为管式换热器。

原料罐：ϕ300mm×400mm，立式；高位槽：200mm×100mm×200mm；产品罐：ϕ150mm×260mm，立式；釜液罐：ϕ150mm×260mm，立式；塔顶冷凝器：ϕ89mm×600mm，卧式；塔釜冷却器：ϕ76mm×200mm，卧式。

8.6.4 实验步骤与注意事项

8.6.4.1 实验步骤

（1）实验步骤（适用于装置一）

1）全回流

① 配制浓度 10%～25%（酒精度，即体积分数）的乙醇水溶液加入原料罐 21 中。

② 打开总电源开关和仪表电源开关。开启塔顶放空阀 12，按照离心泵启动操作规范启动进料泵 19，打开进料管路上的阀门 14 和快速进料阀 17，将料液打入精馏塔塔釜，至塔釜液位计刻度约 30～35cm 处，关闭进料阀门 14 和 17，并关闭进料泵。

③ 关闭塔顶出料管路和塔釜出料管路，启动电加热管电源，调节加热电压至适当位置，使塔釜温度缓慢上升。

④ 在蒸汽上升至塔顶之前打开冷却水进水阀 8，调节冷却水流量计 11 至合适的冷却水流量。

⑤ 待塔顶出现冷凝液后，将回流液流量计阀门 27 全开，启动回流泵 25，使精馏塔处于全回流状态。

⑥ 调节塔釜加热功率，使回流量处于适当状态（观察塔板上汽液接触状态），当塔顶温度和塔釜温度稳定后，记录各实验数据并分别从塔顶取样阀 28 和塔釜取样阀 5 处取样，分析塔顶组成 x_D 和塔釜组成 x_W。

⑦ 自上而下第 1 块、第 2 块、第 9 块和第 10 块塔板处分别取样测定单板效率，取样用注射器从对应塔板的取样口中缓缓抽出液体，取 1mL 左右注入洗净烘干的取样瓶中，各个样品尽可能同时取样。

2）部分回流

① 待精馏塔全回流操作稳定时，打开进料阀 13 或 14，通过进料流量计 16 将进料量调节至适当的值。

② 控制回流泵 25 和产品泵 29 的行程并通过流量计 27 和 31 调节回流液和塔顶馏出液的流量，使回流比保持在某一设定值。

③ 调节流量计 18 控制塔釜残液的排出量，维持全塔进出物料的平衡。

④ 当塔顶及各塔板温度稳定后，记录各实验数据并分别从塔顶取样阀 28、塔釜取样阀 5 和原料罐排净口及取样阀 22 处取样，分析塔顶浓度 x_D、塔釜浓度 x_W 和进料浓度 x_F。

⑤ 实验完毕，停止塔釜加热，关闭进料泵、进料阀，关闭出料泵、出料阀，待塔内无蒸汽上升且塔温下降至适当温度后，关闭冷却水流量计阀门和进水阀。

（2）实验步骤（适用于装置二）

1）全回流

① 配制一定浓度的乙醇-正丙醇溶液加入原料罐 1 中。

② 打开总电源开关和仪表电源开关。按照离心泵启动操作规范启动进料泵 2，打开原料泵循环阀 17 和塔釜快速进料阀 18，以及塔釜放空阀 26，将料液打入精馏塔塔釜，至塔釜液位 2/3 处，关闭阀门 17、18 和 26，并关闭进料泵。

③ 关闭塔顶出料管路和塔釜出料管路，启动塔釜加热电源，调节加热电压至适当位置，使塔釜温度缓慢上升，同时打开压力表阀门。

④ 在蒸汽上升至塔顶之前打开冷却水阀 34，调节冷却水流量计 13 至合适的冷却水流量（80～120L/h）。

⑤ 待塔顶出现冷凝液后，调节塔釜加热功率，使回流量处于适当状态（观察塔板上汽液接触状态），当塔顶温度和塔釜温度稳定后，记录各实验数据并分别从塔顶产品取样阀 31 和塔釜取样阀 25 处取样，分析塔顶组成 x_D 和塔釜组成 x_W。

⑥ 根据需要在精馏塔相应塔板处取样测定单板效率，各个样品尽可能同时取样。

2）部分回流

① 待精馏塔全回流操作稳定时，启动进料泵 2，打开阀门 17、19，将原料液打入高位槽 3，待玻璃观测罐中观察到有稳定流体通过时，选择开启进料阀 22、23 和 24 中之一，而后开启进料总阀 21，调节进料流量计 4 至适当的值。

② 开启回流比控制器，使回流比保持在某一设定值。

③ 开启塔顶产品罐放空阀 32 以及塔釜残液罐放空阀 29，开启塔釜残液出料阀 28 并调节塔釜残液流量计 11，维持全塔进出物料的平衡及塔釜磁翻板液位 14 稳定。

④ 当塔顶及各塔板温度稳定后，记录各实验数据并分别从塔顶产品取样阀 31、塔釜取样阀 25 和原料液取样阀 20 处取样，分析塔顶浓度 x_D、塔釜浓度 x_W 和进料浓度 x_F。

⑤ 实验完毕，停止塔釜加热，关闭进料泵、进料阀，关闭出料阀，待塔内无蒸汽上升且塔温下降至适当温度后，关闭冷却水流量计阀门和进水阀，关闭其他阀门。

8.6.4.2　注意事项

① 确认塔顶放空阀打开，否则无法保证精馏塔在常压条件下操作，也易使塔内压力过大而导致危险。

② 料液一定要加到设定液位 2/3 处方可打开塔釜加热电源，否则塔釜液位过低会导致电加热丝干烧而损坏。

③ 进料泵运行中，进料泵与原料罐之间的循环管路阀门应打开，便于维持原料罐组成均匀。

④ 进料量、回流量、产品量、釜液的排出量的调节要合适，使进出塔中的物料保持平衡。

⑤ 实验过程注意塔釜加热功率缓慢调节，避免发生爆沸。

⑥ 开车时要先接通冷却水再向塔釜供热，或者先向塔釜供热并在蒸汽进入塔顶前接通冷却水，停车时应先关闭塔釜加热电源，待系统冷却后关闭塔顶冷却水。

8.6.5　思考题

① 板式塔汽液两相的流动特点是什么？

② 塔板效率受哪些因素影响？

③ 精馏塔操作中，塔釜压力为什么是一个重要操作参数？塔釜压力与哪些因素有关？

④ 什么叫回流比？精馏中为什么要引入回流比？试说明回流的作用，如何根据实验装置情况，确定和控制回流比？

⑤ 操作中提高回流比的方法是什么？能否采用减少塔顶出料量 D 的方法？

⑥ 其他条件不变，只改变回流比，对精馏塔分离性能会产生什么影响？

⑦ 进料位置是否可以任意选择？它对精馏塔分离性能会产生什么影响？

⑧ 精馏塔在操作过程中，由于塔顶采出率太大而造成产品不合格，恢复正常的最快、最有效的方法是什么？

⑨ 为什么该体系的精馏采用常压操作而不采用加压精馏或真空精馏？

⑩精馏塔的常压操作是怎样实现的？如果要改为加压或减压操作，又怎样实现？

⑪测定全回流和部分回流全塔效率与单板效率时各需测哪几个参数？取样位置在何处？

⑫全回流时测得板式塔上第 n、$n-1$ 层液相组成后，如何求得 x_n^*；部分回流时，又如何求 x_n^*？

⑬在全回流时，测得板式塔上第 n、$n-1$ 层液相组成后，能否求出第 n 层塔板上的汽相单板效率？

⑭精馏塔的塔顶馏出液量应如何确定比较合理？若实际塔顶馏出液量与回流量和塔釜加热量是否需要相应调整？如何调整？

⑮什么是灵敏板和灵敏板温度？灵敏板温度对精馏塔的操作有何意义？

⑯操作过程中若塔体温度整体上升，则可能的原因是什么？需采取什么措施？

⑰操作过程中若塔釜液位持续下降，可能的原因有哪些？要恢复正常，需相应采取什么措施？

⑱若操作过程中出现液泛或严重漏液，则需如何操作使精馏塔恢复正常？

⑲当进料量和进料组成一定时，若要求精馏塔塔顶的组成不低于 x_D^0，则精馏塔的塔顶采出量应如何确定？

8.7 干燥实验

8.7.1 实验目的

① 了解洞道式干燥器的结构和干燥工艺流程。

② 掌握干燥设备的开车、停车及稳定运行操作方法。

③ 掌握干燥曲线和干燥速率曲线的测定方法及其影响因素。

④ 了解干燥速率曲线在工业干燥器设计中的意义。

8.7.2 实验原理

干燥是一种利用热量去湿（常见湿分为水分）的单元操作，它不仅涉及气、固两相间的传热与传质，而且涉及湿分以气态或液态的形式自物料内部向表面传质机理。由于物料所含水分性质和物料形状结构的差异，以及干燥条件的不同，物料中水分干燥速率差别很大。总体而言，干燥速率受到物料性质、结构及其含水性质、干燥介质（通常为湿空气）

的状态（如温度、湿度）、流速、干燥介质与物料的接触方式等各种因素的影响。因为干燥速率的影响因素多而复杂，所以干燥速率主要通过实验测定。

在恒定干燥条件下，即干燥介质的温度、湿度、流速及干燥介质与湿物料的接触方式恒定不变，将湿物料置于干燥介质中测定被干燥湿物料干基含水量、温度随干燥时间的变化，以及干燥速率随干燥时间的变化，即可分别得到如图 8-18、图 8-19 所示的干燥曲线和干燥速率曲线。在恒定干燥条件下，干燥过程可分为三个阶段，即预热阶段、恒速干燥阶段、降速干燥阶段（加热阶段）。湿物料进入干燥器与湿空气接触被预热升温，在这一预热阶段（$A \rightarrow B$），物料温度逐渐上升，同时干燥速率也逐渐提高，在实际干燥过程中这一阶段的时间较短，在工程计算中预热阶段常被并入恒速干燥阶段；若物料表面能够保持湿润状态，经过预热段后，干燥将进入恒速干燥阶段（$B \rightarrow C$），在这一阶段湿物料温度恒定于空气状态所对应的湿球温度，此时传热、传质的推动力恒定不变，湿空气传给湿物料的热量全部用于湿物料中水分的蒸发汽化，其干燥速率较大且恒定不变，故称为恒速干燥阶段，恒速干燥阶段的干燥速率只与湿空气的状态、湿空气流动状况及其与物料接触方式有关，而与湿物料的性质无关，该阶段又称为“表面蒸发汽化控制阶段”；当湿物料内部水分向物料表面扩散的速率低于物料表面水分蒸发速率时，湿物料将无法继续保持湿润状态，物料表面出现部分“干区”，干燥速率下降，此时空气传给物料的热量较水分汽化所需潜热多，导致物料被加热升温，干燥过程进入降速干燥阶段（$C \rightarrow D$），当表面水分蒸发干以后，干燥速率将进一步显著下降（$D \rightarrow E$），直至干燥速率为零，物料的水分含量达到平衡含水量，该阶段的干燥速率主要取决于物料本身的性质，物料的结构形状大小，而与干燥介质的性质无关，故降速干燥阶段又称为“内部水分汽化过程扩散控制阶段”。

单位时间内，单位的物料表面积所蒸发的水分为干燥速率，即：

$$U=\frac{\mathrm{d}W}{A\,\mathrm{d}\theta} \tag{8-46}$$

式中，U 为干燥速率，kg/（m^2 · h）；W 为干燥过程蒸发水分量，kg；A 为干燥面积，m^2；θ 为干燥时间，s。

因为 $\mathrm{d}W=-G_C\mathrm{d}X$，且为了实验测量的可操作性，用差商 $\Delta X/\Delta\theta$ 代替导数 $\mathrm{d}X/\mathrm{d}\theta$，所以

$$U=-\frac{G_C\mathrm{d}X}{A\,\mathrm{d}\theta}=-\frac{G_C\Delta X}{A\,\Delta\theta} \tag{8-47}$$

式中，G_C 为绝干物料量，kg；X 为干基含水量，kg 水/ kg 绝干物料。

干燥过程中，通过测量一定时间间隔 $\Delta\theta$ 内起始和终了湿物料的质量，即可确定该时间间隔内的平均干燥速率。

$$U=-\frac{G_C\Delta X}{A\,\Delta\theta}=\frac{G_C}{A\,\Delta\theta}\left(\frac{G_{S,i}-G_C}{G_C}-\frac{G_{S,i+1}-G_C}{G_C}\right)=\frac{G_{S,i+1}-G_{S,i+1}}{A\,\Delta\theta} \tag{8-48}$$

式中，$G_{S,i}$ 为时间间隔 $\Delta\theta$ 起始时刻湿物料的质量，kg；$G_{S,i+1}$ 为时间间隔 $\Delta\theta$ 终了时刻湿物料的质量，kg。

在时间间隔 $\Delta\theta$ 内，恒速干燥阶段的干燥速率恒定不变，而降速干燥阶段的干燥速率则随时间逐渐减小，所以在时间间隔 $\Delta\theta$ 内，湿物料的干基含水量宜采用 $\Delta\theta$ 内干基含水量的平均值表示。

$$\overline{X}=\frac{X_i+X_{i+1}}{2}=\frac{1}{2}\left(\frac{G_{S,i}-G_C}{G_C}+\frac{G_{S,i+1}-G_C}{G_C}\right)=\frac{G_{S,i+1}+G_{S,i+1}}{2G_C}-1 \tag{8-49}$$

根据式（8-48）和式（8-49）计算的干燥速率U和平均干基含水量$\bar{X}$，可获得干燥速率曲线，若将每个时间间隔$\Delta\theta$的起始和终了时刻的干燥时间θ_i和其干基含水量X_i（$X_i = G_{S,i}/G_C - 1$）作图，即可得到干燥曲线。

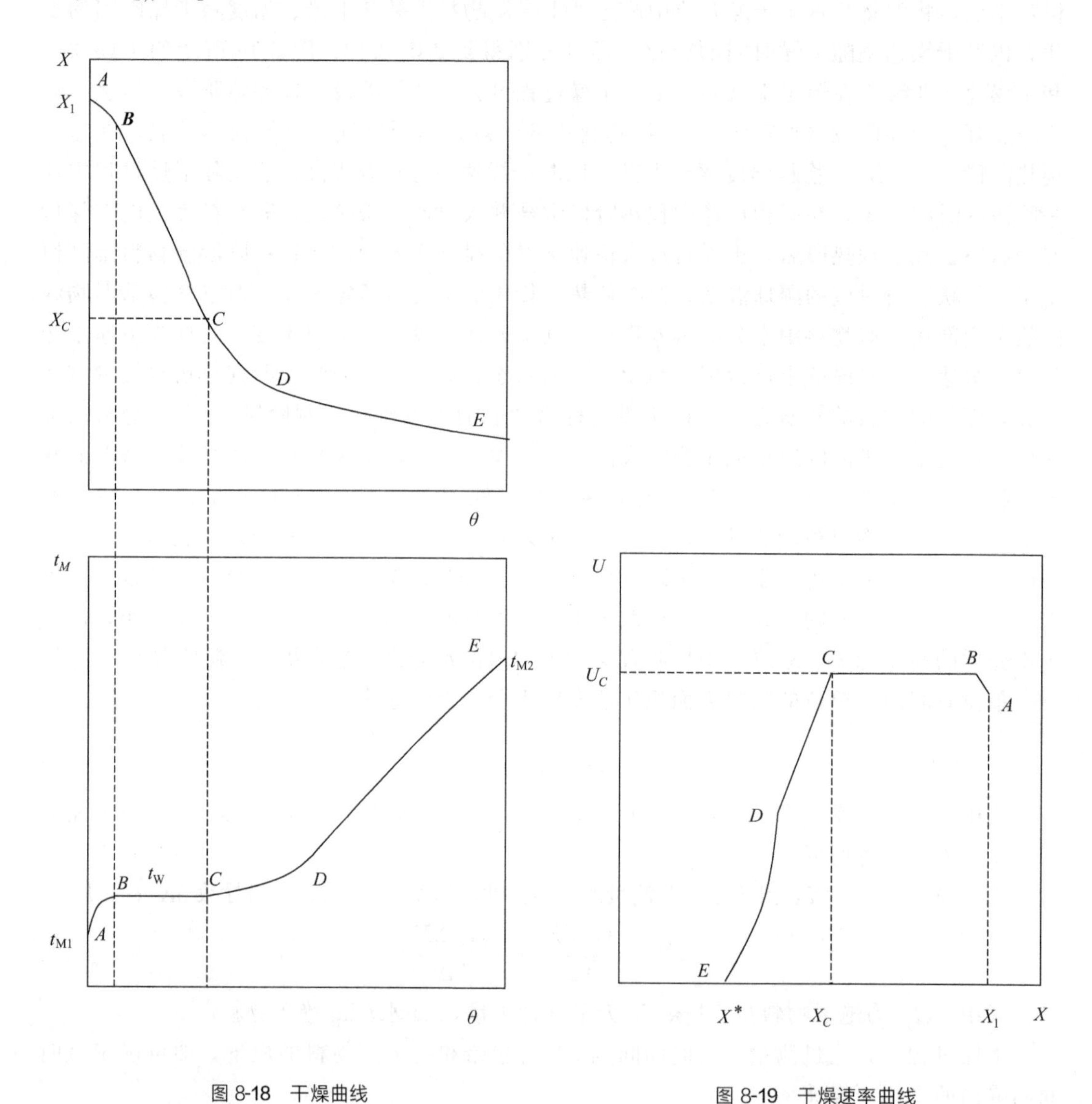

图 8-18 干燥曲线

图 8-19 干燥速率曲线

8.7.3 实验装置与流程

本实验的干燥设备为洞道式循环干燥器，在恒定干燥条件下干燥块状物料。实验装置流程如图 8-20 所示。空气由风机 1 送入管路，经孔板流量计 3 计量和电加热器 4 加热后进入干燥室，加热干燥室中置于试样架上的湿物料后，经排出管道部分返回到风机入口，部分通入大气中。干燥室前方装有干球温度计和湿球温度计。随着干燥过程的进行，物料失去的水分量由称重传感器转化为电信号，并显示于智能数显仪表。空气的风量和温度通过智能数显仪表调节、控制和显示。

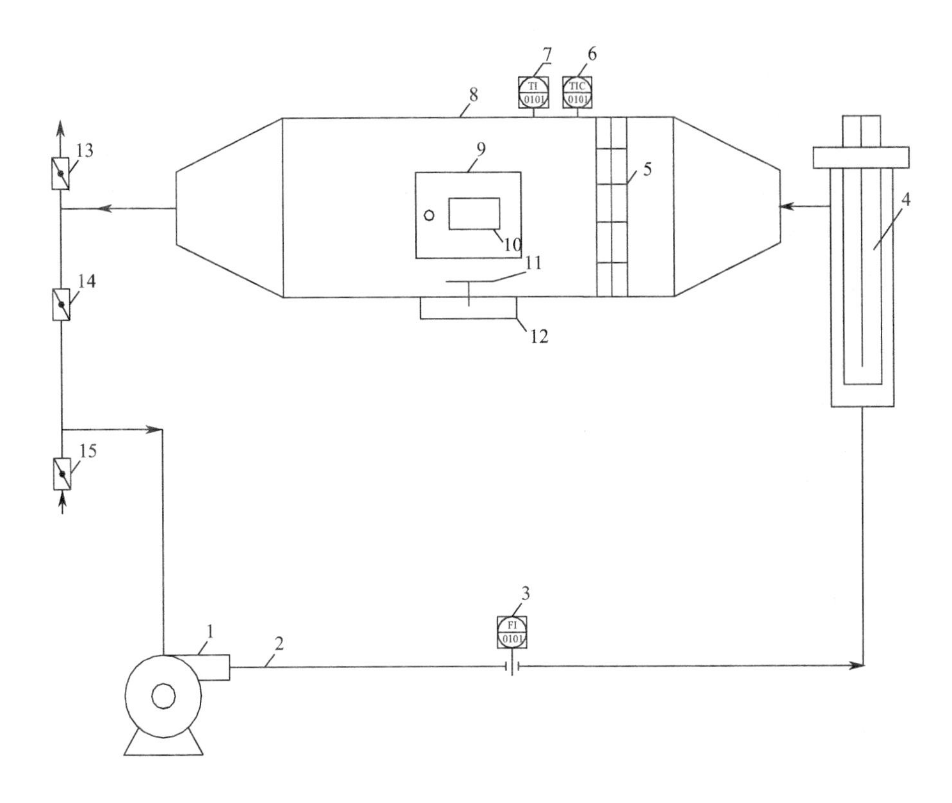

图8-20　干燥实验装置流程图

1—风机；2—管道；3—孔板流量计；4—电加热器；5—气流均布器；6—干球温度计；
7—湿球温度计；8—洞道式干燥器；9—干燥室门；10—玻璃视镜；11—试样架；
12—称重传感器；13～15—蝶阀

设备规格如下。

① 离心式低噪声中压吹风机：CZT-75，370W；

② 电加热器：额定功率4.5kW；

③ 干燥室：180mm×180mm×1250mm；

④ 干燥物料：硬纸板；

⑤ 称重传感器：L6J8型，0～500g。

8.7.4　实验步骤与注意事项

8.7.4.1　实验步骤

① 实验前量取试样尺寸确定干燥面积，并称量绝干试样的质量。

② 打开总电源和仪表电源开关，启动风机，风量大小可通过智能数显仪表调节、控制和显示。

③ 打开加热电源开关，加热器通电加热，加热温度可通过智能数显仪表调节、控制和显示。

④ 在设备背面的U形漏斗中加入一定量的水。

⑤ 调节蝶阀13～15的开度，以保持空气流速、温度和湿度的恒定。

⑥ 关注干球温度，达到预定干燥温度（例如70℃）并稳定后方可进行实验。

⑦ 将试样放入水中浸泡片刻，让水分均匀润湿整个试样，注意含水量不能过多或过少。

⑧ 当干燥室温度恒定时，打开干燥室门，将夹有湿试样的试样架小心放置于称重传感器上，关闭干燥室门，记录湿试样的初始质量。

⑨ 记录每蒸发一定质量水分（例如2g）所需要的干燥时间，直至试样接近平衡水分为止。

⑩ 实验结束，先关闭加热电源开关，待系统冷却后依次关闭风机、仪表电源和总电源，小心取下干燥试样架；收拾整理实验现场。

8.7.4.2 注意事项

① 必须先开风机，后开加热器，否则加热管可能会被烧坏；实验结束时必须先关闭加热器，待系统冷却后再关闭风机。

② 放置干燥物料时应特别注意不能用力下压，因称重传感器的测量上限仅为500g，用力过大容易损坏称重传感器。

③ 实验过程中，不要拍打、碰撞设备，以免引起试样架晃动而影响测量结果。

8.7.5 思考题

① 测定干燥速率曲线有何意义？它对设计干燥器及指导生产有些什么帮助？

② 何为对流干燥？干燥介质在对流干燥过程中的作用是什么？

③ 为什么说干燥过程是一个传热和传质过程？

④ 如何判断干燥条件已稳定，可以开始进行实验？

⑤ 为什么要先开吹风机送气，而后再开电加热器？

⑥ 测定湿球温度时，若水的初始温度不同，对测定的结果是否有影响？为什么？

⑦ 实验过程中干球温度计、湿球温度计是否变化？为什么？

⑧ 若实验过程中湿球温度计中的水干了，则湿球温度计的读数如何变化？为什么？

⑨ 若提高空气流量或空气温度，恒速干燥速率、临界湿含量、平衡含水量如何变化？为什么？

⑩ 影响干燥速率的因素有哪些？若要提高干燥强度，应采取哪些措施（结合本实验装置进行考虑）？

⑪什么是临界含水量？简述干燥中的临界含水量受哪些因素的影响？

⑫本实验是否有废气循环，废气循环对干燥过程有什么影响？干燥热敏性物料或易变形、开裂的物料为什么多使用废气循环？

⑬在70～80℃的空气流中干燥，并经过相当长的时间，能否得到绝干物料？为什么？

⑭有一些物料在热气流中干燥，希望热气流相对湿度要小，而有一些物料则要在相对湿度较大些的热气流中干燥，这是为什么？

⑮如何判断实验可以结束了？

⑯试分析恒速干燥阶段与降速干燥阶段的干燥机理。

⑰恒定干燥条件是什么？请结合干燥速率$U=(\alpha/r_w)(t-t_w)$说明为什么在恒定干燥条件下恒速干燥阶段的干燥速率为常数。

⑱把管路上的蝶阀14的开度关小，对实验所测量的干燥速率曲线有什么影响？

⑲根据实验结果分析不同厚度物料的干燥过程（干燥速率、临界含水量、平衡含水量）有何区别。

第9章 化工原理自组装实验

9.1 自组装实验目的

目前为止化工原理的演示实验，设计型、研究型、综合型、提高型实验等均是在固定的实验装置上完成相对固定的实验项目，实验任务、实验内容、操作步骤等相对固定，甚至实验报告中对实验数据的分析和讨论内容也相对固定。这在一定程度上制约了学生学习主体作用的发挥和实践动手能力、创新能力、工程思维的培养。随着新工科、一流课程等建设工作的推进，我国经济、科技的快速发展对人才提出了更高的要求。基于此，开设了流体阻力、泵性能测试、传热过程强化、列管换热四个自组装实验。其特点和目的在于以下几个方面。

（1）无指定的实验课题

学生根据自己的学习需求，独立自主地制订实验研究课题，鼓励学生在传统实验项目之外树立学习主体地位、培养自我学习能力。

（2）无固定的实验装置

实验室提供实验必要的实验装置基本框架和实验所需的输送机械、管道、管件、测量仪表等配件和安装工具等。由学生根据研究课题的需要自行完成实验装置的组装，提高学生对化工设备、仪器、仪表的认识，提升学生实践动手能力。

（3）无具体的教学讲义

实验教学讲义通常会提供实验内容、实验装置流程、实验原理、实验步骤、实验注意事项等信息，学生可按照讲义按部就班地开展实验工作，过程较为机械，缺乏独立思考，实验过程对学生综合素质与能力的考察较弱。自组装实验的实验课题、实验装置的设计、实验研究方案的制订、实验工作的实施、实验过程突发问题的解决、实验研究报告内容与撰写等各个环节均由学生独立自主完成，可强化学生工程研究素质与能力的综合训练，强调实验教学的过程性评价。

9.2 自组装实验要求

① 自组装实验需要在完成教学大纲要求的实验项目后进行；

② 实验之前，学生需进入实验室充分了解实验室具备的条件（包括实验设备仪器及相关配件等），仔细阅读实验室安全操作规程和注意事项，并了解各种工具的安全使用方法；

③ 确定实验研究课题、研究目标、完成实验装置的设计、实验研究方案的制订、实验过程可能出现问题的解决预案、团队分工协作情况等，经指导教师批准后方可进入实验室开展实验工作；

④ 实验过程严格遵守实验室各项规定，特别注意设备组装和拆卸过程的安全；

⑤ 实验完成后，提交完整的实验研究报告。

9.3 自组装实验主要设备、仪表与配件

9.3.1 流体流动阻力自组装实验

流体流动阻力自组装实验主要设备、仪表和配件如表 9-1 所列。

表 9-1 流体流动阻力自组装实验主要设备、仪表与配件

设备、仪表或配件名称	规格
不锈钢框架	2.50m×0.55m×1.85m
离心泵	MS100/0.55SSC 型，不锈钢，0.55kW，7.2m^3/h，H=12.2m
光滑管	不锈钢，DN8，1.7m
	不锈钢，DN10，1.7m
粗糙管	不锈钢，DN15，1.7m
	不锈钢，DN25，1.7m
	不锈钢，DN32，1.7m
	不锈钢，DN40，1.7m
闸阀	不锈钢，DN15
球阀	不锈钢，DN15
截止阀	不锈钢，DN15
针阀	不锈钢，DN15
蝶阀	不锈钢，DN15
转子流量计	LZB－40
	LZB－25
	VA10－15F
涡轮流量计	LWGY 型，DN25
孔板流量计	d_o=15mm
文丘里流量计	d_o=15mm
变频器	0～50Hz
玻璃倒 U 形管压差计	—
压差传感器	WNK3051 型，0～200.0 kPa
温度数字显示仪	—
流量数字显示仪	—
压差数字显示仪	—
玻璃管液位计	—
热电阻温度计	Pt100
水箱	400mm×420mm×600mm

续表

设备、仪表或配件名称	规格
水箱	780mm×420mm×500mm
管路	不同长度与直径的直管等
阀门	测压阀等
其他配件	管路直接、弯头、三通、软管、垫圈等

9.3.2 泵性能测试自组装实验

泵性能测试自组装实验主要设备、仪表和配件如表 9-2 所列。

表 9-2　泵性能测试自组装实验主要设备、仪表与配件

设备、仪表或配件名称	规格
不锈钢框架	2.50m×0.55m×1.85m
清水泵	MS100/0.55SSC 型，不锈钢，0.55kW，6.0m^3/h，H=14m
	MS60/0.55SSC 型，不锈钢，0.55kW，3.6m^3/h，H=19.5m
	1/2DB70 型，不锈钢，0.55kW，3.0m^3/h，H=70m
灌水装置	—
粗糙管	不锈钢，DN15
	不锈钢，DN25
	不锈钢，DN32
	不锈钢，DN40
压力表	0～0.25MPa，精度 1.5 级
真空表	−0.1～0MPa，精度 1.5 级
功率变送器	PS−139 型
涡轮流量计	LWGY 型，DN40
变频器	0～50Hz
电功率数字显示仪	—
温度数字显示仪	—
流量数字显示仪	—
热电阻温度计	Pt100
水箱	780mm×420mm×500mm
隔板	—
管路	不同长度与直径的直管等
阀门	闸阀、球阀、单向阀等
其他配件	管路直接、弯头、三通、软管、垫圈等

9.3.3 传热过程强化自组装实验

传热过程强化自组装实验主要设备、仪表和配件如表 9-3 所列。

表 9-3 传热过程强化自组装实验主要设备、仪表与配件

设备、仪表或配件名称	规格
不锈钢框架	2.2m×0.55m×1.77m
加热釜	DZF-3 型，3.0 kW，不锈钢
普通传热管	*DN*15，1.2m
槽纹管	*DN*15，1.2m
横纹管	*DN*15，1.2m
缩放管	*DN*15，1.2m
翅片管	*DN*15，1.2m
螺旋线圈强化传热管	*DN*15，1.2m(6 种不同螺距、丝径可选择)
旋涡风机	XGB－12 型，100m^3/h
孔板流量计	—
热电阻温度计	Pt100
热电偶温度计	铜-康铜
压差传感器	WNK3051 型，0～10.00 kPa
热电阻温度显示仪	—
热电偶温度显示仪	—
压差数字显示仪	—
水箱	240mm×240mm×240mm
管路	不同长度与直径的直管等
阀门	闸阀、球阀、泄压阀等
其他配件	管路直接、弯头、三通、软管、垫圈等

9.3.4 列管换热自组装实验

列管换热自组装实验主要设备、仪表和配件如表 9-4 所列。

表 9-4 列管换热自组装实验主要设备、仪表与配件

设备、仪表或配件名称	规格
不锈钢框架	2.2m×0.55m×1.77m
列管换热器	不锈钢，*DN*159，1.5m
热水泵	不锈钢
冷水泵	不锈钢
水箱	1000mm×500mm×600mm
加热器	ES60V-U1(E)型，3kW

续表

设备、仪表或配件名称	规格
转子流量计	VA10-15F
热电阻温度计	Pt100
压差传感器	WNK3051 型，0～10.00 kPa
封头	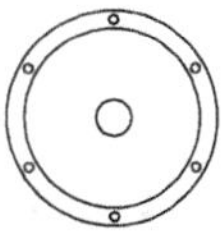
	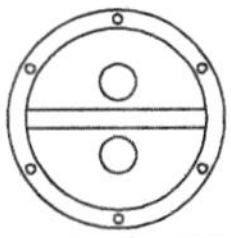
	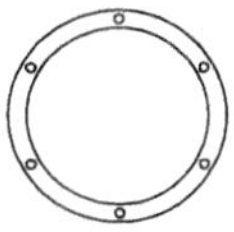
	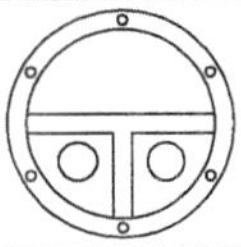
	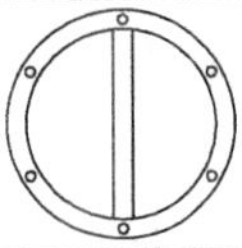
压力表	—
管路	不同长度与直径的直管等
阀门	闸阀、球阀、单向阀等
其他配件	管路直接、弯头、三通、软管、垫圈等

第10章 化工原理提高型实验与实训

10.1 液液萃取提高型实验

10.1.1 实验目的

① 了解搅拌萃取塔、填料萃取塔、喷洒萃取塔的基本结构和萃取工艺流程，操作方法以及各类型萃取塔流体力学性能和传质性能的差异。

② 了解萃取过程的主要影响因素，探索萃取操作条件对萃取过程的影响。

③ 掌握萃取塔滞留分数、泛点等流体力学性能参数的测定方法。

④ 掌握萃取塔传质单元数、传质单元高度、总传质系数的实验测定方法。

10.1.2 实验原理

萃取是分离和提纯物质的重要单元操作之一，是利用混合物中各个组分在萃取剂中溶解度的差异而实现组分分离的单元操作。萃取塔中两种液体在塔内做逆流流动，其中一种液体作为分散相，以液滴形式通过另一种连续流动的液体即连续相，两液相的浓度沿萃取塔高度方向连续变化，并依靠两液相的密度差在塔的两端实现液-液两相的分离。当轻相作为分散相时，相界面出现在塔的上端；反之，当重相作为分散相时，则相界面出现在塔的下端。对微分逆流萃取塔的塔高，也可采用类似于吸收单元操作中填料层高度的计算方法，即采用传质单元数和传质单元高度计算萃取塔高度。其中传质单元数表示萃取分离程度的难易，传质单元的高低表示萃取设备传质性能的好坏。

$$h_0 = H_{OE} N_{OE} \tag{10-1}$$

式中，h_0 为萃取塔的有效传质高度，m；H_{OE} 为以萃取相为基准的总传质单元高度，m；N_{OE} 为以萃取相为基准的总传质单元数，无量纲。

总传质单元数为：

$$N_{OE} = \int_{y_a}^{y_b} \frac{dy}{y^* - y} \tag{10-2}$$

式中，y 为萃取塔内某截面处萃取相的组成，kg BA/kg 水；y^* 为萃取塔内某截面处与萃余相平衡的萃取相组成，kg BA/kg 水；y_a 为萃取剂进入萃取塔时的组成，kg BA/kg 水；y_b 为萃取相离开萃取塔时的组成，kg BA/kg 水。

总传质单元数 N_{OE} 中的 y^* 可由萃取平衡关系式（10-3）计算，而式（10-3）中的萃余相组成 x 则可通过萃取塔的操作线方程求得。萃取实验中以水为萃取剂萃取煤油中的苯甲酸（BA），溶质苯甲酸在原溶剂和萃取剂中的溶解度均很低，且水与煤油之间的溶解度较低，因此两液相在萃取塔内的质量流量可视为常数。但是液液相平衡方程中萃取相与萃余相间呈现非线性关系，如25℃时苯甲酸在水（萃取剂）和煤油（原溶剂）中的相平衡方

程为

$$y^*=4.443060\times10^{-6}+1.247730x-4.440948\times10^{2}x^{2}+5.048579\times10^{4}x^{3}+6.414260\times10^{6}x^{4} \tag{10-3}$$

上式体现了萃取相平衡组成 y^* 与萃余相组成 x 的关系，而 x 与 y 符合物料衡算关系。对萃取塔进行全塔物料衡算可得

$$Fx_b+Sy_a=Ey_b+Rx_a \tag{10-4}$$

式中，F 为原料液流量，kg/h；S 为萃取剂流量，kg/h；E 为萃取相流量，kg/h；R 为萃余相流量，kg/h；x_a 为萃余相离开萃取塔时的组成，kg BA/kg 煤油；x_b 为混合液进入萃取塔时的组成，kg BA/kg 煤油。

对于稀溶液萃取 $F\approx R$、$E\approx S$，且 $y_a=0$，所以

$$y_b=\frac{F}{S}(x_b-x_a) \tag{10-5}$$

同理，在萃取塔内任一截面处进行物料衡算可得萃取塔的操作线方程

$$y=\frac{F}{S}(x-x_a) \tag{10-6}$$

根据式（10-3）和式（10-6）即可将式（10-2）中的 y^* 表达为 y 的函数，而后利用数值积分获得 N_{OE}，最后根据萃取塔有效传质高度 h_0 可求 H_{OE} 和总传质系数。

$$H_{\mathrm{OE}}=\frac{S}{K_{\mathrm{y}}aA} \tag{10-7}$$

式中，A 为萃取塔截面积，m^2；$K_{\mathrm{y}}a$ 为萃取相总体积传质系数，kg/（m^3 · h · kg/kg）。

10.1.3 实验装置与流程

实验装置流程如图 10-1（脉冲填料萃取塔、未装填填料则为喷洒萃取塔）、图 10-2（液液搅拌萃取塔）所示，装置主要由萃取塔、轻相槽、重相槽、输液磁力泵、搅拌电机、空压机、脉冲发生器等设备和部件构成。本装置操作时应先通过重相泵将重相槽内的水输送入萃取塔充满全塔形成连续相，然后开启轻相泵将轻相槽中的煤油输送入塔分散为液滴，形成分散相。分散相（轻相）在塔内向上运动，连续相（重相）向下流动，两相逆流接触传质。分散相在塔顶汇集聚并形成一定厚度的轻液层，在萃取塔上部轻相与重相之间形成明显的界面，界面位置则通过连续相出口的 π 形管闸阀调节，使之稳定于一定高度，轻相由塔顶采出进入萃余相回收槽，重相由塔底部采出并通过 π 形管导出。当轻相作为连续相、重相作为分散相时，流程与上述类似，只是轻重相的分界面在萃取塔的下部。

设备主要技术数据如下。

① 萃取塔：塔体总高 1280mm，内径 50mm。

② 搅拌桨：如图 10-3 所示，有 8 种形式。

③ 填料：如图 10-4 所示，有 11 种形式。

10.1.4 实验步骤与注意事项

(1) 实验步骤

① 将煤油配制成含苯甲酸的混合物（苯甲酸质量分数约为 0.0015～0.002），然后把

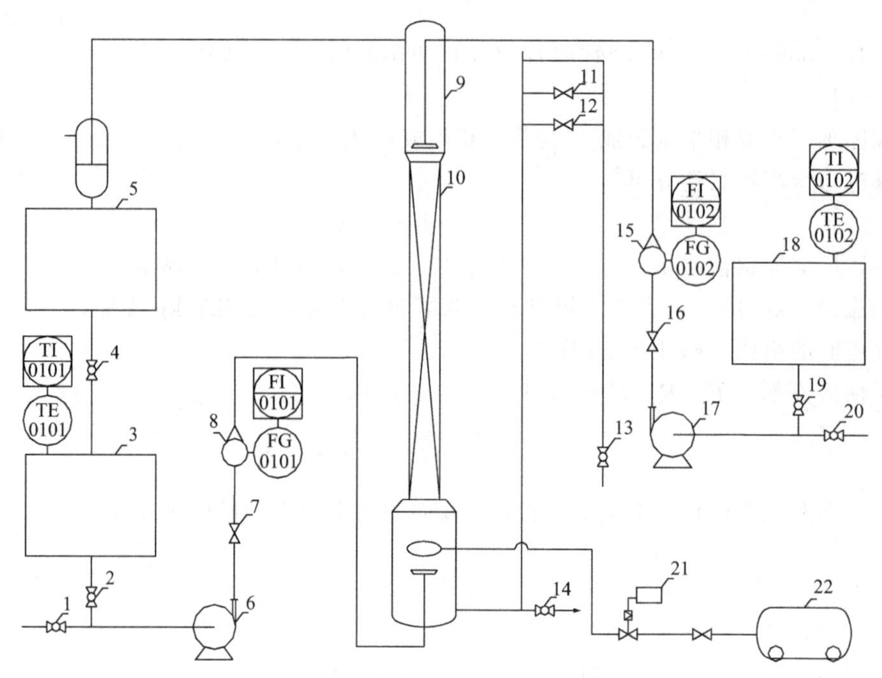

图 10-1　脉冲填料萃取实验装置流程图

1—轻相槽排净阀；2—轻相磁力泵进口阀；3—轻相槽；4—萃余相回收槽与轻相槽联通阀；5—萃余相回收槽；6—轻相磁力泵；7—轻相磁力泵出口阀；8—轻相浮子流量计；9—填料萃取塔；10—填料；11—π形管闸阀 1；12—π形管闸阀 2；13—重相出口阀；14—萃取塔放净阀；15—重相浮子流量计；16—重相磁力泵出口阀；17—重相磁力泵；18—重相槽；19—重相磁力泵进口阀；20—重相槽排净阀；21—脉冲调节器；22—空气压缩机

它灌入轻相槽内；打开轻相磁力泵出口阀和入口阀进行灌泵，开启轻相槽排净阀排空管道空气；关闭轻相槽排净阀、轻相磁力泵出口阀。

② 接通水管，将水灌入重相槽内，打开重相磁力泵入口阀和出口阀进行灌泵，开启重相槽排净阀排空管道空气；关闭重相槽排净阀、重相磁力泵出口阀。

③ 如图 10-5 和图 10-6 所示，打开控制台上的总电源开关，打开仪表电源开关。

④ 打开重相磁力泵开关，缓慢开启磁力泵出口阀，将水送入萃取塔内，直至接近满塔，并开启 π 形管闸阀和萃取相导出阀。

⑤ 开启轻相磁力泵开关，缓慢开启磁力泵出口阀，将煤油输送入塔，煤油分散于萃取塔中向上流动，并在塔顶汇集。

⑥ 对搅拌萃取塔，开启搅拌器，控制适当转速；对脉冲萃取塔，开启脉冲电源。

⑦ 调节两相流量约 10～15L/h，注意对煤油浮子流量计测量的流量进行校正

$$V_{h实际}=V_{h测量}\sqrt{\frac{(\rho_f-\rho_{实际})\rho_{标定}}{\rho_{实际}(\rho_f-\rho_{标定})}}$$

式中，$V_{h实际}$ 为煤油实际体积流量，L/h；$V_{h测量}$ 为煤油浮子流量计指示流量，L/h；ρ_f 为煤油浮子流量计浮子密度，7800kg/m^3；$\rho_{实际}$ 为实际被测流体密度，kg/m^3；$\rho_{标定}$ 为浮子流量计标定条件下标定流体（水）的密度，1000 kg/m^3。

⑧ 通过重相出口 π 形管上闸阀的开度调节萃取塔内轻重两相的界面处于塔顶轻相出口和重相入口之间，并保持稳定。

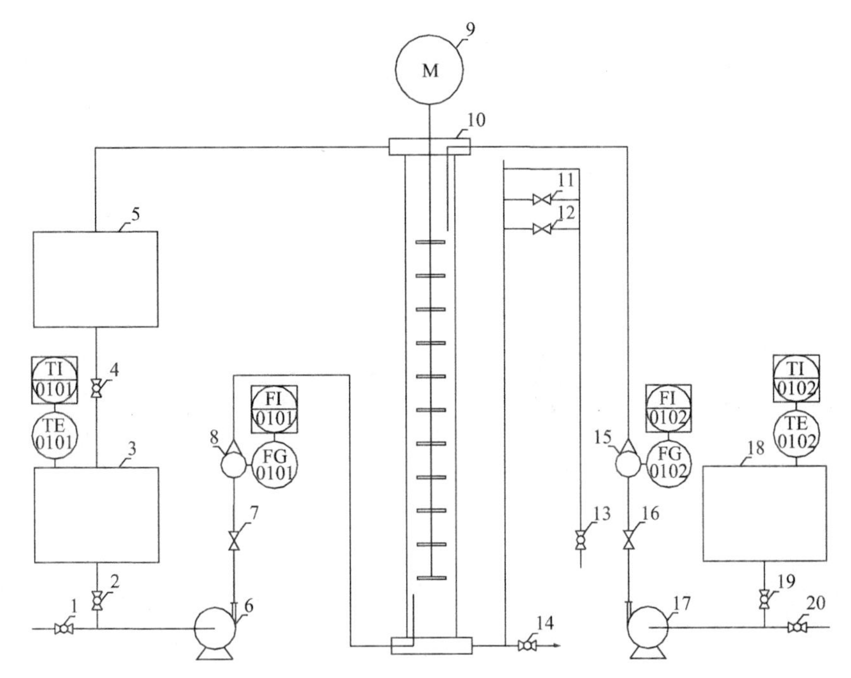

图 10-2　液液搅拌萃取实验装置流程图

1—轻相槽排净阀；2—轻相磁力泵进口阀；3—轻相槽；4—萃余相回收槽与轻相槽联通阀；
5—萃余相回收槽；6—轻相磁力泵；7—轻相磁力泵出口阀；8—轻相浮子流量计；
9—搅拌电机；10—机械搅拌萃取塔；11—π 形管闸阀 1；12—π 形管闸阀 2；13—重相出口阀；
14—萃取塔放净阀；15—重相浮子流量计；16—重相磁力泵出口阀；17—重相磁力泵；
18—重相槽；19—重相磁力泵进口阀；20—重相槽排净阀

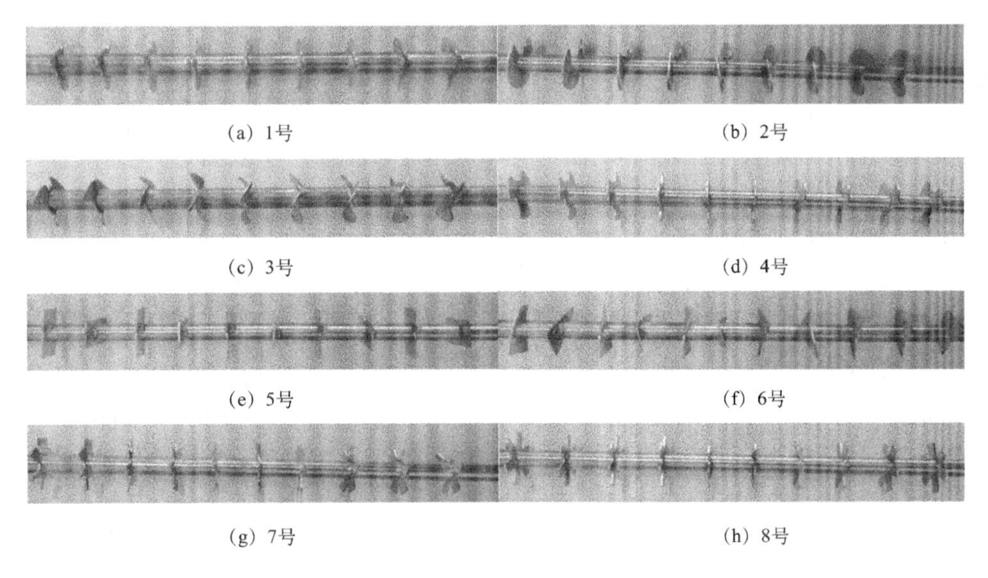

图 10-3　搅拌桨形式

⑨ 萃取塔操作稳定后，对煤油进口（原料液）、煤油出口（萃余相）、水出口（萃取相）分别取样 60～80mL 备用，根据下述方法分析组成。

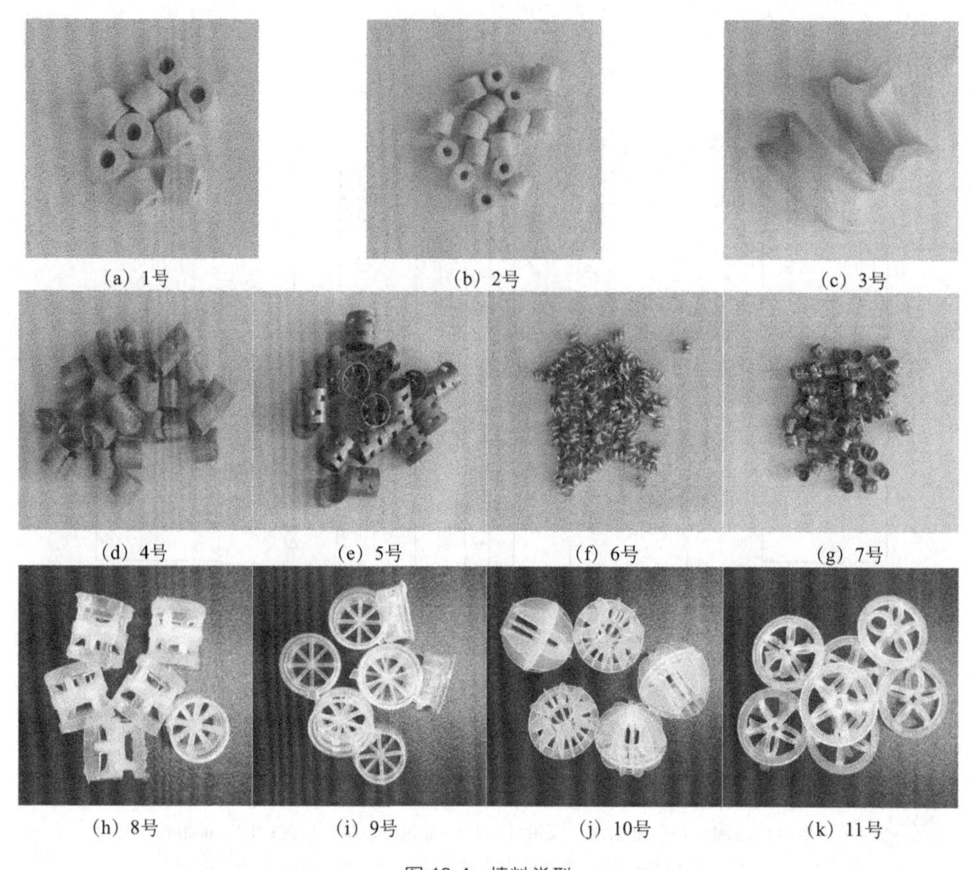

图 10-4　填料类型

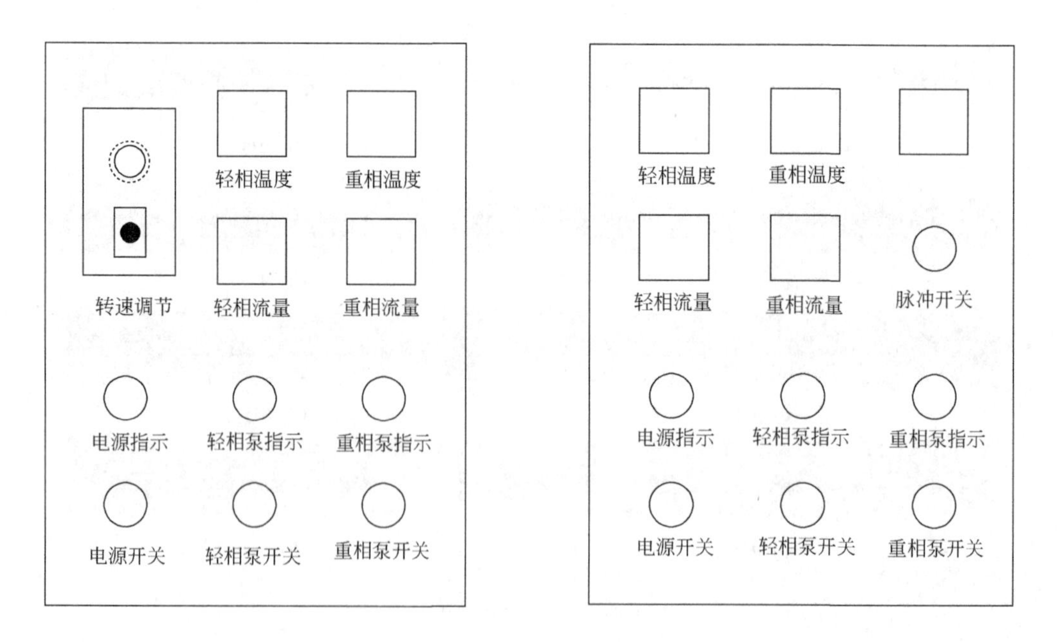

图 10-5　搅拌萃取塔控制面板

图 10-6　脉冲填料/喷洒萃取塔控制面板

本实验采用 NaOH 的乙醇溶液，以酚酞为指示剂，对取样的原料液、萃取相、萃余相中苯甲酸的组成进行酸碱滴定分析，具体步骤如下。

ⅰ. 用标定好组成的 NaOH 乙醇溶液润洗滴定管，而后在滴定管内装填适量 NaOH 乙醇溶液。

ⅱ. 用移液管取待测样品 10mL 或 20mL 于锥形瓶中，滴加 1～3 滴酚酞指示剂。

ⅲ. 用 NaOH 乙醇溶液滴定至终点，则所测量组成为

$$x=\frac{122.12N\Delta V}{800V_{样品}}$$

式中，N 为 NaOH 乙醇溶液当量浓度，mol/L；ΔV 为滴定消耗的 NaOH 乙醇溶液体积，mL；$V_{样品}$ 为样品的体积，mL。

⑩ 测量萃取塔的传质高度即轻重相界面位置与轻相入口之间的垂直距离，同时关闭轻相、重相的进出口阀门，关闭两相磁力泵，关闭搅拌器或脉冲，待全塔内轻、重相完全分离后再次测量界面位置与轻相入口之间的垂直高度，获得滞留分数。

⑪关闭设备总电源，开启萃取塔排净阀，放空塔内液体，放空萃余相回收槽、轻相槽、重相槽内液体，清理实验仪器和实验场地。

（2）注意事项

① 勿直接在轻相槽内配制煤油-苯甲酸溶液，防止未溶解固体颗粒堵塞煤油输送泵的入口管路。

② 磁力泵切不可空载运行。

③ 磁力泵出口阀应缓慢开启，避免快速开启阀门时流体冲击电磁浮子流量计，使其浮子卡住无法复位。

④ 酸碱滴定分析过程中，被测样品尤其煤油样品滴入碱液后要充分振荡，以保证准确判断滴定终点。

10.1.5 思考题

① 分析比较萃取过程与吸收、精馏过程的异同点。

② 萃取过程的影响因素都有哪些？

③ 脉冲、机械搅拌等外加能量对萃取（如萃取传质系数与萃取率）有何影响。

④ 提高萃取剂用量对萃余相组成、萃取相组成有何影响？

⑤ 通过实验数据分析，机械搅拌萃取塔、喷洒萃取塔、填料萃取塔三种萃取设备的性能有何区别？

⑥ 实验中滞留分数如何测定？滞留分数对萃取过程有何影响？

⑦ 分散相液滴越小对萃取过程是否越有利？为什么？

⑧ 试解释萃取塔的液泛现象？

⑨ 测定原料液、萃取相、萃余相的组成可用哪些方法？采用酸碱中和滴定法时，标准碱为什么选用 NaOH 乙醇溶液，而不选用 NaOH 水溶液？

⑩ 若以煤油为连续相，水为分散相，则萃取塔应如何操作？

⑪ 萃取相出口管路 π 形管的作用是什么？

⑫ 煤油进入萃取塔的管路是一倒 U 形管路，为什么如此设置？

⑬ 喷洒萃取塔与填料萃取塔相比，滞留分数、分散相液滴大小有何定性关系？为什么？

⑭ 分界面若向下移动，要恢复原样，该如何调节？

⑮ 分散相流量增大，则塔内滞留分数如何变化？若连续相流量增大，则滞留分数是否变化？

10.2 多效蒸发综合实训

10.2.1 实训目的

① 了解蒸发器的基本结构和工作原理，掌握多效蒸发的工艺流程和操作方法。

② 训练并掌握蒸发生产装置的操作方法，熟练掌握仪表参数的调控，确保多效蒸发装置正常稳定运行。

③ 训练并掌握多效蒸发生产装置的手动和自动操作，熟悉 DCS 控制系统的使用。

④ 训练对异常现象分析，判断故障种类、产生原因并采取适当措施进行排除处理的能力。

⑤ 掌握多效蒸发过程的实验研究方法和相关性能参数的测定方法。

⑥ 培养学生安全、规范、环保、节能的生产意识及敬业爱岗、严格遵守操作规程的职业道德和团队合作精神。

10.2.2 基本原理

蒸发是将不挥发性溶质的稀溶液加热沸腾，使部分溶剂汽化，实现浓缩的单元操作，实现这一功能的设备称为蒸发器。为保证蒸发过程持续稳定进行，必须给蒸发器提供持续不断的热源供给溶剂汽化所需的热量，同时要连续地将汽化蒸汽排出蒸发器。因此，蒸发器通常由加热室和蒸发室构成。

蒸发操作要将大量的溶剂汽化，需要消耗大量热能，因此，蒸发是一个高能耗的单元操作，提高加热蒸汽利用率等节能方法是提高蒸发操作经济性的重要措施。多效蒸发由多个蒸发器串联，一效蒸发器产生的二次蒸汽引入二效蒸发器作为加热蒸汽，从而提高生蒸汽的利用率，实现节能的目的。

本实训装置为双效蒸发，蒸发器为外热式蒸发器。它由加热室和蒸发室组成，加热室位于蒸发器的下部，由加热管束组成，管外的加热蒸汽使管内的溶液沸腾汽化；蒸发器的上部为蒸发室，汽化产生的蒸汽在其中与夹带的液沫分离，而后进入二效蒸发器作为加热热源或进入冷凝器冷凝，浓缩后的溶液一部分循环回加热室继续蒸发浓缩，一部分则从蒸发器的底部排出。

10.2.3 工艺流程、主要设备及仪表

原料液从原料罐 V0101 经原料泵 P0101 输送进入一效加热器 E0101，蒸汽发生器 R0101 产生的生蒸汽在一效加热器 E0101 壳程冷凝加热管程原料液，料液在一效加热器 E0101 和一效蒸发室 E0102 之间自然循环，汽化产生的二次蒸汽在一效蒸发室 E0102 中与夹带的液沫分离，二效蒸发器负压操作，一效蒸发器完成液自动流入二效蒸发器，来自一效蒸发室 E0102 的二次蒸汽则进入二效加热器 E0103 壳程作为二效蒸发器的加热热源，加热管程的料液，料液在二效加热器 E0103 和二效蒸发室 E0104 之间循环蒸发浓缩，二效蒸发器的完成液经完成液泵 P0102 打入产品罐 V0102 贮存。二效蒸发室 E0104 分离出的二次蒸汽则进入冷凝器 E0105 冷凝除去。一效、二效加热器中的冷凝水经疏水阀排除。具体流程如图 10-7 所示。

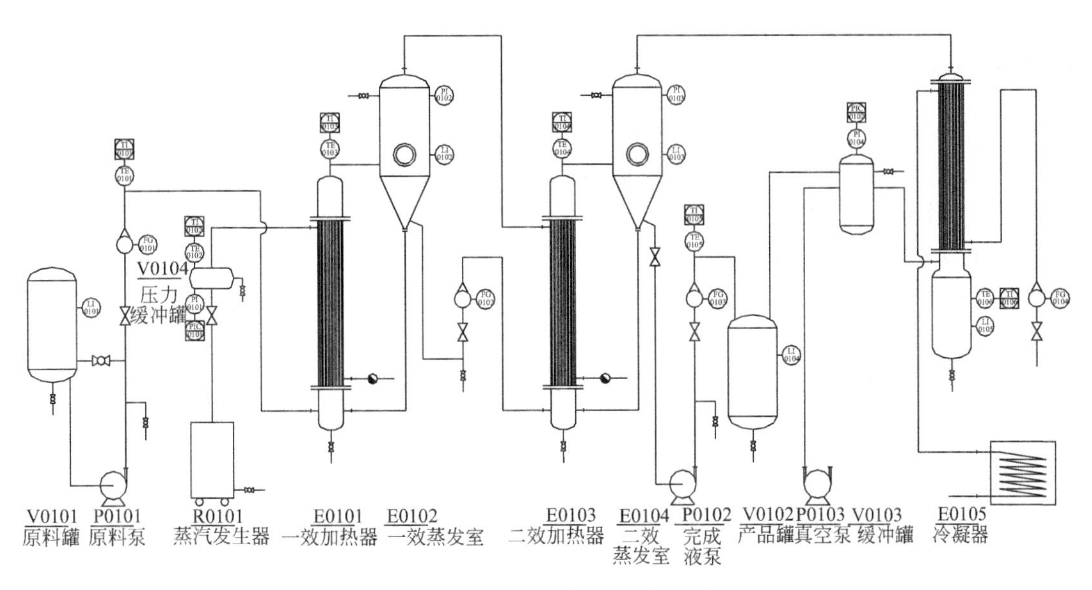

图 10-7　多效蒸发综合实训装置流程图

装置控制仪表面板如图 10-8 所示。

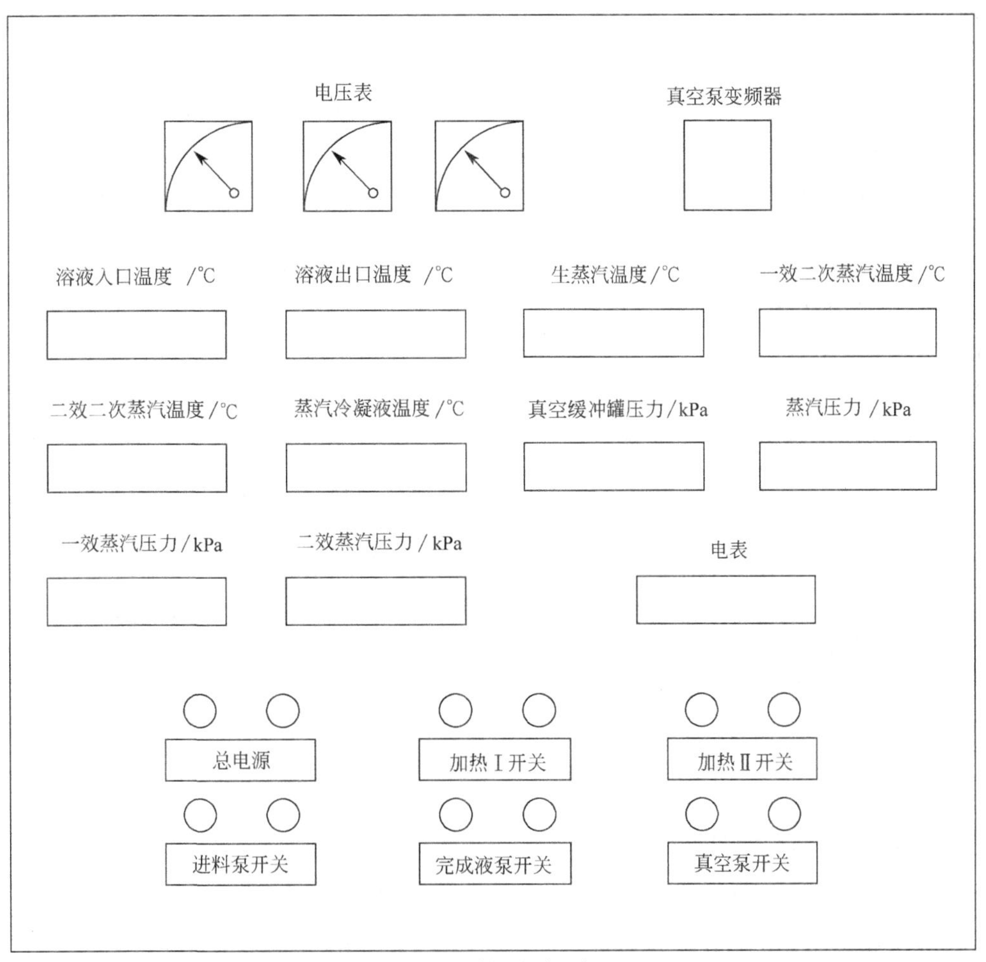

图 10-8　控制仪表面板

10.2.4 实训项目

（1）实训项目一：识图技能和装置实物识别训练

① 识读蒸发装置的工艺流程图，对照装置能独立、熟练详述流程，明确装置中各设备、阀门的作用。

② 识读蒸发装置的测量与控制点，对照装置熟悉各测量点和仪表位置，掌握仪表的调控操作及参数控制。

③ 根据装置实际情况制订多效蒸发装置的详细操作规程。

（2）实训项目二：开车前的动、静设备检查

① 检查各设备是否处于可开车状态。

② 检查管路、管件、阀门连接是否完好，阀门是否灵活好用并处于正确位置。

③ 检查整个装置是否存在“跑、冒、滴、漏”。

④ 检查流体输送设备是否完好，了解离心泵铭牌上的标识内容，检查离心泵安装高度是否合适，离心泵是否需要灌泵，检查离心泵等输送机械进、出口阀门是否处于正常开车状态。

（3）实训项目三：原料以及水、电、气等公用工程和仪表的检查

① 检查原料、水等贮罐液位是否正常。排净加热器、蒸发室、产品罐、冷凝器等内残存液体，向原料罐内加入原料液，控制液位在4/5左右。

② 对装置上的所有测量仪表包括流量测量仪表（转子流量计、孔板流量计、文丘里流量计、涡轮流量计），压力、液位测量仪表（压力表、真空表、磁翻转液位计、玻璃管液位计），温度测量仪表（热电阻温度计），以及控制仪表和相关部件（电动调节阀、传感器、变频仪）等进行识别。

③ 检查仪表柜电源连接是否正常，打开总控开关，检查仪表柜上所示电压表指示是否为380V，总电源指示红灯是否亮起。启动总电源仪表上电，稳定3min后检查各仪表指示是否处于正常范围。

④ 掌握现场控制台仪表和计算机远程控制系统DCS系统的正确操作和监控方法。

（4）实训项目四：动设备试车

① 离心泵、真空泵等动设备启动前需要进行盘车检查方可通电，开车前检查泵的出入口管线、阀门、压力表接头有无泄漏，地脚螺丝及其他连接处有无松动。

② 盘车检查转子是否轻松灵活，泵体内是否有金属碰撞的声音。

③ 开启离心泵入口阀门，打开出口放空阀门，使液体充满泵体，排除泵内积存空气后关闭放空阀，做好开车前准备。

（5）实训项目五：离心泵正常开、停车操作

开车操作：

① 完全开启离心泵入口阀门，关闭所有出口阀门，然后启动电机，当泵运转后全面检查离心泵工作状况。

② 检查电机和泵的旋转方向是否正确。

③ 检查电机和泵是否有杂音、是否异常振动、是否有渗漏。

④ 检查电机电流是否小于额定值，当显示超负荷时应立即停车检查。

⑤ 调整出口流量调节阀，使泵的工作点处于工艺要求状态。

停车操作：

① 逐渐关小离心泵出口阀门至全关。

② 当全部出口阀门完全关闭后，关停电机。

③ 当泵停止运转后，关闭离心泵入口阀门。

（6）实训项目六：多效蒸发操作

① 系统进料：关闭装置中所有阀门，开启一效蒸发器放空阀，按照离心泵的开车规程启动原料泵，由原料罐向一效蒸发器送液，当原料液液位接近一效蒸发器液位观测孔中间位置时，开启缓冲罐压力调节阀，启动真空泵，开启效间流量调节阀，调节缓冲罐压力调节阀，在真空作用下，使一效蒸发室液体进入二效蒸发器，当二效蒸发器中液体液位达到蒸发器液位观测孔中间位置时，关闭原料泵出口流量调节阀停止进料，关闭效间流量调节阀，开启缓冲罐压力调节阀破真空，关闭真空泵，原料泵停车，关闭一效蒸发器放空阀。

② 预热操作：检查蒸汽发生器液位是否正常，打开蒸汽发生器出口阀门，打开蒸汽发生器加热开关，开始加热，一效蒸发器产生的二次蒸汽进入二效加热器，当二效加热器中液体温度达到一定值时，开启真空泵，调节缓冲罐压力至工艺要求值，调节冷却水流量调节阀至一定值。

③ 蒸发操作：当二效蒸发器有二次蒸汽产生后，启动原料泵以一定流量进料，同时调节效间流量调节阀使之流量为一定值（低于进料量），启动完成液泵，调节完成液出口阀门使之流量为一定值（低于效间流量）；调节系统压力、流量等参数，保持系统温度、压力、液位稳定，监测并记录系统温度、压力、液位等数据，稳定操作一定时间后，取样，进行组成分析。

④ 停车操作：关闭蒸汽发生器加热电源；关闭进料流量调节阀、效间流量调节阀和完成液流量调节阀，停止进料和出料；全开缓冲罐压力调节阀，待系统恢复常压状态后关停真空泵，系统冷却后关闭冷却水，关闭总电源，系统内液体温度降至接近常温后，放空系统内积液，清理现场。

说明：在实训之前，应对装置进行全面的认识和了解，熟悉装置中所有设备、管路、阀门、仪表，列出所有设备、阀门、仪表清单，明确其原理和作用；在各实训训练之前应制订详细的操作规程，操作规程通过后方可进行实训操作；多效蒸发操作实训在其他实训项目均完成的前提下方可进行。

10.3　吸收解吸综合实训

10.3.1　实训目的

① 了解吸收塔、解吸塔的基本结构和工作原理，掌握吸收解吸的工艺流程和操作方法。

② 训练并掌握吸收解吸生产装置的操作方法，熟练掌握仪表参数的调控，确保吸收解吸装置正常稳定运行。

③ 训练并掌握吸收解吸生产装置的手动和自动操作，熟悉DCS控制系统的使用。

④ 训练对异常现象分析判断故障种类、产生原因并采取适当措施进行排除处理的能力。

⑤ 掌握吸收解吸过程的实验研究方法和相关性能参数的测定方法。

⑥ 培养学生安全、规范、环保、节能的生产意识及敬业爱岗、严格遵守操作规程的职业道德和团队合作精神。

10.3.2 基本原理

气体吸收简称吸收，它是通过混合气体与液体接触传质，使其中可溶组分溶于液体，将其从混合气体中分离出来的单元操作。气体吸收主要用于分离混合气体，脱除气体中的有害或无用组分，回收有用组分或制备气液反应生成物等。而将液体中溶解的气体脱除的操作称为解吸，解吸是吸收的逆过程，其传质方向和吸收相反。为实现吸收剂的回用或吸收组分的回收和提纯，通常将吸收和解吸过程联用。吸收、解吸设备有填料塔和板式塔两大类。

吸收解吸实训装置包括吸收塔和解吸塔，两塔均采用鲍尔环填料，吸收塔中用水吸收空气中二氧化碳组分，解吸塔则采用空气解吸吸收液中的二氧化碳。二氧化碳在水中溶解度很小，其吸收、解吸过程可按低浓度处理，且其传质过程为液膜控制。

10.3.3 工艺流程、主要设备及仪表

空气由风机 P0101 提供，二氧化碳（溶质）由钢瓶 X0101 提供，空气和二氧化碳混合后经 π 形管进入吸收塔 T0101 底部，由下向上流动通过填料层，与吸收剂（来自解吸塔解吸后的吸收液）在塔内逆流接触，二氧化碳被水吸收，吸收后的尾气放空。吸收后的吸收剂从吸收塔塔底进入贮罐 V0101，由离心泵 P0103 输送，经加热器 E0101 预热后进入解吸塔 T0102 塔顶，与来自风机 P0104 的新鲜空气在解吸塔内逆流解吸，经解吸再生的液体由冷却器 E0102 冷却后进入贮罐 V0102，而后通过离心泵 P0102 输送从吸收塔塔顶进入作为吸收剂，解吸尾气从解吸塔顶放空。具体流程如图 10-9 所示。

装置控制仪表面板如图 10-10 所示。

10.3.4 实训项目

（1）实训项目一：识图技能和装置实物识别训练

① 识读吸收解吸装置的工艺流程图，对照装置能独立、熟练详述流程，明确装置中各设备、阀门的作用。

② 识读吸收解吸装置的测量与控制点，对照装置熟悉各测量点和仪表位置，掌握仪表的调控操作及参数控制。

③ 根据装置实际情况制订吸收解吸装置的详细操作规程。

（2）实训项目二：开车前的动、静设备检查

① 检查各设备是否处于可开车状态。

② 检查管路、管件、阀门连接是否完好，阀门是否灵活好用并处于正确位置。

③ 检查整个装置是否存在“跑、冒、滴、漏”。

④ 检查流体输送设备是否完好，了解离心泵、风机铭牌上的标识内容，检查离心泵安装高度是否合适，离心泵是否需要灌泵，检查离心泵、风机等输送机械进、出口阀门是否处于正常开车状态。

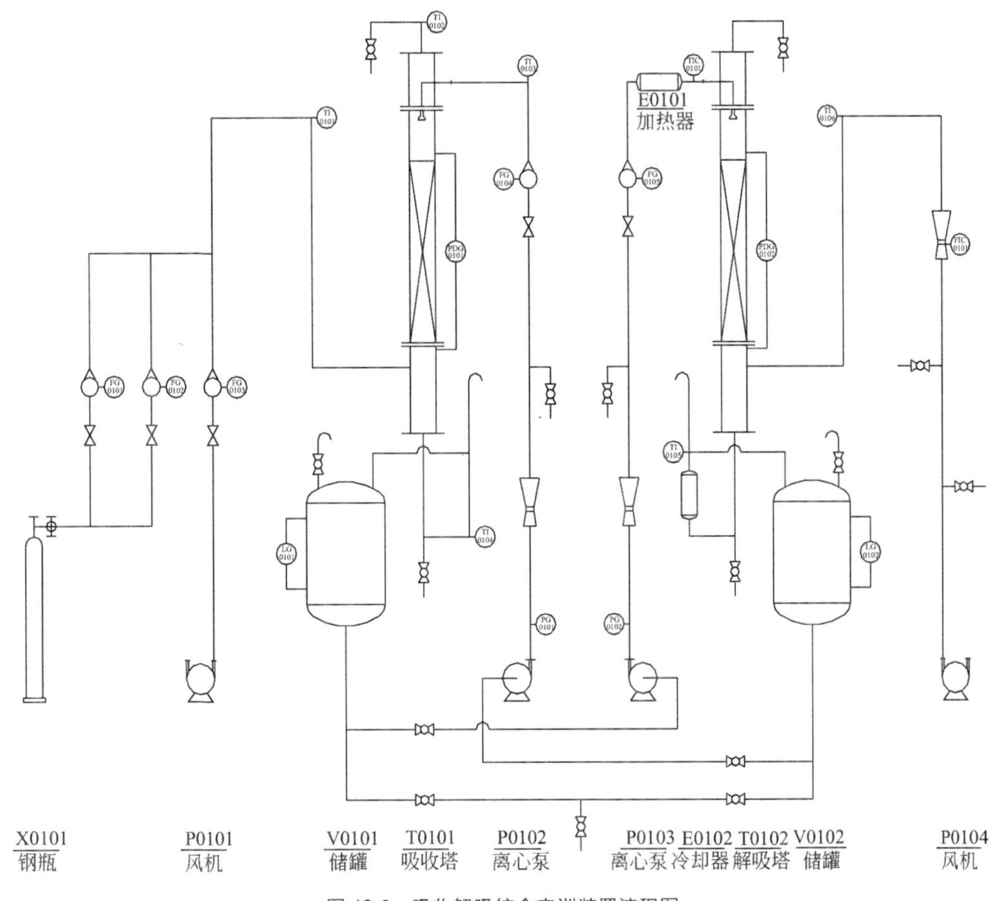

图 10-9　吸收解吸综合实训装置流程图

(3) 实训项目三：原料以及水、电、气等公用工程和仪表的检查

① 检查吸收液、解吸液贮罐液位是否正常，二氧化碳钢瓶贮量是否合适，减压阀是否正常。

② 对装置上的所有测量仪表包括流量测量仪表（转子流量计、文丘里流量计），压力、液位测量仪表（压力表、玻璃管液位计），温度测量仪表（热电阻温度计），以及控制仪表和相关部件（电动调节阀、传感器、变频仪）等进行识别。

③ 检查仪表柜电源连接是否正常，打开总控开关，检查仪表柜上所示电压表指示是否为 380V，总电源指示红灯是否亮起。启动总电源仪表上电，稳定 3min 后检查各仪表指示是否处于正常范围。

④ 掌握现场控制台仪表和计算机远程控制系统 DCS 系统的正确操作和监控方法。

(4) 实训项目四：动设备试车

① 离心泵、风机等动设备启动前需要进行盘车检查方可通电，开车前检查泵的出入口管线、阀门、压力表接头有无泄漏，地脚螺丝及其他连接处有无松动。

② 盘车检查转子是否轻松灵活，泵体、风机内部是否有金属碰撞的声音。

③ 开启离心泵入口阀门，打开出口放空阀门，使液体充满泵体，排除泵内积存空气后关闭放空阀，做好开车前准备。

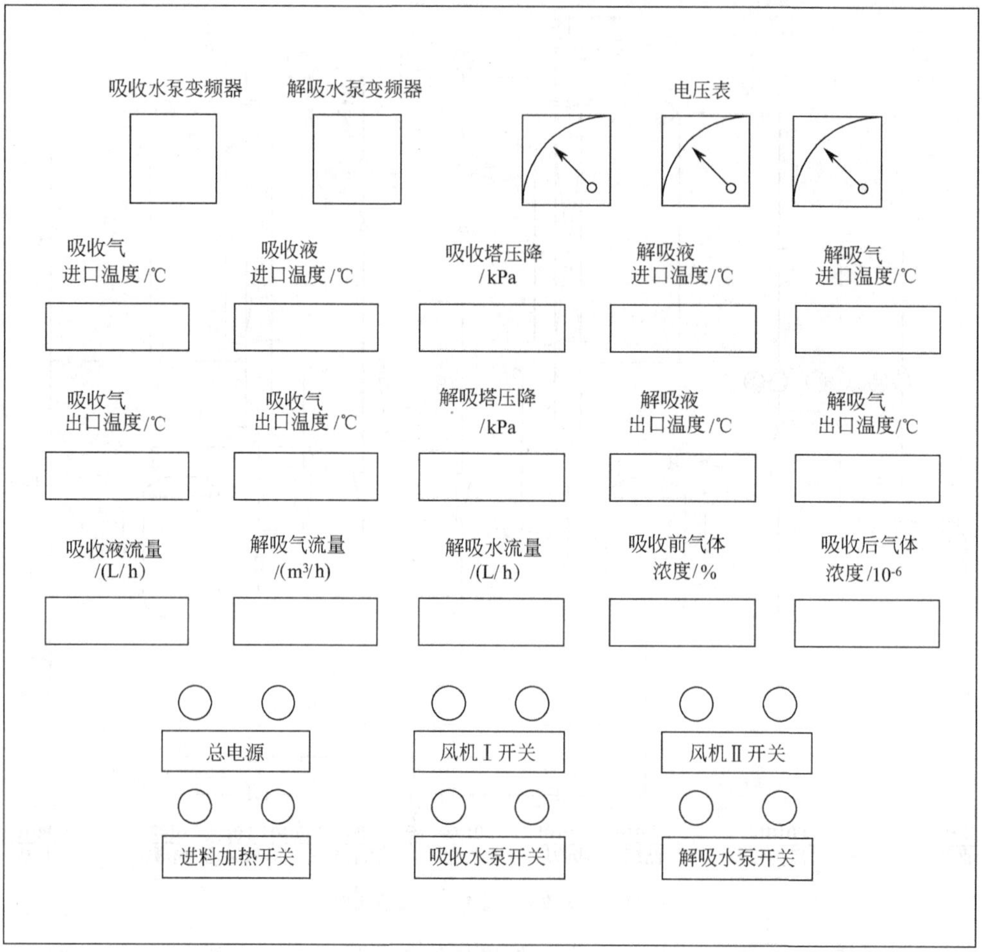

吸收水泵变频器 解吸水泵变频器 电压表

吸收气进口温度/℃ 吸收液进口温度/℃ 吸收塔压降/kPa 解吸液进口温度/℃ 解吸气进口温度/℃

吸收气出口温度/℃ 吸收气出口温度/℃ 解吸塔压降/kPa 解吸液出口温度/℃ 解吸气出口温度/℃

吸收液流量/(L/h) 解吸气流量/(m^3/h) 解吸水流量/(L/h) 吸收前气体浓度/% 吸收后气体浓度/10^{-6}

总电源 风机Ⅰ开关 风机Ⅱ开关

进料加热开关 吸收水泵开关 解吸水泵开关

图 10-10 控制仪表面板

（5）实训项目五：离心泵正常开、停车操作

开车操作：

① 完全开启离心泵入口阀门、关闭所有出口阀门，然后启动电机，当泵运转后全面检查离心泵工作状况。

② 检查电机和泵的旋转方向是否正确。

③ 检查电机和泵是否有杂声、是否异常振动、是否有渗漏。

④ 检查电机电流是否小于额定值，当显示超负荷时应立即停车检查。

⑤ 调整出口流量调节阀，使泵的工作点处于工艺要求状态。

停车操作：

① 逐渐关小离心泵出口阀门至全关。

② 当全部出口阀门完全关闭后，关停电机。

③ 当泵停止运转后，关闭离心泵入口阀门。

（6）实训项目六：旋涡风机正常开、停车操作

开车操作：

① 完全开启旋涡风机旁路阀门，启动电机，当风机运转后全面检查风机工作状况。

② 检查电机和风机的旋转方向是否正确。

③ 检查电机和风机是否有杂音、是否异常振动。

④ 调整风机旁路阀门，使风机的工作点处于工艺要求状态。

停车操作：

① 逐渐开大风机旁路阀门至全开。

② 关停电机。

（7）实训项目七：吸收解吸操作

① 开车操作：按照离心泵操作规程启动吸收液离心泵，向吸收塔输入吸收剂，充分润湿吸收塔填料。启动混合气空气风机，开启二氧化碳钢瓶调节二氧化碳流量与空气混合，控制空气与二氧化碳流量比例，使吸收塔入口混合气组成达到一定数值。按照旋涡风机操作规程启动解吸风机，启动解吸液离心泵，向解吸塔输送解吸液，与解吸空气接触进行解吸，根据操作情况适当开启吸收液冷却器和解吸液加热器。

② 稳定运行：调节混合气、液体和解吸气流量，保持混合气组成、气液温度和流量、吸收塔和解吸塔压降等参数稳定，监测并记录流量、温度、压降、液位等数据，待系统稳定后，取样、进行组成分析。改变气、液流量，重新获得系统稳定，测量不同条件下系统的各个参数，从而获得吸收、解吸过程的流体力学和传质性能。

③ 停车操作：关闭二氧化碳钢瓶总阀，关闭减压阀，关闭空气风机，按照离心泵和旋涡风机停车操作程序关闭吸收液离心泵，关闭解吸液离心泵，关闭解吸气风机。关闭总电源，放空系统内积液，清理现场。

说明：在实训之前，应对装置进行全面的认识和了解，熟悉装置中所有设备、管路、阀门、仪表，列出所有设备、阀门、仪表清单，明确其原理和作用；在各实训之前应制订详细的操作规程，操作规程通过后方可进行实训操作；吸收解吸操作实训应在其他实训项目均完成的前提下方可进行。

Introduction

The experiment of chemical engineering principles is a required course for chemical engineering, pharmaceutical engineering, environmental engineering, food engineering, biological engineering, and related majors. It is a very important practical teaching part in the chemical engineering principles course system, as well as a bridge between theory and practice. Belonging to the category of engineering experiments, each project in the experiment of chemical engineering principles is equivalent to a unit operation in chemical production, and certain engineering concepts could be established through these experiments. At the same time, a large number of engineering problems will be encountered during the experimental process. Therefore, the principles and testing methods of engineering experiments will be learned more effectively here. It would be possible for students to establish a preliminary understanding on the structure of each chemical unit operation equipment, to master the equipment operation methods such as starting, stopping and keeping a stable operation of it; to discover the relationship between complicated real processes and the mathematical models that describe them; to use modern tools such as advanced instruments for analysis and testing, computer software for data processing; as well as to master the basic skills of how to apply engineering research methods such as dimensional analysis method, parameter combination method, equivalent method to solve complex engineering problems. During the whole course, students have been cultivated and improved in basic theory, engineering quality, problem analysis, experimental research, thinking methods and innovative ability, laying a solid foundation for future study and work.

0.1 Teaching objectives

The teaching objectives are mainly as follows.

(1) To consolidate and deepen theoretical knowledge

On the basis of theory of chemical engineering principles, a series of experimental research could be carried out through some typical chemical processes and equipment operations that have been or will be widely used, thus further consolidate and deepen the theoretical knowledge learned previously.

(2) To provide an opportunity to combine theory with practice

Each experimental equipment is a miniaturized factory operation unit. Students can personally be in touch with engineering equipment and apply the theoretical knowledge to solve various practical problems encountered.

(3) To develop the ability on engineering experimental research

① The ability to design experimental programs using engineering research methods to complete specific research projects.

② The ability to observe and analyze experimental phenomena and solve problems encountered during the experiments.

③ The ability to properly select and use measuring instruments and the ability to complete experiments autonomously through teamwork.

④ The ability to select and apply appropriate methods and softwares to process raw data to obtain results.

⑤ The ability to write reports correctly and completely using text, charts, diagrams, etc.

(4) To improve one's overall quality

Through a certain number and different levels of experimental training, a variety of experimental skills should be mastered, the methods to obtain new knowledge and information through experiments should be learned, and one's overall quality should be improved.

0.2 Teaching characteristics

This course emphasizes practicality and engineering concepts, and trains ability and quality throughout the entire process. Demonstration experiments, virtual simulation experiments, as well as design, research, and comprehensive experiments are set up around the basic theories in "Principles of Chemical Engineering" . This will not only train students to master experimental research methods, but also develop an ability to think independently, analyze and solve problems comprehensively.

In addition to completing basic experimental teaching, experimental places as well as basic mechanical parts are also provided for students with learning potential to conduct self-designed experiments. Students can also make appointments to participate in advanced experiments or practical experiments. The ability to independently complete engineering experimental research, practice and inquiry learning is encouraged for further developed.

Some experimental reports in this course are required to be written in the form of research paper. The writing in this form is important to improve students' writing ability, comprehensive knowledge application ability and scientific research ability. A solid foundation could be laid for the writing of graduation papers and the scientific research papers required for future work.

0.3 Teaching content and methods

Engineering experiment is not only a technical work, but also an important technical subject, with its characteristics and systems. Therefore, we especially set up a course for the experiment of chemical engineering principles, consisting of three parts, which are experimental theory teaching, demonstration experiments and virtual simulation experiments, and unit operation experiments.

(1) Experimental theory teaching

It mainly describes the purposes, requirements and methods of the experiment, the characteristics and research methods of the experiments, the error analysis and processing methods of experimental data, as well as the safe operation procedures.

(2) Demonstration experiments and virtual simulation experiments

The demonstration experiments and virtual simulation experiments are with open teaching methods, which are completed independently by students after school. The virtual simulation experiments include three parts: simulation operation, data processing and experimental evaluation.

(3) Unit operation experiments

Unit operation experiments include required and elective experiments. The required experiments include fluid flow resistance experiment, centrifugal pump experiment, filtration experiment, heat transfer experiment, absorption experiment, rectification experiment, drying experiment, as well as advanced liquid-liquid extraction experiment. Each experiment is scheduled for two links of preview and in-site operation. Being highly engineering, the course requires many problems to be considered and analyzed in advance, and necessary preparations must be made. Therefore, it is important to conduct a preview and complete demonstration experiments and virtual simulation experiments before the on-site operation.

Elective experiments are available for students who have the ability or interest for further study, including fluid flow resistance experiment, fluid transport mechanical characteristics experiment, heat transfer enhancement experiment, shell-and-tube heat exchanger self-designed experiment, absorption-desorption experiment, and multi-effect evaporation comprehensive training experiment. The research goals, content, and operating procedures are all set by students, and the experimental projects will be completed independently after those preliminary work being reviewed by teachers.

0.4 Teaching requirements

The teaching of course includes: preview (including demonstration experiments and virtual simulation experiments), experimental operations, data testing and processing, and reports writing. The specific requirements for each part are as follows.

(1) Preview

To complete all the experimental tasks, it is not enough only to know the principles. The following points must also be done.

① Read the experimental handouts carefully, review the relevant content on the text book of "chemical engineering principles" as well as other references, write a preview report, try to ask questions about each experiment, and take the preview class with questions in mind.

② Go to the laboratory site to be familiar with the structure and process of the equipment.

③ Clarify the operating procedures and parameters to be measured, understand the types and methods of related instruments, as well as parameter adjustments and distribution of test points.

④ Complete the demonstration experiments and virtual simulation experiments.

(2) Operations

Experiments are usually carried out in groups of 3 to 4 students. Before entering the laboratory, students must understand the relevant knowledge of electricity safety, as well as prevention of fire, explosion and poisoning. After entering the laboratory, the location of the distribution box, fire extinguisher, protective equipment and emergency exit should be properly observed. There must be both division of labor and cooperation among team members. Each team member has their own responsibilities, and it is necessary to carry out rotation work at an appropriate time. In this way, each student can obtain comprehensive training on the premise of ensuring the overall quality of experimental training. The operation precautions are as follows.

① The start-up operation of the equipment should be carried out item by item according to the procedures described in the textbook. Before starting up, students must check whether the status of each valve is as required. The operation can only be started when the inspection is qualified according to the instructor.

② If there is an abnormality in the equipment or instrument during the operation, immediately stop the equipment according to the shut-down procedures and report to the instructor. At the same time, make full use of this opportunity to study the cause of the abnormality, since it would be an excellent opportunity to develop the ability to analyze and solve problems.

③ Observe the changes in the indicator value all the time to ensure that the operation is performed under stable conditions. When the phenomenon that does not conform to the law appears, attention must be well paid to observation, research, and analyze the cause. No problems should be let go easily.

④ Pay attention to observe the various phenomena and the changes of all measured data, carefully consider the internal laws and reasons for the changes, and make a correct, complete and standardized data record.

⑤ When shutting-down, strictly follow the instructions, turn off the relevant air source, water source, power source, etc. , and restore each valve to the status (open or closed) before the experiment.

(3) Data obtaining

1) Decide which data to test All parameters related to experimental results or necessary for data processing should be tested one by one. Before the experiment, the raw data record table should be designed, which should include the nature of the working medium, operating conditions, equipment geometry, and atmospheric conditions.

2) Distribution of data points　In general, although there are many data to be measured during the experiment, one of the data is often selected as the independent variable to control, the other data affected or controlled by it as the dependent variable. For example, when drawing the characteristic curve of a centrifugal pump, the flow rate is taken as an independent variable. Then data such as heads, shaft power, and efficiency that are related to the flow rate are used as dependent variables. In data processing, it is often necessary to mark the data on various coordinate systems. In order to make the data points form a uniform distribution curve on the coordinates, a consideration of how to choose experimental measurement points for uniform segmentation is required. The most commonly used coordinates for experiments of chemical engineering principles are Cartesian and Double-logarithmic coordinate systems, and the division methods are different between them. The relationship between its segmentation value x and the number of measurements n predetermined in the experiment, and its maximum and minimum control amounts x_{max}, x_{min} are as follows.

① Cartesian coordinate system

$$x_i = x_{min} \qquad \Delta x = \frac{x_{max} - x_{min}}{n-1} \qquad \Delta x_{i+1} = x_i + \Delta x$$

② Double-logarithmic coordinate system

$$x_i = x_{min} \qquad \lg \Delta x = \frac{\lg x_{max} - \lg x_{min}}{n-1}$$

$$\Delta x = \left(\frac{x_{max}}{x_{min}}\right)^{\frac{1}{n-1}} \qquad x_{i+1} = x_i \cdot \Delta x$$

3) Data reading and recording

① Data can be read and recorded only after all parts of the equipment are operating normally and stably. How to judge whether the system has reached a stability? Generally, when the values of two determinations are the same or very close, it can be regarded as stable. After changing the operating conditions, it takes a certain period for each parameter to stabilize. Therefore, it is necessary to wait for stabilization before reading. Otherwise, the experimental results may appear irregular or even abnormal.

② Under a certain operating condition, different data should be read by several people at the same time. If one operator has to read several data at the same time, the reading should be done as fast and accurate as possible.

③ Each reading should be compared with other relevant data and the previous data to see if the correlation is reasonable. If not, find the cause, retest, and indicate on the record.

④ The data should be recorded as the original value, instead of as the result of a calculation. For example, when stopwatch reads 1 minute 23 seconds, it should be recorded as 1′23″ instead of 83″.

⑤ The data should be read according to the accuracy of the meter to a single digit below the minimum division of the meter, which is an estimated value. For example, the minimum scale of a mercury thermometer is 0. 1℃. If the mercury column reads exactly 22. 4℃, it should be recorded as 22. 40℃. Note that it is meaningless to take more than necessary digits of the estimated value.

If some parameters fluctuate greatly during the reading process, try to reduce these fluctuations. In the case that the fluctuation cannot be completely eliminated, the highest point and the lowest point of the fluctuation can be recorded and the average of the two can be taken.

⑥ Do not modify the data by subjective speculation, and do not discard any data easily. For suspicious data, unless there are obvious reasons as misreading or miswriting, it should generally be checked and analyzed during data processing.

⑦ After the data are recorded, check them carefully to see if there are any omissions or mistakes. Pay special attention to the measurement unit of the meter. When the experiment is completed, the original data tables must be submitted to the instructor for inspection and signature, and the experiment can be terminated after the instructor's confirming.

(4) Data processing

The original records can only be collated, and must not be modified at will. Incorrect data that are determined to be caused by negligent errors must be specified before it is excluded from the result. The data processing should be done with calculation examples, and the processing results can be displayed by the tabular method, graphical method or regression analysis method, with the experimental conditions indicated. For the three methods listed above, see Chapter 3 Experimental Data Processing in details.

(5) Writing experimental reports

The experimental report is a comprehensive and systematic summary of experimental work, and is an indispensable part of the practical procedure. With significant engineering characteristics, the experiments of chemical engineering principles are dealing with complex practical and engineering problems. Therefore, the report can be written in traditional experimental report format or research paper format.

1) Traditional experimental report format

① Cover: The main content includes the name of the experiment, the name of the reporter/author, student number and class of the reporter, and the names of teammates, the location and dates of the experiment, and the instructor.

② Purpose: Concisely explain why this experiment was performed and what problems it was designed to solve.

③ Experimental principle: Which is the theoretical basis of the experiment. Briefly explain the basic principles on which the experiment is based, including the main concepts involved in the experiment, the important laws, formulas and deriving results. It is required to be accurate and sufficient.

④ Experimental equipments and flow chart: Briefly draw the equipment flow chart and the specific locations of the test points and control points in it. Mark the numbers of devices, instruments and control valves, write the name of the chart as well as the names of the devices and instruments corresponding to the numbers below flow chart.

⑤ Experimental steps: The operation process is divided into several steps according to the actual operation order, and a sequence number should be added in front of each step to make it clearer. The description of the operation process should be simple and clear.

⑥ Precautions: For operations that are likely to cause the dangerous event of damage to equipments or instruments, and some operations that may have a significant effect on the experimental results should be noted in the precautions to attract attention.

⑦ Raw data records: The tabular method is usually used to record the values read from the measuring instruments during the experiment.

⑧ Data processing: Data processing is a key point of the experiment report, and the sorting and calculating of the raw data are required here. The data calculation process should be illustrated, which is, take a set of raw data as an example, list each calculation process (which explains how the numbers in the result table are obtained), and make the calculation result into a table or graph that is convenient for analysis and discussion, or return the relationship between variables. The table should be easy to show the change law of the data and the correlation of various parameters, and the graph should be able to express the correlation between variables intuitively.

⑨ Analysis and discussion of experimental results: It can reflect the experimenter's theoretical level, and it is an important part of engineering experiment reports, with a comprehensive analysis and research on experimental methods and results. Its main contents include the following aspects.

ⅰ. Analyze and explain the experimental results theoretically, explain its inevitability.

ⅱ. Analyze and discuss the phenomena in the experiment, explaining the internal causes and the main influencing factors.

ⅲ. Analyze the magnitude and cause of the deviation, point out measures to improve the quality of the experiment.

ⅳ. Compare experimental results with those of predecessors and others, explain the similarities or differences.

ⅴ. The value and significance of the results of this experiment in practice, prediction of effect of application and promotion, etc.

ⅵ. Based on experimental results, propose further research directions or suggest improvements to experimental methods or devices.

⑩ Experimental results: The conclusion is the final judgment made based on the experimental results. It must be based on reality, with a theoretical basis, and be concise.

⑪ References: Same as the format described in the following research paper.

2) Research paper format Scientific papers are with unique writing formats, and often includes the following parts: title, author, organization, abstract, key words, foreword (or introduction, preface), text, conclusion, acknowledgment, references, etc. As for specific requirements, *Presentation of theses and dissertations* (GB/T 7713.1—2006) could be referred to.

① Title: The title is also called the headline. It is the general outline of the paper, the basis for document retrieval, the essence of the entire article, a guide for readers to judge whether to read or not. Therefore, the title is required to accurately reflect the main content of the paper.

② Author and organization: The authors are limited to those who select research topics and formulate research plans, directly participate in all or part of the research work and make major contributions. Those who participate in writing papers and are responsible for the content, should be ranked according to their contribution. The organization is written under the author's name.

③ Abstract: The purpose of writing abstract is to make it easier for readers to understand what problems are studied in this paper, what methods are used, what results are obtained, and what is the significance of these results. The abstract is a general statement without explaining or commenting on the content of the paper, and it is a high concentration of the full text. The length of abstracts is generally within 300 words.

④ Key words: Key words are words or terms selected from the paper, which play a key role, mostly illustrate the issue, or represent the content of the paper, and are easy for document retrieval. Usually 3 to 8 key words could be chosen.

⑤ Preface: The preface is the beginning of the main body of the paper, briefly explaining the purpose and scope of the research work, previous work and knowledge gaps in related fields, theoretical foundations and analysis, research assumptions, research methods (to be mentioned, yet no detailed description needed), expected results and significance. It should be concise and different from the abstract. Shorter papers use a short paragraph for brief explanations, without the words "foreword" or "preface."

⑥ Text: This is the core part of the paper, which occupies the main space, mainly including experimental methods, equipments and materials used, experimental and observation results, calculation methods, data, processed charts, formed experimental results, and analysis and discussions of experimental results. The form of this part is mainly determined by the author's intention and the content of the article. The form of this part is mainly determined by the author's intention and the content of the article. It is impossible and should not be stipulated to write in a unified form, However, it is required to be realistic, objective, true, accurate, logical, concise and readable. This section can be divided into several titles according to the content of the paper.

What needs to be emphasized is that the analysis and discussion of experimental results are the main points of the text and the basis on which conclusions are derived. The experimental results need to be further sorted out, from which data or phenomena that mostly reflect the nature of things should be selected and made into graphs or tables that are easy to analyze and discuss. The results and analysis must include the results obtained, but also explain the credibility, reproducibility, error, comparison with the theoretical or analytical results, and the establishment of empirical formulas.

⑦ Conclusion: The conclusion is the viewpoint analyzed and summarized in the theoretical analysis and calculation results (experimental results) of the paper. It is based on the results and discussion (or experimental verification) as the premise and the final judgment made by strict logical reasoning. The conclusion is the essence of the entire research process, from which the level of research results can be seen. Suggestions, research ideas, suggestions for improving equipments and instruments, and issues to be resolved could also be mentioned in conclusions or discussions.

⑧ Acknowledgement: The purpose of acknowledgement is to show respect for the work of all partners. The acknowledgment should be made to persons who have contributed to the research work and paper writing except for the author, such as experts and professors who have supervised the research, those who have helped to collect and organize the data, and those who have revised research and paper writing.

⑨ References: The references reflect the author's scientific attitude and the basis for his research work, as well as the starting point and depth of the paper. Readers can be reminded to consult the original literature, while also the author's respect for the work of others is expressed. The bibliographic description method is sequential coding. The sequential coding system refers to the documents cited by the author in the paper in the order they appear in the article, using Arabic numerals and square brackets to continuously encode, depending on the specific situation, the serial number as the upper corner or as a part of the sentence is marked, and in the reference table after the article, each article is listed in the order of the article number appearing in the paper. For details, please refer to the National Standard *Information and Literature Reference Rules* (GB/T 7714—2015) .

For example, when the cited document is a single article of a journal paper, the description format is: " [No.] Primary responsible person. Document title [J] . Journal name, year, volume (issue) : start and end page number. " .

When the cited literature is a book, the description format is: " [No.] Primary responsible person. The document title [M]. Version (the first version without mark). Place of publication: publisher, publication year: start and end page number. " .

When the cited document is an electronic document, the description format is: " [No.] Primary responsible person. Electronic document title [electronic document and carrier type identification] . Publication or access to address of electronic literature, Publication update date/citation date.

⑩ Appendix: The appendix is at the end of the paper as a supplementary item, and is not required. For some large quantities of important raw data, calculation procedures, materials that are too large to be used as text, and materials that are of reference value to professional colleagues, they can be placed at the end of the paper as appendices (after the reference). Each appendix should be on a separate page.

⑪ Foreign summary: For published papers, some journals require abstracts in foreign languages, in which Chinese title, abstract and key words are usually translated into English. Its location varies from publication to publication.

Writing an experimental reports of Chemical Engineering Principles in the form of a paper is an important mean of developing comprehensive quality and ability. This form of experimental reports should be promoted, especially for research-type experiments, self-designed experiments, and practical experiments. No matter what kind the experimental reportis, it should be academic, scientific, theoretical, normative, creative, and exploratory.

0.5 Basic safety knowledge of the experiment

The equipments used in experiments of chemical engineering principle integrates electrical appliances, meters and mechanical transmission equipments, and chemicals are mostly used. Therefore, before entering the laboratory, the student must understand the safe use of water, electricity and gas, as well as the safety knowledge of fire, explosion and poison prevention, and the experimental rules must be strictly observed.

0.5.1 Experimental rules

① Students are required to go to designated laboratories for experiments on time. Being late, leaving early, and being absent from class without justification are not permitted.

② Before the experiment, students should accomplish a preliminary review, clarify the purposes and requirements of the experiment, master the content, methods, and steps of the experiment, understand the basic situation of the equipments, and receive the instructor's inquiries and check. After the inspection, the experiment can be carried out.

③ Students must pay attention to the dress code for the laboratory (such as not wearing vests, shorts, skirts, and slippers, etc.). Those with long hair must tie up their hair. During the experiment, activities unrelated to the experiment must not be performed, irrelevant equipment or instrument switches must not be touched, and keeping quiet is required throughout the experimental process.

④ It is necessary to operate correctly in accordance with the experimental procedures, observe carefully, record the experimental data in real, do not modify the original records, do not copy the data of others, and do not leave the operation post or interfere with others without authorization.

⑤ During the experiment, students must pay attention to safety, and should choose appropriate protective equipments (such as goggles, gloves, etc.) according to the nature of the experiment. In the event of an accident, keep calm and take prompt and appropriate measures (such as cutting off the power or gas source, promptly rinse with water for chemical burns, etc.) to prevent the accident from expanding and report to the instructor immediately.

⑥ During the experiment, if any equipment is found to be faulty or damaged, it should be reported to the instructor in time to handle it, and the problematic equipment should not be allowed to continue to run anymore.

⑦ For open laboratories, the students should make an appointment in advance (online or direct appointments) to report their experimental purpose, content and required equipment. After the approval, it can be carried out within the time arranged by the laboratory management staff.

⑧ Take care of the equipment, save water, electricity, gas and experimental materials. After the experiment is completed, the experimental data should be signed and confirmed by the instructor, the equipment used should be arranged, the experimental site should be cleaned up, the power and gas sources should be cut off. It is not permitted to leave the laboratory without the instructor's inspection.

0. 5. 2 Basic safety knowledge

Water, electricity, gas and chemicals are generally used in the experiment of chemical engineering principles. With the existence of all these unsafe factors, any omissions or errors may lead to security accidents. Therefore, all personnels in the laboratory must be familiar with necessary relevant safety knowledge and accident prevention technology as well as certain accident handling skills. The most basic safety basics related to chemical engineering principle experiments are listed in this section. Experiments with special safety requirements will be listed in the precautions of each experiment.

0. 5. 2. 1 Electricity safety

(1) Regulations for electricity safety

There are many electrical equipments used in the experiment of chemical engineering principles. Therefore, paying attention to electricity safety is very important. The following rules must be strictly abided by before, during and after the experiment.

1) Before the experiment, the following must be done.

① Know the location of the main switch in the laboratory so that the power can be cut off in time in case of an electrical accident.

② Check the electrical equipments and circuits for compliance. Check the electrical wires for exposure, leaks, whether there are protective grounding or zeroing measures.

③ Before turning on the power, the operating sequence of the start-up and shut-down of the equipments and the methods of emergency shut-down must be clearly understood.

2) During the experiment the following must be done.

① Keep electrical equipments dry and clean. All electrical equipments must not be wiped with a wet cloth when they are charged, and no water should fall on it. The operator's hands must also be dry.

② If there is a trip phenomenon or a fuse blowout during the experiment, an immediately check of whether there is any problem with the equipment should be immediately performed. After changing the fuse, no switch on until further check, or damage could be caused to the equipment.

③ The fuse on the power supply or equipment should be used according to the specified current standard. It is strictly forbidden to use thick fuses or take other metal wires as substitute.

④ If the experimental equipment occurs overheating or burnt smell, the power should be cut off immediately.

⑤ The equipments must be powered off during maintenance.

3) After the experiment, the following must be done.

① Turn off the heating switch, pump or fan switch, meter switch, main power switch of the equipment in order. Pay attention to cutting off the power switch of the heating electrical equipments.

② Before leaving the laboratory, pull down the main switch.

③ In the event of a power outage, all electrical switches must be pull down to prevent the operator from running on without monitoring when sudden power restoration.

(2) Principles for first aid of electric shock

1) Quickly disconnect from power supply To remove the injured person from the power supply: a. cut off the power switch; b. if the power switch is far away, use dry wooden sticks, bamboo poles to disconnect the electric wires or live equipment from the shock person; c. cover hands with several layers of dry clothes or stand on a dry wooden board, then pull the shock person's clothes off the power. Note: no direct touch before the electric shocker is disconnected from the power supply.

2) In-place first aid After the shocker is disconnected from the power supply, observe whether his mind is sober. Those who are conscious should lie flat on the ground and be closely observed, no standing or moving. If they are unconscious, lie on their backs on the ground and ensure that the airway is unobstructed, call or pat the shocker for consciousness every 5 seconds. Do not shake the shocker's head to call. Contact the medical department.

3) Use artificial respiration accurately Check the breathing and heartbeat of the shocker. In case of breath stopping or cardiac arrest, the correctly implementation of cardiopulmonary resuscitation (CPR and artificial respiration) should be done immediately, and contact the medical department for treatment as soon as possible.

4) Persist in rescue Persistence is the hope of the resurrection of the shocker, and the most effort should be made even for the least hope.

0.5.2.2 Chemicals safety

1) Before entering the laboratory for experiment, students must know the name of the chemicals used in this experiment. According to the name, go online search or check the relevant manual to understand the nature of the chemicals and precautions for use. For example, check the MSDS (Material Safety Data Sheet), which provides items including physical and chemical parameters, explosion performance, health hazards, safe use and storage, leak response, first aid measures and related laws and regulations. The correct use of chemicals and first aid measures must be mastered by studying relevant materials.

2) Each laboratory is equipped with eyewash devices and emergency spray devices. The correct using methods should be studied before the experiment.

3) The laboratory should be ventilated.

4) According to the nature of the chemical used, choose the appropriate protective equipments, such as goggles and protective gloves. However, it should be noted that no one kind of gloves can protect all chemicals. The natural rubber has good protection against dilute aqueous solutions, but it is easily penetrated by oil, grease and many organic solvents. Nitrile

rubber gloves can be used to protect oils and fats, but they cannot protect aromatic substances or halogen-containing solvents. Therefore, a reasonable choice is required.

5) In case of chemicals emergency, the following methods should be used.

① Strong acids, strong bases and some other chemical substances have strong irritability and corrosivity. When chemical burns occur, they should be washed with much flowing water, then with low concentrations (2% ~ 5%) of weak bases (caused by strong acids) or weak acid (caused by strong bases) to neutralize and then seek medical treatment.

② If acid (or base) accidentally gets into your eyes, immediately flush them with plenty of water or saline immediately. When rinsing, place your eyes above the eyewash device. Rinse your eyes with water upwards. The rinsing time should not be less than 15 minutes, and do not close your eyes due to pain. After these treatments, go to the eye hospital for further examination and treatment.

③ If sore throat occurs in the experiment, it may be caused by the inhalation of irritating gas . The patient should immediately be transferred to a safe place, the collar button unfastened, and breathing of fresh air made unobstructed. Seek medical treatment in time.

0. 5. 2. 3 High-pressure gas cylinder

High-pressure gas cylinders are high-pressure containers that store various compressed or liquefied gases. Standard high-pressure gas cylinders are manufactured in accordance with national standards and can only be used after strict inspection by relevant departments. Gases in cylinders are generally divided into flammable gases (including hydrogen, methane, ethylene, propylene, acetylene, methyl ether, liquid hydrocarbons, methyl chloride, carbon monoxide, etc.), combustion-supporting gases (including oxygen, compressed air, chlorine, etc.), non-combustible gases (including nitrogen, carbon dioxide, neon, argon, etc.) and toxic gases (including ammonia, chlorine, hydrogen sulfide, etc.) . The outer surface of each types of cylinders is coated with a certain color of paint, in order to make it capable of quickly distinguishing the type of gas to avoid confusion. *Cyclinder Color Mark* (GB/T 7144—2016) stipulates the requirements for the coating color, typeface, character color, color ring, ribbon and inspection color mark on the outer surface of the cylinder. Table 0-1 lists the colors and labels of cylinders commonly used in experiments of chemical engineering principles.

Table 0-1 Colors and labels of common gas cylinders

Gas type	Cylinder color	Text	Text color	Valve outlet thread
oxygen	light blue	oxygen	black	positive
hydrogen	light green	hydrogen	red	negative
nitrogen	black	nitrogen	white	positive
helium	silver gray	helium	dark green	positive
compressed air	black	compressed air	white	positive
carbon dioxide	aluminum white	carbon dioxide	black	positive
ammonia	light yellow	ammonia	black	positive

The main dangers of using a cylinder are explosion or leak. The main reason for the explosion of an inflated cylinder is that the gas expands due to heat and the pressure exceeds the maximum load of the cylinder; or if the bottleneck thread is damaged and the internal pressure rises, the cylinder will fly at high speed in the opposite direction of the gas emitted, or the cylinder will explode when it falls or hits a hard object, which can cause great damage and casualties. Therefore, pay attention to the following when using it.

① Different types of gas cylinders must be stored in strict accordance with national standards or internal industry standards. For example, flammable and combustion-supporting cylinders must be stored in separate compartments. It is strictly prohibited to put them together (such as hydrogen cylinders and oxygen cylinders) ; hydrogen cylinders should be stored separately, and it is best to place them in outdoor cabins. The number of hydrogen bottles must not exceed 2 bottles and the maximum hydrogen content in the air must not exceed 1% (volume) . The cylinder should be placed without exposed to sunlight and strong vibrations. It should not be near heat sources and open flames. The cylinder body should be kept dry.

② Places involving toxic, flammable and explosive gases must be equipped with ventilation facilities as well as monitoring and alarming devices, posted safety warning signs. The distance between flammable cylinders and combustion-supporting cylinders and open flames shall not be less than 10 meters.

③ All cylinders must be inspected regularly. On the shoulder of the cylinder, there are marks printed with steel stamps: manufacturer, date of manufacture, cylinder model, working pressure, air pressure test pressure, air pressure test date and next inspection date, gas volume, cylinder weight, etc. If serious corrosion or serious damage is found during the inspection, it should be inspected in advance. In the event of a bottle valve failure, do not remove the bottle valve or its parts without permission. In addition, the gas cylinders used must have a certificate of compliance provided by the inflatable manufacturer.

④ When moving the cylinder, a cylinder cap and a shockproof ring must be placed well. It is best to carry it with a special stretcher or trolley. It can also be lifted by hand or turned vertically, but it must not be moved with hands on the cylinder valve.

⑤ When the cylinder is placed vertically, it should be fixed securely. According to the shape of the cylinder, appropriate safety devices and anti-tip devices should be adopted.

⑥ All gas cylinders must be equipped with pressure reducing valves, the screws must be tightened during installation to prevent leakage. The selected pressure reducing valves should be used exclusively. The high-pressure chamber of the pressure reducing valve is connected to the gas cylinder, and the low-pressure chamber is the gas outlet and leads to the use system. The indicated value of the high-pressure meter is the pressure of the gas stored in the cylinder, the outlet pressure of the low pressure meter can be controlled by the adjusting screw. Turn the pressure adjustment screw of the low-pressure gauge clockwise to compress the main spring and drive the diaphragm, spring pad and ejector lever to open the valve. Turn the adjustment screw to change the height of the valve opening, so as to adjust the flow of high-pressure gas and achieve the pressure required. In this way, the gas enters the low-pressure chamber from the high-pressure chamber through throttling and decompression, then passes through the outlet to the use system.

⑦ Before opening the main valve of the cylinder, remember to unscrew the adjusting screw of the pressure reducing valve counterclockwise first. When in use, first turn on the main valve of the gas cylinder, then open the pressure reducing valve. After use, close the main valve first, then close the pressure reducing valve after exhausting the remaining gas. Never close the pressure reducing valve without closing the main valve.

⑧ When opening or adjusting the high-pressure gas cylinder valve, the operator should stand at a position perpendicular to the gas cylinder interface instead of in front of the gas outlet to prevent the valve or decompression gauge from hurting people. It is strictly forbidden to knock or bump during operation. Frequently check the pressure gauge readings and leaks.

⑨ Oxygen or hydrogen cylinders are strictly forbidden to come into contact with oil. Avoid mixing flammable gas into the oxygen cylinder. Operators should not wear clothes or gloves that are contaminated with various greases or are sensitive to static electricity, as this may cause combustion or explosion.

⑩ The residual pressure of the cylinder should be above 0. 05MPa after use. The flammable gas should be 0. 2～0. 3MPa, the hydrogen should be kept at 2MPa to prevent danger when it is refilled. The gas cylinders that need to be returned to the factory must be labeled.

0. 5. 2. 4 Fire knowledge

(1) Common fire equipment

The laboratory is equipped with a certain number of fire-extinguishing equipments in accordance with regulations. The experimenters should be familiar with the storage location, the types and use methods of these fire-extinguishing equipments. Moreover, escape routes in different directions should be understood. It should be particularly emphasized that in the same broad category of fire-extinguishing equipments due to the different models, styles, contents, etc., their use methods will also be different. For example, chemical foam fire extinguishers should be used upside down, while air foam fire extinguishers cannot. For dry powder fire extinguishers, there are external pressure storage type, built-in gas storage cylinder type, opening methods such as lifting ring type, hand wheel type, pressure handle type, etc., so experimenters must learn to use it before entering the laboratory.

1) Fire Sandbox Sand isolates the flame from the air and lowers the flame temperature to extinguish the fire. Flammable liquids and other dangerous goods that cannot be extinguished by water can be extinguished with sand when they catch fire. It should be noted that the sand must not be mixed with flammable debris and it must be dry, because the wet sand may cause the burning matter to splash due to the evaporation of water during the fire. Due to the limited amount of sand stored in the sandbox, it can only be used to extinguish local small-scale fire.

2) Asbestos, fire blanket or wet cloth The principle is to isolate the air. It is suitable for quickly extinguishing small area fires, and it is also a common method for extinguishing clothes on fire.

3) CO_2 fire extinguisher The carbon dioxide fire extinguisher is filled with liquid carbon

dioxide. Carbon dioxide gas can exclude air and surrounded the surface of burning objects or be distributed in a more confined space, reducing the oxygen concentration around. In addition, when carbon dioxide is ejected from the storage container, it will rapidly vaporize from liquid to gas and absorb part of the heat from the surrounding.

When in use, first pull out the safety pin, and then press the handle to point the nozzle at the root of the flame. Note: Keep hands on the wooden handle of the cylinder to prevent frostbite, and prevent personnel suffocation on the spot.

It is mainly used to extinguish the initial fire of valuable equipments, archives, instruments, electrical equipments below 600 volts and oil.

4) Foam fire extinguisher Portable chemical foam fire extinguishers are commonly used in laboratories. The outer shell is made of iron. It contains a mixed solution of sodium bicarbonate and a foaming agent and a glass bottle liner, which is filled with an aluminum sulfate aqueous solution. The ejected large amount of carbon dioxide gas foam can adhere to the combustible materials to isolate it from the air, reduce the temperature to achieve the purpose of extinguishing the fire.

When in use, do not invert too early to prevent unexpected ejection. When it is about 10 meters away from the ignition point, the barrel can be turned upside down. Hold the lifting ring with one hand and hold the bottom ring of the barrel with the other. Shake it a few times and aim the jet at the burner. The fire extinguisher should always be kept upside down, otherwise the spray will be interrupted.

It can be used to extinguish solid material fires such as wood, cotton, paper, etc., liquid fires such as gasoline, diesel, etc., but not for water-soluble flammable liquid fires, such as alcohol, ester, ether, or ketone. It cannot be used for gas, metal and precision instrument fires, nor for live equipment, because the foam is conductive, which may cause electric shock to fire extinguishers. The power should be cut off before extinguishing.

5) Dry powder fire extinguisher The dry powder fire extinguisher is provided with fire extinguishing agent such as ammonium phosphate. With easy flowability and drying property, this agent is composed of inorganic salts and pulverized, dried additives. The dry powder fire extinguisher uses carbon dioxide gas or nitrogen gas as the power to spray the dry powder in the cylinder to extinguish the fire. The fire extinguishing principle is that the volatile decomposition products of the inorganic salts in the dry powder and the free radicals or active groups generated by the fuel during the combustion process chemically inhibit and negatively catalyze, which interrupts the chain reaction of the combustion. A chemical reaction takes place outside the surface of the combustibles, and a glass-like covering layer is formed under the action of high temperature, thereby blocking oxygen, and suffocating the fire. In addition, diluting the oxygen and cooling down are also effective.

The most common method for opening a dry powder fire extinguisher is to press the handle. After raising the fire extinguisher to an appropriate position from the fire source, first turn it upside down a few times to loosen the dry powder in the cylinder, and then point the nozzle at the most violent burning point and remove the safety pin. Press the handle, the fire extinguishing agent will spray out. If using an external portable fire extinguisher, hold the nozzle with one hand and lift the lifting ring with the other hand, then the dry powder can be sprayed.

Dry powder fire extinguishers can extinguish common fires, as well as the initial fires of flammable liquids such as petroleum and organic solvents, flammable gases and electrical equipment.

(2) Fire emergency treatment

① Local fire: according to the cause of the fire, immediately use the right fire extinguishing equipment to put out the fire; if there is a large-scale fire, the laboratory staff should notify all personnel to evacuate along the fire passage and report it to the fire department immediately and report to the school leadership.

② When injuries happen, report it to the medical department immediately and ask for support. After the personnel evacuated to the predetermined location, the experimental instructors, laboratory staff and student cadres should immediately count the number of people, confirmed the location of those being absent.

③ If heavy smoke is encountered during escaping, nose and mouth should be covered with a multi-layer wet cloth. Try to crawl as low as possible and never walk upright, in order to avoid being choked by thick smoke.

④ Do not use the elevator when escaping. Leave the fire along the direction of the emergency exit as soon as possible and meet in the open space.

0.6 Tasks

0.6.1 Resistance experiment

Experiment Assignment (1)
Variation of Friction Resistance Loss of Straight Tube with Re

Mission requirements:

① Design the experimental scheme for measuring the resistance loss of water in laminar flow and turbulent flow in circular straight tubes (including smooth tubes and rough tubes) .

② Discuss the regularity of the loss of friction resistance and friction coefficient λ with Re based on experimental results.

③ The experimental results are compared with the Moody friction factor diagram, the rationality is discussed through error analysis.

④ Analyze the cause of frictional resistance loss of straight tubes. Then puts forward the significance and method of reducing friction loss in engineering.

⑤ Measure the local resistance loss of fluid flowing through the tubes, understand the cause of local resistance; master the measurement and calculation methods of local resistance loss.

Experiment Assignment (2) Discussion on the Mechanism of Local Resistance Loss and Several Issues on Reducing Local Resistance Loss

Mission requirements:

① Design the experimental scheme for measuring the local resistance loss of tube fitting under different Re in a certain state.

② Analyze the causes of local resistance. Discuss the significance and methods of reducing local resistance loss in engineering according to experimental results.

③ Compare the calculation method and results of this experiment with the resistance coefficients and equivalent length data of tube fittings and valves found in the relevant manuals, analyze the reasons for the errors.

④ Measure the friction loss of smooth or rough straight tubes; understand the causes of straight tube resistance, master the method of measuring and calculating the straight tube resistance loss.

Experiment Assignment (3) Determining the Roughness of the Tube Wall by an Indirect Method

Mission requirements:

① Design the experimental scheme for measuring the resistance loss of water in laminar flow and turbulent flow in circular straight tubes (including smooth tubes and rough tubes) .

② Analyze the causes of the frictional loss of straight tube and the change of frictional loss of straight tube with Re.

③ Explore the significance of measuring tube wall roughness. The basis and method for indirect measurement of tube wall roughness are proposed.

④ The feasibility of this method is illustrated by comparing the indirect measurement of the wall roughness with the actual one.

⑤ Measure the local resistance loss of fluid flowing through the tube, understand the cause of local resistance, master the measurement and calculation methods of local resistance loss.

0.6.2 Centrifugal pump experiment

Experiment Assignment (1) The Flow Rate Regulation and Energy Consumption Analysis of Centrifugal Pump

In the chemical process, the flow of the pump needs to be adjusted according to changes in production tasks and process conditions. The flow regulation modes of centrifugal pump mainly include outlet valve regulation and motor speed regulation, as well as series and parallel regulation, etc. In addition to their advantages and disadvantages, the energy loss of various adjustment methods are also different. This experiment aims to find out a suitable and energy-saving flow regulation mode through the following experimental assignments.

Mission requirements:

① Design an experimental scheme for measuring the characteristic curve of a single centrifugal pump and the pipeline characteristic curve of high and low resistance pipelines.

② Design an experimental scheme to determine the head curves of two centrifugal pumps of the same model in series and parallel operation at the same speed.

③ Analyze the applicable situations, advantages, disadvantages of outlet valve adjustment and motor variable speed adjustment, series and parallel operation of centrifugal pump.

④ According to the experimental results, the energy consumption is analyzed and a conclusion is drawn through drawing and calculation.

Experiment Assignment (2)
Determination and Adjustment of the Working Point of Centrifugal Pump

When the working point of the centrifugal pump is not in the high efficiency zone or the flow cannot meet the requirements, it is necessary to adjust the working point of the centrifugal pump. In the actual production, the following methods are usually used to adjust the working point: one is to change the characteristic curve of the pipeline, such as adjusting the opening of the outlet valve; the other is to change the characteristic curve of the centrifugal pump, such as changing the speed of the pump, series and parallel operation of the pump, etc. The following tasks are to be completed through experiments.

Mission requirements:

① Design an experimental scheme for measuring the characteristic curve of a single centrifugal pump and the pipeline characteristic curve of high and low resistance pipelines.

② Design an experimental scheme to determine the head curves of two centrifugal pumps of the same model in series and parallel operation at the same speed.

③ Through experiments, analyze and compare the characteristics of outlet valve adjustment, motor variable speed adjustment, series and parallel operation of centrifugal pump.

④ Taking this experimental device as an example, if the working point of the centrifugal pump is not in the high efficiency area for a given pipeline, which way do you think is more reasonable? Why?

Experiment Assignment (3)
The Characteristics of Series and Parallel Operation of Centrifugal Pump

When one centrifugal pump can not meet the requirements of flow or head in actual production, two or more pumps are often combined for operation. Master the characteristics of dual pump operation through experimental research on the characteristics of series and parallel operation of centrifugal pump.

Mission requirements:

① It is necessary to analyze and study the series and parallel operation characteristic curve of the centrifugal pump, master the characteristics of series and parallel pump characteristic curve and the connection with single pump characteristic curve, master the influence of series and parallel pump on pipeline flow and head, etc.

② According to the research content, determine which curves need to be measured, and formulate experimental schemes to complete the experiment.

③ It is necessary to send 20℃ clean water from the storage tank to the water tower. It is known that the water level in the tower is 5m higher than the water level in the tank. The water level of the water tower and the storage tank is constant and connected with the atmosphere. Conveying tube is a ϕ32mm×2mm steel tube with a total length of 10m (or 40m) (including local resistance, λ is taken as 0. 02 in the calculation) . Trial analysis (including calculations, analysis steps and related charts) :

ⅰ. What is the flow rate if a single pump is used for delivery? What is the power and efficiency of the pump consumes during operation? Is it reasonable in economic perspective?

ⅱ. Which operation method (single pump, series connection, parallel connection) is more reasonable to increase the flow rate?

ⅲ. What conclusions can be drawn from experimental studies and case studies?

0. 6. 3 Filtration experiment

Experiment Assignment (1)
The Best Filtration Time and Maximum Production Capacity of Plate and Frame Filter

In one operation cycle of the plate and frame filter, the stages of filtration, washing, slag unloading and reloading are carried out in order. The auxiliary time of slag unloading but reloading is fixed, while the time of filtration and washing can be selected. If the filtering time is too long or too short, the production capacity will be reduced. Therefore, there is an optimal operation cycle to maximize production capacity. According to the equipment of the laboratory, complete the research of the subject.

Mission requirements:

① Design the experimental scheme according to the research theme.

② The parameters necessary for the research on the optimal filtering time and the maximum production capacity of the plate and frame filter are determined through experiments, such as the filtration constant, compressibility index and filtration rate curve.

③ The necessary analysis of the experimental results is made by combining the experimental phenomena and data.

④ Use differential or integral methods to process experimental data, use the graphic method or analytical method to determine the optimal filtering time, production capacity and maximum production capacity of the plate and frame filter, and discuss the influencing factors.

Experiment Assignment (2) The Influence of Filtration Pressure on the Filtration Process of $CaCO_3$ Suspension

Filtration pressure is a very important operating parameter in the filtration process, which has an important influence on filtration rate, filtration constant, production capacity of filter and structure of filter cake. Therefore, the influence of filtration pressure on the filtration process was explored using the laboratory plate and frame filter.

Mission requirements:

① Design the experimental scheme according to the research object.

② Filter parameters such as filter constant and compressibility index are to be determined by experiments.

③ According to the experimental phenomena, the parameters such as filter cake porosity, water content and the internal structure of the filter cake were analyzed under different filtration pressures.

④ The integral method or differential method is used to process the experimental data. the influence of the filtration pressure on the filtration rate, filtration constant and production capacity is explored in conjunction with the graph and table.

Experiment Assignment (3) Design of Industrial Rotary Vacuum Filter

It is necessary to measure the filtration parameters before designing an industrial filter. This work is usually carried out with the same suspension in a small filtration experimental equipment.

A factory needs to filter the suspension containing 5. 0% ~5. 5% $CaCO_3$, while the filtration temperature is 25℃, the density of solid $CaCO_3$ is 2930kg/m^3, the operating vacuum degree of industrial rotary vacuum filter is required to be 0. 08MPa and the production capacity of filtrate is 0. 001m^3/s.

Please use the laboratory plate and frame filter press to conduct experiments. Through the measurement of the relevant filtering parameters, the rotating speed of the rotary vacuum filter n, the immersion degree of the rotary drum φ, the diameter D and the length L of the tumbler are determined.

It can be considered that the resistance of the filter cloth does not change with the filtration pressure and the volume of the filter cake generated from each $1m^3$ of the filtrate obtained does not change significantly, when the design work is carried out.

0. 6. 4 Heat transfer experiment

Experiment Assignment (1)
Enhanced Heat Transfer Process of Spiral Coil

Inserting a spiral coil in the tube is an important measure to enhance the convective heat transfer of the fluid in the tube. What is the enhancement effect of inserting a spiral coil in the tube? What effect does it have on the flow resistance of the fluid in the tube? The double-tube heat exchangers will be used to complete the research on the heat transfer enhancement process of spiral coils.

Mission requirements:

① Consult the data and make a brief review of the current situation and development of enhanced heat transfer by inserting spiral coils into the tube.

② The experimental scheme is designed to determine the heat transfer coefficient K and the convective heat transfer coefficient α_1 for the ordinary double-tube heat exchanger and the double-tube heat exchanger with a spiral coil inserted in the tube.

③ To obtain the correlation formula $Nu = ARe^m Pr^{0.4}$ of the air convective heat transfer coefficient in the two kinds of heat exchange tubes above by means of drawing method or least square method, and to determine the constant A and m.

④ According to the experimental data processing results, the heat transfer effect of ordinary heat transfer tube and the heat transfer tube with spiral coil inserted is compared, and the mechanism of heat transfer enhancement by the spiral coil is discussed.

⑤ Which structural dimensions of the spiral coil in the tube will affect the heat transfer and fluid flow resistance? The advantages and disadvantages of using spiral coil to enhance heat transfer should be also explained.

Experiment Assignment (2)
Measurement of Convection Heat Transfer Coefficient of Fluid in Tube and Design of Tubular Heat Exchanger

The tubular heat exchanger is a kind of heat exchanger widely used in industrial production. Now it is necessary to design a tubular heat exchanger to heat the air to 90℃ with water vapor at 110℃. The air comes from the environment with a flow of $5000m^3/h$, $8000m^3/h$ and $10000m^3/h$ respectively. Please based on the experiment data to make a design of the heat exchanger.

Mission requirements:

① Select a suitable heat exchanger in the laboratory, formulate an experimental scheme, then measure the convective heating coefficient of air in the double-pipe heat exchanger with a spiral coil.

② The empirical correlation of the convective heat transfer coefficient in the tube should be determined according to the experimental results.

③ The temperature of the inlet air and appropriate air flow rate should be confirmed according to the local climate conditions. The design of tubular heat exchanger is based on the relational expression of heat transfer coefficient of convection measured above, and the key structure parameters of the tubular heat exchanger including tube number, tube diameter, wall thickness, tube length, specific type, etc. , can be achieved. Additionally, the air flow resistance should also be checked.

Experiment Assignment (3)
Measurement of Convection Heat Transfer Coefficient Fluid in Tube and Design of double-tube Heat Exchanger

The double-tube heat exchanger has the characteristics of simple structure and large heat transfer driving force, but its heat transfer area is small. The spiral coil is inserted into the tube to enhance the heat transfer, and the heat exchanger is used to preheat the air. The air is heated to 110℃ with 120℃ water vapor, and the air comes from the environment with a flow rates of $80m^3/h$, $90m^3/h$ and $100m^3/h$ respectively.

Mission requirements:

① Select a suitable heat exchanger in the laboratory, formulate an experimental scheme, then measure the convective heating coefficient of air in the inserted spiral coil tube.

② The empirical correlation of convective heat transfer coefficient in the tube is determined according to the experimental results.

③ The temperature of the inlet air and appropriate air flow rate can be confirmed according to the local climate conditions. The design of double-tube heat exchanger is based on the relational expression of heat transfer coefficient of convection measured above, and the key structure parameters of the double-tube heat exchanger, including tube number, tube diameter, wall thickness, tube length, etc. , should be achieved. Additionally, the air flow resistance should be also checked.

0. 6. 5 Absorption experiment

Experiment Assignment (1)
Hydrodynamic Properties and Mass Transfer Performance of Packed Absorption Tower

The absorption experimental device consist of ceramic Raschig ring absorption tower and stainless-steel Bauer ring desorption column. The following hydrodynamic properties and mass transfer performance parameterg are tested by using the device.

Mission requirements:

① In the desorption tower, to measure the pressure drop curve of the packing layer of the dry column, which is the relationship between the superfical velocity u and the pressure drop $\Delta p/z$ of the packing per meter , the u-$\Delta p/z$ function relationship should be related and compared with the theoretical value.

② In the desorption column, measure the pressure drop curve of the packing layer of the wet column when the spray volume is L=100L/h, determine the location of the loading point and flooding point under the spray volume and the division of the three flow areas, then discuss the difference of the hydrodynamic properties between the dry column and the wet column and the engineering significance of measuring the pressure drop curve of the packing layer.

③ In the absorption tower, the flow rate of absorbent is controlled at 80~120L/h, and the concentration of carbon dioxide in the mixed gas entering the tower is controlled at 10%~20% (volume fraction) . Under the premise of the normal operation of the absorption column, take two different liquid flow rates. The height of mass transfer unit H_{OL}, the number of mass transfer units N_{OL}, the total volume absorption coefficient of the liquid phase $K_x a$, the absorption factor A, the composition of gas phase outlet Y_a and the recovery rate of the absorption column η should be measured respectively, The parameters measured under different flow rates of absorbent and the results of material balance should be analyzed and discussed.

Experiment Assignment (2) Hydrodynamic Properties of Packed Columns—Pressure Drop Law and Flooding Law

Using the existing stainless steel Bauer ring packing desorber in the laboratory, the hydrodynamics properities of packing column, pressure drop law and flooding law, were studied.

Mission requirements:

① In the desorption tower, the pressure drop curve of the packing layer of the dry tower, which is the relationship between the superfical velocity u and the pressure drop $\Delta p/z$ of the packing per meter, is to be measured, and the u-$\Delta p/z$ function relationship is related and compared with the theoretical value.

② In the desorption tower, several pressure drop curves of packing layer in the range of spray volume L=0~200L / h should be measured to determine the location of the loading point and flooding point under the spray volume and the division of the three flow areas, then discuss the difference of the hydrodynamic properities between the dry column and the wet column and the engineering significance of measuring the pressure drop curve of the packing layer.

③ The flooding data should be compared with the general correlation diagram of flooding velocity of Eckert packed column, and the main causes of the errors should be analyzed.

④ Complete the measurement of the total mass transfer coefficient $K_x a$ and recovery rate η

of the absorption tower under certain conditions.

Experiment Assignment (3) Suitable Operation Conditions and Mass Transfer Performance of Absorption-Desorption Packed Column

Using a laboratory absorption and desorption device, determine the appropriate operating conditions, the total mass transfer unit height of the liquid phase of the absorption column mass transfer coefficient of the liquid phase and adsorption rate under the conditions that the absorptivity is not less than 5% , $V_{CO_2}/V_{Air} = 0.2$, and the appropriate wetting rate of the packing.

Mission requirements:

① Initially determine the operating conditions of the absorption and desorption tower through theoretical analysis.

② Formulate experimental schemes for the hydraulic performance of desorption columns and absorption columns, measure the pressure drop curves of each tower, and determine the suitable gas and liquid operating ranges of the absorption columns and desorption columns, which lays the foundation for mass transfer experiment.

③ Determine the appropriate operating conditions for mass transfer performance experiments based on the theoretical analysis and hydraulic performance experimental results obtained above.

④ Under suitable operating conditions, determine the parameters of the absorption tower $K_x a$, H_{OL}, N_{OL}, A, Y_a, η and other parameters, discuss the mass transfer performance of the absorption tower, whether or not to achieve the expected goals, and how to adjust parameters if the goals are not achieved.

Experiment Assignment (4) The Characteristics of the Mass Transfer Rate of Carbon Dioxide Absorption in Water

Determine the mass transfer coefficient and recovery rate of carbon dioxide in air absorbed by water under different gas and liquid flow rates by ceramic Raschig ring packed absorption columns. Analysis and discuss the basic characteristics of the mass transfer process of carbon dioxide in the air absorbed by water and the main ways to enhance the mass transfer process.

Mission requirements:

① When the flow rate of mixed gas is $0.8 m^3/h$, the volume fraction of carbon dioxide is 20% , and the absorbent flow rate is 80L/h, 100L/h and 120L/h respectively, measure and calculate the mass transfer unit height H_{OL}, mass transfer unit number N_{OL}, total volume absorption coefficient of the liquid phase $K_x a$, absorption factor A, gas-phase outlet composition Y_a and recovery rate η . The function equation of $K_x a$ and L should be regressed, and the influence of the flow rate of absorbent on the absorption and mass transfer process should be analyzed.

② When the flow rate of the absorbent is 100L/h, the volume fraction of carbon dioxide is 20% , and the flow rate of the mixed gas is 0. 8m^3/h, 1. 0m^3/h, and 1. 2m^3/h respectively, measure and calculate the mass transfer unit height H_{OL}, mass transfer unit number N_{OL}, total volume absorption coefficient of the liquid phase $K_x a$, absorption factor A, gas-phase outlet composition Y_a and recovery rate η. Analyze the influence of mixed gas flow on absorption and mass transfer.

③ Discuss the characteristics of the mass transfer rate and the rate control steps in the process of absorbing carbon dioxide in the air with water based on the results obtained above.

④ Discuss the effective way to enhance the absorption and mass transfer rate of carbon dioxide in water .

0. 6. 6 Distillation experiment

Experiment Assignment (1)
The Effect of Feed Concentration on Operating Conditions and Separation Ability of Distillation Tower

For a certain distillation tower, the feed concentration of cold liquid changes due to the reason of the previous process, which directly affects the distillation operation. Please complete the following tasks based on laboratory equipment and materials.

Mission requirements:

① In a given distillation tower, when the feed concentration changes and other operating conditions do not changed, how will this change affect the quality of the product at the tower top and the tower kettle.

② According to the existing conditions in the laboratory, develop a method for changing the feed concentration and formulate an experimental scheme (including experimental operating conditions, experimental operating methods and precautions, etc.) .

③ Under the condition of total reflux and stable operation, measure the number of theoretical plates, the efficiency of single plate and the overall efficiency.

④ Experiments are performed under continuous distillation and stable operating conditions at a certain reflux ratio. Measure the theoretical number of plates of the whole tower, single plate efficiency and overall efficiency.

⑤ Measure the theoretical number of plates, the efficiency of single plate and the overall efficiency under the condition of different feed concentration.

⑥ According to the experimental results, discuss the influence of feed concentration on the overall efficiency and single plate efficiency, what can be taken to ensure the quality of products in top and bottom of the tower when feed concentration changed.

Experiment Assignment (2)
The Effect of Reflux Ratio on Operating Conditions and Separation Ability of Distillation Tower

For a certain distillation tower, reflux ratio is a parameter which has great influence on product quality and output and is easy to adjust. Please complete the following tasks based on laboratory equipment and materials.

Mission requirements:

① In a given distillation tower, when the feed concentration changes and the operating conditions are not changed, how will this change affect the quality of the product at the top and bottom of the tower.

② According to the existing conditions in the laboratory, develop a method for changing the feed concentration and formulate an experimental scheme (including experimental operating conditions, experimental operating methods and precautions, etc.).

③ Under the condition of total reflux and stable operation, measure the number of theoretical plates, the efficiency of single plate and the efficiency of the overall tower.

④ Experiments should be performed under continuous reflux and stable operating conditions at a certain reflux ratio. Measure the theoretical number of plates of the overall tower, single plate efficiency and overall tower efficiency.

⑤ Measure the theoretical number of plates, the efficiency of single plate and the whole tower efficiency by changing the feed concentration.

⑥ According to the experimental results, discuss the influence reflux ratio on the overall tower efficiency and single plate efficiency, what can be taken to ensure the quality of products in top and bottom of the tower when reflux ratio changed.

⑦ Determine the minimum reflux ratio under one set of operating conditions and calculate the relationship between the minimum reflux ratio and the actual reflux ratio.

Experiment Assignment (3)
The Effect of Feed Position on Operating Conditions and Separation Ability of Distillation Tower

The most suitable feed plate position refers to the feed plate position with the maximum separation ability under the same number of theoretical plates and the same operating conditions, or the feed plate position with the least number of theoretical plates required under the same operating conditions.

In the chemical industry, most distillation towers are equipped with more than two feed plates. The adjustment of the feed position is based on the change of feed components. Please complete the following tasks based on laboratory equipment and materials.

Mission requirements:

① Analyze the effect of changing feed locations on distillation operations and separation ability.

② According to the existing conditions in the laboratory, develop a method for changing the feed concentration and formulate an experimental scheme (including experimental operating conditions, experimental operating methods and precautions, etc.) .

③ Under the condition of total reflux and stable operation, measure the number of theoretical plates, the efficiency of single plate and the whole tower efficiency.

④ Under continuous distillation and stable operating conditions at a certain reflux ratio, measure the theoretical number of full plates, single plate efficiency and overall tower efficiency.

⑤ Measure the theoretical number of plates, the efficiency of single plate and of the whole tower efficiency under the condition of different feeding location.

⑥ According to the experimental results, discuss the influence of feeding location on the overall tower efficiency and single plate efficiency, what can be taken to ensure the quality of products in top and bottom of the tower when feeding location changed.

⑦ Where do you think is the best feed location under the experimental condition? When the feed position changes, how to adjust the operation parameters of the distillation tower to ensure the product quality?

0.6.7 Drying experiment

Experiment Assignment (1)
The Effect of Drying Conditions on Drying Characteristic curve

Compare the constant drying rate of pulp board with certain moisture content, drying area and absolute dry mass under different drying conditions by using the tunnel dryer in the laboratory, and discuss the influence of different drying conditions on the drying rate curve.

Mission requirements:

① Develope the experimental scheme for determining a constant drying rate curve, including experimental procedures, experimental raw data record tables, experimental precautions, etc.

② Different drying conditions are as follows:

ⅰ. The same medium flow rate and different medium temperature (two temperatures) .

ⅱ. Same medium temperature and different medium flow rate (two flow rates) .

ⅲ. Same medium temperature and medium flow rate, different pulp board thickness.

③ According to the experimental results (figures, tables), discuss the effects of different drying conditions on the drying characteristic parameters, such as constant drying rate, reduced drying rate, critical moisture content and equilibrium moisture content, and the engineering significance of measuring the drying rate curve.

Experiment Assignment (2)
Determination of Drying Time Under Constant Drying Conditions

A factory wants to dry pulp boards with a mass of 200kg in a tunnel dryer. The wet content was dried from 1. 76 (kg water/kg absolute dry) to 0. 22 (kg water/kg absolute dry), and the dry surface area was 0. 025 (m^2/kg absolute dry). The air velocity in the tunnel is 2. 6m/s and the temperature is 75℃. In order to determine the residence time of pulp board in the tunnel dryer, a small tunnel dryer and other related testing instruments are provided to measure the constant drying characteristic parameters of the material.

Mission requirements:

① Develope the experimental scheme, including experimental procedures, experimental raw data record tables, experimental precautions, etc.

② Using the measured data from the experiment, explain how do you determine the drying time with the formulas and charts.

③ Discuss the experimental results and explain the engineering significance of measuring the drying rate curve.

0. 6. 8 Liquid-liquid extraction experiment

Experiment Assignment (1)
Hydrodynamic Properities and Mass Transfer Performance of Extraction Tower

The extraction tower has a significant amplification effect, so its hydrodynamic properities and mass transfer performance are very important research objects in the process of amplification. Complete the project research by using spray extraction tower, mechanical agitation extraction tower, packing extraction tower and other devices in the laboratory.

Mission requirements:

① Select a certain type of extraction tower and formulate the experimental scheme (including experimental procedures, experimental raw data record tables and experimental precautions, etc.).

② Measure the retention fraction and flood point of the extraction tower under the conditions of different light and heavy phase flow.

③ Measure the parameters of the height of the mass transfer unit, the number of mass transfer units, the mass transfer coefficient of the extraction tower under different light and heavy phase flow.

④ Explore the effects of light or heavy phase flow on the retention fraction, flood point and mass transfer coefficient of the extraction tower based on the experimental results. Summarize the influence factors and laws of the hydrodynamic properities and mass transfer performance of the extraction tower.

Experiment Assignment (2)
The Effect of Additional Mechanical Energy on the Performance of Extraction Tower

In order to enhance the mass transfer of extraction tower, it can be achieved by external mechanical energy. Common external mechanical energy includes the pulse, mechanical agitation, vibration, etc. Therefore, there are mechanical agitation extraction tower, pulsed packing extraction tower and vibrating sieve plate extraction tower. So, the mode and intensity of me chanical energy input have an important influence on the extraction process. Under the condition of laboratory, research on the effect of mechanical energy on the performance of extraction tower.

Mission requirements:

① Select a certain type of extraction tower as the research object from the mechanical agitation extraction tower, spray extraction tower and packing extraction tower, formulate the experimental scheme (including the experimental steps, the experimental original data record tables and the experimental precautions, etc.) .

② Determine the retention fraction, flooding point, height of mass transfer unit, number of mass transfer units, mass transfer coefficient and other parameters of the extraction tower without additional mechanical energy.

③ Under different agitation speeds and different pulse frequencies, determine the parameters of the mass transfer unit height, the number of mass transfer units, and the mass transfer coefficient of the extraction tower.

④ Explore the effects of the presence or absence of additional mechanical energy on the hydrodynamic properties and mass transfer performance of the extraction tower based on the experimental results. Investigate the influence of mechanical agitation or pulse intensity on the hydrodynamic properties and mass transfer performance of the extraction tower.

Experiment Assignment (3)
The Effect of the Structure and Size of Granular Packing on the Performance of Extraction Tower

The packed tower is simple in structure and easy to operate. It is one of the important mass transfer equipment in the extraction process. However, the material and structure size of the packing affect the wetting performance of the continuous phase and the fragmentation and coalescence behavior of the dispersed phase. Based on the packed extraction tower in laboratory, research on the effects of different materials and sizes of packing on the hydrodynamic properities and mass transfer performance of the extraction tower.

Mission requirements:

① Among the granular packings such as Raschig ring, Bauer ring, θ ring, cascade ring, rosette, etc. in laboratory, select different materials, different structure and different sizes of fillers to develop an appropriate experimental scheme.

② Various selected packings are packed in the packing extraction tower, respectively, and finish the hydrodynamic properities and mass transfer performance experiments.

③ According to the experimental results, analyze the effects of different materials, different structures and different sizes of fillers on the fluid-liquid extraction fluid mechanical properties and mass transfer performance of the experimental system.

The above tasks are only for "three type" experiments and enhanced experiments. A few typical tasks were listed here for reference by experimenters. Teachers and experimenters can formulate corresponding experimental tasks according to actual needs.

Chapter 1
Research Methods of Chemical Engineering Principle Experiment

Chemical engineering principle experiment belongs to the category of engineering experiments. Engineering experiments are different from those of basic courses. The methods of basic course experiments are theoretical and rigorous, and the objects of study are usually simple, basic and even ideal. However, engineering experiments are faced with complex experimental and engineering problems. The difficulty lies in the large number of variables, which involves the materials are ever-changing, the equipment structure and size are very different. Therefore, in addition to summarize production experience, experimental research is an important basis for the establishment and development of disciplines. There are three research methods for med in the process of chemical engineering principle experiment: direct experiment method, dimensional analysis method and mathematical model method.

1.1 Direct experiment method

Direct experiment method means that experiments are directly conducted objects to obtain the law of the relationship between the relevant parameters. It is the most basic method to solve practical engineering problems with strong pertinence and reliable results. However, this research method usually obtains the regular relationship between some parameters with less variables. The experimental results can only be applied to specific experimental conditions and equipments, or extended to the same experimental conditions. Therefore, it has great limitations.

1.2 Dimensional analysis method

For a multi-variable engineering problem, in order to study the law of the process, the grid method is often used to plan experiments, that is, to fix other variables in order and change the target value of a certain variable. For example, the main factors affecting fluid resistance are pipe diameter d, pipe length l, average flow velocity u, fluid density ρ, fluid viscosity μ and tube wall roughness ε. If each variable change 10 times, 10^6 experiments are needed. It is easy to see that the number of variables appears on the power, the more variables involved, the number of experiments required will increase dramatically. Therefore, it is necessary to find a way to reduce the workload and make the results universal. Dimensional analysis is an experimental method which can solve the above problems and is widely used in engineering research.

The basic theory of dimensional analysis is the principle of dimensional consistency and the π theorem of Buckingham. The principle of dimensional consistency is that all physical equations derived from physical laws must have the same dimensions. π theorem is: the number of independent groups of factor numbers N obtained by dimensional analysis is equal to the difference between the number of variables n and the number of basic factors m, that is $N=n-m$.

The method of dimensional analysis is to sort out the multivariable functions into the relation of some dimensionless groups. Then the coefficients and exponents in the quasi-relational expression are obtained through experiments, which greatly reduces the experimental workload, and it is also easy to apply the experimental results to engineering calculations and designs.

It should be clear that factors are different from units when using the dimension analysis. Dimension refers to the type of physical quantity, and the unit is the standard used to compare the size of the same type of physical quantity. For example, the length can be expressed in meters (m), centimeters (cm) and millimeters (mm). However, the types of units belong to the same category of the length. If L、M、T respectively represent length, quality and time, then $[L]$, $[M]$, $[T]$ respectively represent the dimensions of length, quality and time.

There are two types of factors. One is the basic factor, which is independent of each other and cannot be derived from each other. The other is the derived factor, which is derived from the basic factor. For example, in the international system of units, there are seven basic dimensions: length, mass, time, thermodynamic temperature, amount of substance, current intensity and light intensity. The commonly used dimensions of length, mass and time can be represented respectively by [L], [M], [T]. The dimensions of other physical quantities can be derived from these seven basic dimensions and written in the form of power exponential product.

The derived dimension of a physical quantity is Q, $[Q]=[M^a L^b \theta^c]$, where a, b, c are constants. If the exponents of the basic factors are all zero, this physical quantity is called dimensionless group, such as, the Reynolds number reflecting the state of fluid flow is a dimensionless group.

$$Re=du\rho/\mu \tag{1-1}$$

Tube diameter d, basic dimension: $[d]=[L]$;

Flow rate u, derived dimension: $[u]=[LT^{-1}]$;

Density ρ, derived dimension: $[\rho]=[ML^{-3}]$;

Viscosity μ, derived dimension: $[\mu]=[ML^{-1}T^{-1}]$;

$$[Re]=[d][u][\rho]/[\mu]=[L][LT^{-1}][ML^{-3}]/[ML^{-1}T^{-1}]=[L^0M^0T^0]$$

1.2.1 Specific steps of dimensional analysis

① Find the independent variables that affect the process;

② Determine the basic dimensions of independent variables;

③ The function expressions of the dependent variables and independent variables are constructed, which are usually expressed in the form of the exponential equation;

④ Represent the factors of all independent variables with basic factors, write the factor formula that includes each independent variable;

⑤ The quasi-number equation is obtained according to the dimensional consistency principle of the physical equation and the π theorem;

⑥ Summarize the specific functional formula of the quasi-number equation through experimental induction.

1.2.2 Dimensional analysis examples

As an example, the relationship between the friction coefficient and the resistance of fluid flowing in the tube is obtained. The loss per unit volume of fluid Δp_f due to internal friction is related to six factors according to the nature of friction resistance and relevant experimental research. The functional relationship is as follows:

$$\Delta p_f = f(d, l, u, \rho, \mu, \varepsilon) \tag{1-2}$$

We do not know what form of this function. But mathematically, it is advisable to approach any non-periodic function in the form of the power function. Therefore, it is generally changed to the form of the following power function in the chemical industry.

$$\Delta p_f = K d^a l^b u^c \rho^d \mu^e \varepsilon^f \tag{1-3}$$

Although the power exponent of each physical quantity in the above formula is unknown, according to the principle of dimensional consistency, it can be known that the factor on the right side of the equation must be the same as the factor of Δp_f. So how to combine it into several groups of indifference times to meet the requirements? From equation (1-2), the number of variables $n=7$ (including Δp_f) represents the basic factors $m=3$ (length [L], time [T], mass [M]) of these physical quantities. Therefore, the number of indifference degree groups is $N=n-m=4$ according to the π theorem of Buckingham.

Variables are dimensionless through dimensional analysis. The factors of physical quantities in formula (1-3) are as follows:

$$[\Delta p_f] = [ML^{-1}T^{-2}] \qquad [d] = [l] = [L] \qquad [u] = [LT^{-1}]$$

$$[\rho] = [ML^{-3}] \qquad [\mu] = [ML^{-1}T^{-1}] \qquad [\varepsilon] = [L]$$

Substituting the dimensions of each physical quantity into formula (1-2), the dimensions at both ends are:

$$[ML^{-1}T^{-2}] = K[L]^a[L]^b[LT^{-1}]^c[ML^{-3}]^d[ML^{-1}T^{-1}]^e[L]^f$$

According to the principle of dimensional consistency, the exponents of the basic quantities on both sides of the equal sign in the above formula must be equal, and the equations can be obtained:

Basic dimension [M]: $d+e=1$

Basic dimension [L]：a+ b+ c−3d−e+ f=−1

Basic dimension [T]：−c−e=−2

This system of equations includes three equations，but there are six unknowns. If we use three unknowns b，e，f to express a，d，c to solve this system of equations，we can know：

$$\begin{cases} a=-b-c+3d+e-f-1 \\ d=1-e \\ c=2-e \end{cases} \qquad \begin{cases} a=-b-e-f \\ d=1-e \\ c=2-e \end{cases}$$

Substituting a，d，c into equation (1-3)，we can get：

$$\Delta p_f = K d^{-b-e-f} l^b u^{2-e} \rho^{1-e} \mu^e \varepsilon^f \tag{1-4}$$

Combining the physical quantities with the same exponent：

$$\frac{\Delta p_f}{u^2\rho} = K\left(\frac{l}{d}\right)^b \left(\frac{du\rho}{\mu}\right)^{-e} \left(\frac{\varepsilon}{d}\right)^f \tag{1-5}$$

$$\Delta p_f = 2K\left(\frac{l}{d}\right)^b \left(\frac{du\rho}{\mu}\right)^{-e} \left(\frac{\varepsilon}{d}\right)^f \left(\frac{u^2\rho}{2}\right) \tag{1-6}$$

Since the friction loss Δp should be proportional to the tube length l，so in the formula $b=1$，the formula is used to calculate the friction resistance of the fluid in the pipe (Fanning formula)

$$\Delta p_f = \lambda \frac{l}{d}\left(\frac{u^2\rho}{2}\right) \tag{1-7}$$

By comparison，the relationship of friction coefficient λ is obtained：

$$\lambda = 2K\left(\frac{du\rho}{\mu}\right)^{-e} \left(\frac{\varepsilon}{d}\right)^f \tag{1-8}$$

or

$$\lambda = \Phi\left(Re, \frac{\varepsilon}{d}\right) \tag{1-9}$$

It can be seen from the above analysis that under the guidance of dimensional analysis，the calculation of fluid resistance in a complex multi-variable tube is simplified to the study and determination of friction coefficient λ. It is based on the correct judgment of the influencing factors of the process and the logical processing of the number group. The above example can only tell us：λ is the function of Re and ε/d，as for the specific form between them，it still has to be determined experimentally. Obtain empirical formulas or calculation drawings through experiments and use them to guide engineering calculations and engineering design. The friction coefficient diagram of Moody is the curve of the relationship between friction coefficient λ and Re，ε/d，which is the result of this experiment. Many experiments have studied the formulas for calculating the friction coefficient λ under various specific conditions，among which Blasius formula is more famous：

$$\lambda = \frac{0.3164}{Re^{0.25}} \tag{1-10}$$

Other research results can be found in relevant textbooks and manuals.

Note the two points of the dimensional analysis method.

① The form of the resulting number group is related to the method of solving simultaneous equations. How to combine variables into useful quasi-numbers is a problem that researchers must pay attention to. In the previous example, if you do not use b, e, f to represent a, d, c but instead use d, e, f to represent a, d, c the resulting number group form will be different. However, these different number groups can be transformed into the four number groups obtained in the previous example by multiplying and dividing each other.

② It is necessary to have an essential understanding of the problems of the process under study. If an important variable is omitted or an irrelevant variable is introduced, it will lead to incorrect results and even lead to false conclusions. Therefore, the application of dimensional analysis must be cautious.

From the above analysis, it can be seen that the dimension analysis method is to reduce the number of independent variables and greatly reduce the number of experiments by combining variables into a dimensionless groups. Another advantage is that if the experiment is carried out according to formula (1-2), in order to change ρ and μ, a variety of fluids must be changed in the experiment; in order to change d, the experimental device (tube diameter) must be changed. When the equation (1-6) from the dimensional analysis is used to guide the experiment, to change $du\rho/\mu$, only the flow rate needs to be changed. To change l/d, only the distance of the measuring section needs to be changed. Therefore, the experimental results of water and air can be extended to other fluids, and the experimental results of small-sized models can be applied to large-scale experimental devices. The dimensionless work before the experiment is an effective method for planning the experiment, and it is widely used in chemical experiments.

1.3 Mathematical model method

1.3.1 Main steps of mathematical model method

The mathematical model method is based on a full understanding of the research problems, the work is carried out according to the following main steps.

① Simplify complex problems reasonably without distortion and propose a physical model that approximates the actual process, which is easy to describe with mathematical equations;

② Make a mathematical description of the obtained physical model, that is, establish a mathematical model, then determine the initial and boundary conditions of the equation, and solve the equation;

③ The rationality of the mathematical model is tested by experiments, the parameters of the model are determined.

1.3.2 Mathematical model method example

Take the determination of the pressure drop of the fluid through the fixed bed as an example. The pores between particles in fixed bed form many small channels for fluid to pass through. These channels are tortuous and cross-linked, and the cross-section size and shape of these channels are very irregular, it is difficult to calculate the pressure drop of fluid through such a complex channel theoretically, but it can be solved by the mathematical model method.

(1) Physical model

The flow of fluid through the particle layer is mostly creeping. The surface area of unit volume bed plays a decisive role in the flow resistance. In this way, in order to solve the pressure drop problem, the actual flow process in the particle layer can be greatly simplified as follows on the premise of ensuring the same unit volume surface area, so that it can be described by mathematical equations.

The irregular channels in the bed are simplified into a set of parallel thin tubes of length L_e. The inner surface area of the thin tubes is equal to the entire surface of the bed particles, and the total flow space of the thin tubes is equal to the void volume of the particle bed.

The equivalent diameter d_e of these virtual thin tubes can be obtained based on the above assumptions:

$$d_e=\frac{4\times \text{Cross-sectional area of channel}}{\text{Wetting perimeter}} \tag{1-11}$$

The numerator and denominator are multiplied L_e, then

$$d_e=\frac{4\times \text{Flow space of bed}}{\text{All inner surfaces of the tube}} \tag{1-12}$$

The flow space of the bed is ε based on the bed volume of 1m^3, and the particle surface of each 1m^3 bed is the specific surface of the bed α_B. If the reduction of the exposed particle surface due to particle contact is ignored, the relationship between the specific surface α and α_B of the particle is $\alpha_B=\alpha\ (1-\varepsilon)$, therefore:

$$d_e=\frac{4\varepsilon}{\alpha_B}=\frac{4\varepsilon}{\alpha(1-\varepsilon)} \tag{1-13}$$

According to this simplified physical model, the pressure drop of a fluid through a fixed bed can be equivalent to the pressure drop of a fluid through a set of thin tubes of equivalent diameter d_e and length L_e.

(2) Mathematical model

The simplified physical model described above has simplified the pressure drop of a fluid through a bed with complex geometric boundaries into a pressure drop through a uniform circular tube. To this, the existing theory can be used to make the following mathematical description:

$$h_f=\frac{\Delta p}{\rho}=\lambda\frac{L_e}{d_e}\frac{u_1^2}{2} \tag{1-14}$$

Where, u_1 is the flow rate of the fluid in the tube. u_1 can be taken as the flow velocity between the particle voids in the actual packed bed, and its relationship with the flow velocity of the empty bed is as follows:

$$u=\varepsilon u_1 \tag{1-15}$$

Substituting formula (1-13), (1-15) into formula (1-14):

$$\frac{\Delta p}{L}=\left(\lambda\frac{L_e}{8L}\right)\frac{(1-\varepsilon)\alpha}{\varepsilon^3}\rho u^2 \tag{1-16}$$

The length of the tube L_e is not equal to the actual length L, but it can be considered that L_e is directly proportional to the actual bed height L, that is $L_e=kL$, the coefficient k is incorporated into the friction coefficient, so

$$\frac{\Delta p}{L}=\lambda'\frac{(1-\varepsilon)\alpha}{\varepsilon^3}\rho u^2 \tag{1-17}$$

Where, $\lambda'=\frac{\lambda}{8}\frac{L_e}{L}$.

The above formula is a mathematical model of the pressure drop of fluid through a fixed bed, which includes an unknown coefficient λ' to be determined. λ' is called the model parameter, which can also be called the flow coefficient of friction of the fixed bed in its physical sense.

(3) Testing of models and estimation of model parameters

The simplified treatment of the bed is only a hypothesis, its validity must be tested by experiments, and the parameters of the model λ' must also be determined by experiments. Both Kozeny and Ergun have carried out experimental research about this, and obtained the correlations of λ' and Re' in different ranges under different experimental conditions. Due to space limitations, please refer to other related books for details.

1.3.3 Comparison of mathematical model method and dimensional analysis method

For the mathematical model method, the key to determining success or failure is the reasonable simplification of complex processes, that is, whether a simple enough physical model that can be represented by mathematical equations without distortion is obtained. Only by having a deep understanding of the inherent laws of the process, especially the particularity of the process, and using it according to the specific research purpose, which can be possible to greatly and reasonably simplify the real complex process while maintaining it on a specified equivalent. When the above example is simplified, only the physical model is equivalent to the actual process on the side of resistance loss.

For dimenstonal analysis, the key is to be able to list the main factors that affect the process. It doesn't require an in-depth understanding of the laws of the process itself, as long as several factorial analysis experiments are performed to investigate the degree of influence of each variable on the experimental results. Under the guidance of dimensional analysis，the experimental research can only get the external relations of the process，but not the internal laws of the process. However，this is a major feature of dimensional analysis，which makes it a general method applicable in principle to various research objects.

Both the mathematical model method and the dimensional analysis method need to solve the problem through experiments，but the purpose of the experiments is quite different. The experimental purpose of the mathematical model method is to test the rationality of physical model and determine a small number of model parameters；the experimental purpose of dimensional analysis is to find the functional relationship between the dimensionless variables.

Chapter 2
Error Analysis of Experimental Data

Errors exist in all experiments and measurements. Furthermore, errors are caused by extremely complex factors, which influence the nature of errors. Therefore, the goal of error analysis is to evaluate the accuracy of experimental data, identify the sources of errors and their impacts, and eliminate or reduce errors, so as to improve the quality of experimental results. This chapter presents a brief introduction to the basic concepts and estimation methods of some errors encountered in chemical engineering principle experiments.

2.1 True values and the mean

A true value is a definite value that theoretically exists for a physical quantity, also called a theoretical value or defined value. In the strictest sense, true values cannot be measured because measuring instrument, measuring method, environment, human observation, measurement procedure can not be perfect, thus they only represent desired values. In scientific experiments, the definition of true value is as follows: when the number of measurements is infinitely large, the probability of positive and negative errors will be theoretically equal according to the law of distribution of errors, thus the measurements can be summed and averaged to provide a value extremely close to the true value when there is no systematic error. Therefore, in practice, a "true value" is the mean value obtained when the number of measurements is infinitely large (or a so-called "established value" accepted in the literature and manuals) . However, the number of measurements is limited for engineering experiments, thus, a limited number of observed data are averaged to provide a so-called optimal value, which approximates the true value. This optimal value is referred to as a mean value—or a mean for simplicity. There are several types of means commonly used in practice, which are as follows.

(1) The arithmetic mean $\overline{x}$

This is the most commonly used mean. When measured values follow a normal distribution, the principle of least squares can verify that the arithmetic mean is the optimal value or the most reliable value in a set of measured data with equal precision.

$$\overline{x}=\frac{x_1+x_2+\cdots+x_n}{n}=\frac{1}{n}\sum_{i=1}^{n}x_i \tag{2-1}$$

where, each of x_1, x_2, $\cdots$, x_n is an observed value; n is the number of measurements.

(2) Root mean square $\overline{x}_{rms}$

$$\overline{x}_{rms}=\sqrt{\frac{x_1^2+x_2^2+\cdots+x_n^2}{n}}=\sqrt{\frac{1}{n}\sum_{i=1}^{n}x_i^2} \tag{2-2}$$

(3) Weighted mean $\bar{w}$

When measuring the same physical quantity via different methods or when different people measure the same physical quantity, reliable values are sometimes multiplied by weights, summed and averaged to obtain a weighted mean.

$$\bar{w}=\frac{w_1x_1+w_2x_2+\cdots+w_nx_n}{w_1+w_2+\cdots+w_n}=\frac{\sum_{i=1}^{n}w_ix_i}{\sum_{i=1}^{n}w_i} \tag{2-3}$$

where, each of x_1, x_2, $\cdots$, x_n is a measured value; w_1, w_2, $\cdots$, w_n are the respective weights of the measured values, which are empirically determined in many cases.

(4) Geometric mean $\bar{x}_G$

$$\bar{x}_G=\sqrt[n]{x_1x_2x_3\cdots x_n} \tag{2-4}$$

(5) Logarithmic mean x_m

$$x_m=\frac{x_1-x_2}{\ln x_1-\ln x_2}=\frac{x_1-x_2}{\ln\frac{x_1}{x_2}} \tag{2-5}$$

If $1<\frac{x_1}{x_2}<2$, the arithmetic mean can be used instead of the logarithmic mean, which leads to an error within 4%.

It is necessary to select one from the above types of means to best approximate the true value for a set of measured values. The selection of the best mean for a set of observed data depends on the distribution that the data follow most closely. In chemical engineering experiments, data mostly follow a normal distribution, thus, an arithmetic mean is generally used.

2.2 Error classification

For any measurement, irrespective of how precise the instrument is, how perfect the method is, how carefully the experiment is conducted, results measured at different times are not necessarily the same and are subject to certain errors and deviations. In the strictest sense, an error refers to the difference between an experimentally measured value (a directly or indirectly measured value) and the true value (a definite value that objectively exists), the deviation refers to the difference between the experimentally measured value and the mean. In practice, the terms "error" and "deviation" are used interchangeably. According to the natures and causes of errors, errors can be divided into three types: systematic errors, random errors and gross errors.

(1) Systematic errors

Systematic errors are the errors that remain constant or vary in a certain manner with the conditions when measuring the same physical quantity multiple times. The term trueness can be used to characterize the magnitude of systematic errors. The smaller the systematic error, the higher the trueness, and vice versa.

Systematic errors are generally caused by the following reasons.

1) Measuring instrument The precision of the instrument fails to meet the requirements or the instrument has zero offset.

2) Reagent quality The reagent is impure or the quality fails to meet the requirements.

3) Measurement method An approximate measurement method is used for measurement or a simplified calculation formula is used for calculation.

4) Environmental factors External factors such as temperature, pressure and humidity undergo changes.

5) Personal habits Such as consistently over-reading or under-reading experimental data, or consistently using a darker or a lighter solution color for titration endpoint detection.

Systematic errors are important errors and should be eliminated as much as possible when conducting measurements. Generally, systematic errors are regular, their causes are often known or can be eliminated after identification. Regarding systematic errors that cannot be eliminated, they should be determined or estimated.

(2) Random errors

Random errors, also known as accidental errors, are caused by some random factors that are not easy to control. They mainly affect the dispersion of measurement results but completely obey a statistical law. Random errors can be described probabilistically. As the number of measurements increases, the random error of the mean can be reduced, but never be eliminated. In error theory, the term "precision" is often used to characterize the magnitude of random errors. The larger the random error, the lower the precision, and vice versa.

(3) Gross errors

Gross errors are evidently inconsistent with the actual situation, they are mainly caused by carelessness by the experimenter, such as misreading, mismeasurement and misrecording. Measured values containing gross errors are called bad values and should be eliminated according to relevant criteria when collating the data.

In summary, systematic errors and gross errors can always be avoided, whereas random errors are inevitable. Thus, the best experimental results should contain only random errors.

2.3 Error representation methods

(1) Absolute error $D(x)$

The absolute value of the difference between the measured value x and true value A is called an absolute error, which can be expressed as

$$D(x)=|x-A| \tag{2-6}$$

In engineering calculations, a true value is often approximated by an arithmetic mean $\bar{x}$ or relative true value, which is measured using high-precision standard instruments. Therefore, the absolute error can be expressed as

$$D(x)=|x-\bar{x}| \tag{2-7}$$

(2) Relative error $E_r(x)$

The ratio of the absolute error to the true value is called a relative error, which can be expressed as

$$E_r(x)=\frac{D(x)}{|A|} \tag{2-8}$$

Similar to the absolute error, the true value is often approximated by the arithmetic mean $\bar{x}$. Therefore, the relative error can be expressed as

$$E_r(x)=\frac{D(x)}{|\bar{x}|} \tag{2-9}$$

Relative errors reflect the accuracy of measurement and are generally expressed as a percentage or a permillage. Therefore, relative errors can be compared among different physical quantities, and they are related to the magnitudes of measured values and absolute errors.

(3) Arithmetic mean error δ

The arithmetic mean error of a physical quantity measured n times can be expressed as

$$\delta=\frac{\sum_{i=1}^{n} d_i}{n}=\frac{\sum_{i=1}^{n}|x_i-\bar{x}|}{n}, \quad i=1,2,\cdots,n \tag{2-10}$$

where, n is the number of measurements; d_i is the deviation of a measured value from the mean, $d_i=x_i-\bar{x}$.

The disadvantage of arithmetic mean error is that it cannot reveal the degree of consistency among measured values.

(4) Standard error σ

Standard errors are also known as root mean square errors, which can be expressed as

$$\sigma=\sqrt{\frac{\sum_{i=1}^{n} d_i^2}{n}}=\sqrt{\frac{\sum_{i=1}^{n}(x_i-\bar{x})^2}{n}} \tag{2-11}$$

A standard error is more sensitive to larger or smaller errors in a set of measured values, thereby allowing it to well represent accuracy. Formula (2-11) is suitable for an infinite number of measurements. In practice, the number of measurements is not infinite. Thus, formula (2-11) can be rewritten as

$$\sigma=\sqrt{\frac{\sum_{i=1}^{n}(x_i-\bar{x})^2}{n-1}} \tag{2-12}$$

Standard errors are not the actual errors of measured values. The magnitude of σ only indicates the degree of dispersion of any one of a set of measured values of the same precision with respect to the arithmetic mean value under a certain condition. A small σ suggests that the set of measured values is dominated by small errors and the dispersion of any measured value with respect to the arithmetic mean is small, namely, the measurement reliability is high.

Among the various error representation methods mentioned above, relative errors and standard errors are superior to others when used to compare the precision of various measurements or the quality of the measurement results, Moreover, standard errors are more commonly used in the literature.

2.4 Precision, trueness and accuracy

The quality and level of measurement can be described by errors and by trueness. To indicate the sources and natures of errors, the following three concepts are generally used.

(1) Precision

Precision measures the consistency among several measurements of the same physical quantity, that is, it measures the repeatability and reflects the degree of random errors, with high precision meaning small random errors.

(2) Trueness

Trueness refers to the combination of all systematic errors in the measurement under the given conditions. It reflects the degree of the overall systematic error. The higher the trueness, the smaller the overall systematic error.

(3) Accuracy

Accuracy refers to the degree of approximation between a measured value and the true value. It reflects the degree of the overall magnitude of combined systematic errors and random errors. The higher the accuracy, the smaller the systematic errors and random errors.

The difference between the three can be metaphorized as shooting a target. As shown in Figure 2-1, A's systematic error is small, while its random error is large, that is, it has high trueness and low precision. B has a large systematic error with a small random error, that is, it has low trueness and high precision. Finally, both C's systematic error and random error are low, indicating high accuracy. How-

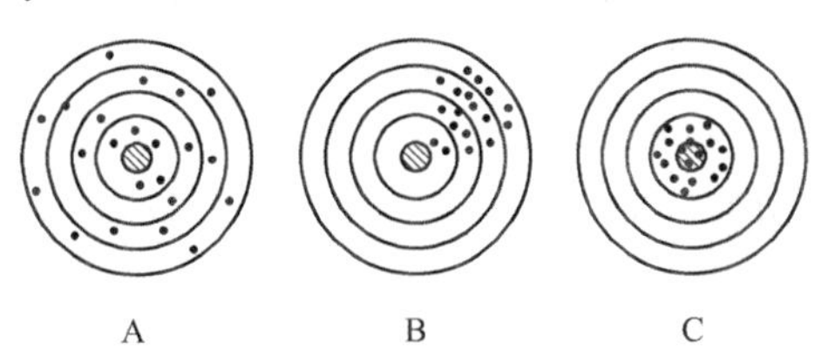

Figure 2-1 Schematic illustration of precision, trueness, and accuracy

ever, experimental measurements do not have a true value as straightforward as the bullseye, and the experimenter is tasked with determining the unknown true value.

For experimental measurements, high precision does not necessarily mean high trueness, and vice versa. However, high accuracy always indicates high precision and trueness.

2.5 Effective digits of experimental data

In experiments, it is necessary to retain the appropriate significant digits when measuring and calculating values. It is not simply the case that the more digits after the decimal point in a measured or calculated value, the more accurate the value. The first reason is that any measured value to be recorded should contain an estimated digit in the last place, and such digit is of uncertainty. The accurate digits obtained by direct reading are referred to as reliable digits, and those obtained by estimation as inaccurate digits. In a measured value, all the digits including one inaccurate digit that can represent the magnitude of the measured physical quantity are called significant digits. The position of the decimal point in the value is only related to the unit of the measurement. For example, the two values of 28.3 mm and 0.0283 m have the same accuracy and the same significant digits except for the position of the decimal point. The second reason is that the measuring instrument used under the given experimental conditions can only achieve a certain accuracy, thus, measured or calculated values cannot and should not exceed the accuracy range allowed by the instrument. In the above length measurement, for example, the smallest scale division is 1 mm, and the reading can be recorded to 0.1mm (the digit 3 in the above two values is an estimated digit), thus, there are three significant digits in the above values.

(1) The role of "0" in significant digits

All leading zeros before the first non-zero digit, counted from the left, only serve to determine the position of the decimal point. The zeros between non-zero digits and the trailing zeros are all significant digits except for the trailing zeros in positive integers. The number of significant digits in positive integers ending with zeros is uncertain. For example, 350g does not necessarily represent 350.0g, as the former may have 2 or 3 significant digits, while the latter has 4 significant digits. In contrast, 0.0350 g has 5 digits, but only 3 significant digits.

In scientific research and calculation, to clearly demonstrate the accuracy of a value, scientific notations can be used. The steps are to note the significant digits, add a decimal point after the first significant digit, and use 10 raised to an integer power to indicate the order of magnitude of the value. For example, to indicate that the number of significant digits of 35800 is 4, it can be written as 3.580×10^4. If it is to be presented with only 3 significant digits, it can be written as 3.58×10^4.

(2) Rounding rules for significant digits

For rounding of values, a common practice is to round the last to-be-retained significant digit to the nearest integer. However, on some occasions with higher precision requirements, rounding of significant digits should be conducted as per the standard "Rules of Rounding off for Numerical Values & Expression and Judgement of Limiting Values." (GB/T 8170—2008). That is, after the significant digits of a value are determined, only one inaccurate digit is generally retained when presenting the value, whereas the excess digits are processed as per rounding off rules. In this national standard, the rules of rounding off are as follows.

① If the leftmost digit of the to-be-discarded digits is smaller than 5, the to-be-discarded digits are discarded while maintained the remaining digits constant.

② If the leftmost digit of the to-be-discarded digits is larger than 5, the digit immediately before the to-be-discarded digits is increased by 1 while discarding the to-be-discarded digits.

③ If the leftmost digit of the to-be-discarded digits is 5 and it is followed by at least one non-zero digit, the digit immediately before the to-be-discarded digits is increased by 1 while discarding the to-be-discarded digits.

④ If the leftmost digit of the to-be-discarded digits is 5, it is not followed by any digit or all the tailing digits are zeros; further, if the last significant digit to be retained is an odd digit (1, 3, 5, 7, 9), then the last significant digit to be retained should be increased by 1 while discarding all the to-be-discarded digits. If the first two conditions are met and the last significant digit to be retained is an even digit (0, 2, 4, 6, 8), it is kept constant while discarding all the to-be-discarded digits.

⑤ When rounding a negative value, the steps are to convert the negative value to its absolute value, round the absolute value, and then place a negative sign before the rounded-off result.

For example, the following values are rounded to 3 significant digits:

12.1498 ⟶ 12.1　　rule①

12.1698 ⟶ 12.2　　rule②

12.2598 ⟶ 12.3　　rule③

12.2508 ⟶ 12.3　　rule③

12.2500 ⟶ 12.2　　rule④

12.3500 ⟶ 12.4　　rule④

(3) Calculation rules of significant digits

1) Summation and subtraction　Before summation and subtraction, all values are rounded to the same number of decimal places as the value with the fewest decimal places.

For example, when suming the three values 12.2508, 17.4 and 0.083, the calculation formula should be written as

$$12.3 + 17.4 + 0.1 = 29.8$$

2) Multiplication and division　Before multiplication and division, all values are rounded to the same number of significant digits as the value with the fewest significant digits, and the calculation results should also be rounded to the smallest number of significant digits as above.

For example, when multiplying three values 13.786, 1.034 and 0.128, the multiplication formula should be written as

$$13.8\times1.03\times0.128=1.82$$

3) Logarithmic operation The number of significant digits for a logarithm should be the same as the antilogarithm.

For example, lg12.34=1.091, ln12.34=2.513.

(4) Significant digits of directly measured values

The number of significant digits for a directly measured value is determined by the accuracy of the measuring instrument. In general, the last significant digit should represent tenths of the smallest scale division of the measuring instrument, that is, the value should contain one estimated digit. For example, the value in Figure 2-2 (a) can be recorded as 1.71, where the last digit 1 is estimated. The estimated digit may vary slightly depending on the experimenter's reading habits. However, if the smallest scale division of the measuring instrument does not represent $1\times10^{\pm n}$ where n is an integer, the estimated digit should be at the same decimal place as that of the significant digit of the smallest scale division of the measuring instrument. For example, the smallest scale division in Figure 2-2 (b) is 0.2, thus, the estimated digit may be 7, and the reading may be recorded as 1.7.

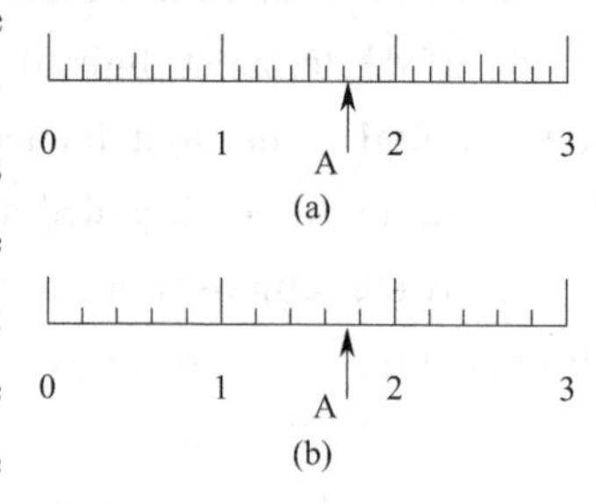

Figure 2-2 Readings under different scale division conditions

(5) Significant digits of values not directly measured

In experiments, some values are not directly measured by the instrument and insteadly they must be derived through calculation. Chemical engineering principle experiments belong to the category of engineering experiments. In principle, calculations in chemical engineering principle experiments must be in accordance with the calculation rules of significant digits, however, the actual situations must also be considered.

① In calculation, for constants such as π and g as well as some factors such as $\sqrt{2}$ and 1/3, the actual number of significant digits is determined by the numbers of significant digits in the original values on which the calculation was performed. Assuming that the largest number of significant digits in the original values is n, the above constant values may be rounded to $n+2$ significant digits to avoid larger errors due to the introduction of constant values.

② When calculating a value, to consider the precision of the results and the convenience of the calculation, all intermediate calculation results in the engineering field are generally rounded to 5 or 6 significant digits.

③ Error values should generally be rounded to 1 or 2 significant digits. Given that errors serve to reflect the degree of accuracy of values in question, it is necessary to avoid over-optimism about the result accuracy and ensure sufficient redundancy for errors. Therefore, when retaining 1 or 2 significant digits of an error, the last digit of the retained significant digits can be increased by 1 without the need to consider the aforementioned rounding-off rules. For example, if the error is 0. 5612, it can be written as 0. 6 or 0. 57.

2. 6 Error handling methods

(1) Systematic errors

Systematic errors cannot be corrected solely by the error theory, and the correction should be based on a thorough understanding of the measurement methods and principles of the physical quantities in question. When analyzing and processing systematic errors, before error elimination or correction is conducted, the identification of the significant systematic errors in the measured values is crucial.

Systematic errors can be divided into two categories: constant systematic errors and variable systematic errors. They can be simply discriminated based on observation or comparison. Commonly used methods for eliminating or reducing systematic errors include source elimination, correction value-based elimination, substitution elimination, cancelation of positive and negative errors, exchanged elimination, symmetric elimination, an even number of measurements every half period (for periodic systematic errors) and regression elimination. Proper treatment of systematic errors depends largely on familiarity with the measurement technology and on the richness of experiences with various measurement technology problems. For more details about the elimination and correction of systematic errors as well as the criteria for the degree of elimination, readers may refer to relevant professional books.

(2) Random errors

Random errors are inevitable in experimental measurements. How to determine the most reliable measurement results from a set of experimental data with random errors is a basic issue to address for experimental data processing.

Random errors follow a normal distribution in most cases, and they are unimodal, bounded, symmetric and able to cancel each other out. The ability to cancel each other out refers to the fact that when the number of measurements approaches infinity, the arithmetic mean of the errors approaches zero, which is the most essential statistical property of random errors. In other words, all compensable errors can be treated as random errors in principle. According to the statistical law of random errors, measurement results are often expressed with their arithmetic mean and the standard deviation.

(3) Gross errors

Experimental results sometimes contain several values with particularly large deviations. Retention of these values would cause a great impact on the mean and random errors. However, random removal of these data to ensure consistency among experimental results would result in a risk of losing useful information and thus is evidently inappropriate. If these values are generated by gross errors in the measurements, they are generally called questionable or bad values and must be eliminated. In contrast, if these values are generated by random errors, they are not bad values and thus should not be eliminated. Therefore, if the technical and physical cause of the outliers is unknown and it is impossible to determine whether the values contain gross errors or random errors, the following statistical methods may be used to judge the cause and eliminate the outliers as appropriate.

(4) Judgment and elimination of questionable values

The basic working principle of the statistical method is that a critical value at a given significance level is determined for a certain distribution. For any measured value, if its error exceeds the critical value, the error is considered not to be a random error but a gross error, and the measured value should be eliminated. There are three commonly used statistical judgment criterias, which are as follows.

1) Pauta criterion The Pauta criterion, also known as the 3σ criterion, is based on the premise that the number of measurements is sufficiently large. In general, the number of measurements is relatively small, so the 3σ criterion is only approximately valid.

For a series of measured values x_i ($i=1\sim n$), if each measured value x_i contains only random errors, according to the normal distribution law of random errors, the probability that the deviation d_i falls outside $\pm 3\sigma$ is approximately 0.3%. The σ can be calculated according to formula (2-12). If the deviation of a measured value in the series is greater than 3σ, as shown below

$$|d_i| > 3\sigma \tag{2-13}$$

the measured value contains a gross error and should be eliminated.

According to the 3σ criterion, it is rational to eliminate all values with a deviation greater than 3σ and then recalculate σ for the remaining values, followed by application of the 3σ criterion for further elimination of outliers.

The greatest advantage of this method is simple calculation and convenient application, but when the number of experimental values is small, it is difficult to remove the bad values. In the case of $n \leqslant 10$, it is impossible to observe $|d_i| > 3\sigma$. For this reason, when the number of measurements is small, it is better not to use this criterion.

2) Grubbs criterion Suppose a number of independent measurements are made on a certain quantity to obtain a series of measured values x_i ($i=1\sim n$). When x_i follows a normal distribution, it is possible to calculate $\bar{x}$ and σ according to formula (2-1) and formula (2-12), respectively. To verify whether there are gross errors in the series, the measured values x_i are arranged in ascending order,

$$x_{(1)} \leqslant x_{(2)} \leqslant \cdots \leqslant x_{(n)}$$

Grubbs derives the distribution of $g_{(1)}$ and $g_{(n)}$:

$$g_{(1)} = \frac{\bar{x} - x_{(1)}}{\sigma} \tag{2-14}$$

$$g_{(n)} = \frac{x_{(n)} - \bar{x}}{\sigma} \tag{2-15}$$

For a given significance level α, the corresponding critical value $g_{(0)}(n, \alpha)$ can be obtained from Table 2-1 according to the number of experiments n. If the formula

$$g_{(i)} \geqslant g_{(0)}(n, \alpha) \tag{2-16}$$

holds for measured value x_i, the measured value is considered to contain a gross error and should be eliminated.

Table 2-1 Grubbs critical values $g_{(0)}(n, \alpha)$

n	Significance level α			n	Significance level α		
	0.05	0.025	0.01		0.05	0.025	0.01
3	1.15	1.16	1.16	17	2.47	2.62	2.79
4	1.46	1.48	1.49	18	2.50	2.65	2.82
5	1.67	1.71	1.75	19	2.53	2.68	2.85
6	1.82	1.89	1.94	20	2.56	2.71	2.88
7	1.94	2.02	2.10	21	2.58	2.73	2.91
8	2.03	2.13	2.22	22	2.60	2.76	2.94
9	2.11	2.21	2.32	23	2.62	2.78	2.96
10	2.18	2.29	2.41	24	2.64	2.80	2.99
11	2.23	2.36	2.48	25	2.66	2.82	3.01
12	2.28	2.41	2.55	30	2.74	2.91	3.10
13	2.33	2.46	2.61	35	2.81	2.98	3.18
14	2.37	2.51	2.66	40	2.87	3.04	3.24
15	2.41	2.55	2.70	45	2.92	3.09	3.29
16	2.44	2.59	2.75	50	2.96	3.13	3.34

3) t-test criterion It is proved by the mathematical statistical theory that when the number of measurements is small, the random variable follows a t-distribution. The t-distribution is not only related to the measured values but also the number of measurements n. When $n>10$, the t-distribution is very close to a normal distribution. Therefore, when the number of measurements is small, it is more reasonable to judge the presence of gross errors according to the t-test criterion established for the t-distribution. A prominent characteristic of the t-test criterion is first removing a questionable measured value and then determining whether this value should be removed according to the t-test criterion.

Suppose that a certain physical quantity is measured multiple times to obtain a series of values x_i $(i=1\sim n)$. If the measured value x_j is considered questionable, it is eliminated, then the mean of the remaining values is calculated as follows:

$$\bar{x} = \frac{1}{n-1}\sum_{\substack{i=1 \\ i \neq j}}^{n} x_i \tag{2-17}$$

The standard error σ of the remaining measured values is calculated according to formula (2-18) (excluding $d_i = x_i - \bar{x}$).

$$\sigma=\sqrt{\frac{1}{n-2}\sum_{\substack{i=1\\i\neq j}}^{n}d_i^2} \tag{2-18}$$

According to the number of measurements n and the selected significance level α, the t-test coefficient $K(n, \alpha)$ can be found from Table 2-2. If the formula

$$|x_j-\bar{x}|>K(n,\alpha)\sigma \tag{2-19}$$

holds for the measured value x_j, x_j is considered to contain a gross error and should be eliminated, otherwise, it does not contain a gross error and should be retained.

Table 2-2 $K(n, \alpha)$ table of t-distribution

n	Significance level α		n	Significance level α	
	0.05	0.01		0.05	0.01
4	4.97	11.46	18	2.18	3.01
5	3.56	6.53	19	2.17	3.00
6	3.04	5.04	20	2.16	2.95
7	2.78	4.36	21	2.15	2.93
8	2.62	3.96	22	2.14	2.91
9	2.51	3.71	23	2.13	2.90
10	2.43	3.54	24	2.12	2.88
11	2.37	3.41	25	2.11	2.86
12	2.33	3.31	26	2.10	2.85
13	2.29	3.23	27	2.10	2.84
14	2.26	3.17	28	2.09	2.83
15	2.24	3.12	29	2.09	2.82
16	2.22	3.08	30	2.08	2.81
17	2.20	3.04			

Among the above three criterias, the t-test criterion is generally used when the number of measurements is very small. The 3σ criterion is suitable for a large number of measurements. However, since it is easy to use without the need to refer to tables, it is also often used in scenarios where the number of measurements is relatively small with a relatively less rigorous requirement.

2.7 Error estimation of directly measured values

(1) Error estimation of a single measured value

In experiments, when the conditions do not permit or the requirements for measurement results are not rigorous, the physical quantity in question may be measured once. In such a scenario, it is possible to make a rational estimation of the errors of the measured value according to the actual situation.

The following presents how to estimate the errors of a single measured value according to the measuring instrument.

1) Estimation of measurement errors for instruments with specified accuracy grades (such as electrical instruments and rotameters) The accuracy of a measuring instrument is often expressed in terms of the maximum fiducial error and the accuracy grade of the instrument. The maximum fiducial error of the instrument is defined as

$$\text{maximum fiducial error}=\frac{\text{absolute error of instrumental reading}}{\begin{array}{c}\text{Absolute measurement range of the}\\ \text{instrument at a given accuracy grade}\end{array}}\times 100\% \tag{2-20}$$

In formula (2-20), the absolute error of instrumental reading for a physical parameter refers to the maximum absolute difference between the measured values and standard values of the physical parameter under a certain condition. For multi-scale instruments, the absolute error of displayed values and the measurement range vary among the scales.

Formula (2-20) shows that in the case of a fixed absolute error of an instrumental reading, the larger the measurement range, the smaller the maximum fiducial error.

For example, there are 7 kinds of accuracy grades for electrical instruments in China, namely 0. 1, 0. 2, 0. 5, 1. 0, 1. 5, 2. 5 and 5. 0, there are 3 kinds of accuracy grades for metal tube rotameters, namely 1. 5, 2. 5 and 4. The smaller the number, the higher the accuracy. If the accuracy grade of an instrument is 2. 5, the maximum fiducial error of the instrument is 2. 5%.

A question arises as to how to estimate the absolute error and relative error of a measured value when using an electrical instrument. Suppose the instrument has an accuracy grade P and its maximum fiducial error is P%. If the measurement range of the instrument is x_n and the instrumental reading is x_i, the absolute error and relative error of the reading value can be calculated using formula (2-20):

$$D(x)\leqslant x_n\times P\% \tag{2-21}$$

$$E_r(x)=\frac{D(x)}{x_i}\leqslant\frac{x_n}{x_i}\times P\% \tag{2-22}$$

Formulas (2-21) and (2-22) indicate the following:

① When the accuracy grade P and measurement range x_n of the instrument are fixed, the larger the measured value x_i, the smaller is the relative error of the measurement.

② When choosing an instrument, it is discouraged to blindly emphasize a high accuracy grade, as the relative error of the measurement is also related to $\frac{x_n}{x_i}$, thereby making it necessary to consider both the accuracy grade of the instrument and $\frac{x_n}{x_i}$.

2) Estimation of measurement errors for instruments without specified accuracy grades (such as electronic balances) For example, the national standard "Electronic balance"(GB/T 26497—2011) divides electronic balances into four grades of accuracy, namely special, high, medium and ordinary accuracy, which are denoted by symbols Ⓘ、ⒾⒾ、ⒾⒾⒾ and ⒾⒾⒾⒾ, respectively. The accuracy of such instruments can be expressed as follows:

$$\text{accuracy}=\frac{0.5\times\text{nominal scale division}}{\text{measurement range}} \tag{2-23}$$

The nominal scale division in the formula refers to the smallest scale division of the measuring instrument.

If there is an electronic balance with a nominal scale division of 0.1 mg and a measurement range of 0 to 200 g, its accuracy can be calculated as follows:

$$\text{accuracy}=\frac{0.5\times 0.1}{(200-0)\times 10^3}=2.5\times 10^{-7}$$

When the accuracy of an instrument is known, the nominal scale division can also be determined according to formula (2-23).

When using the instrument, the absolute errors and relative errors of the above type of measurements can be determined as follows, respectively.

$$D(x)\leqslant 0.5\times \text{nominal scale division} \tag{2-24}$$

$$E_r(x)\leqslant \frac{0.5\times \text{nominal scale division}}{\text{measured value}} \tag{2-25}$$

(2) Error estimation of repeated measurement values

If a physical quantity is measured multiple times, the measurement error can be estimated using the standard error. Suppose a physical quantity in question is repeatedly measured n times and the measured values are x_1, x_2, …, x_n, whose mean is $\bar{x}=(x_1+x_2+\cdots+x_n)/n$ and standard error is $\sigma=\sqrt{\sum(x_i-\bar{x})^2/(n-1)}$. The absolute error and relative error of $\bar{x}$ can be calculated using formula (2-26) and (2-27), respectively.

$$D(\bar{x})=\frac{\sigma}{\sqrt{n}} \tag{2-26}$$

$$E_r(\bar{x})=\frac{D(\bar{x})}{|\bar{x}|} \tag{2-27}$$

(3) Actual errors of measured values

Measurement errors determined using the above methods are always much smaller than the actual errors of measured values. This is because the instrument may fail to be adjusted to the ideal state, such as failing to be placed vertically or horizontally or undergoing improper zero adjustment, which will introduce errors. The actual working conditions of the instrument may fail to meet the specified normal working conditions and thus introduce errors. After long-term use, the instrument may undergo component wear and assembly condition changes, thus introducing errors. The operator's habits and bias may also introduce errors. The signal captured by the instrument may not actually be the same as the signal to be detected and the instrument circuit may be subject to interference, which will introduce errors as well.

In short, the actual errors of measured values are affected by many factors. To obtain measurement results with higher accuracy, it is necessary to use better instruments with scientific attitudes and methods and possess solid theoretical knowledge and practical experience.

2.8 Error propagation in indirect measurements

In many experiments and studies, some results are not from direct instrumental measurements but are calculated by substituting some experimentally measured values into a calculation formula, that is, they are indirectly measured values. For example, $Re=du\rho/\mu$ is an indirectly measured value. Because directly measured values are always subject to some errors, it is inevitable that indirectly measured values also contain some errors. That is, the direct measurement errors inevitably propagate into the indirectly measured values and thus form indirect measurement errors. The basic formulas for error propagation are described below.

Suppose the indirectly measured value y is a function of the directly measured values $x_1, x_2, \cdots, x_n$:

$$y=f(x_1,x_2,\cdots,x_n) \tag{2-28}$$

The differential of y is as follows:

$$dy=\frac{\partial y}{\partial x_1}dx_1+\frac{\partial y}{\partial x_2}dx_2+\cdots+\frac{\partial y}{\partial x_n}dx_n \tag{2-29}$$

Δy, Δx_1, Δx_2, $\cdots$, Δx_n are substituted for dy, dx_1, dx_2, $\cdots$, dx_n, respectively. When the direct measurement errors (Δx_1, Δx_2, $\cdots$, Δx_n) are very small and considering the most unfavorable situation that each partial derivative of y in formula (2-29) and each error of the directly measured values are non-negative, it is possible to derive the error propagation formulas for the maximum absolute error and maximum relative error of the indirectly measured value y.

The formula for calculating the maximum absolute error that propagates to y is as follows:

$$\Delta y=\left|\frac{\partial y}{\partial x_1}\right|\cdot|\Delta x_1|+\left|\frac{\partial y}{\partial x_2}\right|\cdot|\Delta x_2|+\cdots+\left|\frac{\partial y}{\partial x_n}\right|\cdot|\Delta x_n| \tag{2-30}$$

or

$$\Delta y=\sum_{i=1}^{n}\left|\frac{\partial y}{\partial x_i}\cdot\Delta x_i\right| \tag{2-31}$$

In formula (2-31), $\frac{\partial y}{\partial x_i}$ is error propagation coefficient; Δx_i is absolute error of directly measured value; Δy is maximum absolute error of indirectly measured value.

The formula for calculating the maximum relative error that propagates to y is as follows:

$$E_r=\frac{\Delta y}{y}=\frac{1}{f(x_1,x_2,\cdots,x_n)}\left(\left|\frac{\partial y}{\partial x_1}\right|\cdot|\Delta x_1|+\left|\frac{\partial y}{\partial x_2}\right|\cdot|\Delta x_2|+\cdots+\left|\frac{\partial y}{\partial x_n}\right|\cdot|\Delta x_n|\right) \tag{2-32}$$

or

$$E_r(y)=\frac{\Delta y}{y}=\sum_{i=1}^{n}\left|\frac{\partial y}{\partial x_i}\cdot\frac{\Delta x_i}{y}\right| \tag{2-33}$$

Formulas (2-30) to (2-33) are used for the sumative combination of errors, and such combined errors represent the maximum possible errors, which are very unlikely to occur in practice. According to probability theory, a geometric combination of errors is more in line with the actual situation. The following are general formulas of geometric combination.

The absolute error of the indirectly measured value can be calculated as

$$\Delta y=\sqrt{\left(\frac{\partial y}{\partial x_1}\Delta x_1\right)^2+\left(\frac{\partial y}{\partial x_2}\Delta x_2\right)^2+\cdots+\left(\frac{\partial y}{\partial x_n}\Delta x_n\right)^2} \tag{2-34}$$

The relative error of the indirectly measured value now becomes

$$E_r(\lambda)=\frac{\Delta y}{y}=\sqrt{\left(\frac{\partial y}{\partial x_1}\cdot\frac{\Delta x_1}{y}\right)^2+\left(\frac{\partial y}{\partial x_2}\cdot\frac{\Delta x_2}{y}\right)^2+\cdots+\left(\frac{\partial y}{\partial x_n}\cdot\frac{\Delta x_n}{y}\right)^2} \tag{2-35}$$

The standard error of the indirectly measured value is

$$\sigma_y=\sqrt{\left(\frac{\partial y}{\partial x_1}\sigma_{x_1}\right)^2+\left(\frac{\partial y}{\partial x_2}\sigma_{x_2}\right)^2+\cdots+\left(\frac{\partial y}{\partial x_n}\sigma_{x_n}\right)^2} \tag{2-36}$$

In formula (2-36), σ_{x_1}, σ_{x_2}, …, σ_{x_n} are standard errors of the directly measured values; σ_y is standard error of the indirectly measured value.

Simple formulas of additive combination and geometric combination of errors for some commonly used functions can be found in relevant books.

2.9 Application examples of error analysis

In addition to calculating the accuracy of measured values, error analysis can also be used to evaluate experimental designs. After identifying the main sources of errors and the error magnitude due to each source, useful improvement suggestions may be made for the experimental scheme and instruments, as illustrated by the following three examples.

【Example 2-1】 It is necessary to conduct a resistance test in a small copper tube (Dg6; nominal diameter of 6 mm). Because the inner diameter of the copper tube is too small, it cannot be measured by a general vernier caliper and thus is indirectly measured using a volumetric method. That is, a tube segment with a height of 400 mm (absolute error of ± 0.5 mm) is cut and the volume of water inside it is measured, so as to calculate the mean inner diameter. The measuring gauge is a pipette, whose volume scale is accurate and systematic errors can be neglected. The volume is measured 3 times as 11.31mL, 11.26mL and 11.30mL. The problem is to find the arithmetic mean $\bar{x}$, mean absolute error $D(x)$ and relative error $E_r(x)$ of the volume.

Solution: arithmetic mean $\bar{x}=\frac{\sum x_i}{n}=\frac{11.31+11.26+11.30}{3}=11.29\text{mL}$

mean absolute error $D(x)=\frac{|11.29-11.31|+|11.29-11.26|+|11.29-11.30|}{3}=0.02\text{mL}$

relative error $E_r(x)=\frac{D(x)}{|\bar{x}|}=\frac{0.02}{11.29}=0.18\%$

【Example 2-2】 The tube used for measuring the friction coefficient λ in fluid flow resistance experiments is a stainless-steel tube with a diameter of 8 mm. The spacing between pressure measurement points is $l=1.7$m. The laminar flow tube is now replaced by a small copper tube (Dg6) as described in Example 2-1. The spacing between pressure measurement points remains unchanged, and the flow rate is measured with a 500mL graduated cylinder. It is expected that the accuracy of the friction coefficient λ will not be lower than 4.5% at Re=2000. Is it possible to meet the accuracy requirement for λ when using the copper tube? Does the spacing between pressure measurement points need to be adjusted? Is the flow error introduced using a 500mL graduated cylinder within a reasonable range?

Solution: the function form of λ is $\lambda=\frac{2g\pi^2}{16}\cdot\frac{d^5(R_1-R_2)}{lV_s^2}$

Where, R_1, R_2 are pressure gauge readings (liquid column) at the two pressure measurement points, m; V_s is flow rate, m^3/s; l is spacing between the two pressure measurement points, m.

The relative error is

$$E_r(\lambda)=\frac{\Delta\lambda}{\lambda}=\pm\sqrt{\left[5\left(\frac{\Delta d}{d}\right)\right]^2+\left[2\left(\frac{\Delta V_s}{V_s}\right)\right]^2+\left(\frac{\Delta l}{l}\right)^2+\left(\frac{\Delta R_1+\Delta R_2}{R_1-R_2}\right)^2}\times100\%$$

It is required that $E_r(\lambda)<4.5\%$. Given that the relative error $\frac{\Delta l}{l}$ is less than $\frac{E_r(\lambda)}{10}$, it can be neglected. The remaining three sub-errors can be equivalently allocated to the remaining three components, Accordingly, the relationship between each sub-error m_i and the total error is expected to be as follows:

$$E_r(\lambda)=\sqrt{3m_i^2}=4.5\%$$

each sub-error:

$$m_i=\frac{4.5}{\sqrt{3}}\%=2.6\%$$

(1) Estimation of sub-error m_1 for the flow rate component

First, V_s is determined:

$$V_s=Re\frac{d\mu\pi}{4\rho}=2000\times\frac{0.008\times10^{-3}\times\pi}{4\times1000}m^3/s=1\times10^{-5}m^3/s=10mL/s$$

Such a small flow rate can be measured with a 500mL graduated cylinder. The systematic error of the graduated cylinder is very small and can be neglected. The reading error is $\Delta V=\pm5$mL, the systematic error of the stopwatch used for timing can also be neglected. The random error introduced by the starting and stopping of the stopwatch is estimated to be $\Delta\tau=\pm0.1$s. When Re=2000, if the water volume measured each time is approximately 450mL, it will take a time τ of approximately 48 s. Therefore, the maximum error of the measured flow rate is

$$\frac{\Delta V_s}{V_s}=\pm\left(\frac{\Delta V}{V}+\frac{\Delta\tau}{\tau}\right)=\pm\left(\frac{5}{450}+\frac{0.1}{48}\right)=\pm(0.011+2.08\times10^{-3})$$

The values in the formula indicate that the relative error $\frac{\Delta V}{V}$ is relatively large and $\frac{\Delta\tau}{\tau}$ is very small, so that the latter can be neglected. Therefore, the sub-error of the flow rate component is calculated to be 2.2% as follows:

$$m_1 = 2\frac{\Delta V_s}{V_s} = 2\times 0.011\times 100\% = 2.2\%$$

which does not exceed the upper limit set for the sub-error of each component.

(2) Estimation of sub-error m_2 for the tube diameter component

As shown by Example 2-1, the tube diameter d can be indirectly measured using the volumetric method:

$$V = \frac{\pi}{4}d^2 h\text{, then } d = \sqrt{\frac{V}{h}\times\frac{4}{\pi}}$$

The tube height is $h = 400$ mm, and the absolute error is $\pm$ 0.5 mm. To ensure that error estimation is in line with actual conditions, the relative error of d is calculated using geometric combination:

$$\frac{\Delta d}{d} = \frac{1}{2}\left(\frac{\Delta V}{V} + \frac{\Delta h}{h}\right)$$

As shown by Example 2-1, the relative error $\frac{\Delta V}{V}$ is 0.18%, and it is substituted into the above formula to calculate m_2 to be 0.8%:

$$m_2 = 5\frac{\Delta d}{d} = \frac{5}{2}\left(\frac{\Delta V}{V} + \frac{\Delta h}{h}\times 100\%\right) = \frac{5}{2}\times\left(0.18\% + \frac{0.5}{400}\times 100\%\right) = 0.8\%$$

which does not exceed the upper limit set for the sub-error of each component.

(3) Estimation of sub-error m_3 for the differential pressure component

The single-tube differential pressure gauge is measured using a ruler with a scale division of 1mm. The systematic error can be neglected, and the absolute random error ΔR of the reading is $\pm$0.0005m.

$$m_3 = \frac{\Delta R_1 + \Delta R_2}{R_1 - R_2} = \frac{2\Delta R_1}{R_1 - R_2} = \frac{2\times 0.0005}{R_1 - R_2}$$

The measured differential pressure $R_1 - R_2$ is proportional to the distance l between the two pressure measurement points as follows:

$$R_1 - R_2 = \frac{64}{Re}\cdot\frac{l}{d}\cdot\frac{u^2}{2g} = \frac{64}{2000}\cdot\frac{l}{0.006}\cdot\frac{\left(\frac{9.4\times 10^{-6}}{0.785\times 0.006^2}\right)^2}{2g} = 0.031$$

Where u is the mean flow velocity, m/s.

The impact of the changes in l on the error of the differential pressure component can be calculated as summarized in Table 2-3.

Table 2-3 Impact of changes in l on the error of differential pressure component

l/m	$R_1 - R_2$/m	$\frac{2\Delta R_1}{R_1 - R_2}$/%
0.500	0.015	6.7
1.000	0.030	3.3
1.500	0.045	2.2
2.000	0.060	1.6

As shown in Table 2-3, $l \geqslant 1.500$m meets the error requirement. For $l = 1.500$m, the sub-error m_3 of the differential pressure component is

$$m_3 = \frac{\Delta R_1 + \Delta R_2}{R_1 - R_2} = \frac{2\Delta R_1}{R_1 - R_2} = \frac{2 \times 0.0005}{0.03 \times 1.500} \times 100\% = 2.2\%$$

The overall error is

$$E_r(\lambda) = \frac{\Delta\lambda}{\lambda} = \pm\sqrt{m_1^2 + m_2^2 + m_3^2} \times 100\% = \pm\sqrt{(2.2\%)^2 + (0.8\%)^2 + (2.2\%)^2} \times 100\% = \pm 3.2\%$$

The above error analysis indicates the following:

① When using the copper tube instead of the steel tube, the accuracy requirement for λ can be met.

② The original distance between the pressure measuring points meets the requirement. If $l >$ 1.500m, the error can be further reduced.

③ Although the sub-error of the diameter component has a relatively large error propagation coefficient (i. e., 5) with a large impact on the total error, the indirect diameter measurement through a volumetric method is reasonable with high accuracy, thereby reducing the overall error.

④ The sub-error of the flow rate component is within a reasonable error range, that is, it is appropriate to use a 500mL graduated cylinder for flow rate measurement. If a graduated cylinder with a higher accuracy is used, the reading error will be reduced, thus the experimental results will be more accurate.

【Example 2-3】 In the experiment of Example 2-2, when Re=300, what is the relative error of measured λ? (experimental conditions: $l = 1.7$ m, water temperature is 20℃, $R_1 - R_2 = 6.8$mm; it takes 319 s for 450mL of water to flow through the tube)

Solution: as shown in Example 2-2, $m_1 = 2.2\%$, $m_2 = 0.8\%$

$$m_3 = \frac{2\Delta R_1}{R_1 - R_2} = \frac{2 \times 0.0005}{0.0068} \times 100\% = 14.7\%$$

$$E_r(\lambda) = \pm\sqrt{m_1^2 + m_2^2 + m_3^2} = \pm\sqrt{(2.2\%)^2 + (0.8\%)^2 + (14.7\%)^2} \times 100\% = \pm 14.9\%$$

The result shows that due to the decline of differential pressure, its relative error rises, resulting in an increase in the relative error of measured λ. When Re=300, the theoretical value of λ is $\frac{64}{Re} = 0.213$. If the experimental result deviates from this value (for example, $\lambda = 0.181$ or $\lambda = 0.245$), it does not necessarily mean that the measured λ is inconsistent with the theoretical value as the magnitude of deviation needs to be considered as well. The deviations of the values in parentheses with respect to the theoretical value are caused by low measurement precision. By increasing the precision of differential pressure measurement or increasing the number of measurements and taking the mean of measured values, it is possible for the measured λ to be consistent with the theoretical value. In summary, the above examples fully illustrate the importance of error analyses in experiments.

Chapter 3

Experimental Data Processing

One of the important purposes of chemical engineering principle experiments is to process the raw data from experiments to obtain a quantitative relationship between the variables. On this basis, further analyze the experimental phenomena and summarize the laws, verify the consistency between the experimental results and the theory. Furthermore, basic data and theoretical guidance could be provided for production and design. The processing of experimental data is a process that infer the true value of measurement, based on the concept and state of the research object, using the mathematical operation as a tool, and finally derive some whole conclusions with regular conclusions.

There are three commonly used methods for processing experimental data: tabular method, graphic method and mathematical equation representation method.

3.1 Tabular method

The data are listed in a certain order according to the relationship between the independent and dependent variables of the experimental data to form a data table, which is the tabular method. There are many advantages for tabular method, such as the original data record table will make the data easy to compare, the form is compact, the relationship between several variables can be expressed in the same table. It is usually the first step in organizing data, laying the foundation for plotting graphs or organizing mathematical formulas.

3.1.1 Classification of data tables

Experimental data tables are generally divided into two categories: original data record tables and experimental data processing result tables. The measurement of the relationship between λ-Re in fluid resistance experiment is taken as an example for illustration.

(1) Original data record tables

The original data record table is designed according to the specific content of the experiment to clearly record all the data to be tested. The table must be designed before the experiment. A raw data record table is shown in Table 3-1.

(2) Experimental data processing result table

The experimental data processing result table can be subdivided into intermediate calculation result tables (representing the calculation results of the main variables of the experimental process), comprehensive result tables (expressing the conclusions obtained from the experiment) and error analysis

Table 3-1 Fluid resistance experiment original data record table

Equipment Number No. ____ Experiment starting temperature______℃ Experiment end temperature______℃ Experiment time______

Inner diameter of smooth pipe d=______m Distance between pressure measuring points l=______m

Inner diameter of rough pipe d=______m Distance between pressure measuring points l=______m

No.	Flow V_s/(L/h)	Smooth pipe differential pressure Δp_f			Flow V_s/(L/h)	Rough pipe differential pressure Δp_f		
		kPa	cmH_2O			kPa	cmH_2O	
			left	right			left	right
1								
2								
⋮								
n								

tables (expressing the error range between experimental values and reference values or theoretical values), etc. Which table to use in the experiment report should be determined based on the specific conditions of the experiment. The data processing result table of fluid resistance experiment is shown in Table 3-2, and the error analysis result table is shown in Table 3-3.

Table 3-2 Fluid flow resistance experimental data processing result table (smooth pipe)

No.	Flow V_s/(m^3/s)	Average velocity u/(m/s)	Δp_f/kPa	Re	λ	λ-Re Relationship
1						
2						
⋮						
n						

Table 3-3 Fluid resistance experimental error analysis result table

No.	Re	Experimental λ-Re Relationship	$\lambda_{Experimental}$	$\lambda_{Theoretical}$(or $\lambda_{Empirical}$)	$\lambda_{Theoretical}$(or $\lambda_{Empirical}$)	Relative error/%
1						
2						
⋮						
n						

3.1.2 Considerations for designing experimental data tables

① The tables should be designed to be concise and easy to read and use. Items of record and calculate should meet the needs of the experiment. Various experimental conditions can be written between the title of the table and the table.

② The name, symbol and unit of measurement of the physical quantity are listed in the header. The symbol and the unit of measurement are separated by slashes. Slashes cannot be reused continuously, and parentheses can be used when appropriate. Units of measurement should not be mixed in numbers, which can easily cause indistinguishability.

③ Pay attention to the number of significant digits, that is, the number recorded should match the accuracy of the measuring instrument, no more and no less.

④ When the value of the physical quantity is too large or too small, it should be expressed in scientific notation. Enter it in the header in form of "symbol of physical quantity$\times 10^{\pm n}$/ unit of measurement" . Note: the relationship between $10^{\pm n}$ in the header and data in the table should be consistent with following formula.

$$\text{Actual value of physical quantity} \times 10^{\pm n} = \text{data in the table}$$

⑤ In order to facilitate the reference, the serial number and title of a table should be indicated on top of it. The serial numbers of the tables should be written in the order in which they appear in the report and explained in the text. A table should not spread across pages. When the pages must be spread across, the "continued table" must be noted on the table on the following pages.

⑥ The data should be written clearly and neatly. When revising, it is advisable to use a single line to cross out the mistakes and write the correct ones below. The name of the recorder and the question to be remarked can be written as a "table note" below the table.

3.2 Graphic method

The graphic method is the most commonly used method to represent the relationship between variables in the experiment. Collected experimental data or calculation results are plotted on appropriate coordinates, then data points are connected into smooth curves or straight lines. The advantage of this method is that it is intuitive and clear, and easy to compare and see the characteristics in the data such as extreme points, turning points, periodicity, rate of change, etc. The accurate graph can also be used for calculus calculation without knowing the mathematical expression, and it is therefore widely used.

In engineering experiments, the correct drawing must follow the following basic principles, in order to obtain a smooth curve (or straight line) with the smallest deviation from the experimental points.

3.2.1 Basic principles of coordinate system selection

The coordinate systems commonly used in the experiment of chemical engineering principles are cartesian coordinate system, single logarithmic coordinate system and double logarithmic coordinate system. The single logarithmic coordinate system is shown in Figure 3-1. One axis is a common coordinate axis with uniform division, the other axis is a logarithmic coordinate axis with uneven division. The double logarithmic coordinate system is shown in Figure 3-2, where axes are logarithmic axes.

(1) Cartesian coordinate system

The functional relationship between variables x and y is:

$$y = a + bx$$

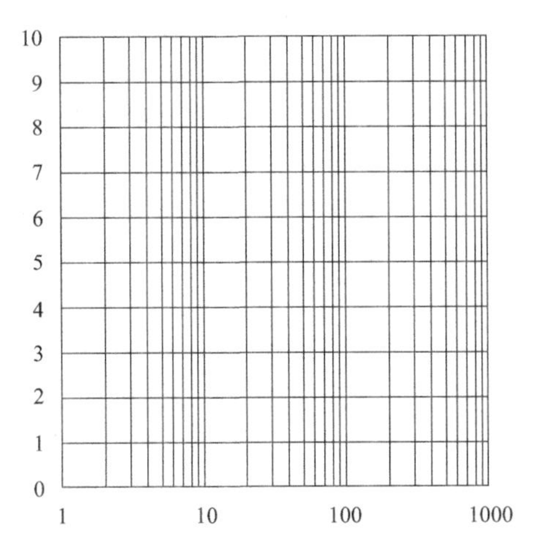

Figure 3-1 Single logarithmic graph

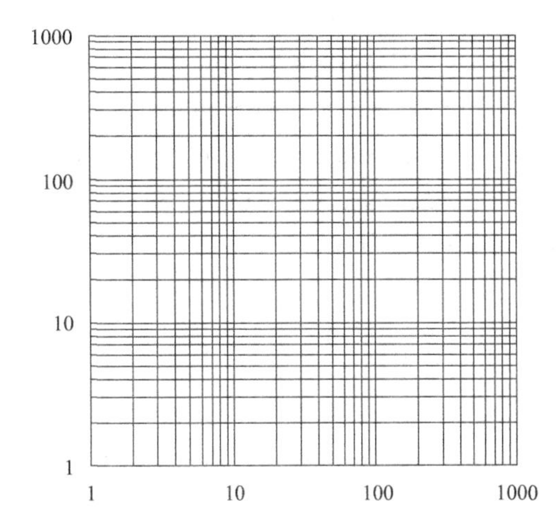

Figure 3-2 Double logarithmic graph

This is a straight line function. The variable x, y are plotted on a rectangular coordinate paper to obtain a straight line graph, and the intercept a and slope b can be obtained.

(2) Single logarithmic coordinate system

In the following cases, it is recommended to use a single logarithmic coordinate system.

① One of the variables has changed by several orders of magnitude within the scope of the study.

② In the initial stage where the independent variable gradually increases from zero, when a small change in the independent variable causes a large change in the dependent variable, using a single logarithmic coordinate can extend the maximum change range of the curve and make the graph outline clearer.

③ When you need to transform a non-linear relationship into a linear relationship, single logarithmic coordinates could be chosen. For example, in the formula of exponential function $y=ae^{bx}$ or logarithmic function $y=a+b\lg x$, a, b are undetermined coefficients. See part 3. 3. 1 for details.

(3) Double logarithmic coordinate system

In the following cases, it is recommended to use the double logarithmic coordinates system:

① The variables x and y have changed by several orders of magnitude in value.

② The beginning of the curve needs to be divided into an expanded form.

③ When you need to transform a non-linear relationship into a linear relationship, double logarithmic coordinates could be used. For example, in the formula of the power function $y=ax^b$, a, b are undetermined coefficients. See part 3. 3. 1 for details.

3. 2. 2 Determination of coordinate division

Coordinate indexing is determined according to the physical quantity that each coordinate axis can represent, that is, select an appropriate coordinate scale. The method of determining the coordinate division is as follows.

① In order to get better graphics, when the error $D(x)$, $D(y)$ of x, y are known, the scale should be chosen to make the side length of the experimental point to be $2D(x)$, $2D(y)$ (approximately square), and to make $2D(x)=2D(y)=1\sim2mm$, if $2D(x)=2D(y)=2mm$, then their scale should be:

$$M_y=\frac{2mm}{2\Delta y}=\frac{1}{\Delta y}mm \tag{3-1}$$

$$M_x=\frac{2mm}{2\Delta x}=\frac{1}{\Delta x}mm \tag{3-2}$$

If the temperature error is known as $D(t)=0.05℃$, then

$$M=\frac{1mm}{0.05℃}=20mm/℃$$

At this time, the coordinate of temperature of 1℃ is 20mm. If it feels too long, it is advisable to take $2D(x)=2D(y)=1mm$, which makes the coordinate of 1℃ to be as long as 10mm.

② If the error of the measured data is unknown, the division of the coordinates should match the effective number of the experimental data, that is, the most suitable division is to make the experimental curve coordinate reading and the experimental data have the same effective number of digits. Secondly, the ratio between the horizontal and vertical coordinates may not necessarily be the same. It should be selected according to the specific situation, so that the slope of the experimental curve is between 30° and 60°, and the accuracy of such curve coordinate reading is satisfactory.

③ It is recommended to use the proportional constant of the coordinate axis $M=(1,2,5)\times10^{\pm n}$ (n is a positive integer), and do not use the proportional constants as 3, 4, 6, 7, 8, 9, etc., because the latter will cause trouble in drawing graphics and errors are tend to occur.

3.2.3 Considerations for designing experimental data graphs

① For a system with two variables, it is customary to put the independent variable on the horizontal axis and the dependent variable on the vertical axis. The variable name, symbol and unit should be marked on sides of both axes. For example, the horizontal axis of the characteristic curve of the centrifugal pump must be marked with: flow / (m^3/h). Pay special attention to unit, which is often overlooked by beginners.

② Coordinate indexing should be appropriate to make the functional relationship of variables clear. The origin of the Cartesian coordinate is not necessarily selected as the zero point, it should be determined according to the range of the plotted data, so as to make the figure uniformly centered.

The origin of the logarithmic coordinates is not zero point. The values marked on the logarithmic axis are true values, not logarithmic values. For example, several data are obtained during a resistance experiment, such as $Re_1=1.1\times10^4$, $\lambda_1=0.031$, $Re_2=7.8\times10^4$, $\lambda_1=0.019$. Plot them on the double logarithmic coordinates as Figure 3-3 to be Point A and Point B. Since the logarithms of 1, 10, 100, 1000, etc. are 0, 1, 2, 3, etc. , the distance of each order of magnitude on the logarithmic coordinate paper is equal, but the scale within the same order of magnitude is not equal. The coordinate system should be strictly followed when using the coordinate paper.

③ If several sets of measured values are plotted on the same piece of coordinate paper at the same time, different symbols (such as o, Δ, ×, etc.) should be used to show the difference. If n different functions are plotted on a piece of coordinate paper, the function name should be marked on the curve.

④ The graphs must be with serial numbers and names. The serial numbers should be written in the order appearing in the report and explained in the text. If necessary, there should also be an annotation.

⑤ The graph line should be smooth. Use the tools such as curve board to connect the discrete points into a smooth curve or straight line, and make the line pass as many experimental points as possible, or make the points outside the line as close to the line as possible, and make the points on both sides of the line roughly equal .

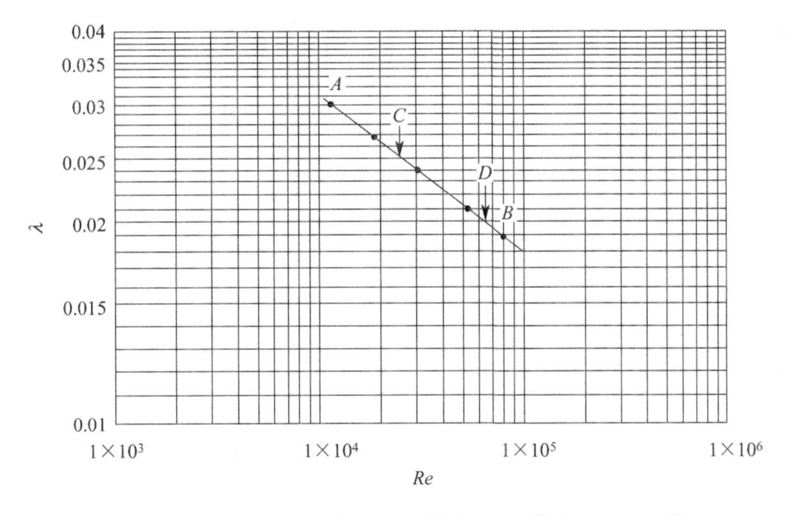

Figure 3-3 Relationship between friction coefficient λ and Re

3. 3 Mathematical equation representation method

In experimental research, in addition to using tables and graphs to describe the relationship between variables, experimental data is often organized into equations to describe the relationship between the independent and dependent variables of a process or phenomenon, that is, to establish a mathematical model of the process. The method is to draw the experimental data into a curve and compare it with the typical curve of the known functional relationship (linear equation, power function equation, exponential function equation, parabolic function equation, hyperbolic function equation, etc.) and determine the appropriate functional relationship, then use the graphic method or regression analysis method to determine the various constants in the functional formula. It is very common to use a computer to regress the experimental data results to mathematical equations for experimental data processing. For example, Excel and Origin are commonly used data processing and drawing software in chemical engineering principle experiment. Whether the obtained function expression can accurately reflect the relationship existing in the experimental data should be confirmed by inspection.

The principle of mathematical equation selection is that it requires both simple forms and fewer constants, it can accurately express the relationship between experimental data. However, it is often difficult to meet both the conditions. Usually, as long as the necessary accuracy is guaranteed, the simple linear relationship is selected or converted into by an appropriate method. For example, the power function and the exponential function are converted into linear equations by taking the logarithm, so that the data processing work is simplified.

3. 3. 1 Graphic analysis method

When the formula is selected, the constants in the equation can be obtained by the graphic method. This section takes power function, exponential function and logarithmic function as examples.

(1) Linear illustration of power function

For the power function $y=ax^b$, a and b are the coefficients to be solved. If drawing directly on the cartesian coordinate system, the line will be a curve, therefore, it is supposed to take the logarithms of both sides of the above formula at the same time, then:

$$\lg y=\lg a+b\lg x \tag{3-3}$$

Let $\lg y=Y$, $\lg x=X$, then formula (3-3) is transformed into:

$$Y=\lg a+bX \tag{3-4}$$

Formula (3-4) is a straight line equation, which is plotted as a straight line on a rectangular coordinate paper, then values of a and b can be obtained through its intercept and slope. However, in order to avoid the trouble of finding logarithmic values for each experimental data, it is advisable to use double logarithmic graph paper for drawing. The method of finding coefficients a and b in double logarithmic coordinates is as follows.

1) How to find the coefficient b　The coefficient b is the slope of the straight line, as shown in Figure 3-3. The method of calculating the slope on logarithmic coordinates is different from that on rectangular coordinates. The slope of the straight line on the double logarithmic coordinate paper needs to be calculated with a logarithmic value. Take any two points on the straight line (note: the points must be taken on the line instead of using the experimental points directly) and substitute them into formula (3-5) to calculate, then:

$$b=\frac{\lg y_2-\lg y_1}{\lg x_2-\lg x_1} \tag{3-5}$$

2) How to find the coefficient a　Take the value of any point on the straight line and the

calculated slope b, substitute them into the original equation $y=ax^b$ and obtain the value a by calculation.

3) Equation-oriented approach The values of a and b can also be solved by taking two points on a straight line and using simultaneous equations.

【Example 3-1】 In the fluid resistance experiment, a smooth tube was used and the data in the turbulent zone were measured. The values of friction coefficient λ and Re calculated based on the experimental data have been plotted on Figure 3-3, and the experimental points have been fitted into a straight line (the specific numbers are omitted here) . Please determine the value of a and b in equation of the relationship between the friction coefficient λ and Re $\lambda=aRe^b$.

Solution One

Take two arbitrary points on the straight line in Figure 3-3: point *C* (2. 6×104, 0. 025) and point *D* (6. 3×104, 0. 020), calculate the value of b according to formula (3-5)

$$b=\frac{\lg y_2-\lg y_1}{\lg x_2-\lg x_1}=\frac{\lg 0.020-\lg 0.025}{\lg(6.3\times10^4)-\lg(2.6\times10^4)}=-0.2521$$

$$a=\frac{\lambda}{Re^b}=\frac{0.025}{(2.6\times10^4)^{-0.2521}}=0.3243$$

Solution Two

Substitute the values of points C and D into the equation $\lambda=aRe^b$ to solve simultaneously:

$$\begin{cases}0.025=a\times(2.6\times10^4)^b\\0.020=a\times(6.3\times10^4)^b\end{cases}$$

The solution is: a=0. 3243, b=−0. 2521

According to the effective number rule, a=0. 32, b=−0. 25, then the relationship between the friction coefficient λ and Re obtained by the experiment is:

$$\lambda=0.32Re^{-0.251} \text{ or } \lambda=\frac{0.32}{Re^{0.25}}$$

(2) Linear illustration of exponential or logarithmic functions

When the functional relationship studied is exponential function $y=ae^{bx}$ or logarithmic function $y=a+b\lg x$, the data plotting on the single logarithmic graph paper forms a straight line. The linearization method is as follows:

Take the logarithm on both sides of the formula $y=ae^{bx}$, then

$$\lg y=\lg a+bx\lg e \tag{3-6}$$

Let $\lg y=Y$, $b\lg e=k$, then formula (3-6) becomes:

$$Y=\lg a+kx \tag{3-7}$$

For the logarithmic function $y=a+b\lg x$, let $\lg x=X$, then:

$$y=a+bX \tag{3-8}$$

The method of seeking coefficients is as follows.

1) How to find the coefficient b For $y=ae^{bx}$, the vertical axis is logarithmic, and the slope is:

$$k=\frac{\lg y_2-\lg y_1}{x_2-x_1} \tag{3-9}$$

$$b=\frac{k}{\lg e} \tag{3-10}$$

For $y=a+b\lg x$, the horizontal axis is logarithmic, and the slope is:

$$b=\frac{y_2-y_1}{\lg x_2-\lg x_1} \tag{3-11}$$

2) How to find the coefficient a The method is basically the same as the one described in the power function. Substitute the coordinate value at any point on the line and the coefficients b already obtained into the function relationship to solve a.

3) Equation-oriented approach The values of a and b can also be solved by taking two points on a straight line and using simultaneous equations.

(3) Graphical solutions of binary linear equations

If the physical quantity of the object under study includes a dependent variable and two independent variables, the following functional formula can be used when they are in a linear relationship:

$$y=a+bx_1+cx_2 \tag{3-12}$$

When illustrating such a functional formula, first of all, one of the independent variables should be kept constant. For example, let x_1 be a constant, then the formula can be rewritten as

$$y=d+cx_2 \tag{3-13}$$

In formula, $d=a+bx_1=\text{constant}$.

Based on the data y and x_2, a straight line can be plotted in rectangular coordinates, as shown in Figure 3-4 (a). The coefficient c of x_2 can be thus determined.

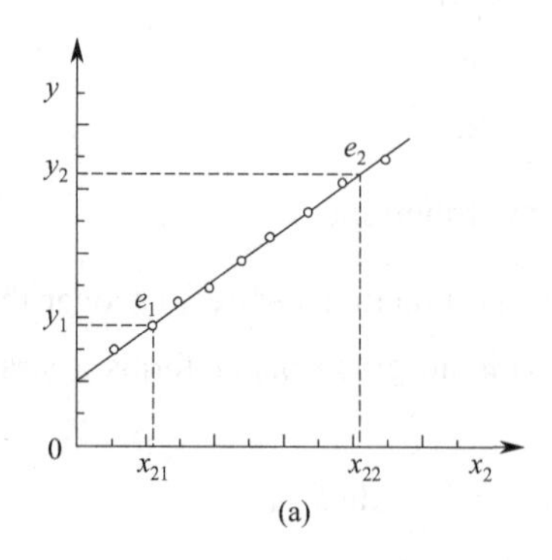

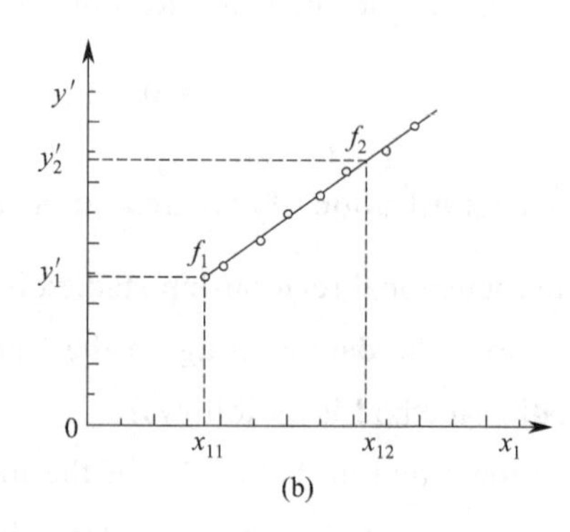

Figure 3-4 Schematic diagram of binary linear equation

Take two points e_1 (x_{21}, y_1), e_2 (x_{22}, y_2) on the line in Figure 3-4 (a), then

$$c=\frac{y_2-y_1}{x_{22}-x_{21}} \tag{3-14}$$

When c is obtained, substitute it into formula (3-12) and rewrite formula (3-12) into the following form:

$$y-cx_2=a+bx_1 \tag{3-15}$$

Let $y'=y-cx_2$, then a new linear equation can be obtained:

$$y'=a+bx_1 \tag{3-16}$$

Calculate y' from the experimental data y, x_2 and c. Draw the straight line in Graph 3-4 (b) from y' and x_1, then take two points f_1 (x_{11}, y'_1), f_2 (x_{12}, y'_2) on the straight line to determine the two constants a and b.

$$b=\frac{y'_2-y'_1}{x_{12}-x_{11}} \tag{3-17}$$

$$a=\frac{y'_1x_{12}-y'_2x_{11}}{x_{12}-x_{11}} \tag{3-18}$$

It should be pointed out that when determining b and a, its independent variable x_1 and x_2 should be changed at the same time, so that its result covers the entire experimental range.

Sherwood used 7 different fluids to study the forced convection heat transfer through a circular straight tube, a large amount of data is obtained and processed in the form of power function, and the formula is:

$$Nu=CRe^mPr^n \tag{3-19}$$

where Nu is Nusselt number; Re is Reynolds number; Pr is Prandtl number; C, m and n are Undetermined constant.

Nu changes with Re and Pr. Take the logarithm of the two sides of the above formula and use variable substitution to make it into the form of a binary linear equation:

$$\lg Nu=\lg C+m\lg Re+n\lg Pr \tag{3-20}$$

Let $y=\lg Nu$, $x_1=\lg Re$, $x_2=\lg Pr$, $a=\lg C$, then the above formula can be expressed as a binary linear equation:

$$y=a+mx_1+nx_2 \tag{3-21}$$

Now rewrite the formula (3-20) into the following form to determine the constant n (let the variable Re be a constant, and the independent variable is reduced by one)

$$\lg Nu=(\lg C+m\lg Re)+n\lg Pr \tag{3-22}$$

Sherwood sets Re to be a constant of 10^4, plots the experimental data of 7 different fluids on a double logarithmic graph paper, then draws the relationship between Nu and Pr to obtain exponent n, of Prandtl number, then solve by the following graphical method:

$$\lg(Nu/Pr^n)=\lg C+m\lg Re \tag{3-23}$$

Taking Nu/Pr^n as the ordinate and Re as the abscissa, drawing on a double logarithmic coordinate paper, the C and m value can be obtained from the slope and intercept. In this way, all undetermined constants C, m and n in the empirical formula are determined.

3.3.2 Regression analysis method

The process of obtaining the empirical formula by graphic method was introduced in part 3.3.1. Although the graphical method has many advantages, since the fitted line is drawn through discrete points during the drawing process, where the arbitrariness is greater, the reading from the coordinate chart will also cause errors. In order to get the quantitative relationship between these experimental data variables to make it as practical as possible, one of the most widely used mathematical statistical methods is regression analysis method, in which some statistical laws that reflect the internal things can be found from a large number of scattered data observed, those laws can be expressed in the form of mathematical models.

Regression is also called fitting. For two variables with a related relationship, if described with a straight line, it is called a unary linear regression, if described with a curve, it is called a unary nonlinear regression. For the three variables with correlation, one dependent variable and two independent variables, if described by plane, it is called binary linear regression, and if by surface description, it is called binary nonlinear regression. By analogy, it can be extended to n-dimensional space for regression, which is called multiple linear regression or multiple nonlinear regression. When dealing with experimental problems, nonlinear problems are often converted to linear problems to deal with, and the most effective method for establishing linear regression equations is linear least squares. Although most of the regression problems encountered during the experiment of chemical engineering principles are mostly binary or more regressions, as the concept of univariate linear regression is easy to understand, this part mainly talks about how to find the linear regression equation of one variable, which lays the foundation for other regression analysis learning and regression analysis with Origin software.

3.3.2.1 Solution of the unary linear regression equation

In the statistical methods of scientific experiments, it is usually necessary to find the functional relationship $y=f(x)$ between the independent variable x_i and the dependent variable y_i from the obtained experimental data $(x_i, y_i, i=1,2,\cdots,n)$. Since the experimental measurement data generally have errors, it cannot be required that all experimental points are on the curve represented by $y=f(x)$, and only needs to satisfy that the residual $d_i=y_i-f(x_i)$ of the experimental points (x_i, y_i) and $f(x_i)$ is less than the given error. This kind of problem that seeks to approximate the function expression $y=f(x)$ of the experimental data relationship is called curve fitting.

To carry out curve fitting, first of all, according to the characteristics of the experimental data, the appropriate function form should be selected, then determine the target function. For example, after obtaining the experimental data of two variables, if each data point is marked on a common rectangular coordinate paper and the distribution is similar to a straight line, linear regression can be considered to find its expression.

Given n experimental points $(x_1, y_1), (x_2, y_2), \cdots, (x_n, y_n)$, the discrete points are shown in Figure 3-5, so the following straight line can be used to represent their relationship.

$$y'=a+bx \tag{3-24}$$

In this formula: y' is the value calculated by the regression formula, called the regression value; a, b are regression coefficients

For each measured value x_i, a regression value y' can be obtained by formula (3-24). The absolute value $d_i=|y_i-y_i'|=|y_i-(a+bx_i)|$ of the difference between the regression value y' and the measured value y_i indicates the degree of deviation of y_i from the regression line. The smaller the deviation between the two, the better the straight line fits the experimental data points. The $|y_i-y_i'|$ value represents the distance of the point (x_i, y_i) along the direction parallel to the y-axis to the regression line, as shown by the vertical lines d_i in Figure 3-6.

The target function for fitting should be determined during curve fitting. The method of selecting the sum of squared residuals as the objective function is called the least square method. This method is a rigorous and effective method of seeking approximate function expressions of experimental data. It is defined as: the most ideal curve is to make the sum of squared residuals of each point on the same curve to be the smallest.

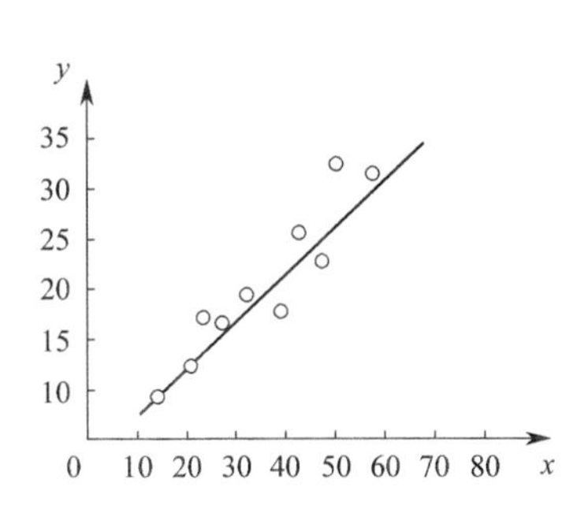

Figure 3-5 Schematic diagram of unary linear regression

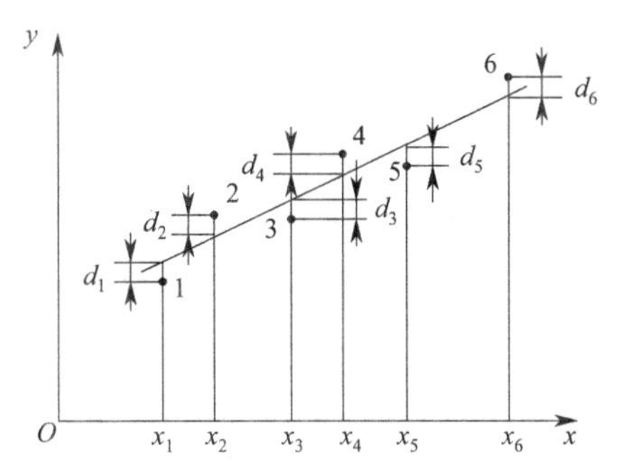

Figure 3-6 Experimental curve diagram

Let the sum of squared residuals Q be:

$$Q=\sum_{i=1}^{n} d_i^2=\sum_{i=1}^{n}[y_i-(a+bx_i)]^2 \tag{3-25}$$

x_i and y_i are known here, so Q is a function of a and b. In order to minimize the Q value, according to the extremum principle, as long as the partial derivatives of a and b $\left(\frac{\partial Q}{\partial a}, \frac{\partial Q}{\partial b}\right)$ are respectively obtained and let to be zero, the value of a and b can be evaluated. This is the principle of least squares, as:

$$\begin{cases}\dfrac{\partial Q}{\partial a}=-2\sum\limits_{i=1}^{n}(y_i-a-bx_i)=0 \\ \dfrac{\partial Q}{\partial b}=-2\sum\limits_{i=1}^{n}(y_i-a-bx_i)x_i=0\end{cases} \tag{3-26}$$

A normal equation could be obtained from equation (3-26):

$$\begin{cases}a+\bar{x}b=\bar{y} \\ n\bar{x}a+\left(\sum\limits_{i=1}^{n}x_i^2\right)b=\sum\limits_{i=1}^{n}x_iy_i\end{cases} \tag{3-27}$$

where

$$\bar{x}=\frac{1}{n}\sum_{i=1}^{n}x_i \quad \bar{y}=\frac{1}{n}\sum_{i=1}^{n}y_i \tag{3-28}$$

Solve equation (3-27), a (intercept) and b (slope) in the regression could be achieved:

$$b=\frac{\sum(x_i \cdot y_i)-n\bar{x}\bar{y}}{\sum x_i^2-n(\bar{x})^2} \tag{3-29}$$

$$a=\bar{y}-b\bar{x} \tag{3-30}$$

【Example 3-2】 The data obtained during the calibration of the rotameter are shown in Table 3-5, please use the least square method to find the experimental equation.

Table 3-4 The reading and flow rate data obtained during the calibration of the rotameter

reading x/grid	0	2	4	6	8	10	12	14	16
flow $y/(m^3/h)$	30.00	31.25	32.58	33.71	35.01	36.20	37.31	38.79	40.04

Solution: $\sum(x_iy_i)=2668.58, \bar{x}=8, \bar{y}=34.9878, \sum x_i^2=816$

$$b=\frac{\sum(x_i \cdot y_i)-n\bar{x}\bar{y}}{\sum x_i^2-n(\bar{x})^2}=\frac{2668.58-9\times8\times34.9878}{816-9\times8^2}=0.623$$

$$a=\bar{y}-b\bar{x}=34.9878-0.623\times8=30.0$$

Therefore, the regression equation is: $y=30.0+0.623x$

3.3.2.2 Verification of regression effect

The relationship between experimental data variables is uncertain, each value of a variable corresponds to the entire value set. When x changes, the distribution of y also changes in a certain way. In this case, the relationship between the variables x and y is called the correlation.

In the calculation process of the regression equation seeking mentioned above, it is not necessary to assume in advance that there must be some correlation between the two variables. As far as the method itself is concerned, even if the discrete points on the graph are completely disorganized, there could still be a straight line to represent the relationship between x and y using the least square method, which is obviously meaningless. In fact, it only makes sense to perform linear regression when the two variables are linearly related. Therefore, the regression effect must be tested.

(1) Correlation coefficient

The correlation coefficient r could be used here to test the regression effect, who is a quantitative index that characterizes the degree of linear correlation between two variables. If the linear equation obtained by regression is $y'=a+bx$, the calculation formula of r is (deduction process ellipsis):

$$r=\frac{\sum(x_i-\bar{x})(y_i-\bar{y})}{\sqrt{\sum(x_i-\bar{x})^2\sum(y_i-\bar{y})^2}} \tag{3-31}$$

The variation range of r is $-1\leqslant r\leqslant 1$, and its signs depend on $\sum(x_i-\bar{x})(y_i-\bar{y})$ and are consistent with the slope of the regression line equation b. The geometric meaning of r can be illustrated by Figure 3-7.

When $r=\pm1$, that is, the n groups of experimental values (x_i, y_i) are all on the straight line, it is said to be completely correlated. See (d) and (e) in Figure 3-7.

When $0<|r|<1$, which is the vast majority of cases, there is a certain linear relationship between x and y. When $r>0$, y increases as x increases, where it was said that y is positively correlated with x, as shown in (b) in Figure 3-7; When $r<0$, y decreases as x increases, where it was said that y is negatively correlated with x, as shown in (c) in Figure 3-7. The smaller $|r|$ is, the farther the scattered points are from the regression line. When $|r|$ is closer to

1, that is, the closer the n sets of experimental values are to $y=a+bx$, the closer the relationship between variables y and x is to a linear relationship. When $r=0$, there is no linear relationship between variables, as shown in Figure 3-7(a). It should be noted that the absence of a linear relationship does not mean that there is no other functional relationship, as shown in Figure 3-7(f).

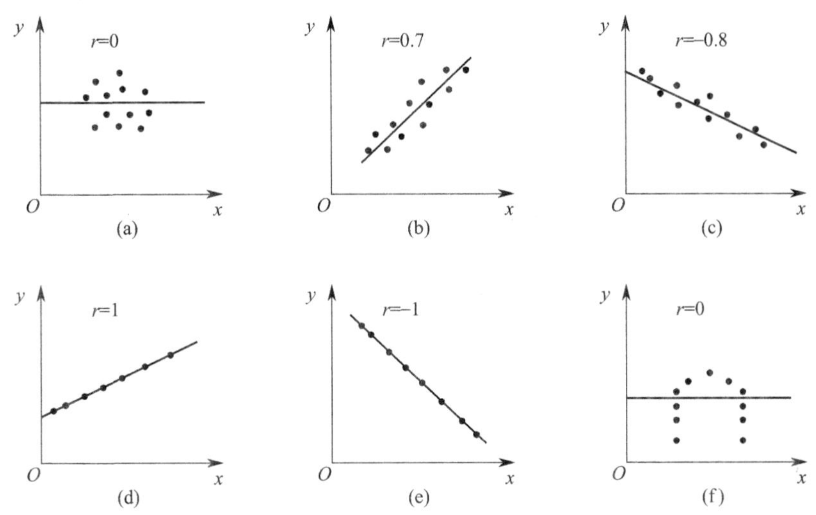

Figure 3-7 The geometric meaning of correlation coefficient

(2) Significance test

As discussed above, the closer the absolute value of the correlation coefficient r is to 1, the more linear the correlation between x and y. But how close is enough to show that there is a linear correlation between x and y? This makes it necessary to test the correlation coefficient for significance. Only when $|r|$ reaches a certain level can the regression line be used to approximate the relationship between x and y, when the correlation is significant. In general, if the correlation coefficient r can reach a significant value of the correlation or not is related to the number of experimental data n. Therefore, only when $|r|>r_{min}$ a linear regression equation can be used to describe the relationship between its variables. The r_{min} value can be found in Table 3-5. With this table, the corresponding r_{min} can be found based on the n and the significant level coefficient α. Generally, α could be taken at 1% or 5%. The smaller α, the higher the significance.

【Example 3-3】 Find the actual correlation coefficient r of the rotameter calibration experiment in Example 3-2.

Solution: $n=9$, $n-2=7$, look up Table 3-5:

$\alpha=0.01$, $r_{min}=0.798$; $\alpha=0.05$, $r_{min}=0.666$

$\bar{x}=8$, $\bar{y}=34.9878$

$\sum(x_i-\bar{x})(y_i-\bar{y})=149.46$, $\sum(x_i-\bar{x})^2=240$, $\sum(y_i-\bar{y})^2=93.12$

$$r=\frac{\sum(x_i-\bar{x})(y_i-\bar{y})}{\sqrt{\sum(x_i-\bar{x})^2\sum(y_i-\bar{y})^2}}=\frac{149.46}{\sqrt{240\times 93.12}}=0.99976>0.798$$

The above calculation results show that the correlation coefficient of this example is highly significant at the level of α 0. 01.

Table 3-5 Correlation coefficient r_{min} test table

$n-2$	α		$n-2$	α	
	0. 05	0. 01		0. 05	0. 01
1	0. 997	1. 000	21	0. 413	0. 526
2	0. 950	0. 990	22	0. 404	0. 515
3	0. 878	0. 959	23	0. 396	0. 505
4	0. 811	0. 917	24	0. 388	0. 496
5	0. 754	0. 874	25	0. 381	0. 487
6	0. 707	0. 834	26	0. 374	0. 478
7	0. 666	0. 798	27	0. 367	0. 470
8	0. 632	0. 765	28	0. 361	0. 463
9	0. 602	0. 735	29	0. 355	0. 456
10	0. 576	0. 708	30	0. 349	0. 449
11	0. 553	0. 684	35	0. 325	0. 418
12	0. 532	0. 661	40	0. 304	0. 393
13	0. 514	0. 641	45	0. 288	0. 272
14	0. 497	0. 623	50	0. 273	0. 354
15	0. 482	0. 606	60	0. 250	0. 325
16	0. 468	0. 590	70	0. 232	0. 302
17	0. 456	0. 575	80	0. 217	0. 283
18	0. 444	0. 561	90	0. 205	0. 267
19	0. 433	0. 549	100	0. 195	0. 254
20	0. 423	0. 537	200	0. 138	0. 181

Chapter 4

Applications of Excel and Origin Software

4.1 Application of Excel in the experimental data processing

Microsoft Excel is a piece of data processing software for personal computers. It has a friendly operation interface, full-featured calculation functions and adequate chart tools, it is an efficient means for experimental data processing in the chemical engineering field. Arithmetic operations and function calculations are mainly involved during the processing of experimental data in the field of chemical engineering. The arithmetic operators are shown in Table 4-1. The priority order of operations is power, multiplication/division, then addition/subtraction. Operators with the same priority (multiplication and division, addition and subtraction) are applied according to their order in the calculation formulas. If an operation order needs to be changed, it can be achieved by adding brackets "( ) ." If "—" is used as a negative sign, it takes precedence over other arithmetic operators.

Table 4-1 Arithmetic operators and priority orders

Arithmetic operator	Meaning	Priority order
+	Addition	3
—	Subtraction	3
*	Multiplication	2
/	Division	2
^	Power	1

The common calculation functions of Excel are shown in Table 4-2.

Table 4-2 Common calculation functions

Function	Meaning
SUM(number1:number2)	Sum of values from cell number 1 to cell number 2
AVERAGE(number1:number2)	Average of values from cell number 1 to cell number 2
ABS(number)	the Absolute value of the number
MAX(number1:number2)	Maximum of values from cell number 1 to cell number 2
MIN(number1:number2)	Minimum of values from cell number 1 to cell number 2
LOG(number,base)	The logarithm of the number to the specified base
LN(number)	The natural logarithm of the number
SQRT(number)	The square root of the number
PI()	π

The above operations and functions are sufficient to complete the experimental data processing of chemical engineering research. For example, in the single-pump characteristic curve of a centrifugal pump, according to the experimentally tested raw data (such as the flow rate, vacuum, outlet pressure and motor power), the flow velocity, head, effective power and efficiency can be calculated. The inlet velocity calculation is used as an example. As shown in Figure 4-1, enter "=B5 / 3600 / PI()/ 0. 035 ^ 2 * 4" in cell I5 (where cell B5 is the flow rate, 0. 035 is the inlet pipe diameter), the head H is calculated based on the experimental data, The relationship between the head H and the flow Q is fitted through the following equation:

$$H=18.8354-0.1712Q^2$$

Therefore, enter "=18. 8354—0. 1712 * B5 ^ 2" in cell P5 to obtain the head value of the fitting equation listed above. Then, compare it with the experimentally determined head value (cell L5). Entering "=ABS((L5-P5) /L5 * 100)" in cell Q5, followed by the error between the fitting equation of the head versus the flow rate and the experimental data are obtained. Other physical quantities that need to be calculated are also processed through a similar approach. After completing the calculation of all physical quantities at data point 1 (corresponding to the data in row 5 in the Excel table), as shown in Figure 4-2, select I5 to Q5 then drag and drop them through the filling handle to obtain the calculation results for each physical quantity, as shown in Figure 4-3.

单泵特性曲线

流量 /(m³/h)	真空度 /MPa	出口压力 /MPa	电机功率 /W	水温 /℃	进口流速 /(m/s)	出口流速 /(m/s)	密度 /(kg/m³)	扬程 /m	有效功率 /W	轴功率 /W	效率	H拟合值 /m	误差 /%
9.12	0.0180	0.050	683.2	33.0	2.63	1.83	994.65	7.08	174.9	649.0	0.269	4.60	35.05
8.17	0.0140	0.073	665.6	33.2									
7.40	0.0115	0.090	649.5	33.4									
6.62	0.0090	0.109	627.3	33.4									
5.87	0.0064	0.120	604.0	33.5									
5.02	0.0040	0.135	574.1	33.6									
4.10	0.0030	0.150	536.6	33.7									
3.12	0.0000	0.165	491.6	33.7									
2.41	0.0000	0.170	455.9	33.7									
1.36	0.0000	0.180	403.5	33.8									
0.58	0.0000	0.185	365.9	33.8									
0.00	0.0000	0.190	328.2	33.8									

Figure 4-1 Example of arithmetic operations

The above method allows most of the functions required for experimental data processing to be conducted in the field of chemical engineering. In addition, the numerical integration calculation of the number of mass transfer units is involved in experiments conducted in this field. The data processing of an extraction experiment is used as an example to illustrate how to use Excel tools for a numerical integration.

In the extraction experiment, the solubility of the solute in the original solvent and extract-

单泵特性曲线

序号	流量 /(m³/h)	真空度 /MPa	出口压力 /MPa	电机功率 /W	水温 /℃	进口流速 /(m/s)	出口流速 /(m/s)	密度 /(kg/m³)	扬程 /m	有效功率 /W	轴功率 /W	效率	*H*拟合值 /m	误差 /%
1	9.12	0.0180	0.050	683.2	33.0	2.63	1.83	994.65	7.08	174.9	649.0	0.269	4.60	35.05
2	8.17	0.0140	0.073	665.6	33.2									
3	7.40	0.0115	0.090	649.5	33.4									
4	6.62	0.0090	0.109	627.3	33.4									
5	5.87	0.0064	0.120	604.0	33.5									
6	5.02	0.0040	0.135	574.1	33.6									
7	4.10	0.0030	0.150	536.6	33.7									
8	3.12	0.0000	0.165	491.6	33.7									
9	2.41	0.0000	0.170	455.9	33.7									
10	1.36	0.0000	0.180	403.5	33.8									
11	0.58	0.0000	0.185	365.9	33.8									
12	0.00	0.0000	0.190	328.2	33.8									

Figure 4-2 Group 1 data calculation settings

单泵特性曲线

序号	流量 /(m³/h)	真空度 /MPa	出口压力 /MPa	电机功率 /W	水温 /℃	进口流速 /(m/s)	出口流速 /(m/s)	密度 /(kg/m³)	扬程 /m	有效功率 /W	轴功率 /W	效率	*H*拟合值 /m	误差 /%
1	9.12	0.0180	0.050	683.2	33.0	2.63	1.83	994.65	7.08	174.9	649.0	0.269	4.60	35.05
2	8.17	0.0140	0.073	665.6	33.2	2.36	1.64	994.59	9.06	200.6	632.3	0.317	7.41	18.23
3	7.40	0.0115	0.090	649.5	33.4	2.14	1.48	994.52	10.57	212.0	617.0	0.344	9.46	10.52
4	6.62	0.0090	0.109	627.3	33.4	1.91	1.33	994.52	12.29	220.5	595.9	0.370	11.33	7.78
5	5.87	0.0064	0.120	604.0	33.5	1.69	1.18	994.49	13.17	209.5	573.8	0.365	12.94	1.78
6	5.02	0.0040	0.135	574.1	33.6	1.45	1.01	994.46	14.48	197.0	545.4	0.361	14.52	0.26
7	4.10	0.0030	0.150	536.6	33.7	1.18	0.82	994.42	15.94	177.1	509.8	0.347	15.96	0.13
8	3.12	0.0000	0.165	491.6	33.7	0.90	0.63	994.42	17.18	145.3	467.0	0.311	17.17	0.08
9	2.41	0.0000	0.170	455.9	33.7	0.70	0.48	994.42	17.70	115.6	433.1	0.267	17.84	0.78
10	1.36	0.0000	0.180	403.5	33.8	0.39	0.27	994.39	18.74	69.1	383.3	0.180	18.52	1.17
11	0.58	0.0000	0.185	365.9	33.8	0.17	0.12	994.39	19.25	30.3	347.6	0.087	18.78	2.47
12	0.00	0.0000	0.190	328.2	33.8	0.00	0.00	994.39	19.77	0.0	311.8	0.000	18.84	4.71

Figure 4-3 Data processing results

ant is extremely low. However, the relationship between the composition of the extraction phase and the raffinate phase in the phase equilibrium equation is non-linear. For example, the phase equilibrium equation of benzoic acid in water (extractant) and kerosene (original solvent) at 25℃ is as follows:

$$y=4.443060\times10^{-6}+1.247730x-4.440948\times10^{2}x^{2}+5.048579\times10^{4}x^{3}+6.414260\times10^{6}x^{4}$$

In formula, x and y are solute components in the raffinate phase and the extraction phase.

Consequently, it is usually difficult to obtain analytic integrals when calculating the number of mass transfer units, and it is necessary to use a graphical or numerical integration method to obtain the number of mass transfer units.

$$N_{OE}=\int_{y_a}^{y_b}\frac{dy}{y^*-y}$$

The upper and lower limits of the integration (the outlet and inlet compositions of the extraction phase) can be obtained through experimental analysis of the compositions. Now, two conditions i. e., $y_a=0$ and $y_b=0.0010$, are used as examples to illustrate how to use Excel software to integrate the above equation.

First, as shown in Figure 4-4, selecting column A in the Excel sheet, setting the integration step size to 0. 0001, and recording the y value in column B starting from $y=0$, the next row of cells will increase by one step size based on the previous adjacent row of cells. The fill handle of the cell is used to pull down and fill in the cells, which is stopped when the value of the bottom cell in column B reaches 0. 0010.

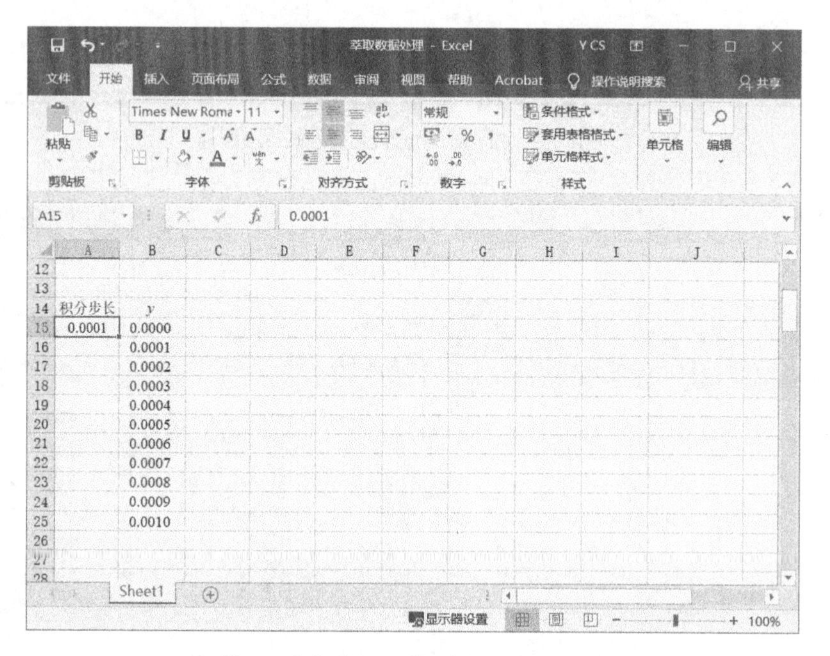

Figure 4-4 Integration interval settings

In this way, several intervals are drawn between the upper and lower limits of the integral. The example shows that ten intervals were obtained from 0 to 0. 0010, these division points must also correspond to the extraction phase composition of a certain section in the extraction tower. Then, the raffinate phase composition x in these sections can be obtained based on the material balance of the whole tower, and y^* can be solved through the raffinate phase composition x according to the phase equilibrium equation, the result of which is shown in Figure 4-5.

Based on the above results, the value of the integrand $1/(y^*-y)$ of each division point can be obtained, as shown in Figure 4-6.

As shown in Figure 4-7, to solve the above integral value, N_{OE} may be obtained by the area of the shape enclosed by the function between $y_a=0$ and $y_b=0.0010$ as well as the horizontal axis.

	A	B	C	D
14	积分步长	y	y	y*
15	0.0001	0.0000	8.2000E-04	7.5971E-04
16		0.0001	9.0333E-04	8.1066E-04
17		0.0002	9.8667E-04	8.5778E-04
18		0.0003	1.0700E-03	9.0133E-04
19		0.0004	1.1533E-03	9.4157E-04
20		0.0005	1.2367E-03	9.7878E-04
21		0.0006	1.3200E-03	1.0132E-03
22		0.0007	1.4033E-03	1.0452E-03
23		0.0008	1.4867E-03	1.0751E-03
24		0.0009	1.5700E-03	1.1031E-03
25		0.0010	1.6533E-03	1.1295E-03

Figure 4-5 Material balance and phase equilibrium calculations

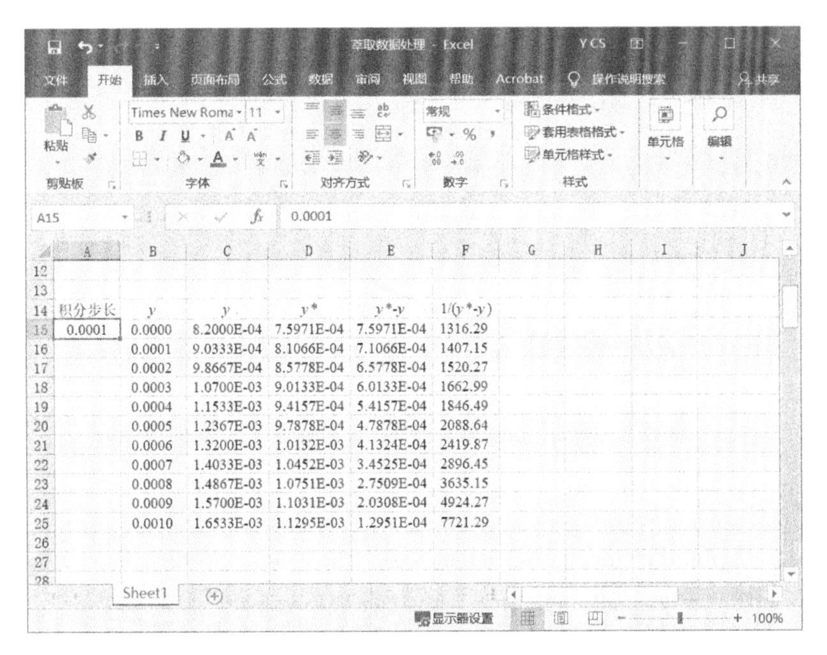

	A	B	C	D	E	F
14	积分步长	y	y	y*	y*-y	1/(y*-y)
15	0.0001	0.0000	8.2000E-04	7.5971E-04	7.5971E-04	1316.29
16		0.0001	9.0333E-04	8.1066E-04	7.1066E-04	1407.15
17		0.0002	9.8667E-04	8.5778E-04	6.5778E-04	1520.27
18		0.0003	1.0700E-03	9.0133E-04	6.0133E-04	1662.99
19		0.0004	1.1533E-03	9.4157E-04	5.4157E-04	1846.49
20		0.0005	1.2367E-03	9.7878E-04	4.7878E-04	2088.64
21		0.0006	1.3200E-03	1.0132E-03	4.1324E-04	2419.87
22		0.0007	1.4033E-03	1.0452E-03	3.4525E-04	2896.45
23		0.0008	1.4867E-03	1.0751E-03	2.7509E-04	3635.15
24		0.0009	1.5700E-03	1.1031E-03	2.0308E-04	4924.27
25		0.0010	1.6533E-03	1.1295E-03	1.2951E-04	7721.29

Figure 4-6 Integrand calculations

Although the shape is irregular, its area can be calculated by summing up the areas of each interval, as shown in Figure 4-8. The area of each interval can be approximated as a rectangle, the area of which is easy to obtain. For example, the area of the interval (0.0001, 0.0002) can be calculated by multiplying the integrand function value $1/(y^*-y)|_{y=0.0001}$ of the interval division point and the interval step size. Figure 4-8 shows that, although there is an error between the rectangular area and the actual area, as long as the interval step size is sufficiently small, the

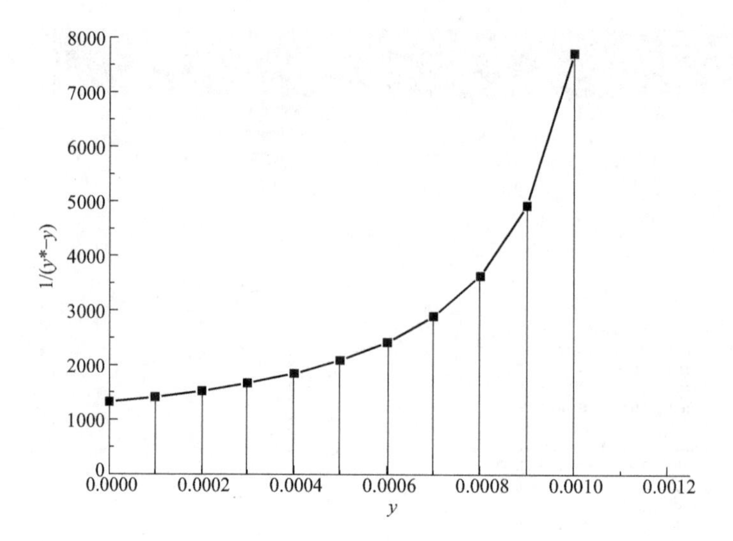

Figure 4-7 Diagram of integrands

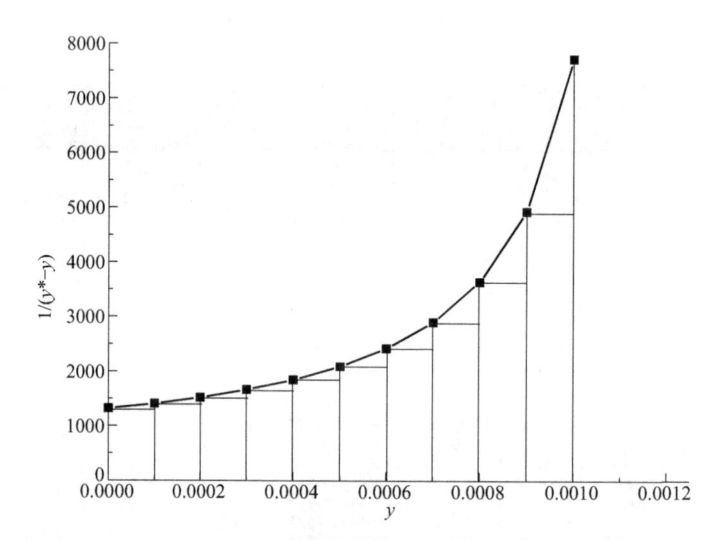

Figure 4-8 Rectangular integration

rectangular area is infinitely close to the actual area. Therefore, in theory, as long as the interval step size is sufficiently small, the actual area can be obtained.

In addition to rectangular integration, trapezoidal integration can also be used. Similar to the above method, trapezoidal integration is applied to approximate the actual area with the trapezoidal area shown in Figure 4-9 during each interval. Using the interval (0.0001, 0.0002) as an example, the trapezoid area is

$$\frac{[1/(y^*-y)|_{y=0.0001}+1/(y^*-y)|_{y=0.0002}]\times 0.0001}{2}.$$

This clearly shows that the results are more in line with the actual situation when using trapezoidal integration under the same conditions.

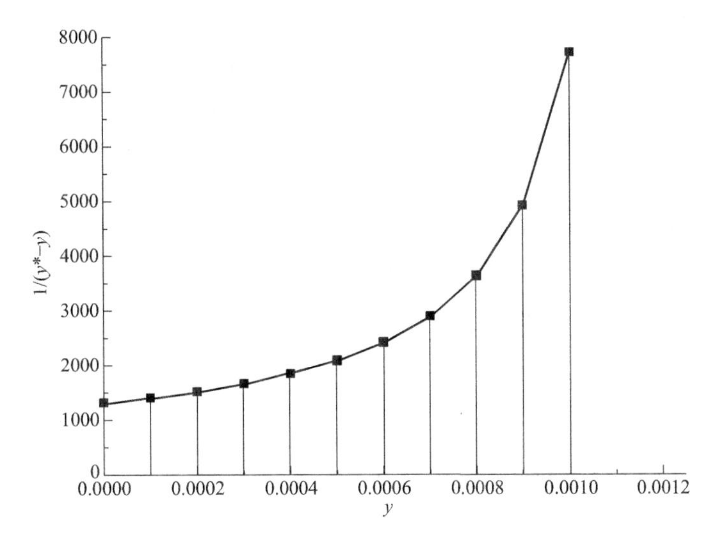

Figure 4-9 Trapezoidal integration

In Excel, the interval (0. 0001, 0. 0002) is also used as an example. As shown in Figure 4-10, the rectangular integral is F14 * $ A$ 15, and the trapezoidal integral is (F14 + F15) * $ A$ 15/2. The area of the other intervals can be calculated by copying through the filling handle, the result of which is shown in Figure 4-10.

G14 | 梯形积分

	A	B	C	D	E	F	G	H
12								
13								
14	积分步长	y	y	y*	y*-y	1/(y*-y)	梯形积分	矩形积分
15	0.0001	0.0000	8.2000E-04	7.5971E-04	7.5971E-04	1316.29	0.1362	0.1316
16		0.0001	9.0333E-04	8.1066E-04	7.1066E-04	1407.15	0.1464	0.1407
17		0.0002	9.8667E-04	8.5778E-04	6.5778E-04	1520.27	0.1592	0.1520
18		0.0003	1.0700E-03	9.0133E-04	6.0133E-04	1662.99	0.1755	0.1663
19		0.0004	1.1533E-03	9.4157E-04	5.4157E-04	1846.49	0.1968	0.1846
20		0.0005	1.2367E-03	9.7878E-04	4.7878E-04	2088.64	0.2254	0.2089
21		0.0006	1.3200E-03	1.0132E-03	4.1324E-04	2419.87	0.2658	0.2420
22		0.0007	1.4033E-03	1.0452E-03	3.4525E-04	2896.45	0.3266	0.2896
23		0.0008	1.4867E-03	1.0751E-03	2.7509E-04	3635.15	0.4280	0.3635
24		0.0009	1.5700E-03	1.1031E-03	2.0308E-04	4924.27	0.6323	0.4924
25		0.0010	1.6533E-03	1.1295E-03	1.2951E-04	7721.29		
26							2.6920	2.3718
27								

Sheet1

循环引用: G26

Figure 4-10 Integration results

The integration results of N_{OE} can be obtained by accumulating the integration results of each interval. The trapezoidal integration and rectangular integration values are shown in G26 and H26, respectively; however, are these results correct? This method requires reducing the step size of the intervals and repeating the above calculation process. If the integration results of different step sizes are similar or the error is less than the allowable value, the integration result is final. Otherwise, the step size needs to be reduced to repeat the calculations. According to the method described above, the step sizes are set to 0. 0001, 0. 00005, 0. 00002 and 0. 00001, the trapezoidal integral results are calculated as 2. 692, 2. 666, 2. 658 and 2. 657, with rectangular integral results of 2. 372, 2. 506, 2. 594 and 2. 625, respectively. During the process of reducing the step size, the trapezoidal integration converges quickly compared to the rectangular integration. When the step size is reduced from 0. 00002 to 0. 00001, the error of the two integrations is 0. 04% , the error of the rectangular integration is 1. 2%. Therefore, $N_{OE}=2.657$ can be regarded as the final accurate solution.

The above-mentioned numerical integration method is calculated using an Excel sheet. It is necessary to continuously change the step size and repeat the calculations until an accurate solution is obtained, the calculation process of which is tedious. Therefore, a polynomial regression can also be applied on the integrand based on Figure 4-7 to obtain the following:

$$\frac{1}{y^*-y}=1261.11846+3.096720\times10^6 y-1.945360\times10^{10}y^2+6.816080\times10^{13}y^3$$
$$-9.377030\times10^{16}y^4+4.828130\times10^{19}y^5+1.448440\times10^{17}y^6$$

Therefore,

$$N_{OE}=\int_0^{0.001}\frac{dy}{y^*-y}=1261.11846y+1.54836\times10^6 y^2-6.48453\times10^9 y^3+1.70402\times10^{13}y^4$$
$$-1.87541\times10^{16}y^5+8.04688\times10^{18}y^6+2.0692\times10^{16}y^7\Big|_0^{0.001}$$
$$=1.2611185y+1.54836-6.48453+17.0402-18.75406+8.0468833+2.069\times10^5$$
$$=2.658$$

This result is similar to the trapezoidal integration result.

4.2 Application of Origin in experimental data regression and plotting

Origin has been recognized as the fastest, most flexible and most convenient engineering drawing software available. Compared with other drawing software, Origin has a simple interface and powerful technical drawing and data processing functions. It is the first choice of drawing tool for scientists and professionals in engineering fields. Origin has numerous functions.. A drawing and the processing of the characteristic curve of the centrifugal pump are used here as an example to illustrate the application of Origin in the drawing and processing of experimental data based on chemical engineering principles. Open the main interface of Origin, as shown in the Figure 4-11, select "Add New Column..." in the menu bar "Column," enter 2 in the dialog box to add two columns, and obtain the four-column data table Book1 of A, B, C and D. Fill-in or copy the centrifugal pump characteristic curve data in this table: flow, head, power and efficiency (as shown in Figure 4-12) .

First, select "symbol" and then "scatter" from "Plot" in the menu bar, open the dialog box, as shown in Figure 4-13, set the data from column A to the x-axis and data from column B

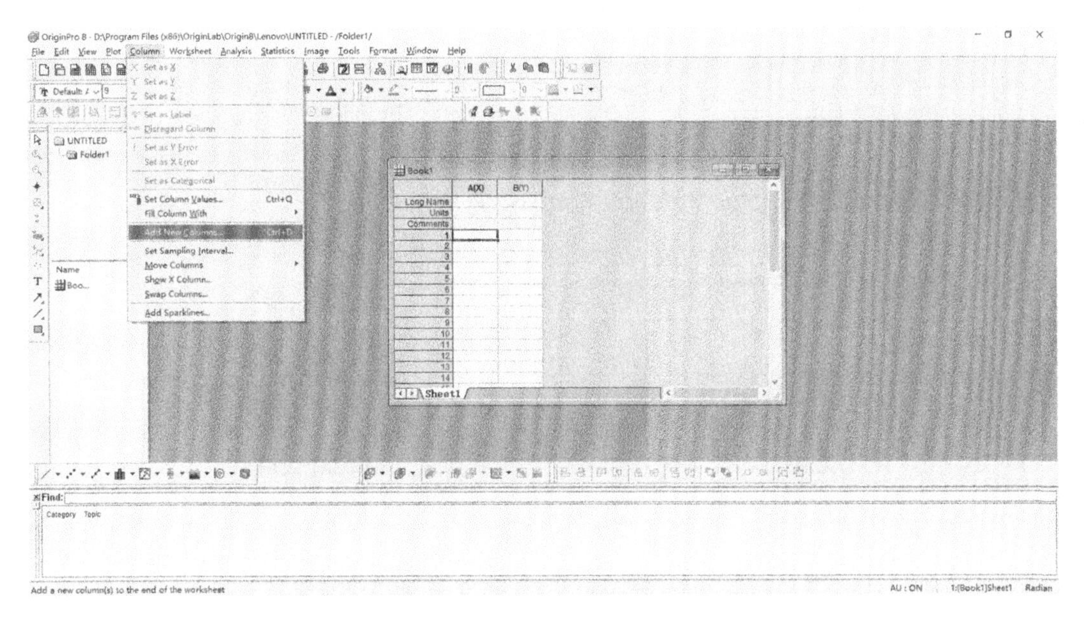

Figure 4-11 The main window of Origin

Book1

	A(X)	B(Y)	C(Y)	D(Y)
Long Name	Q	H	N	η
Units	m^3/s	m	W	
Comments				
1	9.12	7.08	649.04	0.2695
2	8.17	9.06	632.32	0.3173
3	7.4	10.57	617.03	0.3437
4	6.62	12.29	595.94	0.3699
5	5.87	13.17	573.8	0.3651
6	5.02	14.82	545.4	0.3612
7	4.1	15.94	509.77	0.3473
8	3.12	17.18	467.02	0.3111
9	2.41	17.7	433.11	0.267
10	1.36	18.74	383.33	0.1801
11	0.58	19.25	347.61	0.0871
12	0	19.77	311.79	0
13				

Sheet1

Figure 4-12 Data entry

to the y-axis, and click the "Add" button to add the data in the figure below, which is considered to be layer 1 by default. After confirmation, the default data graph shown in Figure 4-14 (the default name of this window is Graph 1) will be generated, which is relatively rough. Any element in the figure can be double-clicked or right-clicked to select "Properties" to open the corresponding property setting dialog box (Figure 4-15). In these dialog boxes, tabs such as "Scale" "Tick Labels" and "Title & Format" can be selected to set up the coordinates, coordinate scales, coordinate names and formats, initial value of the coordinates, coordinate scale form, data point form, legend accordingly in the data map of Figure 4-14. The relationship between the head and the flow is shown in Figure 4-16.

The values of the head, power and efficiency curves in the centrifugal pump characteristic chart are quite different. If these three relationship curves were drawn in a unified coordinate sys-

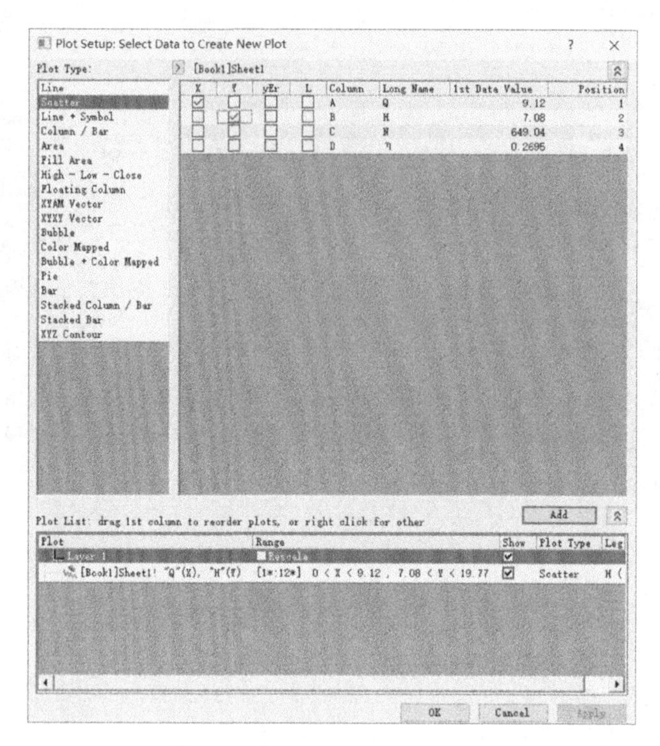

Figure 4-13 Data graph settings

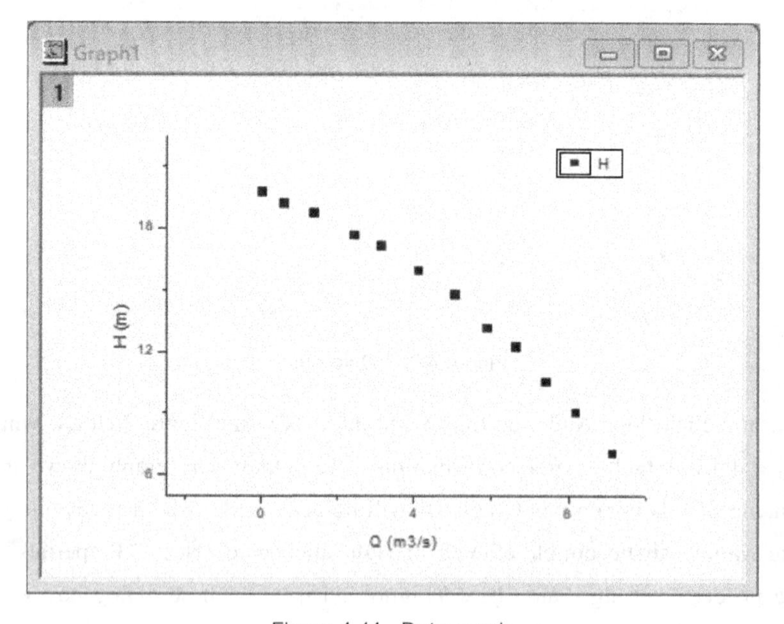

Figure 4-14 Data graph

tem, it would not only affect the aesthetics of the chart, but also inappropriately show certain individual relationship curves owing to the limit of the coordinate range. Therefore, different coordinate axes need to be set for different relationship lines.

With the graph window Graph1 activated, select "New Layer (Axes)" followed by "(Linked) Right Y" from "Graph" in the menu to add a new layer, Layer2 (the layer ID is

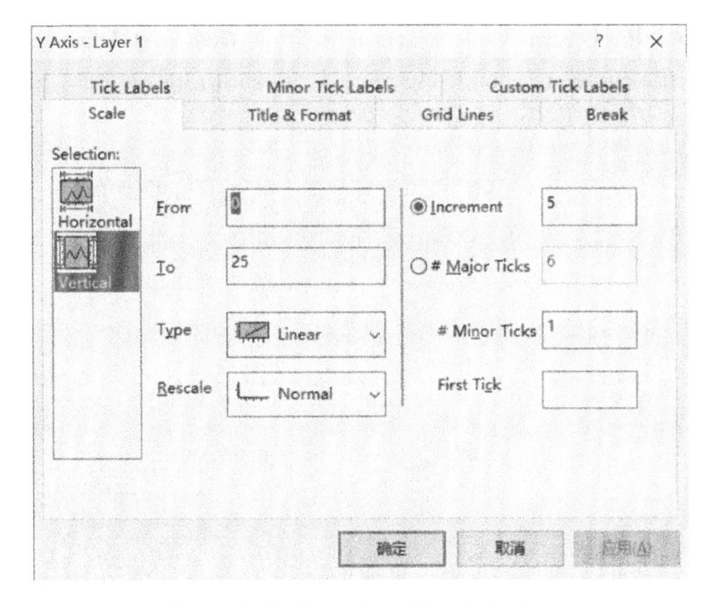

Figure 4-15 Property setting dialog box

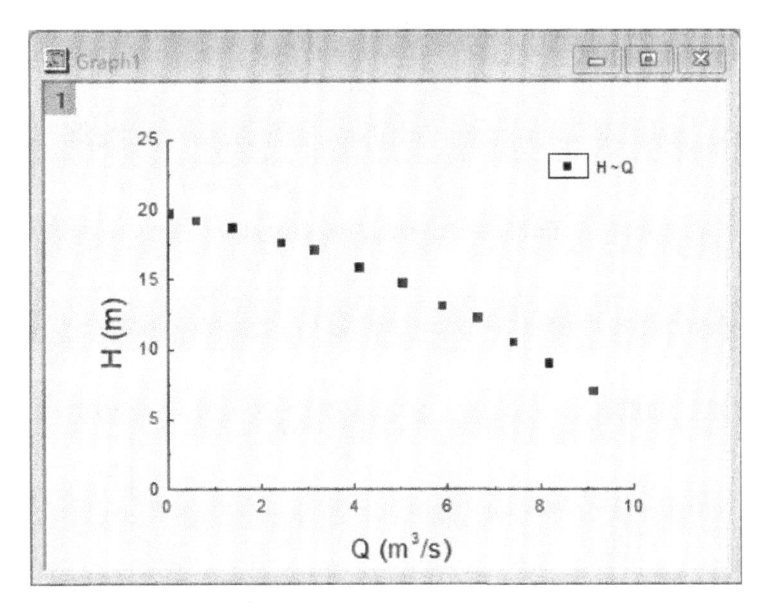

Figure 4-16 Data graph of head versus flow

displayed in the upper left corner of the Graph1 window), set a new y-axis on the right side of the figure, with Layer 1 and Layer 2 sharing the same x-axis. Then, select "Graph" followed by "Add Plot to Layer" and "Scatter," in the pop-up dialog box (Figure 4-17), select the data table Book1, which is required for charting, and set the data columns A and C to the x- and y-axes, and click the "Add" and "OK" buttons to draw the power and flow data graphs. Figure 4-18 is obtained after proper settings. Similarly, set the layer 3, and create a new y-axis on the right side of the graph for efficiency. To avoid an overlapping with the y-axis for power in the layer 2, set the y-axis position in the layer 3 at the position of x = 12. 5 (the axis value in the "Title & Format" tab of the property settings window in Figure 4-15 is changed from "Right" to

"At Position=", and the Percent/Value is set to 12. 5) . Then, all data regarding the centrifugal pump characteristic curve are completely plotted, as shown in Figure 4-19.

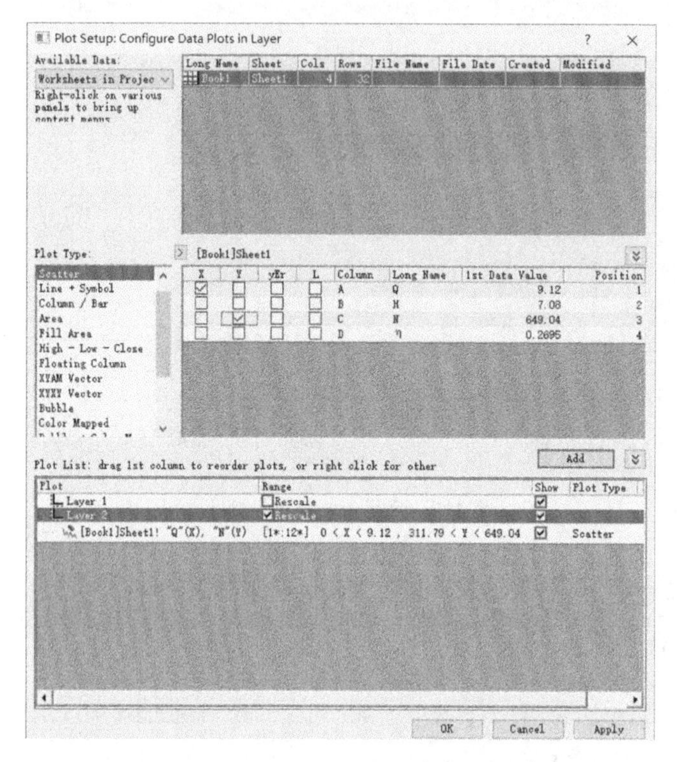

Figure 4-17 Plot setting dialog box

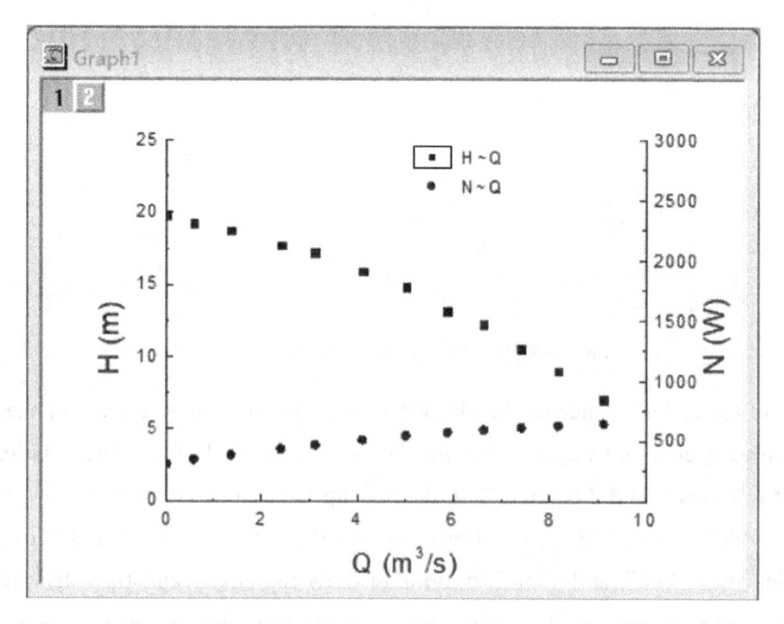

Figure 4-18 Head and power versus flow data graphs

Based on this, the relational equations of the head, power and efficiency versus flow can be obtained by applying a correlation regression on the three characteristic curves in Figure 4-19, the relational

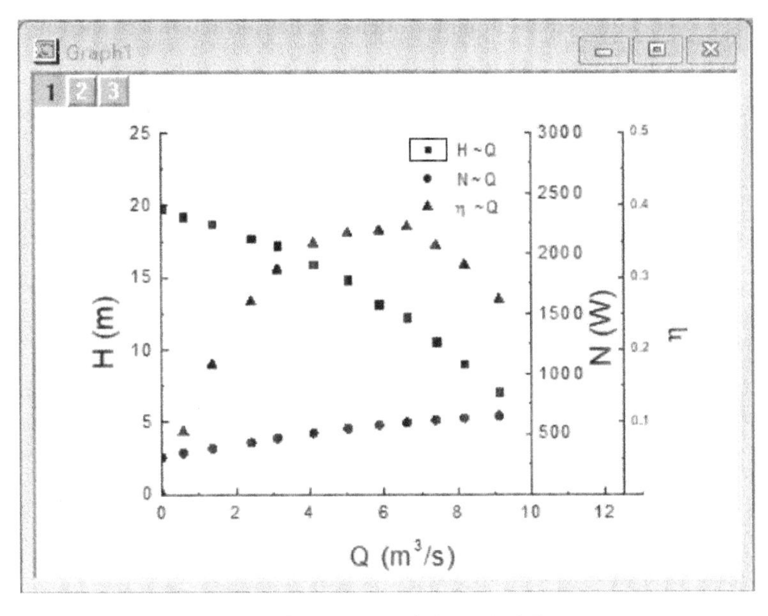

Figure 4-19　Data graph of characteristic curves

equations of N-Q and η-Q can be processed using linear equations and polynomials, respectively. A linear fitting is used for the N-Q relationship. Select Layer 2 where the power data are stored, and click the menu bar "Analysis" "Fitting" and "Fit Linear", followed by "Open Dialog..." sequentially to open the linear fitting setting dialog box (Figure 4-20). A linear fitting can be conducted under

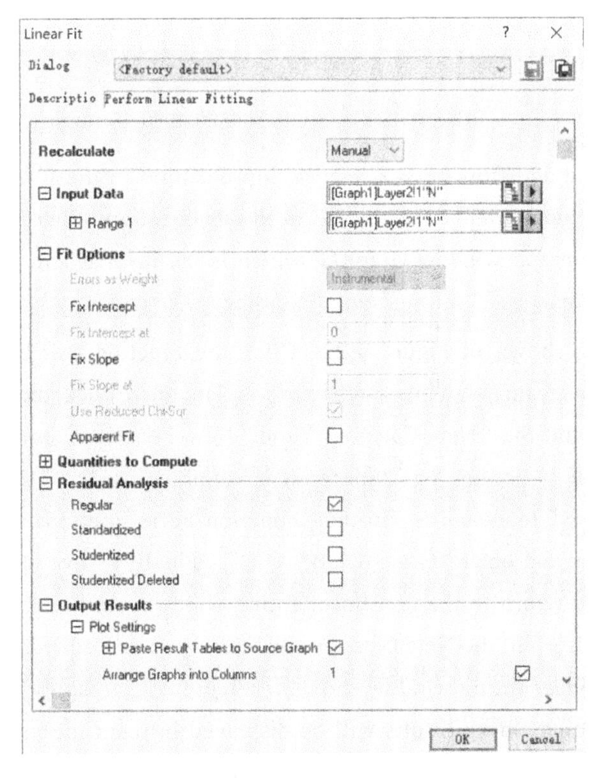

Figure 4-20　Linear fitting dialog box

the default settings. The fitted equation and related error information are shown in the data table Book1 (Figure 4-21), the fitted line is also shown in Graph1 at the same time, which visually demonstrates the result of the fitting. Similarly, select Layer 3 where the efficiency data are located, then click "Analysis" "Fitting" "Fit Polynomial" and "Open Dialog..." sequentially from the menu bar to open the polynomial fitting setting dialog box. In the dialog box, click the option of "Polynomial Order" to set the number of polynomial terms and other parameters can be set as required.

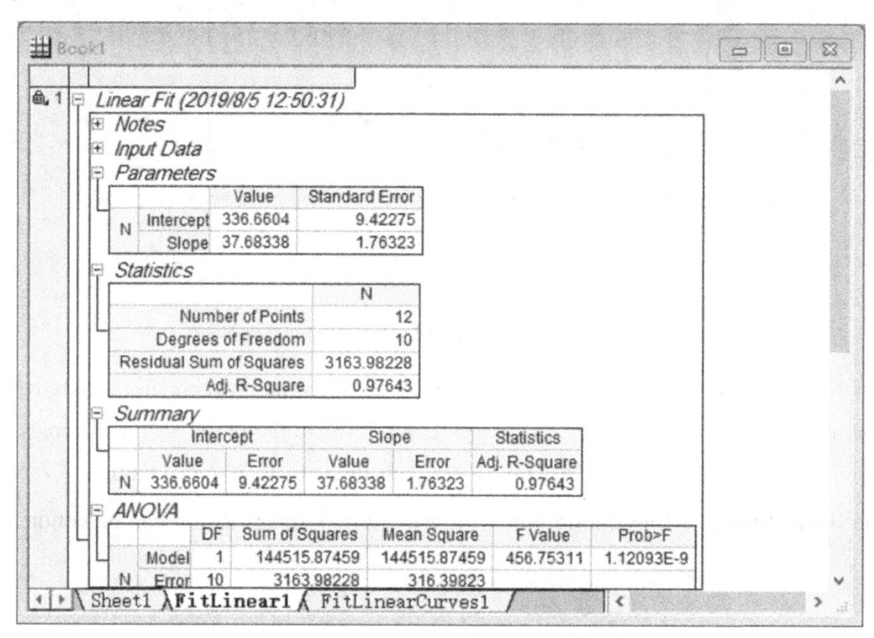

Figure 4-21 Linear fitting results

In theory, the head and flow have a relationship equation in the form of $H = A - BQ^2$. In Origin, a user-defined function must be applied to fit this equation. First, activate Layer 1 in the upper left corner of the graph window Graph1, which is the layer where the head data are located. Second, select "Analysis" "Fitting" and "Nonlinear Curve Fit..." sequentially to open the non-linear fitting dialog shown in Figure 4-22. Click the function edit button in the middle to open the fitting function manager window (Figure 4-23), then click the tree menu "User Defined" on the left side and the "New Category" and "New Function" buttons on the right side. Further edit the function by setting the "Independent Variables" to Q, "Dependent Variables" to H and "Parameter Names" to A and B. In the "Function Form" option, select the "Equations" and "Function" text box and enter "H = A-B* Q^2". Finally, save and exit the fitting function manager window to return to the non-linear fitting window. In "Category" and "Function", select New Function (User) in the New Category directory you just edited. Then, click the iterative button in the non-linear fitting window shown in Figure 4-22 to apply the iterative fitting step by step, the corresponding fitting results will be displayed in real time in the text box below it. Alternatively, directly click the "Fit" button to complete the iterative fitting immediately.

Through the above method, the graph window shown in Figure 4-24 is finally obtained,

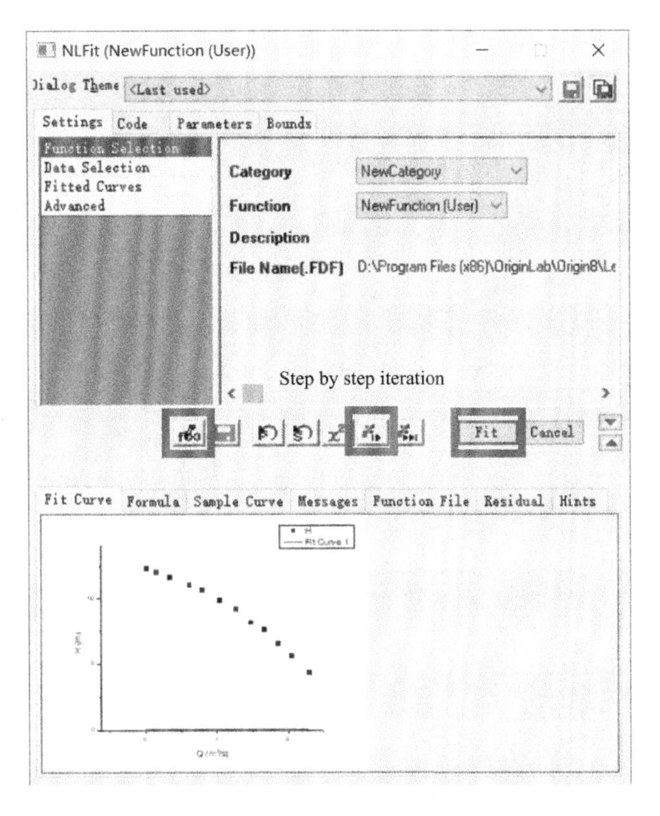

Figure 4-22 Non-linear fitting window

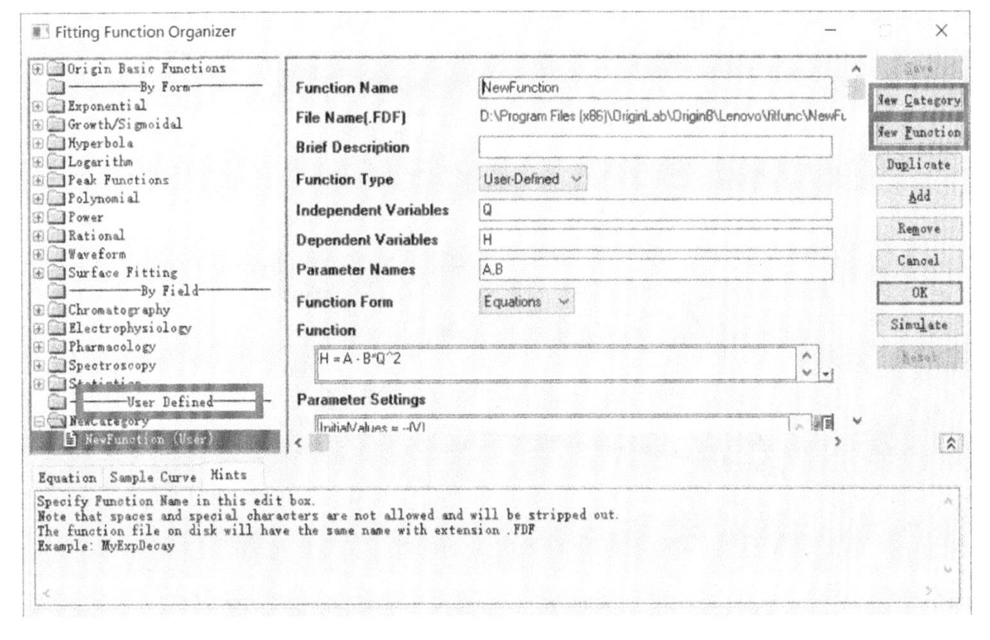

Figure 4-23 Fitting function manager window

where the three centrifugal pump characteristic curves and their respective fitted lines are displayed. The fitting equations and fitting errors can be found in the Book1 window. For other settings and operations, please refer to the related literatures on Origin.

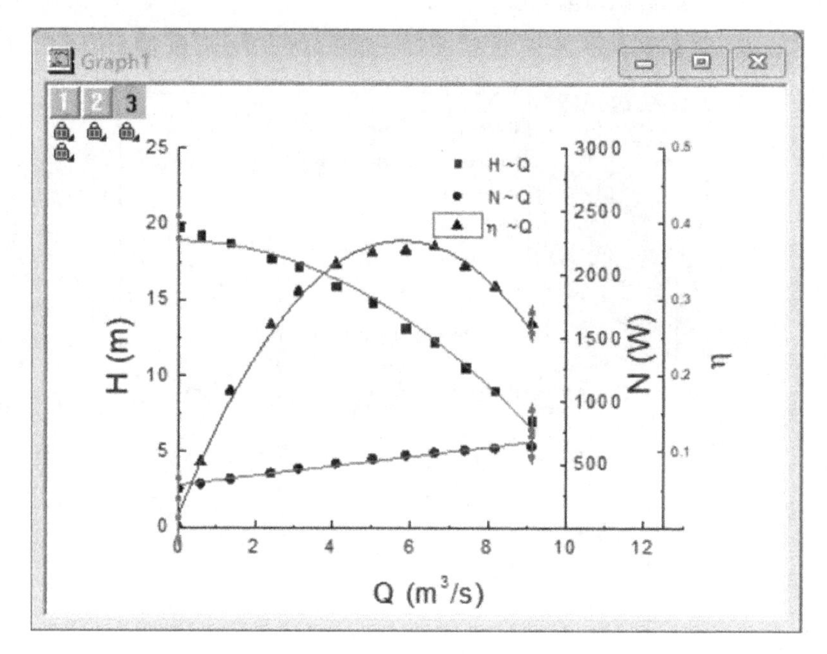

Figure 4-24 Characteristic curve fitting graph window

In the "Edit" menu, select "Copy Page" to copy and paste the centrifugal pump characteristic curve graph in the window shown in Figure 4-24 into any text, as shown in the Figure 4-25. Clearly, the figure also needs to be marked accordingly with the model, speed and other information of the centrifugal pump.

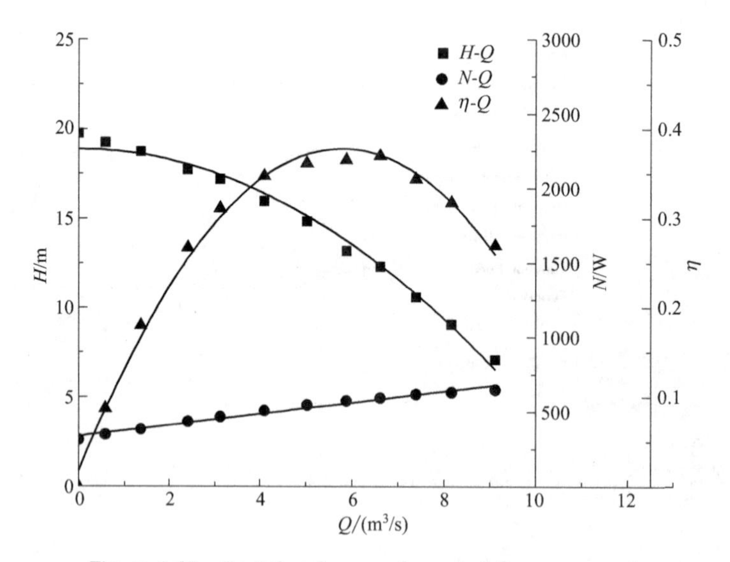

Figure 4-25 Centrifugal pump characteristic curve graph

Chapter 5

Commonly Used Measuring Instruments in the Laboratory

Temperature, pressure, flow and liquid level are the four major parameters that need to be measured in the chemical process. A brief introduction to the commonly used measuring instruments will be given in this chapter.

5.1 Temperature measurement

In chemical production and experiments, temperature is one of the most important parameters to be measured and controlled. Almost every chemical principle experiment equipment is equipped with temperature measurement instruments.

Temperature is a physical quantity that characterizes how hot or cold an object is. Temperature cannot be measured directly, but can be measured indirectly by means of the heat exchange between cold and hot objects, as well as the characteristics of some physical properties of objects (such as expansion, resistance, thermoelectric effect, etc.) that vary with the degree of cold and heat. According to the measuring methods, the temperature measuring instruments can bedivided into contact type and non-contact type. The temperature measured in the experiment of chemical engineering principle can be measured by contact thermometer. The types, advantages and disadvantages of common temperature measuring instruments are shown in Table 5-1. This section focuses on the expansion thermometer, thermocouple thermometer and thermistor thermometer in the contact thermometer.

Table 5-1 Types, advantages and disadvantages of commonly used temperature measuring instruments

Temperature measurement method	Thermometer type		Temperature measurement range /℃	Advantages	Disadvantages
Contant	Expansion	Glass tube	−50～600	Structure simple, easy to use, accurate measurement and low price	The upper limit of measurement and accuracy are limited by the quality of the glass, fragile. Recorded data cannot be transferred remotely
		Bimetallic	−80～600	Structure compact, firm and reliable	Low accuracy, limited measure range and use range

continued

Temperature measurement method	Thermometer type		Temperature measurement range /℃	Advantages	Disadvantages
Contact	Pressure	Liquid Gas Steam	−30～600 −20～350 0～250	Structure simple, shock-proof and explosion-proof, with record and alarm function, low price	Low accuracy, short temperature measurement distance, serious hysteresis
	Thermocouple	Platinum rhodium-platinum Nickel chromium-nickel silicon Nickel-chromium-copper	0～1600 −50～1000 −50～600	Wide temperature measurement range and high accuracy, convenient for long-distance, multi-point, centralized measurement and automatic control	Cold junction temperature compensation is required, low accuracy in the low temperature section
	Thermistor	Platinum Copper	−200～600 −50～150	High measurement accuracy, convenient for long-distance, multi-point, centralized measurement and automatic control	Not suitable for high temperature measurement, with an influence of ambient temperature
Non-contact	Radial	Radial Optical Colorimetric	400～2000 700～3200 900～1700	Without destroying the measured temperature field during measurement	Low accuracy in low temperature measurement, with an influence of environmental conditions on the accuracy
	Infrared ray	Photoelectric detection Thermoelectric detection	0～3500 200～2000	Large temperature measurement range, suitable for measuring temperature distribution, without influencing the measured temperature field, fast response	Easily susceptible to outside interference and difficult to calibrate

5.1.1 Expansion thermometer

(1) Glass tube thermometer

The glass tube thermometer is the most commonly used instrument for measuring temperature. It is with simple structure, low price, convenient reading and high accuracy. The measurement range is −50℃ to 600℃. The disadvantage is that it is fragile and cannot be repaired after damage. The commonly used glass tube thermometers in the laboratory are mercury and organic liquid (such as ethanol) thermometers. The mercury thermometer has a wide measuring range, a uniform scale and accurate readings, but it can cause pollution after damage. The liquid in the organic liquid (ethanol, benzene, etc.) thermometer has obvious reading after coloring, but since the expansion coefficient changes with temperature, the scale is uneven and the reading error is large. Glass tube thermometers are divided into three types: rod type, internal standard type and electric contact type as shown in Table 5-2.

Table 5-2 Commonly used glass tube thermometer

Items	Rod type	Internal standard type	Electric contact type
Characteristics	Most commonly used in the lab $d=6\sim8$mm l = 250mm, 280mm, 300mm, 420mm, 480mm	Commonly used in industry $d_1=18$mm, $d_2=9$mm $l_1=230$mm, $l_2=130$mm $l_3=60\sim2000$mm	Used for control, alarm, etc.; including two types: fixed contact and adjustable contact
Outline drawing	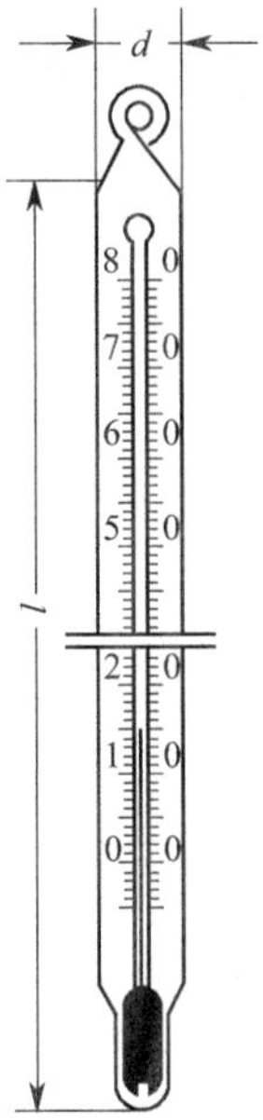	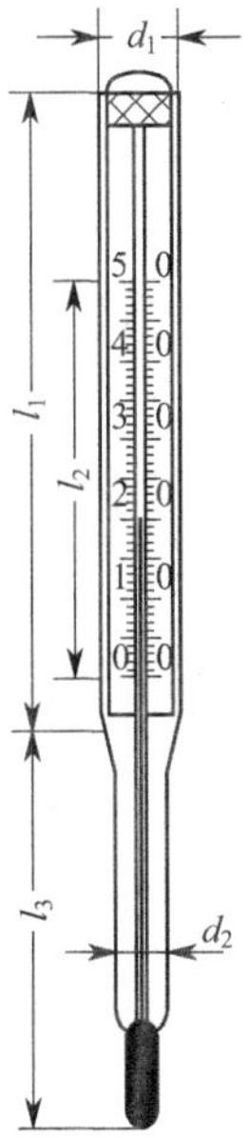	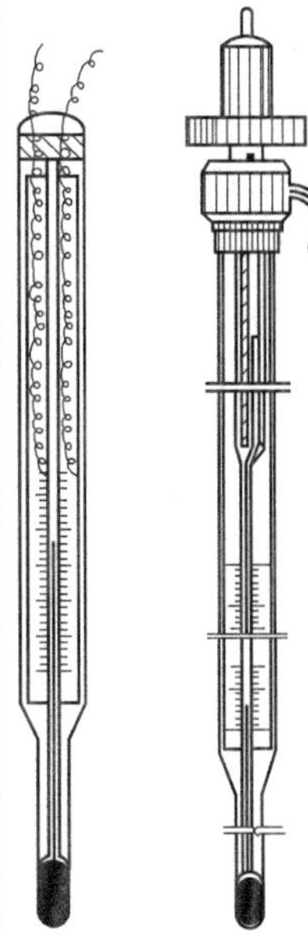 Cable

(2) Calibration of glass tube thermometer

Calibration is required for accurate measurement with a glass tube thermometer. There are two methods: one is to compare with a standard thermometer under the same conditions; the other is to use the phase transition point of pure substances, such as ice-water, water-water vapor system calibration.

When calibrating by the first method, the calibrated glass tube thermometer and standard thermometer (the second-class standard thermometer purchased) can be inserted into the thermostatic bath. After the temperature of the thermostatic bath stabilizes, compare the indicated value of the calibrated thermometer with the standard one. Note that heating-up calibration should be adopted, because the organic liquid has adhesion to the capillary wall. When the temperature drops, some liquid will stay on the capillary wall, affecting the accurate reading. When cooling, mercury thermometers will lag due to friction.

If there is no standard thermometer in the laboratory, the thermometer can be corrected with ice-water or water-water vapor phase transition for the calibration.

1) Ice-water system for 0℃ point calibration Fill a 100mL beaker with crushed ice or ice

cubes, then pour distilled water so that the liquid level reaches 2cm below the ice level. Insert the thermometer so that the thermometer scale is easy to observe or the 0℃ scale is exposed to the ice surface. Stir and observe the change of the mercury column. When the indicating temperature is constant, record the reading, which is the corrected 0℃. Take care not to completely dissolve the ice cubes.

2) Water-steam system for 100℃ point calibration Figure 5-1 is the installation diagram of the calibration thermometer. There should be a gap with the plug to balance the pressure between inside and outside of the test tube. Add a small amount of zeolite and 10mL of distilled water to the test tube. Adjust the thermometer so that the mercury ball is 3 cm above the liquid surface. Heating with a small fire causes the steam to condense on the wall of the test tube to form a ring. Pay attention to controlling the power to maintain the ring at about 2cm above the mercury ball. If there is a droplet on the mercury ball, it indicates that the thermal equilibrium between liquid and gas has been achieved. When the temperature is constant, observe the mercury column readings, record the readings, then correct the air pressure to be the corrected 100℃.

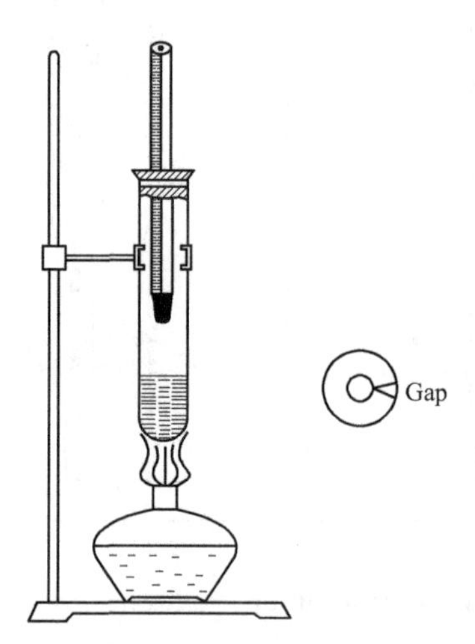

Figure 5-1 Calibration thermometer installation scheme

(3) Installation and use of glass tube thermometer

① The glass tube thermometer should be installed on equipment without large vibration and collisions, especially the organic liquid glass tube thermometer. The liquid column is likely to be interrupted with large vibration.

② The center of the temperature-sensing bulb of the glass tube thermometer should be placed in the most sensitive place of temperature change (such as the position with the largest fluid velocity in the tube).

③ The glass tube thermometer should be installed in a position that is convenient for reading, cannot be inverted and try not to install it at an angle.

④ The mercury thermometer reads at the highest point of the convex surface; the organic liquid glass tube thermometer reads at the lowest point of the concave surface.

⑤ In order to accurately measure the temperature, all the liquid in the thermometer should be in the object to be measured when measuring.

5.1.2 Thermocouple thermometer

Thermocouple thermometer is based on thermoelectric effect. It has a large measuring range, simple structure and reliable temperature measurement, it is easy to use, accurate and convenient for remote transmission of signals and automatic recording and centralized control. Therefore, it is extremely popular in chemical production and experiments.

Thermocouple thermometer consists of three parts: thermocouple (temperature-sensing element), measuring instrument (potentiometer, etc.) and the wire (compensation wire and copper wire) connecting the thermocouple and the measuring instrument. The temperature measurement system of the thermocouple thermometer is shown in Figure 5-2.

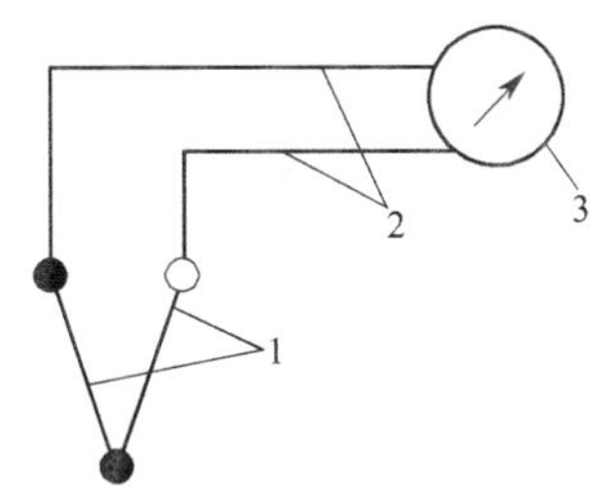

Figure 5-2 Thermocouple thermometer temperature measurement system diagram

1—Thermocouple; 2—Wire; 3—Measuring instrument

(1) Principle of temperature measurement

Connect two different kinds of metal wires or alloy wires A and B into a closed loop. If the two contacts are placed in the heat source with temperature t_0 and t, the electromotive force will be generated in this circuit, as shown in Figure 5-3 (a) . If a DC millivoltmeter is connected into this loop (disconnect the metal B to the millivoltmeter, or disconnect the mV meter at the t_0 connector of the two metal wires), as shown in Figure 5-3 (b), (c), a potential indication would be shown in the millivoltmeter, and this phenomenon is called the thermoelectric effect. The closed circuit composed of different wires is called thermocouple. One end of the closed circuit is inserted into the measured medium to feel the measured temperature, which is called the working end or hot end of the thermocouple, while the other end is connected with the wire, which is called the cold end or free end. Conductors A and B are called thermoelectric poles.

At the contact point of two metals, set $t > t_0$, due to the different junction temperatures, two thermoelectric potentials of different sizes and opposite directions $e_{AB}(t)$ and $e_{AB}(t_0)$ are generated. For the same metal A or B, the kinetic energy of free electrons is different at both ends, and two corresponding electromotive forces $e_A(t, t_0)$ and $e_B(t, t_0)$ are generated. This electromotive force is called the thermoelectric potential. There are both contact potential and temperature difference potential in the thermocouple circuit, so the total potential in the circuit is

$$\begin{aligned} E_{AB}(t,t_0) &= e_{AB}(t)+e_B(t,t_0)-e_{AB}(t_0)-e_A(t,t_0) \\ &= [e_{AB}(t)-e_{AB}(t_0)]-[e_A(t,t_0)-e_B(t,t_0)] \end{aligned} \quad (5\text{-}1)$$

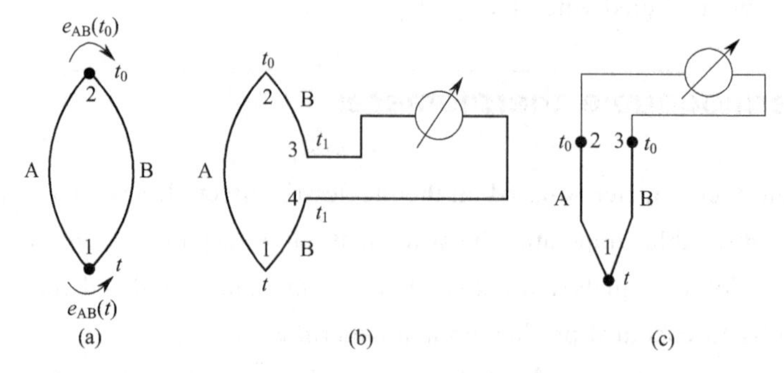

Figure 5-3 Schematic diagram of thermocouple measurement principle

Since the temperature difference potential is much smaller than the contact potential, it can be ignored, so equation (5-1) can be simplified as

$$E_{AB}(t,t_0)=e_{AB}(t)-e_{AB}(t_0) \tag{5-2}$$

When $t=t_0$, $E_{AB}(t,t_0)=0$; when t_0 is constant, $e_{AB}(t_0)$ is constant, then the thermoelectric potential becomes a single-valued function of temperature, regardless of the length and diameter of the thermocouple. In this way, as long as the size of the thermoelectric potential is measured, the temperature of the temperature measurement point can be judged. This is the principle of using thermoelectric phenomena to measure temperature.

(2) Selection of compensation wire

According to the principle of thermocouple temperature measurement, only when the temperature of the cold end of the thermocouple remains unchanged, the thermoelectric potential is a single-valued function of the measured temperature. Thermocouples are generally short (especially precious metals), the working end of the thermocouple is close to the cold end, the cold end is exposed to the space and is easily affected by the surrounding environment, so the temperature of the cold end is difficult to maintain constant. In order to keep the temperature of the cold end constant, a special wire can be used to extend the cold end of the thermocouple. It is also made of two metal materials with different properties, it has the same thermoelectric characteristics with the connected thermocouple within a certain temperature range (0 ~ 100℃). Different compensation wires should be chosen for different thermocouples. Table 5-3 shows the materials and characteristics of the compensation wire used with various thermocouples.

Table 5-3 Material and characteristics of compensation wire used with various thermocouple

Thermocouple type	Compensation wire			
	Positive		Negative	
	Material	Color	Material	Color
Platinum rhodium-platinum	Copper	Red	Copper-nickel alloy	Green
Nickel chromium-nickel silicon Copper-constantan	Copper	Red	Constantan	Brown
Nickel chromium-copper	Nichrome	Brown green	Copper	Yellow

(3) Cold junction temperature compensation

The instrument matched with the thermocouple is calibrated according to the temperature-thermoelectric potential relationship curve of various thermocouples while keeping the temperature of the cold end at 0℃. After the compensation wire is used, although the cold end of the thermocouple is extended from a place with higher temperature and instability to a lower temperature and more stable operation room, the temperature is still often higher than 0℃ and not constant , so the thermoelectric potential generated by the thermocouple must be smaller, and the measured value also changes with the temperature of the cold junction. Therefore, when applying thermocouple temperature measurement, in order to get accurate measurement results, the cold junction temperature must be kept at 0℃ , or a certain corrections should be made, which is called cold junction temperature compensation of thermocouple

The method of cold junction temperature compensation in the laboratory is usually to insert the two cold ends of the thermocouple into a test tube containing insulating oil, then put it into a container containing an ice-water mixture, as shown in Figure 5-4. The temperature at the cold end is therefore maintained at 0℃.

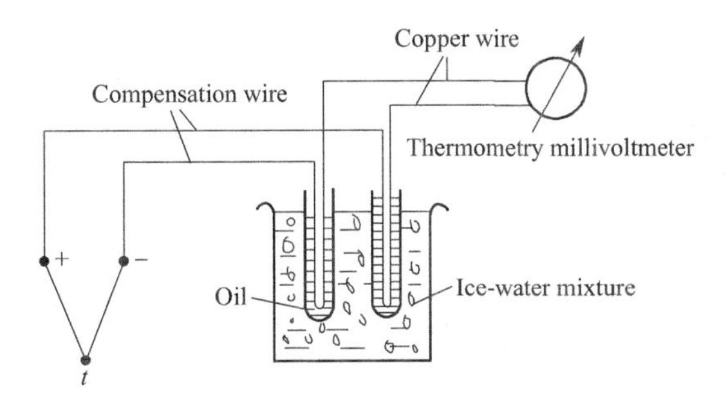

Figure 5-4 The method of keeping the temperature of cold end of thermocouple at 0℃

(4) Several commonly used thermocouples

At present, the following kinds of thermocouples widely used in China.

1) Platinum rhodium-platinum thermocouple The index number is S. The positive electrode of the thermocouple is an alloy wire composed of 90% platinum and 10% rhodium, and the negative electrode is platinum. This kind of thermocouple can be used for a long time in the range below 1300℃ , it can measure 1600℃ high temperature for a short time in a good environment. Because it is easy to obtain high-purity platinum and rhodium, and the thermocouple has high replication accuracy and measurement accuracy, it can be used for precise temperature measurement and as a reference thermocouple. The disadvantage is that the thermoelectric potential is weak and the cost is high.

2) Nickel chromium-nickel silicon thermocouple The index number is K. The positive electrode of the thermocouple is nickel-chromium, and the negative electrode is nickel-silicon. The thermocouple can measure temperatures below 900℃ for long periods in oxidizing or neutral

media, and up to 1200℃ for short-term measurements. With the characteristics of good reproducibility, large generated thermoelectric potential, good linearity, low price and disadvantage that the measurement accuracy is low, it can fully meet the requirements of industrial measurement, and is the most commonly used thermocouple in industrial production.

3) Nickel chromium-copper thermocouple The index number is EA. The positive electrode of the thermocouple is nickel-chromium, and the negative electrode is copper. It is suitable for reductive or neutral media. The long-term use temperature should not exceed 600℃, for short-term measurement it can reach 800℃. This thermocouple is characterized by high electric sensitivity and low price.

4) Copper-constantan thermocouple The index number is CK. The positive electrode of the thermocouple is copper, and the negative electrode is constantan. It is characterized by high accuracy at low temperature, it can measure the low temperature of −200℃, while the upper limit temperature is 300℃. Meanwhile, the price is quite low.

(5) Calibration of thermocouple thermometer

Due to the oxidation, corrosion and recrystallization of the thermocouple material at high temperatures, the thermoelectric characteristics of the thermocouple will change, making the measurement error larger. In order to ensure a certain degree of accuracy, thermocouples must be regularly calibrated to obtain changes in thermoelectric potential. When changes beyond the specified error range, the thermocouple wire can be replaced or the low temperature end of the thermocouple can be cut off a bit, then used after welding. A calibration must be done before use.

According to China's regulations, various thermocouples must be verified at the specified temperature, the maximum error of each temperature cannot exceed the allowable error range, as shown in Table 5-4, otherwise it should not be used.

Table 5-4 Commonly used thermocouple allowable deviation

Model	Material	Checkpoint /℃	Allowable deviation			
			Temperature/℃	Deviation/℃	Temperature/℃	Deviation/%
S	Platinum rhodium-platinum	600,800,1000,1200	0～600	±2.4	> 600	±0.4
K	Nickel chromium-nickel silicon (aluminum)	400,600,800,1000	0～400	±4	> 400	±0.75
E	Nickel chromium-copper nickel (constantan)	300,400,600	0～300	±4	> 300	±0.1

5.1.3 Thermistor thermometer

In addition to thermocouple thermometers, thermistor thermometers are often used in industrial production, using the characteristic that the resistance value of the temperature measuring element changes with the temperature. The thermoelectric thermometer outputs a small thermoelectric potential in the measurement of temperatures below 500℃, which is prone to errors. Therefore, in industrial production, thermistor thermometer are often used in the range of −120~ 500℃. Thermistor thermometer is the most commonly used temperature detector in the middle and low temperature area. Its main features are high measurement accuracy and stable performance. The measurement accuracy of platinum thermistor thermometor is the highest. It is not only widely used in industrial temperature measurement, but also made into a standard reference instrument. Under special circumstances, the lower limit of the thermistor thermometer measurement can be −270℃, while the upper limit can be 1000℃.

The resistivity of pure metals and most alloys increases with increasing temperature. Within a certain range, the relationship between resistance and temperature is linear. If it is known that the resistance of a metal conductor at a temperature of 0℃ is R_0, then the resistance at a temperature of t is

$$R = R_0(1 + \alpha t) \tag{5-3}$$

In formula, α is average temperature coefficient of resistance.

Different metals have different average temperature coefficients of resistance, only metals with larger average temperature coefficients of resistance can be used as thermistor for temperature measurement. The best and most commonly used thermistor material is pure platinum, with a measurement range of −200~500℃. Copper wire thermistor thermometers are also often used in industrial production, with a measurement range of −150~180℃.

In order to reduce the influence of the resistance of the wire on the measurement, the three-wire circuit shown in Figure 5-5 is often used to measure the resistance of the thermistor. Be careful to limit the current through R_t, otherwise a larger error might occur.

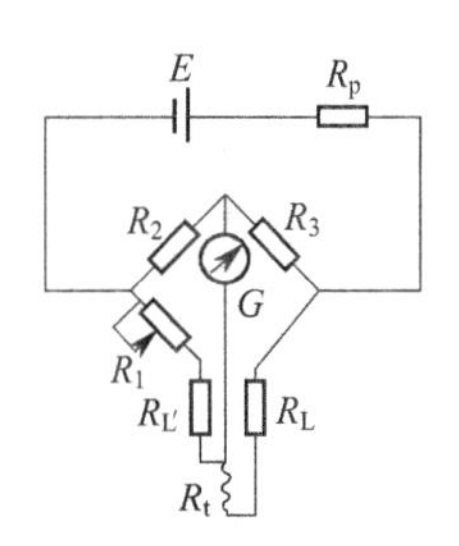

Figure 5-5 Three wire connection line

The thermistor should be calibrated before use and after use for a certain period of time to ensure its accuracy. The calibration of the working standard or standard thermistor is usually carried out at several equilibrium points, such as 0℃ ice, water equilibrium points, etc., but its requirements are high, and the method and equipment are complex. China has certain regulations for the calibration of thermistor. The inspection of industrial thermistor only requires that the values of R_0 (resistance value at 0℃) and R_{100} (resistance value at 100℃) do not exceed the specified range.

5.1.4 Principles of thermometer selection and use

When selecting and using a thermometer, the following points should be considered.

① Whether the temperature of the measured object needs to be indicated, recorded and automatically controlled;

② Temperature measurement range and accuracy requirements;

③ Whether the size of the temperature sensing element will destroy the temperature field of the measured object;

④ When the measured temperature changes, whether the hysteresis performance of the temperature sensing element meets the requirements;

⑤ Whether the measured object and the environmental conditions damage the temperature sensing element;

⑥ When using a contact thermometer, the temperature sensing element must be in good contact with the object to be measured, and there should be no heat exchange with the surrounding environment, otherwise the temperature measured will be different from the real value;

⑦ The temperature sensing element needs to be inserted into the measured medium to a certain depth. In the gas medium, the insertion depth of the metal protection tube is 10 to 20 times that of the protection tube, and the insertion depth of the non-metallic protection tube is 10 to 15 times that of the protection tube.

5.2 Pressure (pressure drop) measurement

In chemical production and experiments, pressure is one of the important parameters. For example, it needs to measure the pressure drop of the fluid flowing through the tube in the tube resistance experiment; to measure inlet and outlet pressures to understand the performance and installation of the pump in the pump performance experiment; to frequently observe the pressure at the top of the tower and the kettle to understand whether the operation of the tower is normal in distillation experiments. In addition, some other parameters, such as level and flow, are often converted by measuring pressure or pressure drop.

There are many instruments for pressure measurement. According to different conversion principles, they can be roughly divided into four types: liquid column pressure gauge, elastic pressure gauge, electrical pressure gauge and piston pressure gauge.

5.2.1 Liquid column pressure gauge

(1) Structure and characteristics of the liquid column pressure gauge

Based on the principle of hydrostatics and converts the measured pressure into the height of the liquid column for measurement, according to the different structural forms, liquid column pressure gauge can be divided into the following categories: U-tube differential pressure gauge, single-tube differential pressure gauge, inclined tube differential pressure gauge and U-tube dual-indicating hydraulic differential gauge. The structures and characteristics are shown in Table 5-5. With a simple structure and being easy to use, its accuracy is affected by factors such as capillary, action, density and parallax of the working fluid. The measurement range is narrow, and it is generally used to measure lower pressure, vacuum or pressure difference. It cannot automatically indicate and record, so the application is limited.

Table 5-5 Structure and characteristics of liquid column pressure gauge

Name	Schematic diagram	Measurement range	Static equation	Remarks
Positive U-tube differential pressure gauge	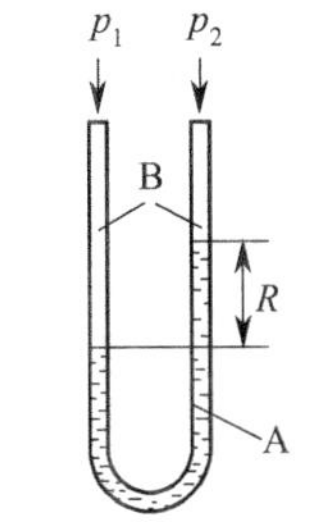 p_1 p_2 B R A	$R \leqslant 800$mm	$\Delta p = Rg(\rho_A - \rho_B)$(liquid) $\Delta p = Rg\rho$(gas)	The zero point is in the middle of the scale. It is often used as a standard differential pressure meter to correct the flow. It is suitable for the case where the indicator density is greater than the measured fluid density.
Inverted U-tube differential pressure gauge	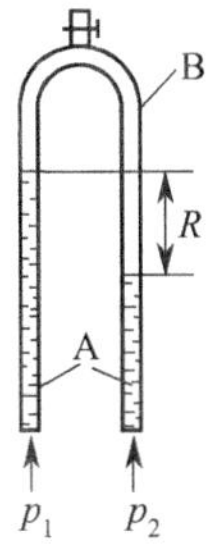 B R A p_1 p_2	$R \leqslant 800$mm	$\Delta p = Rg(\rho_A - \rho_B)$(liquid)	Taking the liquid to be measured as the indicator liquid, it is suitable for the measurement of small pressure difference and the density of the indicator is less than that of the measured fluid.
Single tube differential pressure gauge	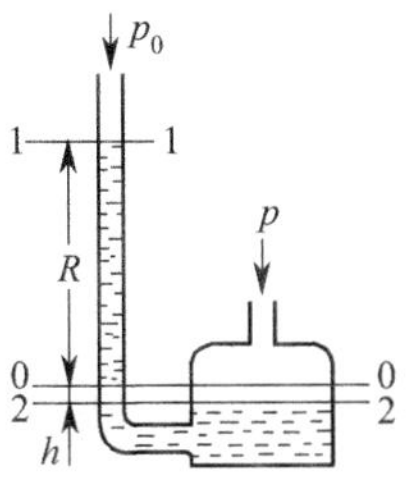 p_0 1 1 R p 0 2 0 2 h	$R \leqslant 1500$mm	$\Delta p = R\rho(1 + S_1/S_2)g$ $S_1 \ll S_2$, $\Delta p = Rg\rho$ S_1: Vertical pipe cross-sectional area S_2: Expansion room cross-sectional area	The zero point is at the lower end of the scale. The zero point needs to be adjusted before use. It can be used as a standard differential pressure gauge.
Inclined tube differential pressure gauge	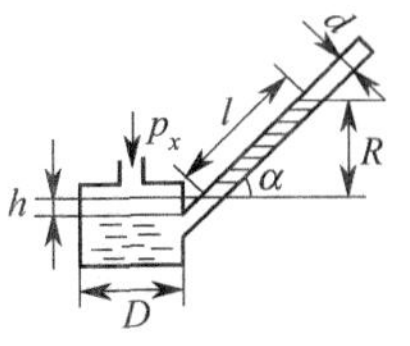 p_x l d R α h D	$R \leqslant 1200$mm	$\Delta p = l\rho g(\sin\alpha + S_1/S_2)$when $S_1 \ll S_2$, $\Delta p = l\rho g \sin\alpha$ S_1: Cross-sectional area S_2: Expansion room cross-sectional area	When α is less than 15°, the measurement range can be adjusted by changing α. The zero point is at the lower end of the scale and needs to be adjusted before use.

continued

Name	Schematic diagram	Measurement range	Static equation	Remarks
U-tube double indicator hydraulic difference meter		$R \leqslant 500\text{mm}$	$\Delta p = Rg(\rho_A - \rho_C)$	The U-tube is equipped with two indicator liquids of similar density, A and C, and an enlarged chamber above the two arms, which is designed to improve the measurement accuracy and is suitable for cases where the pressure difference is small.

(2) Precautions for use of the liquid column pressure gauge

① The measured pressure cannot exceed the measuring range. Sometimes if the pressure increases due to improper operation of the measured object, the working fluid will be washed away, which will not only cause losses, but also pollute the environment.

② The measured medium cannot be miscible with the working fluid or react chemically, if so, the working fluid should be replaced or the isolation fluid should be added. Commonly used isolation fluid is shown in Table 5-6.

Table 5-6 Isolation fluid for certain media

Measuring medium	Isolation fluid	Measuring medium	Isolation fluid
Chlorine	98% concentrated sulfuric acid or fluorine oil	Ammonia	Transformer oil
Hydrogen chloride	kerosene	Water gas	Transformer oil
Nitric acid	Pentachloroethane	Oxygen	Glycerin

③ The installation location should avoid places with excessive heat, cold and vibration.

④ Adjust the working liquid level to the zero line before use.

⑤ When reading the pressure value, the line of sight should be on the liquid column surface, the concave surface should be seen when observing water, and the convex surface should be seen when observing the mercury.

⑥ When the working fluid is water, a little ink or other water-soluble pigments can be added to the water in order for easy observe.

5.2.2 Elastic pressure gauge

The elastic pressure gauge is a pressure measuring instrument that uses various forms of elastic elements to produce elastic deformation after the elastic element is compressed under the action of the pressure of the measured medium. With the advantages of simple structure, reliable use, clear reading, low price, wide measurement range and sufficient accuracy, it is the most widely used pressure measuring instrument. If additional devices are added, such as recording mechanism, electrical conversion device, control element, etc. , it can realize pressure recording, remote transmission, signal alarm, automatic control, etc. The structure and measuring range of the elastic pressure gauge are shown in Figure 5-7. This section focuses on the spring tube pressure gauge.

Table 5-7 Structure and measuring range of elastic pressure gauge

Type	Name	Schematic diagram	measuring range/Pa	
			Minimum	Maximum
Membrane	Flat film	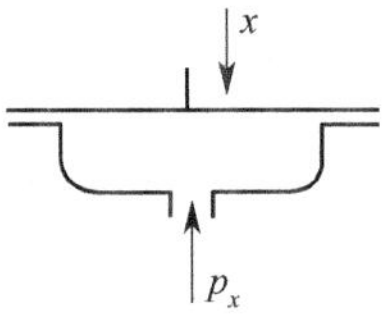 x, p_x	$0\sim10^4$	$0\sim10^8$
	Corrugated membrane	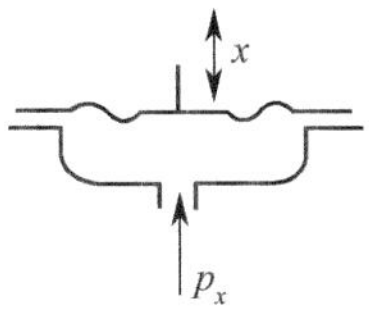 x, p_x	$0\sim1$	$0\sim10^6$
	Flexible membrane	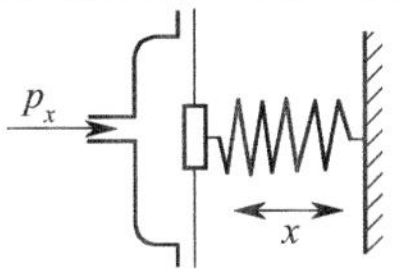 p_x, x	$0\sim10^{-2}$	$0\sim10^5$
Bellows	Bellows	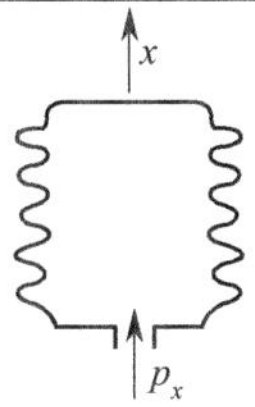 x, p_x	$0\sim1$	$0\sim10^6$
Spring tube	Single-tune spring tube	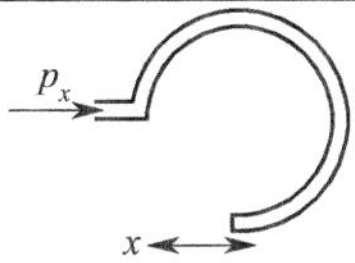 p_x, x	$0\sim10^2$	$0\sim10^9$
	Multi-turn spring tube	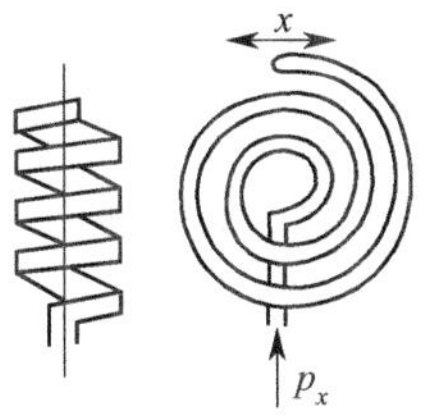 x, p_x	$0\sim10$	$0\sim10^8$

(1) Working principle of spring tube pressure gauge

The structure of the spring tube pressure gauge is shown in Figure 5-6. The spring tube 1 is a measuring element of a pressure gauge. The figure shows a single-turn spring tube, which is a flat, elliptical, hollow metal tube with an arc shape. The free end B of the pipe is closed, and the other end of the pipe is fixed on the joint 9, which is connected to the pressure measuring point. After being compressed, the spring tube deforms elastically, displacing the free end B.

Since the input pressure is proportional to the displacement of the free end B of the spring tube, as long as the displacement at point B is measured, the magnitude of the pressure can be gained. This is the basic measurement principle of the spring tube pressure gauge.

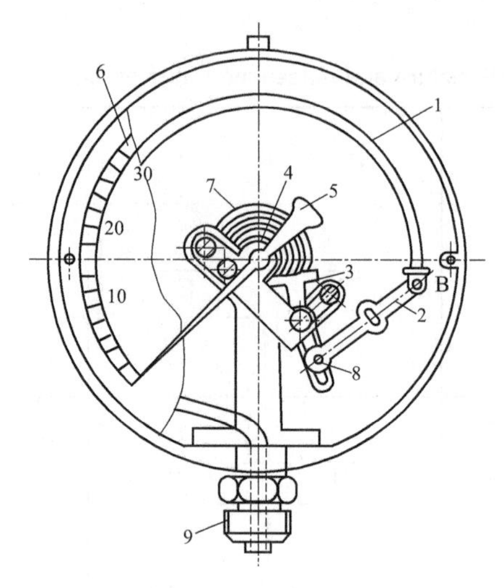

Figure 5-6 The structure of the spring tube pressure gauge

1—Spring tube; 2—Pull rod; 3—Sector gear; 4—Central gear; 5—Pointer; 6—Panel; 7—Balance spring; 8—Adjusting screw; 9—Joint

The displacement of the free end B of the spring tube is generally small, and difficult to display directly, so it must be indicated by a magnifying mechanism. The specific magnification process is as follows: the displacement of the free end B of the spring tube passes through the pull rod 2 to deflect the sector gear 3 counterclockwise, then, the pointer 5 is clockwise deflected by the coaxial central gear 4, and the value of the measured pressure is indicated on the scale of the panel 6. Since the displacement of the free end of the spring tube has a proportional relationship with the measured pressure, the scale of the spring tube pressure gauge is linear.

The balance spring 7 is used to overcome the deterioration of the instrument due to the transmission gap between the sector gear and the central gear. Changing the position of the adjusting screw 8 (that is, changing the magnification factor of the mechanical transmission) can adjust the range of the pressure gauge.

(2) Precautions for use and installation of spring tube pressure gauge

The correct use and installation of the pressure gauge is an important part of ensuring the accuracy of the measurement results and the service life of the pressure gauge.

① The instrument type should be selected correctly according to the process requirements. When measuring the pressure of explosive, corrosive and toxic fluids, special instruments should be used. For example, the spring tubes of ordinary pressure gauges mostly use copper alloys, while the material of ammonia pressure gauge spring tubes is carbon steel, and copper alloys are not allowed, because ammonia gas corrodes copper very strongly. Another example is a pressure gauge for oxygen should prohibit oil from entering, because it is easy to cause an explosion.

② The meter should work within the allowable pressure range. Generally, the maximum value of the measured pressure should not exceed 2/3 of the scale of the instrument. When measuring the pulsating pressure, it should not exceed 1/2 of the upper limit of the measurement. In both cases, the measured pressure should not be less than 1/3 of the scale of the instrument.

③ The distance between the installation place of the instrument and the measuring point should be as short as possible to avoid slow indication.

④ The meter must be installed vertically without leakage.

⑤ When using the instrument under vibration, a shock absorber must be installed. When measuring the steam pressure, a condensate tube should be installed to prevent the high-temperature steam from directly contacting the pressure measuring element. When measuring the pressure of corrosive media, an isolation tank equipped with neutral media should be added.

⑥ When the measured pressure is small, the pressure gauge and the pressure tap are not at the same height, the measurement error caused by the height should be corrected according to $\Delta p = \pm hg\rho$ (h is the height difference in the formula, ρ is the density of the medium in the pressure guide tube, g is the acceleration of gravity) .

⑦ The meter must be checked regularly.

5. 2. 3 Electric pressure gauge

Electric pressure gauge is an instrument that can convert pressure into electrical signal for transmission and display. This instrument has a wide measurement range, can transmit signals over long distances, realize automatic control of pressure, and meets the requirements of continuous improvement of industrial automation.

Electric pressure gauges are generally composed of pressure sensors, measuring circuits and signal processing devices, as shown in Figure 5-7. Common signal processing devices include indicators, recorders, controllers and microprocessors. The pressure sensors in the pressure gauge include piezoelectric magnetic type, piezoelectric type, capacitive type, inductive type and resistance strain type. The following section mainly introduces piezoresistive pressure sensors and capacitive pressure sensors.

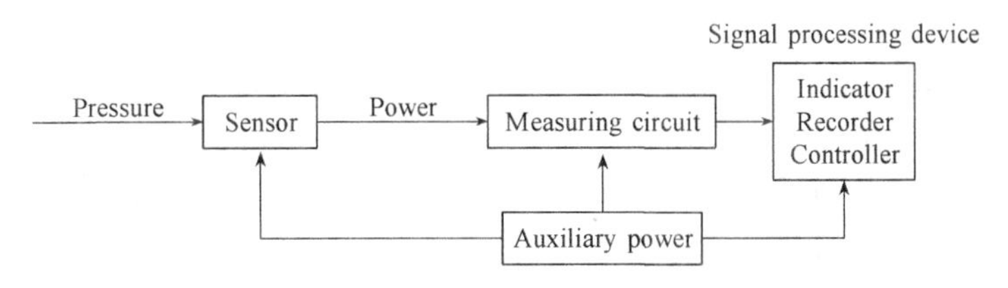

Figure 5-7 Block diagram of electric pressure gauge

(1) Piezoresistive pressure sensor

Piezoresistive pressure sensors are also called solid-state pressure sensors or diffusion-type piezoresistive pressure sensors. Its working principle is the piezoresistive effect of single crystal silicon. The single crystal silicon diaphragm and the strain resistance film are combined together using an integrated circuit process to form a silicon piezoresistive chip, which is then packaged in a sensor case and then connected with an electrode wire. The structure principle of a typical piezoresistive pressure sensor is shown in Figure 5-8. In the figure, there are two cavities on both sides of the silicon diaphragm, usually the upper tube is connected to the atmosphere or other reference pressure source, and the lower tube is connected to the high-pressure cavity which is filled with silicone oil and has an isolation diaphragm to isolate it from the measured object.

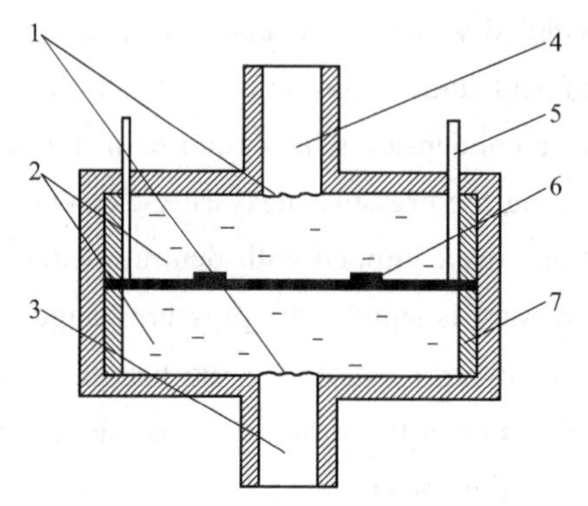

Figure 5-8 Piezoresistive pressure sensor structure principle diagram

1—Isolation diaphragm; 2—Silicone oil; 3—High pressure side; 4—Low pressure side
5—Lead; 6—Silicon diaphragm and strain resistance; 7—Support

When the pressure of the measured object acts on the silicon diaphragm through the high pressure side, the isolation diaphragm and the silicone oil, the silicon diaphragm is deformed, two of the four strain resistors on the diaphragm are compressed and two are pulled extend to change the internal resistance of the Wheatstone bridge, and convert it into the corresponding electrical signal output. The bridge is powered by a constant voltage source or a constant current source, which reduces the influence of temperature on the measurement results. The change value of the strain resistance sheet has a good linear relationship with the pressure, so the accuracy of the piezoresistive pressure sensor can often reach 0.1%.

The piezoresistive pressure sensor has the characteristics of high accuracy, reliable operation, high frequency response, small hysteresis, small size, light weight, simple structure, etc. It can work in harsh environments and is convenient for digital display.

(2) Capacitive pressure sensor

Capacitive pressure sensors are sensors that use two parallel plates to measure pressure, as shown in Figure 5-9. The diaphragm will be displaced with pressure acted on, changing the board distance d, causing the capacitance to change. After the conversion of the measuring circuit, the magnitude of the acting pressure p can be obtained. When ignoring the edge effect, the capacitance C of the plate capacitor is

$$C=\frac{\varepsilon S}{d} \tag{5-4}$$

In formula, ε is dielectric constant; S is overlap area between plates; d is distance between plates.

From formula (5-4), we can see that the capacitance C is related to ε, S and d. When the measured pressure affects any of the three parameters, the capacitance will change. Therefore, capacitive pressure sensors can be divided into three types: variable area type, variable dielectric type and variable inter-electrode distance type.

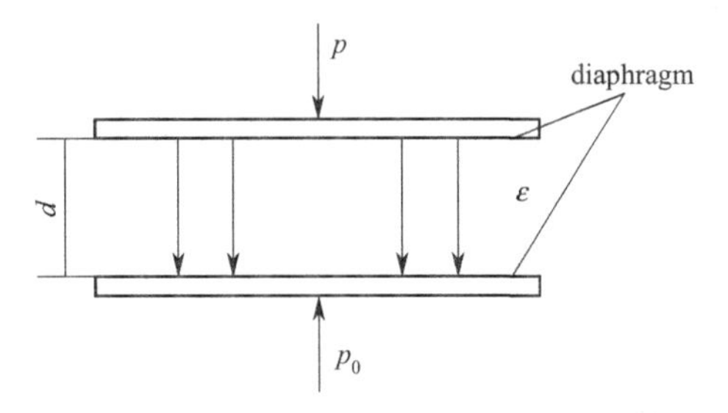

Figure 5-9 Schematic diagram of capacitive pressure

The main characteristics of capacitive pressure sensors are as follows:

① High sensitivity, especially suitable for low pressure and micro pressure measurement.

② No moving parts inside, no energy consumption and small measurement error.

③ The mass of the diaphragm is very small, so it has a higher frequency, thus ensuring a good dynamic response capability.

④ Using gas or vacuum as the insulating medium, the pressure loss is small, and it will not cause temperature change.

⑤ The structure is simple, glass, quartz or ceramic are often used as insulating support, so they can work under harsh conditions such as high temperature and radiation.

In recent years, new materials, new processes and micro integrated circuits have been used, and the signal conversion circuit of the capacitive pressure sensor has been assembled with the sensor, effectively eliminating the effects of electrical noise and parasitic capacitance. The measuring pressure range of the capacitive pressure sensor can be from tens of pascals to hundreds of megapascals, and the range of use are greatly expanded.

5. 2. 4 Piston pressure gauge

Piston pressure gauge is based on the principle of hydraulic transmission pressure of the hydraulic press, it converts the measured pressure into the mass of the balance weight added to the piston. Its measurement accuracy is very high, and the allowable error can be as small as 0. 05% to 0. 02% , but the structure is more complicated and the price is more expensive. It is generally used as a standard pressure measuring instrument to check other types of pressure gauges.

5. 2. 5 Calibration of pressure gauge

The new pressure gauge must be calibrated before leaving the factory to verify whether its technical specifications meet the specified accuracy. The pressure gauge should be calibrated after being used for a period of time, the purpose of which is to determine whether it meets the original accuracy. If the error is confirmed to exceed the specified value, it should be repaired. The repaired pressure gauge still needs to be calibrated before it can be used.

There are generally two types of pressure gauge calibration methods: static calibration method and dynamic calibration method: a. Static calibration determine the static accuracy and the instrument grade, including "standard table comparison method" and "weight calibration method"; b. Dynamic calibration mainly determines the dynamic characteristics of the pressure gauge, such as the transition process, time constant and static accuracy of the instrument. The commonly used method is the "shock tube method".

5.3 Flow measurement

In chemical production and experiments, it is often necessary to measure the flow of various media (liquid, gas, steam, etc.) in order to provide a basis for operation and control. Flow can be divided into instantaneous flow and total amount. Instantaneous flow refers to the amount of fluid flowing through a section of the pipeline per unit time; total amount refers to the sum of the fluid flow through the pipeline in a certain period of time, that is, the cumulative value of the instantaneous flow rate in a certain period of time.

The instantaneous flow and total flow can be expressed by mass or volume. The fluid that flows through a unit of time is called mass flow in terms of mass, and volumetric flow in terms of volume. There are many methods for measuring the flow rate, and their measurement principles and applied instrument structures are different. There are three types of flowmeters: velocity flowmeter, volumetric flowmeter, and mass flowmeter.

5.3.1 Velocity flowmeter

Velocity flowmeter calculates the flow rate by measuring the flow velocity of the fluid in the pipeline as the measurement basis, such as differential pressure flowmeter, rotor flowmeter, electromagnetic flowmeter, turbine flowmeter, weir flowmeter, etc. This section introduces differential pressure flowmeters, rotameters and turbine flowmeters that are commonly used.

(1) Differential pressure flowmeters

Differential pressure flowmeters use the pressure difference generated when a fluid flows through a throttling device or a constant velocity tube to achieve flow measurement. Among them, the differential pressure flowmeter composed of a throttling device and a differential pressure meter is the most widely used flow measurement instrument in industrial production and experimental devices. Common throttling devices include orifice plates (Figure 5-10), nozzles and venturi tubes. The following focuses on the orifice flowmeter.

The orifice flowmeter is a flowmeter that calculates the volume flow of fluid by measuring the pressure change caused by the fluid before and after flowing through the orifice plate. The relationship between the flowmeter reading and the fluid volume flow is:

$$V_s = C_0 A_0 \sqrt{\frac{2gR(\rho_A - \rho)}{\rho}} \tag{5-5}$$

In formula, C_0 is discharge coefficient; A_0 is cross-sectional area of small holes in orifice plate; ρ_A is indicator fluid density; R is liquid level difference of the Indicator fluid.

For orifice plates made according to standard specifications and precision, pressure is taken by the angle joint method (called standard orifice plate), C_0 depends on the cross-sectional ratio A_0/A_1 (A_1 is the tube cross-sectional area) and the Reynolds number in the tube Re_1. It can be seen from Figure 5-11 that after Re_1 exceeds a certain limit, C_0 no longer changes with Re_1 and becomes a constant. Obviously, in the design and use of the orifice plate, it is expected that Re_1 is greater than the limit value.

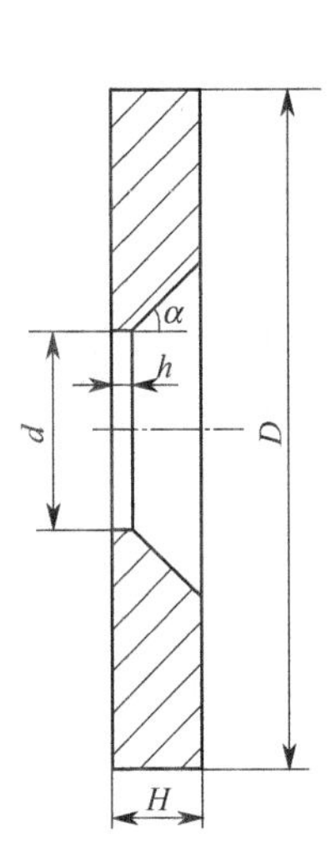

Figure 5-10 Schematic diagram of orifice cross section

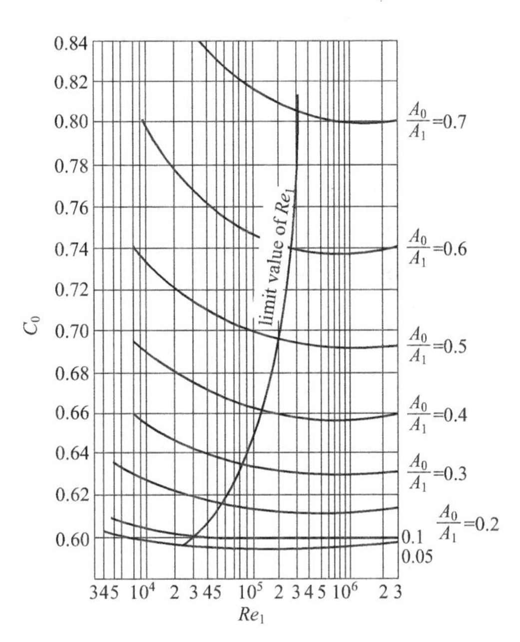

Figure 5-11 Relationship between discharge coefficient C_0, Re_1 and A_0/A_1

The orifice plate has a simple structure. It is easy to process and has a low manufacturing cost. Its main disadvantage is that the loss of resistance is large, which reduces the maximum passing capacity of the fluid.

Attention should be paid to the correct direction when installing the orifice flowmeter, and the requirements are strict and no burrs are allowed during processing, otherwise it will affect the measurement accuracy. Orifice flowmeters should not be used for unclean or corrosive fluids that easily make the orifice dirty, worn and deformed.

(2) Rotameter

The working principle of rotameter is different from the differential pressure flowmeter described above. The differential pressure flowmeter reflects the size of the flow rate with a change in differential pressure under the condition that the throttle area (such as the orifice flow area) is unchanged, so it is also called a constant cross-section variable pressure flowmeter. However, the rotameter uses the change in throttle area to measure the flow rate without changing the pressure drop. This kind of flowmeter is particularly suitable for measuring the flow rate of tubes with a diameter of less than 50mm, and the measured flow rate can be as small as a few liters per hour. Therefore, it is widely used in the laboratory.

1) The working principle of the rotameter The rotameter is mainly composed of two parts, one is a tapered tube that gradually expands from bottom to top (usually made of glass, the cone angle is 40″～ 4°); the other is a rotor placed in the tapered tube and can move freely (made of metal or other materials). When the flow rate is zero, the rotor sinks at the lower end of the tube, when the fluid flows from the bottom to the top, it is pushed up and suspended in the fluid in the tube. The rotor will be suspended at different positions with the flow rate changes, as shown in Figure 5-12.

When fluid flows through the conical tube, the rotor inside receives an upward force, which causes the rotor to float. When this force is exactly equal to the net gravitational force of the rotor immersed in the fluid (the gravity of the rotor minus the buoyancy of the fluid to the rotor), the upper and lower forces acting on the rotor are balanced. At this time, the rotor stops floating at a certain height. If the flow rate of the measured fluid suddenly changes from small to large, the upward force acting on the rotor increases, because the net gravity of the rotor in the fluid is constant, the rotor rises. As the position of the rotor in the conical tube rises, the annular gap between the rotor and the conical tube increases, that is, the flow area increases, and the flow velocity of the fluid flowing through the annulus becomes slow, so the upward force of the fluid acting on the rotor becomes smaller. When the force of the fluid acting on the rotor is again equal to the rotor's net gravity in the fluid, the rotor stabilizes at a new height. In this way, the balance position of the rotor in the conical tube corresponds to the flow rate of the measured medium. This is the basic principle of rotameter to measure flow.

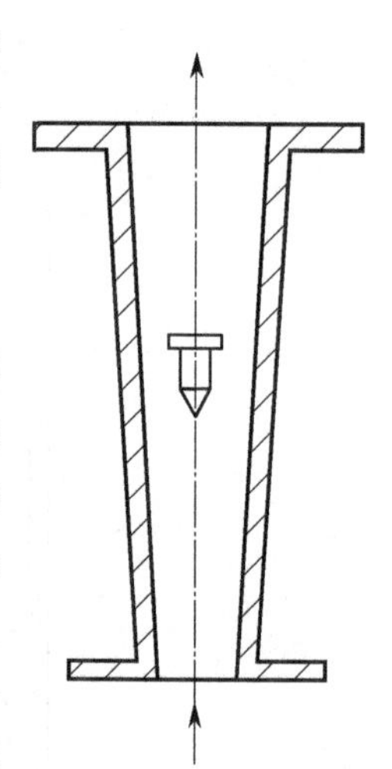
Figure 5-12 Working principle of rotameter

The calculation formula of the volume flow of the rotameter is:

$$V_s = C_R A_0 \sqrt{\frac{2g(\rho_f - \rho)V_f}{\rho A_f}} \tag{5-6}$$

In formula, C_R is flow Coefficient; A_0 is ring gap area; ρ, ρ_f are the density of fluid and rotor; V_f, A_f are the volume and cross-sectional area of the rotor (where the cross-section is the largest)

The value of the flow coefficient C_R mainly depends on the configuration of the rotor, and it is also related to the Reynolds number of the fluid flowing through the annular gap. For the rotor configuration shown in Figure 5-12, when the Reynolds number reaches 10000, C_R is constant at 0. 98.

It can be seen from formula (5-6) that the flow rate is related to the annular gap area A_0.

When the size of the tapered tube and the rotor is fixed, this A_0 depends on the height of the rotor in the tube, so the corresponding flow value is engraved outside the tapered tube, the flow rate can be read directly according to the level of the rotor's equilibrium position.

When reading the flowmeter scales of rotors with different shapes, the maximum cross section of the rotor should be used as the reading reference, as shown in Figure 5-13

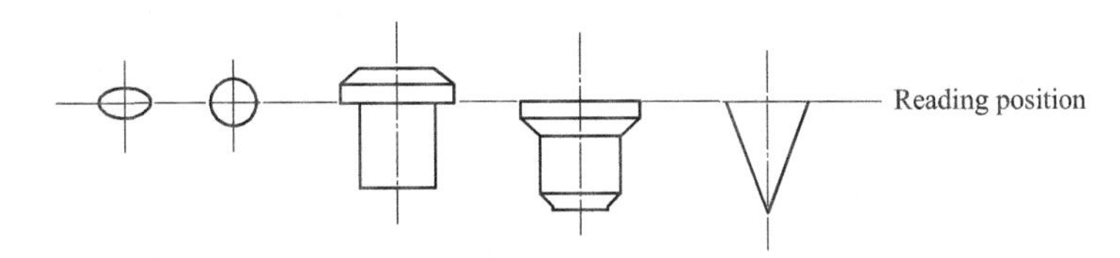

Figure 5-13 The correct reading position of different rotameters

2) Conversion of flow rate when rotameter is used measuring other substances The rotameter is a non-standardized instrument, and each rotameter is accompanied by factory-calibrated flow data. For the liquid flowmeter, the manufacturer uses 20℃ water for calibration; for the gas flowmeter, it uses 20℃, 101. 3kPa air for calibration. Therefore, in use, if the calibration conditions are not met, it needs to be corrected as follows:

For liquid:

$$V_1 = V_0 \sqrt{\frac{(\rho_f - \rho_1)\rho_0}{(\rho_f - \rho_0)\rho_1}} \tag{5-7}$$

In formula, V_1 is the actual flow rate of the liquid under working conditions; V_0 is the reading of rotameter water calibration; ρ_f is the density of the rotor; ρ_1 is density of liquid under working condition; ρ_0 is density of water at factory calibration.

3) Change of rotameter range When the measurement range exceeds the range of the existing rotameter, the measurement range can be changed by changing the rotor density, methods of which are such as hollowing the solid rotor or adding filler to the hollow rotor. In the case that the shape and size of the rotor remain the same, the flow rate can be converted by formula (5-8) :

$$V_0' = V_0 \sqrt{\frac{\rho_f' - \rho_0}{\rho_f - \rho_0}} = V_0 \sqrt{\frac{m_f' - V_f\rho_0}{m_f - V_f\rho_0}} \tag{5-8}$$

In formula, V_0, ρ_0, ρ_f, m_f are the fluid volume flow, fluid density, rotor density and rotor mass before the rotor is changed respectively; V_0', ρ_f', m_f' are the volume flow rate, rotor density, and rotor mass of the rotor after the rotor is changed respectively.

(3) Turbine flowmeter

The turbine flowmeter is designed based on the principle of conservation of momentum. An impeller that can rotate freely is installed in the fluid flowing tube. When the fluid passes through the turbine, the blades of the turbine rotate due to the impact of the flowing fluid, and the rotation speed changes with the change of the flow rate. Under the specified flow range and a certain fluid viscosity, the rotational speed is linearly related to the flow rate. Therefore, by measuring the rotation speed or the number of revolutions of the impeller, the flow or total amount of fluid flowing through the pipeline can be determined. The turbine speed is converted into a pulsed electrical signal by an appropriate device. By measuring the pulse frequency or using an appropriate device to convert the electrical pulse into a voltage or current output, the flow rate is finally measured.

The structure diagram of the turbine flowmeter is shown in Figure 5-14. It mainly consists of the following parts: the turbine 1 is made of stainless steel material with high magnetic permeability. The impeller core is equipped with a spiral blade, the fluid acts on the blade to make it rotate. The deflector 2 is used to stabilize the flow direction of the fluid and support the impeller. The magnetoelectric induction converter 4 is composed of a coil and a magnetic steel, it is used to convert the rotation speed of the impeller into a corresponding electrical signal, which is supplied to the preamplifier for amplification. The entire turbine flowmeter is installed on the Shell 3, which is made of non-magnetic stainless steel, and the two ends are connected to the fluid pipeline.

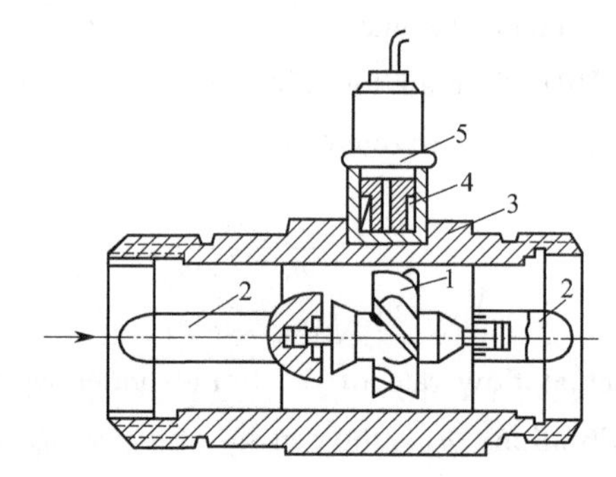

Figure 5-14 The structure diagram of the turbine flowmeter

1—Turbine; 2—Deflector; 3—Shell; 4—Magnetoelectric induction converter; 5—Preamplifier

The working process of the turbine flowmeter is as follows: when the fluid passes through the gap between the turbine blade and the tube, the force generated by the pressure difference between the front and back of the blade pushes the blade to rotate the turbine. Meanwhile, the highly magnetic turbine periodically sweeps through the magnetic steel, causing the magnetic resistance of the magnetic circuit to change periodically, and the magnetic flux in the coil also changes periodically, thereby inducing an AC signal. The frequency of the AC electrical signal is proportional to the speed of the turbine, that is, proportional to the flow rate. This electrical signal is amplified by a preamplifier and sent to an electronic counter or electronic frequency meter to accumulate or indicate flow.

The turbine flowmeter is easy to install, there is no need for sealing or gear transmission mechanism between the magnetoelectric induction converter and the blade, so the measurement accuracy is high and it can withstand high pressure. Due to the principle of magnetic induction conversion, the response is fast and pulsating flow can be measured. The output signal is an electrical frequency signal, which is convenient for long-distance transmission without interference.

The turbine of the turbine flowmeter is easy to wear, so the measured medium should not carry mechanical impurities and generally a filter should be added. It should be installed horizontally, and there must be a certain straight tube section before and after to make the fluid flow more stable. Generally, the length of the inlet straight tube section is more than 10 times the inner diameter of the tube, and the outlet is more than 5 times.

5. 3. 2 Volumetric flowmeter

A volumetric flowmeter calculates the flow rate based on the number of fixed volumes of fluid discharged per unit time. Only the wet flowmeters commonly used in the laboratory are introduced here.

The wet flowmeter is mainly composed of a drum-shaped casing, a rotating drum and a transmission counting mechanism, as shown in Figure 5-15. The rotating drum is composed of a cylinder and four curved blades, and the four blades constitute four cells of equal volume. The lower part of the drum is submerged in water. The water filling level is indicated by the water level gauge 7. When the gas enters into a chamber from the air inlet tube 9 in the middle of the back, then successively discharged from the top, the drum is forced to rotate. The number of rotations of the drum is displayed on the dial through the counting mechanism. With the stopwatch, the gas flow can be directly measured. The wet flowmeter can be used directly to measure gas flow, as well as a standard instrument for the verification of other flowmeters.

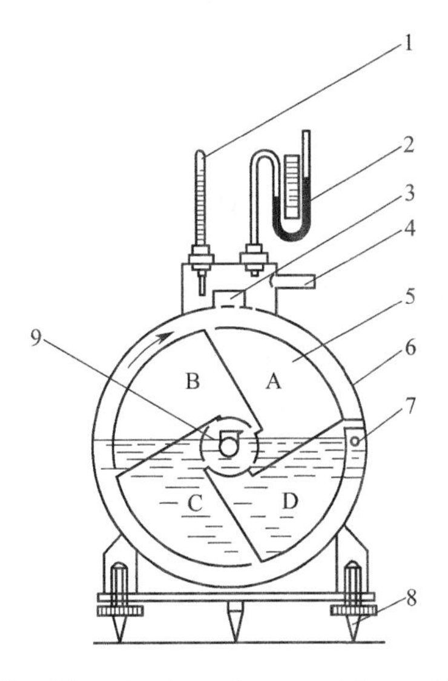

Figure 5-15 The structure diagram of the wet flowmeter

1—Thermometer; 2—Differential pressure gauge; 3—Gradienter; 4—Exhaust pipe; 5—Drum; 6—Shell; 7—Water level gauge; 8—Adjustable supporting feet; 9—Intake pipe

As shown in the Figure 5-15, the gas enters through the inlet tube during operation. While chamber B is inhaling and chamber C starts to inhale, chamber D will exhaust.

Wet flowmeters are generally calibrated with standard volumetric flasks. The volume of the standard volumetric flask is V_v, the volume indication of the wet flowmeter is V_w, then the difference ΔV between the two is $\Delta V = V_v - V_w$. When the pointer of the flowmeter makes one revolution, the total volume on the dial is 5L. Generally, a 1L volumetric flask is used for 5 calibrations. The total volume of the flowmeter is $\sum V_w$, and the average correction factor is

$$C_w = \frac{\sum \Delta V}{\sum V_w} \tag{5-9}$$

Therefore, after calibration, the relationship between the volumetric flow rate of the wet flowmeter V_s and the flowmeter indication V_s' should be:

$$V_s = V_s' + C_w V_s' \tag{5-10}$$

5. 3. 3 Mass flowmeter

This is a flowmeter measuring the mass of fluid flow, including two types: direct and indirect (also called derivation type). Direct mass flow meters directly measure mass flow, including calorimetry, angular momentum, gyro and Coriolis force. The indirect mass flow meter uses the density and volume flow rate to calculate the mass flow rate. With the advantage that the measurement accuracy is not affected by changes in fluid temperature, pressure, viscosity, etc., the mass flow meter is a developing flow measurement instrument.

There are many types of flow meters. With the improvement of industrial production automation, many new flow measurement instruments appeared. Ultrasonic, laser, X-ray and nuclear magnetic resonance and other emerging flow measurement technologies are gradually being applied in industrial production.

5. 3. 4 Flowmeter inspection and calibration

In order to get accurate flow value, the flowmeter must be applied correctly, the structure and characteristics of the flowmeter should be fully understood. In addition, attention must be paid to the correct maintenance and management of the flowmeter in use, while regular calibration is required, too. The flowmeter must be calibrated under the following situations.

① When a flowmeter that has not been used for a long time is to be used;

② When high-precision measurement is required;

③ When there is doubt about the measured value;

④ When the characteristics of the measured fluid do not meet the characteristics of the fluid used to calibrate the flowmeter.

The calibration methods of liquid flowmeter include container type, weighing type, standard volume tube type and standard flowmeter type.

The calibration of gas flowmeters includes container type, sonic nozzle type, soap film tester type, standard flowmeter type, wet flowmeter type, etc. When calibrating a gas flowmeter, special care must be taken to measure the temperature, pressure and humidity of the gas flowing through the calibrated flowmeter and standard flowmeter. In addition, before the calibration work, the characteristics of the gas must be understood, such as whether it is soluble in water, and whether its properties will change with temperature and pressure.

5.4 Liquid level measurement

Liquid level is a measure of how much liquid is stored in a device or container. Liquid level detection can provide decision-making basis for ensuring the normal production process, such as adjusting material balance, mastering material consumption and determining product output.

Liquid level gauges vary depending on the nature of the system, and there are many types. Common liquid level gauges include: direct reading liquid level gauges (glass tube liquid level gauges, glass plate liquid level gauges), differential pressure liquid level gauges (pressure liquid level gauge, blowing method pressure liquid level gauge), buoyancy liquid level gauge (float, buoy and magnetic flap), electrical Liquid level gauge (electric contact type, magnetostrictive type, capacitive type), ultrasonic liquid level gauge, radar liquid level gauge, radioactive liquid level gauge, etc.

In the following, the direct reading level gauge, differential pressure level gauge, and buoyancy level gauge commonly used in the laboratory are going to be introduced.

5.4.1 Direct reading level gauge

(1) Measuring principle

The measurement principle of the direct reading liquid level gauge is to use the meter to directly read the level of the gas or liquid phase in the container. It has simple measurement and intuitive reading, but it is inconvenient to transmit signals remotely, which is suitable for on-site direct reading. The measurement principle is shown in Figure 5-16.

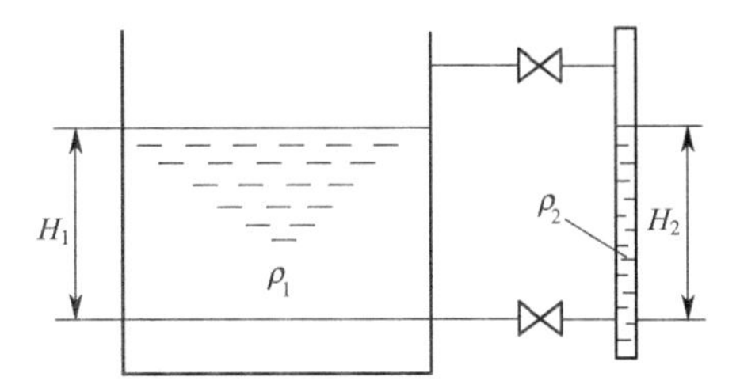

Figure 5-16 Measuring principle of direct reading level gauge

According to the principle of isobaric surface

$$\rho_1 gH_1 = \rho_2 gH_2 \tag{5-11}$$

When $\rho_1 = \rho_2$, $H_1 = H_2$.

When the medium temperature is high, $\rho_1 \neq \rho_2$, an error will occur. However, it is widely used due to its simplicity and practicality, and it is sometimes used for the calibration of the zero level and the highest level of the automatic level gauge.

(2) Glass tube level gauge

The early glass tube level gauges were only used for atmospheric pressure open containers due to structural shortcomings such as fragility and limited length. Now, because the glass tube has been replaced by quartz glass and the protective metal tube is added, the shortcoming of fragility is overcome. In addition, because quartz has the characteristics of high temperature and high pressure resistance, the scope of use of glass tube level gauges has been broadened. With the difference in the refractive index of light in liquid and air, a two-color glass tube liquid level gauge with color filter glass is made. The gas phase is red and the liquid phase is green, making reading rather easy.

The commonly used glass tube level gauges are shown in Figure 5-17. The upper and lower ends are connected to the equipment by flanges and installed with valves. The upper and lower valves are equipped with steel balls. When the glass tube is damaged due to an accident, the steel balls block the passage under the pressure of the container, so that the container is automatically sealed, which can prevent the liquid in the container from continuously flowing out. Steam jackets can also be used for heat tracing to prevent condensable liquids from clogging the pipeline.

(3) Glass plate level gauge

As shown in Figure 5-18, the direct-reading glass plate liquid level gauge has the glass plates on the front and back sides staggered. From the front glass plate of the level gauge, one can see the blind area between the front and back glass plates, and vice versa. In this way, the shortcoming of the blind zone in each measurement could be overcome.

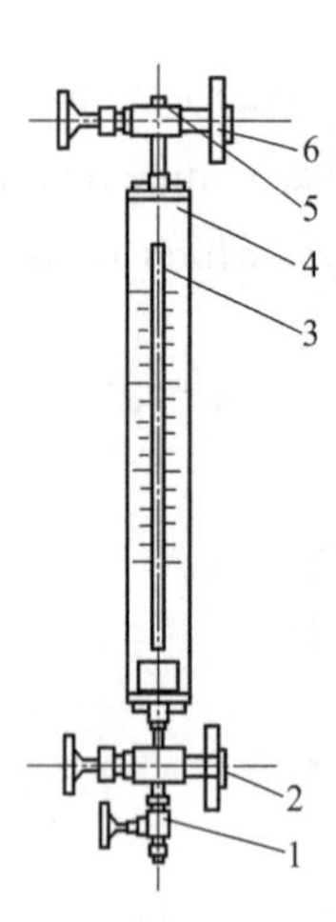

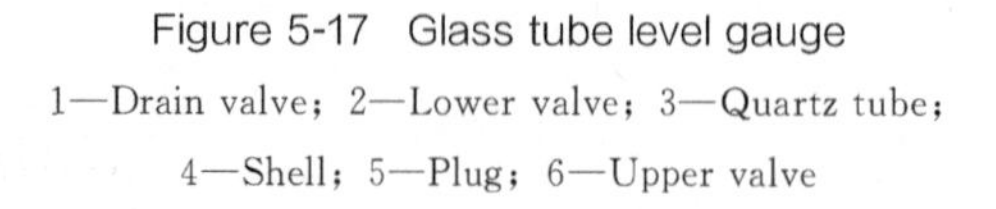

Figure 5-17 Glass tube level gauge

1—Drain valve; 2—Lower valve; 3—Quartz tube; 4—Shell; 5—Plug; 6—Upper valve

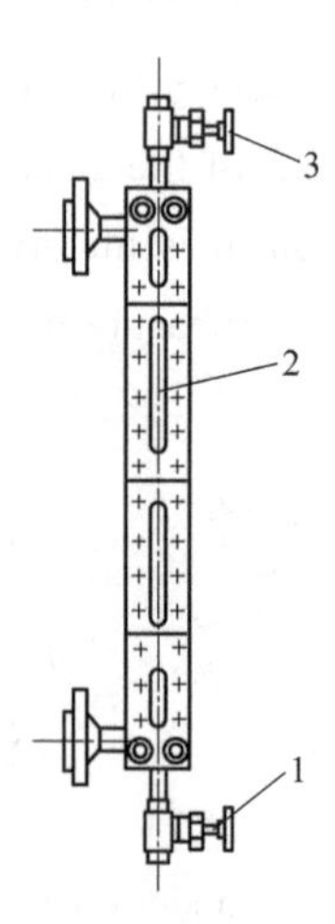

Figure 5-18 The direct-reading glass plate level gauge

1—Lower valve; 2—Glass plate; 3—Upper valve

5.4.2 Differential pressure level gauge

(1) Measuring principle

The liquid level measurement system is shown in Figure 5-19, the differential pressure meter 2 measures the pressure difference between the left and right sides:

$$\Delta p = p_2 - p_1 = (p_0 + \rho gH) - p_0 = \rho gH$$

$$H = \frac{\Delta p}{\rho g} \tag{5-12}$$

In formula, Δp is differential pressure; ρ is measured fluid density; H is liquid level height.

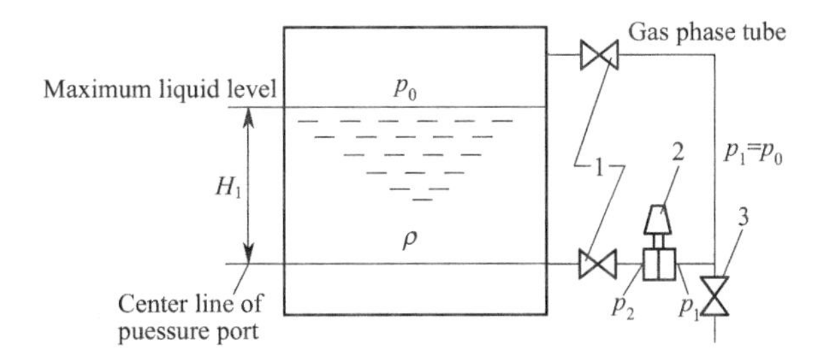

Figure 5-19 Measuring principle of differential pressure level gauge

1—Shut off valve; 2—Differential pressure instrument; 3—Drain valve for gas phase

Since the density ρ of the measured liquid is known, the pressure difference measured by the differential pressure transmitter is proportional to the liquid level height, the liquid level height can be calculated by formula (5-12) .

(2) Differential pressure liquid level measurement principle with positive and negative migration

The measuring principle of differential pressure liquid level measurement principle with positive and negative migration is shown in Figure 5-20. The gas-pressure guiding tube is filled with liquid condensed from vapor instead of gas. Where ρ is the density of the measured fluid, ρ_1 is the density of the medium, h_1 is the height of the condensate. When the gas phase continues to condense, the condensate will automatically overflow from the gas phase port and return to the container under test to keep h_1 unchanged. When the liquid level is at zero position, the negative end of the transmitter is subjected to a pressure of $\rho_1 gh_1$. This pressure must be counteracted in the calculation, which is called negative migration. If the starting measuring level is at H_0, the pressure of ρgH_0 at the positive end of the transmitter should also be offset, which is called positive migration. Therefore, the total migration of the transmitter is $\rho_1 h_1 g - \rho gH_0$.

$$p_2 - p_1 = p_0 + \rho g(H_1 + H_0) - (p_0 + \rho_1 gh_1) = \rho gH_1 + \rho gH_0 - \rho_1 gh_1$$

$$H_1 = \frac{(p_2 - p_1) + (\rho_1 gh_1 - \rho gH_0)}{\rho g}$$

The last item $\rho_1 h_1 g - \rho gH_0$ of the numerator is the total migration amount, so when there is migration, the total migration amount must be offset, and the above formula becomes

$$H_1=\frac{p_2-p_1}{\rho g} \tag{5-13}$$

That is, the range of the instrument is $\Delta p=\rho g H_1$.

When the measured fluid is corrosive or easy to crystallize, a double-flange differential pressure transmitter with isolation diaphragm can be selected. The calculation of migration and instrument range still uses the formula above, however, in the formula, ρ_1 is the density of the silicone oil filled in the capillary, h_1 is the height difference between the centers of the two flanges.

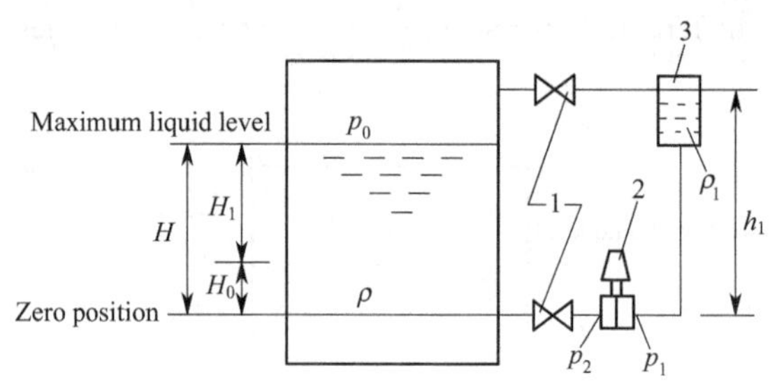

Figure 5-20 Measuring principle diagram of differential pressure level gauge with positive and negative migration

5.4.3 Buoyancy liquid level gauge

Being the earliest type of liquid level instrument, the buoyancy liquid level gauge uses the principle of object buoyancy in liquid to achieve liquid level measurement. Buoyancy liquid level gauges are divided into float type, float ball type and float cylinder type. The first two types are with constant buoyancy, while the latter is variable. Their measurement principles are introduced in the following.

(1) Float liquid level gauge

The measuring principle of float liquid level gauge is shown in Figure 5-21. The float floats on the surface of the liquid by buoyancy, and rises and falls as the liquid level rises and falls. When the liquid level rises, the float floats up, and the steel wire rope is retracted into the watch body by the tension of the spring in the indicator table to keep the balance of the gravity and buoyancy of the float and the tension of the spring. At this time, the liquid level value is transmitted to the indicator through the steel belt and the reduction gear, While the indicator indicates the liquid level value, the transmitter sends a signal proportional to the liquid level height.

According to the structure, the transmitter can be divided into:

① Steel belt gear mechanism and electric transmitter, which can perform local liquid level indication and transmit and output 4 ~ 20mA DC signal;

② Steel belt gear mechanism, as well as a steel belt with coding hole and transmitter with code reading device, which can perform local liquid level indication, and the transmitter outputs

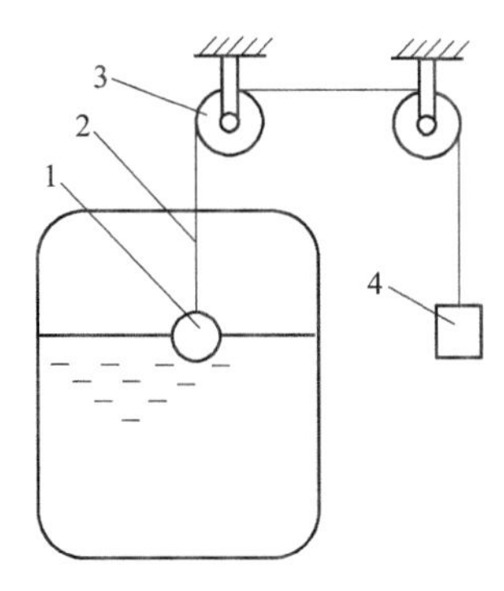

Figure 5-21 Measuring principle diagram of float type liquid level gauge

1—Float; 2—Steel belt; 3—Guide pulley device; 4—Indicating instrument or transmitter

pulse signal to the secondary instrument for indication.

(2) Float ball liquid level gauge

The measuring principle of the float ball liquid level gauge is shown in Figure 5-22. The magnetic float above the liquid level in the container moves up and down with the liquid level. The position of the magnetic float can be measured to obtain liquid level information, the position signal of the magnetic float can also be converted into an electrical signal for remote transmission and control.

When the density of the tested material changes, the floating ball weight can also be changed to ensure the measurement.

(3) Float cylinder liquid level gauge

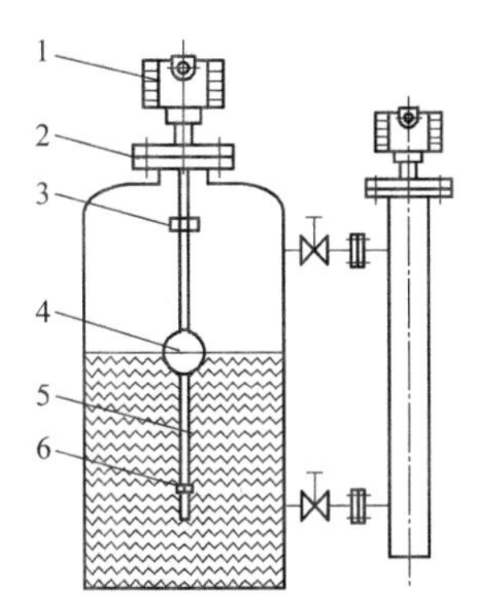

Figure 5-22 Measuring principle diagram of float ball liquid level gauge

1—Indicating instrument or transmitter; 2—Connecting flange; 3—Upper limit 4—Float ball; 5—Guide connecting rod; 6—Lower limit

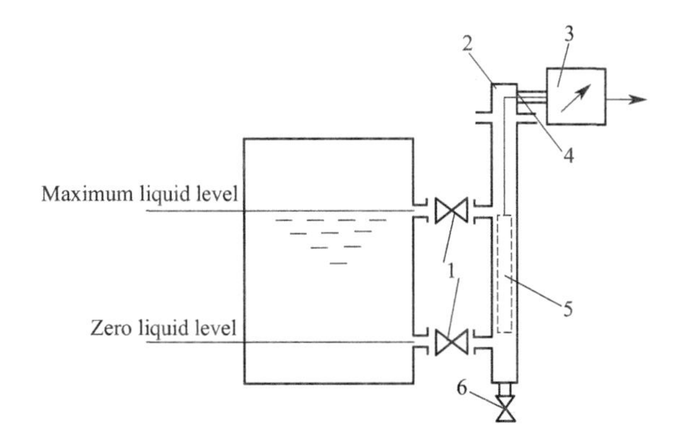

Figure 5-23 Measuring principle diagram of float cylinder liquid level gauge

1—Globe valve; 2—Float cylinder body; 3—Indicating instrument or transmitter; 4—Torsion tube assembly; 5—Float cylinder; 6—Discharge valve

The measuring principle of float cylinder liquid level gauge is shown in Figure 5-23. From the zero position to the highest liquid level, the float cylinder is completely immersed in the liquid. The buoyancy forces the float cylinder to have a small upward displacement. The liquid level is measured by detecting the change in the buoyancy of the float cylinder. When the liquid level is at zero, the torsion tube receives the torsional torque generated by the gravity of the buoy (at this time, the torsional torque is maximum), which is at "zero" degree. When the liquid level gradually rises to the highest, the torsion tube receives the torsion torque generated by the maximum buoyancy (the torque at this time is the smallest) and rotates through an angle, the transmitter changes this rotation angle to 4～20mA DC signal, which is proportional to the measured liquid level.

(4) Magnetic flap liquid level gauge

As shown in Figure 5-24, the magnetic flap liquid level gauge is equipped with a magnetic steel float in the float chamber (made of non-magnetic stainless steel) connected to the container. The scale of the flap is attached to the wall of the float chamber. When the liquid level rises or falls, the float also rises and falls. The flap (half white and half red) is attracted by the magnetic steel in the float and flipped over. The flipped part shows red, non-flipped part shows white, and the boundary indicates the liquid level.

In addition to the indicator scale for local indication, the magnetic flap liquid level gauge can also be equipped with an alarm device and a remote transmission device, which can convert the liquid level into a 4-20mA DC signal and send it to the receiving instrument.

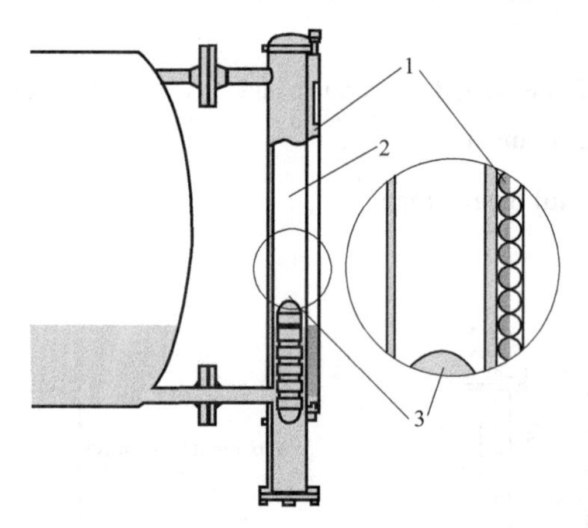

Figure 5-24 Schematic diagram of the magnetic flap liquid level gauge

1—Flap scale; 2—Float; 3—Magetic float

Chapter 6

Virtual Simulation Experiments of Chemical Engineering Principle

The simulation software for laboratory experiments in the subject Principles of Chemical Engineering simulates actual experiments and processes in real time using dynamic mathematical models, and produces observations and results consistent with real experiments through the interaction of 3D simulation of the experimental setups. Using these simulations, students can understand the underlying principles of the experiments, familiarize themselves with the experimental processes, master the steps, observe the outcomes, record the data, obtain the results to deepen their understanding of the theories applicable to chemical engineering and also verify the relevant formula.

This software is developed by Beijing Oubeier Software Technology Development Limited Company. It consists of nine 3D experiment simulation softwares, including 3D simulation modules for the Reynolds demonstration, comprehensive study of chemical flow processes, comprehensive performance measurement of centrifugal pump, filtration under constant pressure, comprehensive study of heat transfer, comprehensive study of rectification, carbon dioxide absorption and desorption, liquid-liquid extraction tower experiment and tunnel drying experiment.

6.1 Overview of virtual simulation experiments

6.1.1 Basic operations

For this illustration, the 3D virtual simulation software for carbon dioxide absorption and desorption and chemical flow process are used as examples.

① Select the experiment on the simulation experiment platform. For example, select "Determine curves of the desorption tower dry packing materials" on the software interface for carbon dioxide absorption and desorption. Click "Start" to enter the interface shown in Figure 6-1. If the user wants more information regarding the experiments, click "Introduction to experiments" (detailed introduction is included in the individual experiments) . To study the simulation experiment in detail, click "Enter System," to reach the interface in Figure 6-2, then click "Start" to go to the interface in Figure 6-3.

If the user selects "Smooth Tube Resistance Measurement Experiment" in the 3D simulation of chemical flow processes, click "Start" to enter the interface in Figure 6-4.

② The name and the status of each part appear when the mouse slides over it. Valves, power button or pumps may be switched on/off by single clicking the left mouse button.

Figure 6-1 Simulation software interface for carbon dioxide absorption and desorption

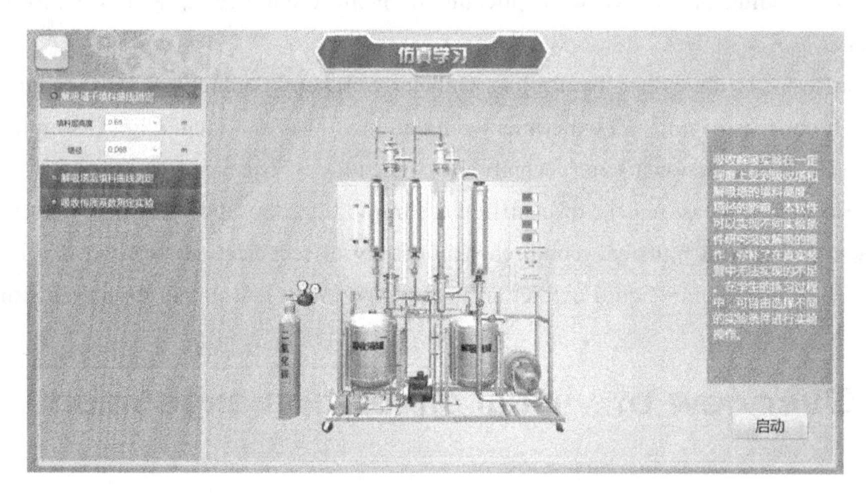

Figure 6-2 Simulation learning interface after entering the system

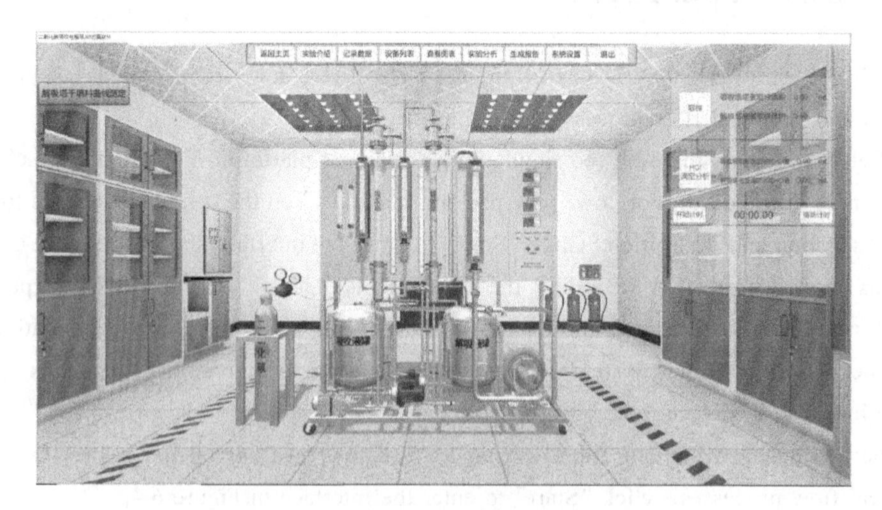

Figure 6-3 Absorption laboratory interface

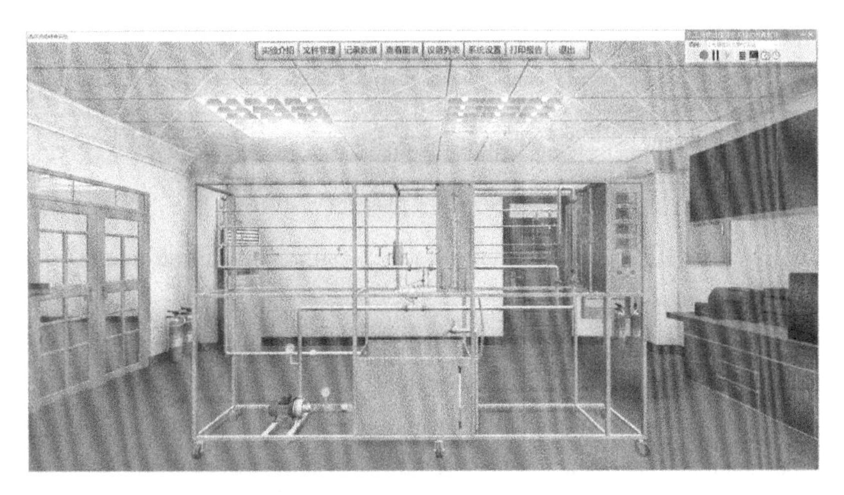

Figure 6-4 Comprehensive chemical flow process laboratory interface

③ Scene control: use the W, S, A, D keys of the keyboard or the up, down, left and right keys (Figure 6-5) to control the movement. Press the right mouse button to rotate the angle of each view.

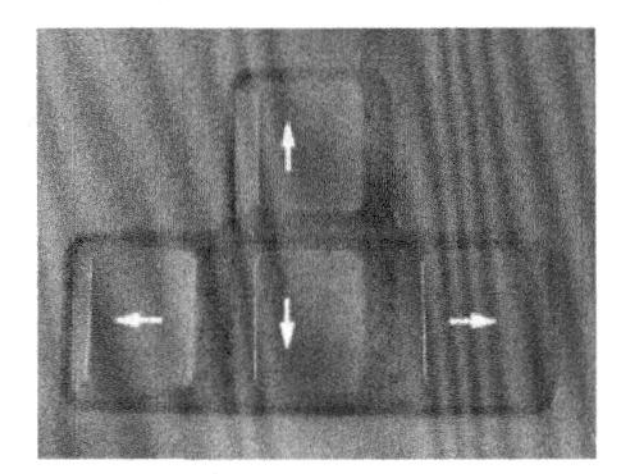

Figure 6-5 Scene control keys

④ Zoom in: select the part to be enlarged and double-click the left mouse button. After zooming in, hold down the right mouse button and drag to move the field of view up, down, left or right; press any key on the keyboard to restore the previous setting.

⑤ Part lookup: click "Device List" on the menu bar, select the name of the part to be viewed, single click the left mouse button to quickly locate the enlarged target, hold down the right mouse button and drag the mouse to move the field of vision up, down, left or right, press any key on the keyboard to restore the previous setting.

⑥ If a device or equipment model has a parameter setting interface, the user can hold down the left mouse button to rotate the view, while the mouse wheel can be scrolled to reduce or enlarge the model.

6.1.2 Description of menu selection functions

Figure 6-3 and Figure 6-4 show the menu selection items that appear at the top of the laboratory interface, as shown in Figure 6-6.

The functions of the menu selection items are as follows:

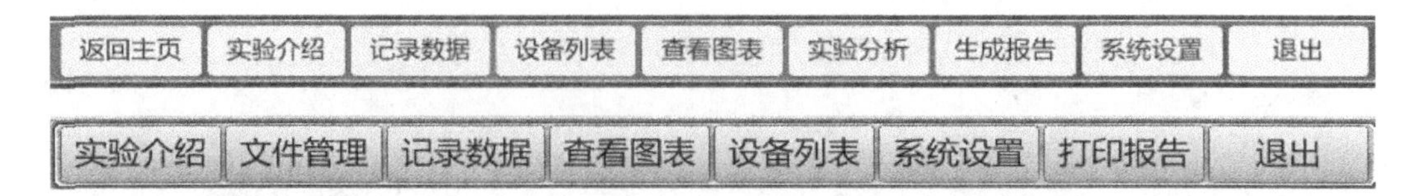

Figure 6-6　Menu selection items

【Return to main menu】　Click to return to the simulation learning interface. After that, a new project will be initiated.

【Introduction to the experiment】　Introduce of the basics of the experiment, such as the purpose and content of the experiment, principles of the experiment, basic layout of the experimental device, experimental methods and procedures, as well as the precautions.

【File management】　the data storage files can be established and set as the current record files. Some software packages have this function. The 3D simulation software for the comprehensive study of chemical flow processes is used as an example for this illustration.

Operation method: click "Save as" at the bottom as shown in Figure 6-7, then the interface in Figure 6-8 will appear. A user can modify the name of the new file and set it as the current record file, then click "Save." If a user clicks "New" again, a new file can be generated.

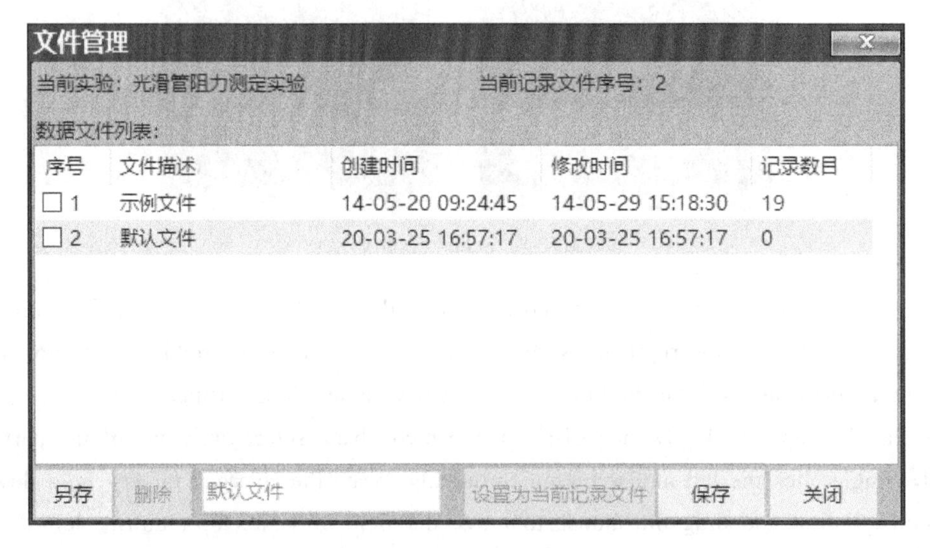

Figure 6-7　File management interface 1

【Record data】　Manage data to enable data recording and data processing functions.

The data management and recording interface of the 3D simulation software for the comprehensive study of chemical flow processes is shown in Figure 6-9.

Operation method:

① Click "Record Data" below to pop up the record data box, fill in the measured data, then press OK.

② After recording the data, check the data to be calculated (if a user wants to process all data, check "select all below") . After selecting the data, click the "Data Processing" button to calculate the results as per the recorded data.

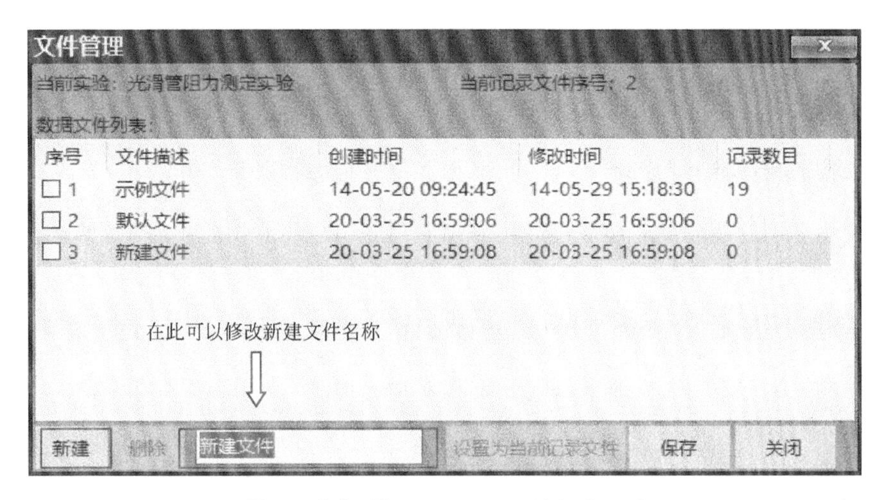

Figure 6-8 File management interface 2

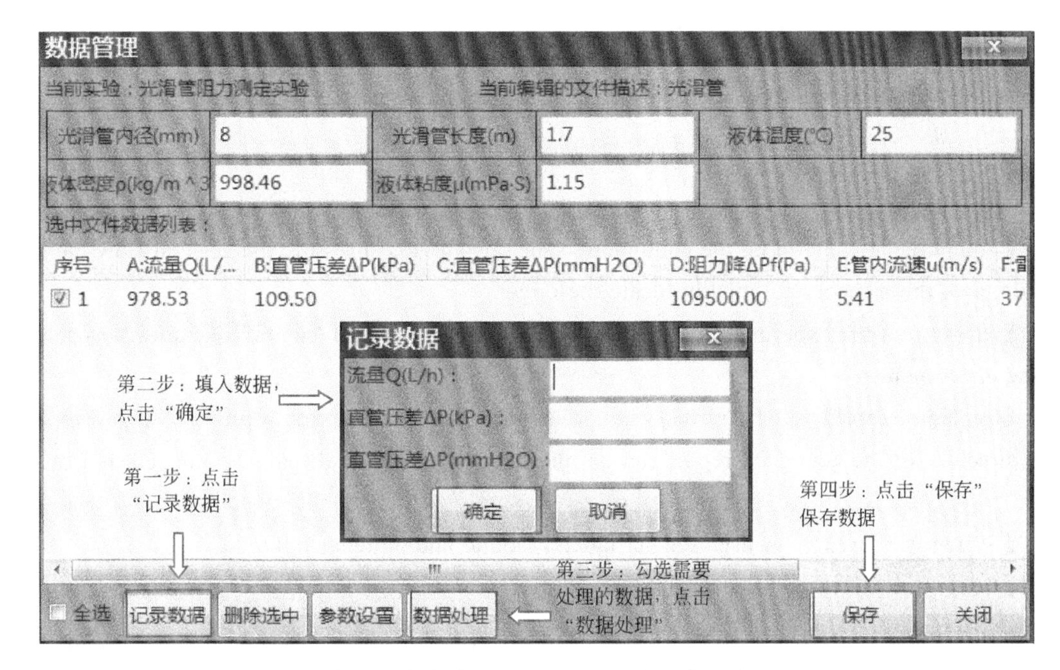

Figure 6-9 Data management interface 1

③ If an error occurs in the data recording, check the relevant group of data, then click "Delete Selected" to delete the selected data with errors.

④ After data processing, click the "Save" button, then close the window.

Figure 6-10 shows the data management interface of the 3D simulation software for carbon dioxide absorption and desorption. In addition to the data to be recorded in the column after each serial number, such as air flow and height difference in the U-tube, the data for the other columns should be computed independently and the results filled in. Finally, click the "Submit" button and close the window.

【View charts】 Based on the recorded experimental data and data processing results, a user can generate a target table or display a relationship curve for the project, then insert it into the

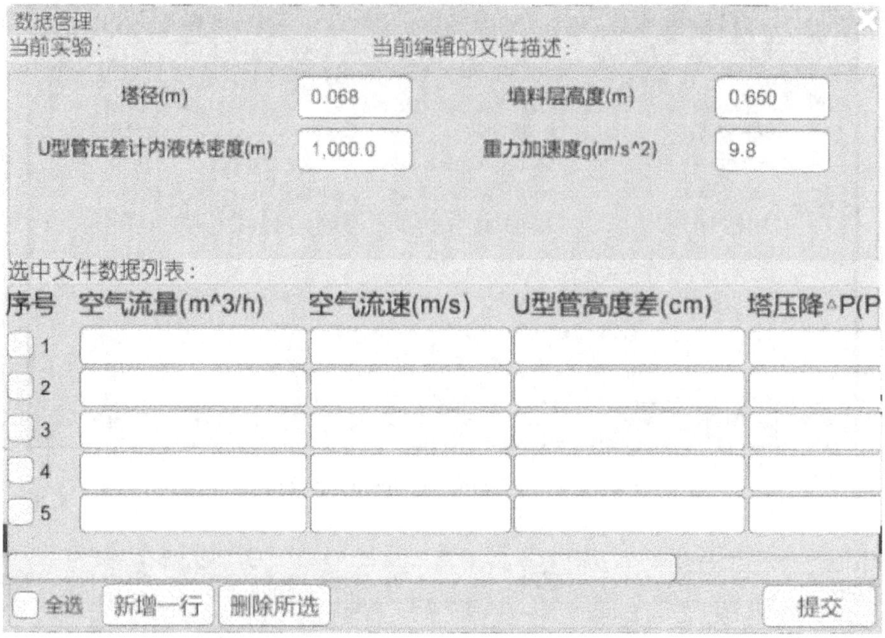

Figure 6-10 Data management interface 2

experiment report.

【Equipment List】 List the equipment, such as valves, pressure gauges and flow meters. Single click the category to quickly locate the target.

【Analysis of the experiment】 Multiple choice questions and true or false questions related to the experiment.

【Generate report】 The virtual simulation software can generate a preview of the report. The directory of the generated report can be placed as per the default path or one selected by the user.

【System settings】 A user can set labels, sound and ambient light.

【Exit】 Click to exit the experiment.

6. 1. 3 Instrument and valve adjustment instructions

(1) Numerical display instrument panel

This type of instrument panel is only for display and does not perform any operations. The corresponding value is displayed directly, as shown in Figure 6-11.

(2) Instrument settings

The value denoted as PV in the top row is the displayed value, while SV in the bottom row is the set value, as shown in Figure 6-12.

For example, if the temperature needs to be adjusted from 25. 0℃ to 65. 0℃, press the  key on the control panel, a flashing number will appear in the SV display window of the in-

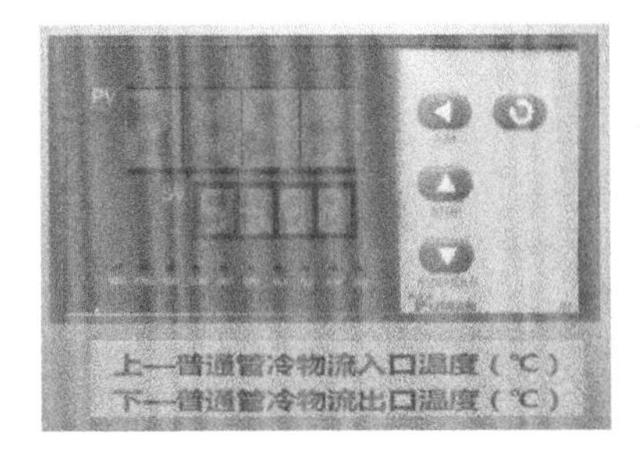

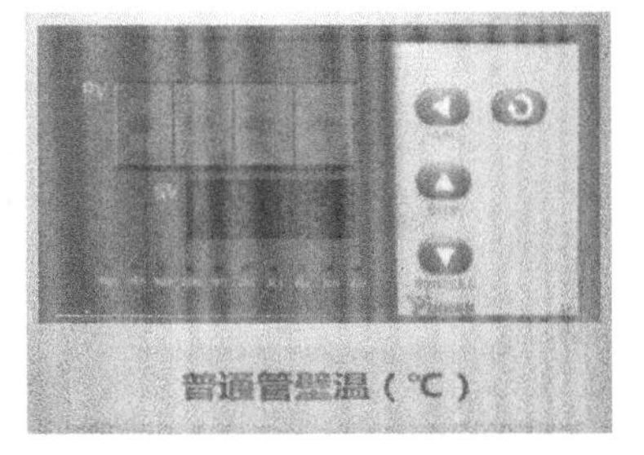

Figure 6-11 Numerical display instrument panel

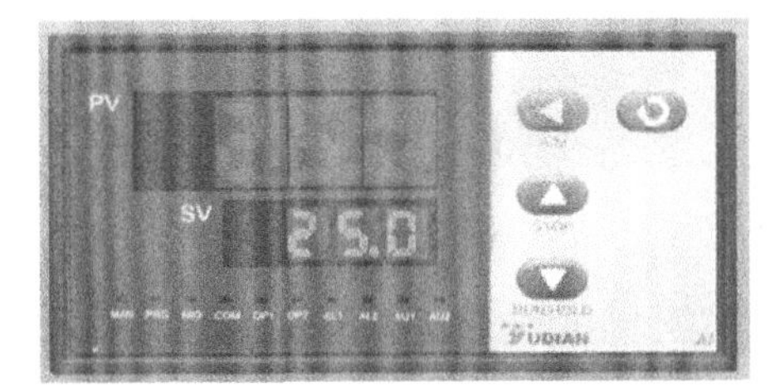

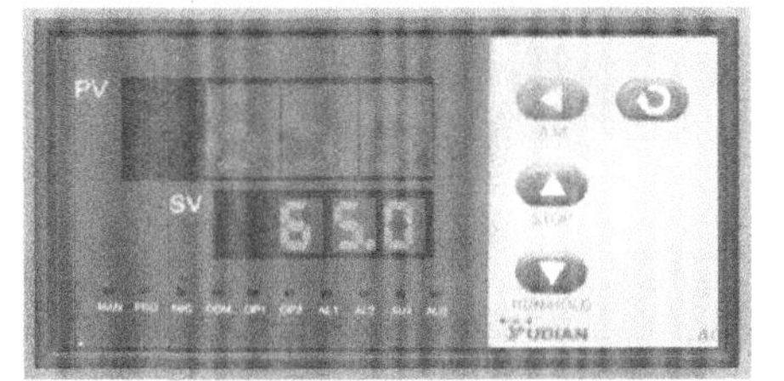

Figure 6-12 Setting instrument panel

strument. Each time the key is pressed, the flashing number will move to the left stepwise. A user can use the and keys to adjust the value of the flashing number and press again to confirm the adjustment.

(3) Use methods of the frequency converter

The frequency converter panel is shown in Figure 6-13. The on/off button of the pump is denoted as RUN / STOP. The method of setting the pump frequency is as follows: when the pump is started, press the RESET button; the value displayed on the panel will flash from the last digit. Continue to press the button to change the location that flashes. The value of the flashing number can be changed by pressing the up or down button. After the setting is performed, press the READ / ENTER button which can be automatically adjusted to the set value.

Figure 6-13 Frequency converter panel

(4) Flow adjustment in the centrifugal pump experiment

The flowmeter display panel is shown in Figure 6-14. The flowmeter on the control cabinet is used to adjust the degree of opening of the electric flow control valve. PV on the flowmeter display shows the current flow, while SV shows the current opening of the electric valve. The opening of the electric flow control valve can be adjusted using the up and down buttons.

(5) Adjustment of valves

The rotameter adjustment is shown here as an example. Click the valve to be adjusted, then an adjustment dialog box appears. Subsequently, click "open" or pull the progress bar to adjust

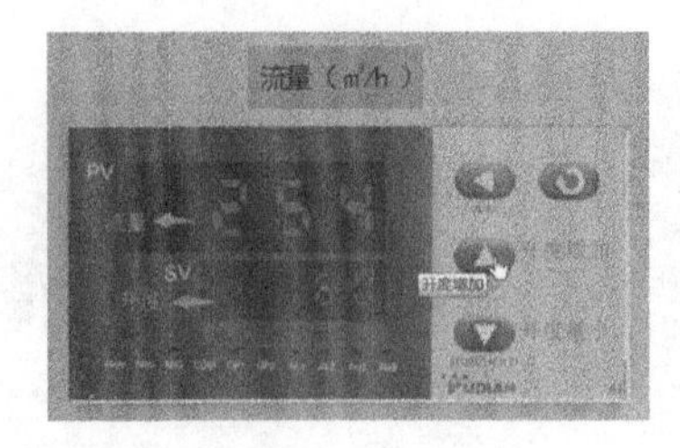

Figure 6-14 Flowmeter display panel

the degree of opening. Move the mouse over the flowmeter to display the value of the flow rate. Adjust the opening until the desired flow rate is achieved , as shown in Figure 6-15.

Figure 6-15 Diagram of valve adjustment

6.1.4 Precautions

① Some simulation experiments require data processing to be completed independent of the software. After clicking "Record Data" in the menu bar, calculate such values and enter in the data management system for submission.

② Some simulation experiments require the calculation of relevant data based on the parameters provided by the software under "View Charts" in the menu bar and fill in the corresponding columns with the calculation results.

③ The rotameter reading needs to be corrected during data processing.

④ The required content corresponding to "Analysis of the Experiment" in the menu bar of the simulation must be completed.

6.2 Simulation experiment of Reynolds demonstration

6.2.1 Main interface of simulation

The main interface of the Reynolds demonstration simulation experiment is shown in Figure 6-16.

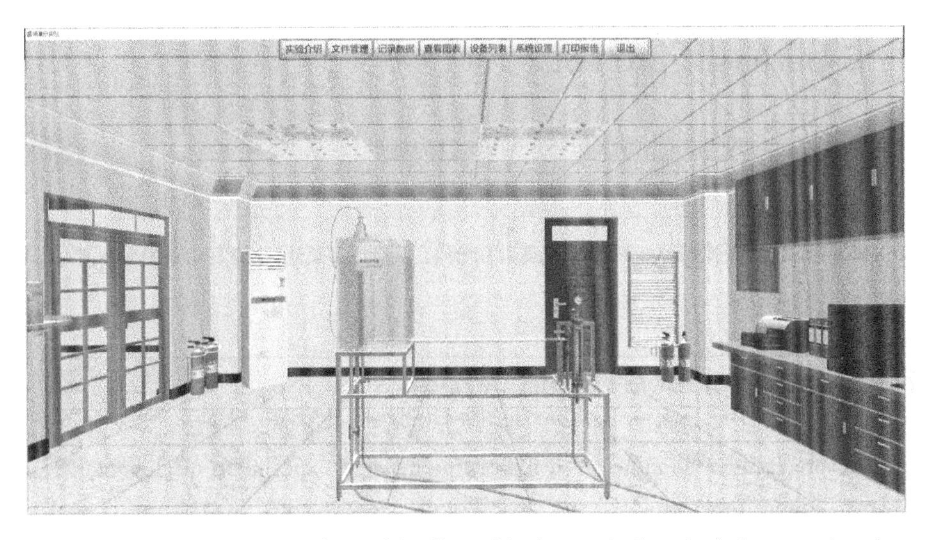

Figure 6-16 The main interface of the Reynolds demonstration simulation experiment

6. 2. 2 Reynolds demonstration experimental device

The diagram of the Reynolds demonstration experimental device is shown in Figure 6-17.

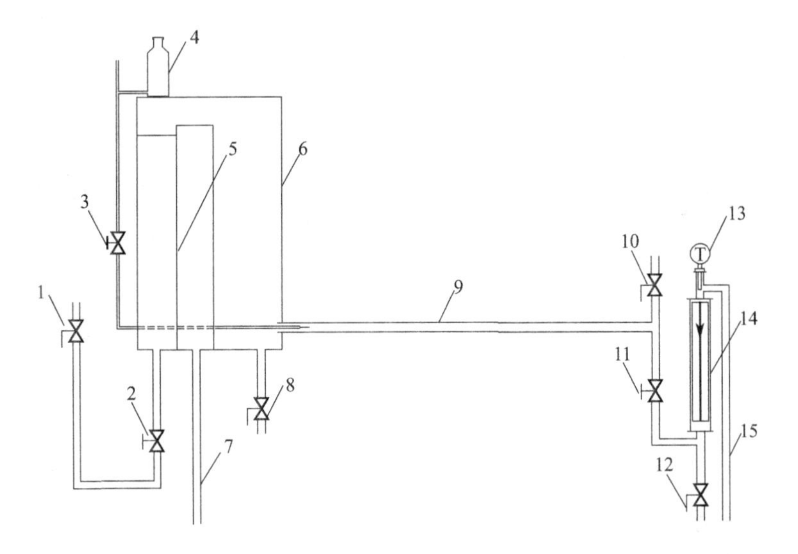

Figure 6-17 Diagram of Reynolds demonstration experimental device

1— Faucet; 2—Feeding water regulating valve; 3—Red ink inlet valve; 4— Red ink bottle; 5—Overflow plate; 6—Water tank; 7—Overflow tube; 8—Drain valve 1; 9—Test tube; 10—Vent valve; 11—Flow regulating valve; 12—Drain valve 2; 13—Temperature display panel; 14—Rotameter; 15—Outlet tube

The main components of the experimental device are the water tank, red ink bottle, test tube and rotameter. The water in the water tank flows slowly through the test tube, while the red ink flows through the center of the tube. The water flow conditions in the experimental tubes can be observed.

6.2.3 Experiment projects

Simulation experiment of Reynolds demonstration.

6.3 Comprehensive experiment of chemical flow processes

6.3.1 Main interface of simulation

The main interface of the simulation for comprehensive experiment of chemical flow processes is shown in Figure 6-18.

Figure 6-18 The main interface of simulation for the comprehensive experiment of chemical flow processes

6.3.2 Experimental device for the comprehensive experiment of chemical flow processes

The diagram of the comprehensive experiment of chemical flow processes is shown in Figure 6-19.

This experimental device mainly consists of water tank, centrifugal pump, flow meters, vacuum gauges, pressure gauges, inverted U-tube and pressure sensors. The water pump draws water from the water storage tank, sends it to the experimental system, measures the flow rate through a flowmeter, then sends it to the testing tubes. Subsequently, the water flows back to the water storage tank through the return tube.

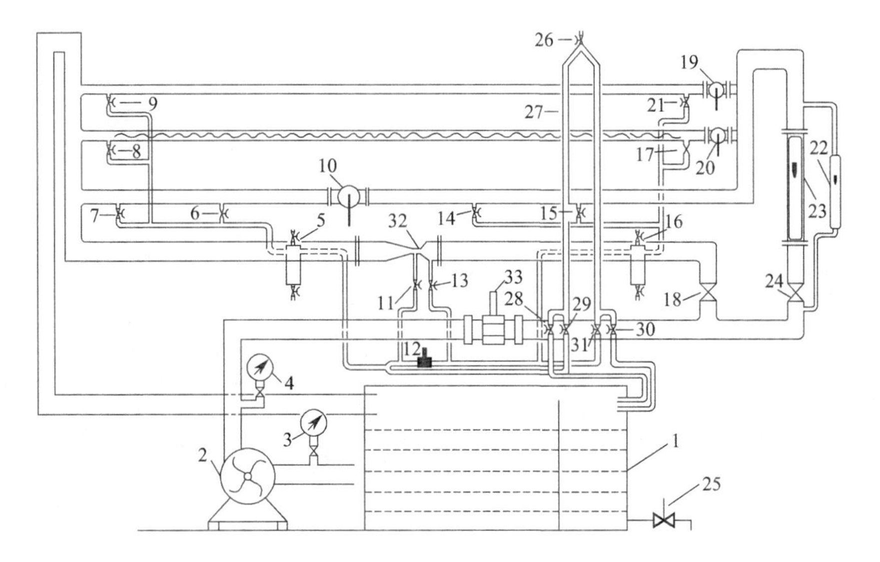

Figure 6-19 Diagram of the comprehensive experiment of the chemical flow processes

1—Water tank; 2—Water pump; 3—Inlet vacuum gauge; 4—Outlet pressure gauge; 5, 16—Buffer tank top valves; 6, 14—Near-end valve for local resistance measurement; 7, 15—Far-end valve for local resistance measurement; 8, 17—Rough tube pressure measurement valve; 9, 21—Smooth tube pressure measurement valve; 10—Local resistance valve; 11—Left valve of the Venturi meter pressure drop sensor; 12—Pressure sensor; 13—Right valve of the Venturi meter pressure drop sensor; 18, 24—Valves; 19—Smooth tube valve; 20—Rough tube valve; 22—Small rotameter; 23—Large rotameter; 25—Water tank drain valve; 26—Inverted U-tube vent valve; 27—Inverted U-tube; 28, 30—Inverted U-tube drain valve; 29, 31—Inverted U-tube balance valve; 32—Venturi meter; 33—Turbine flowmeter

6.3.3 Experiment projects

(1) Determination of resistance of smooth tubes

(2) Determination of resistance of rough tubes

(3) Determination of local resistance

(4) Determination of characteristic curves of centrifugal pump

(5) Determination of pipeline characteristic curves

(6) Flow performance measurement

6.4 Experiment of centrifugal pump comprehensive performance measurement

6.4.1 Main interface of simulation

The main interface of simulation for centrifugal pumb comprehensive performance measurement experiment is shown in Figure 6-20.

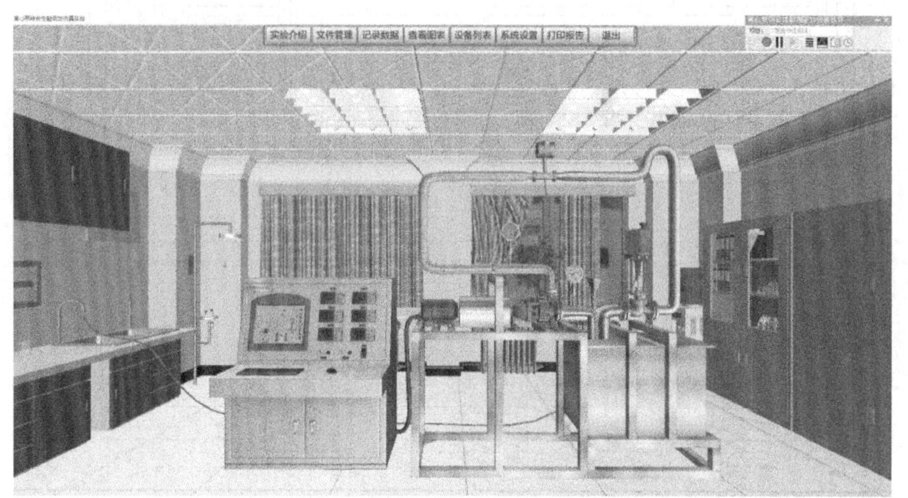

Figure 6-20 The main interface of simulation for centrifugal pumb comprehensive performance measurement experiment

6.4.2 Centrifugal pump experimental device

The diagram of the centrifugal pump experimental device is shown in Figure 6-21.

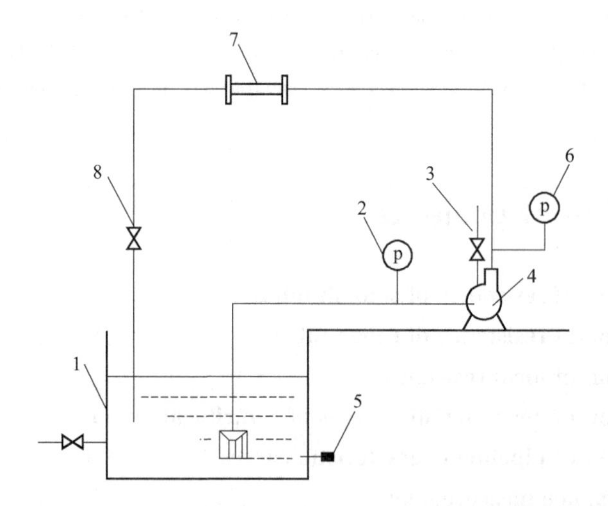

Figure 6-21 Diagram of the centrifugal pump experimental device

1—Water tank; 2—Vacuum gauge; 3—Pump priming valve; 4—Centrifugal pump; 5—Temperature sensor; 6—Pressure gauge; 7—Turbine flowmeter; 8—Electric flow regulating valve

This experimental device is mainly composed of a water tank, centrifugal pump, turbine flowmeter, vacuum gauge, pressure gauge and temperature sensor. The centrifugal pump 4 transfers the water from the water tank 1 to the experimental system; the flow is adjusted by an electric flow regulating valve 8, the fluid is measured by the turbine flowmeter 7 and returns to the water storage tank. Note: the centrifugal pump is installed above the water level of the water tank. Remember to prime the pump before the operation.

6.4.3 Experiment projects

(1) Determination of characteristic curves of the centrifugal pump

(2) Determination of pipeline characteristic curves

6.5 Filtration experiment under constant pressure

6.5.1 Main interface of simulation

The main interface of simulation for Filtration experiment under constant pressure is shown in Figure 6-22.

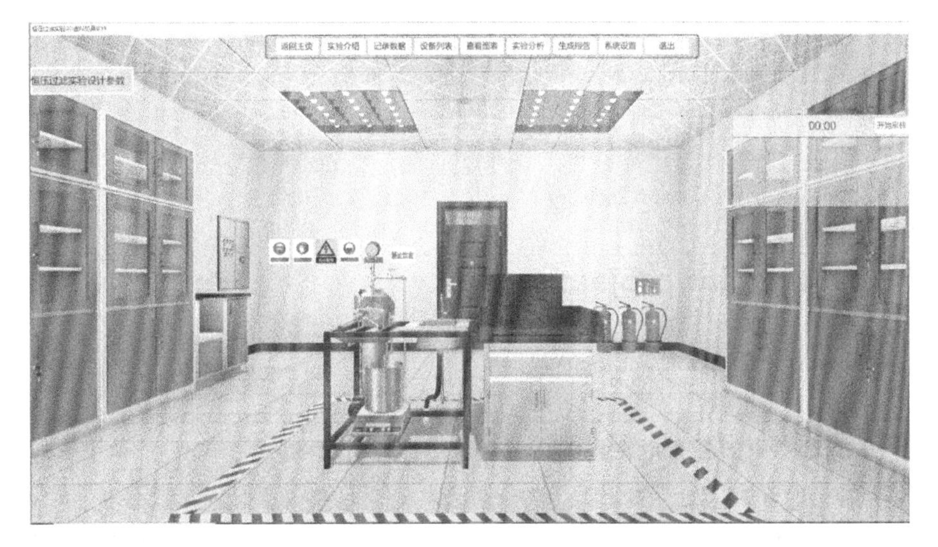

Figure 6-22 The main interface of simulation for filtration experiment under constant pressure

6.5.2 Experimental device of filtration under constant pressure

Experimental device for filtration under constant pressure is shown in Figure 6-23.

The main components of the experimental device are the suspension tank, feed pump, filter, clear liquid tank, electronic scale, air compressor, computer. The plate filter has three layers of filter plates in total, with a canvas filter screen as the filter medium.

The prepared calcium carbonate ($CaCO_3$) suspension is circulated and stirred by feed pump 8 at the bottom of the suspension tank, so that the filter slurry does not precipitate. By adjusting the opening of the pressure regulating valve 7, the feed liquid is passed through the bypass line and the feed valve 11 and sent to the plate filter 13 for filtration. The filtrate flows into the clear liquid tank 16 and is weighed by an electronic scale 17. After the filtration is completed, the fil-

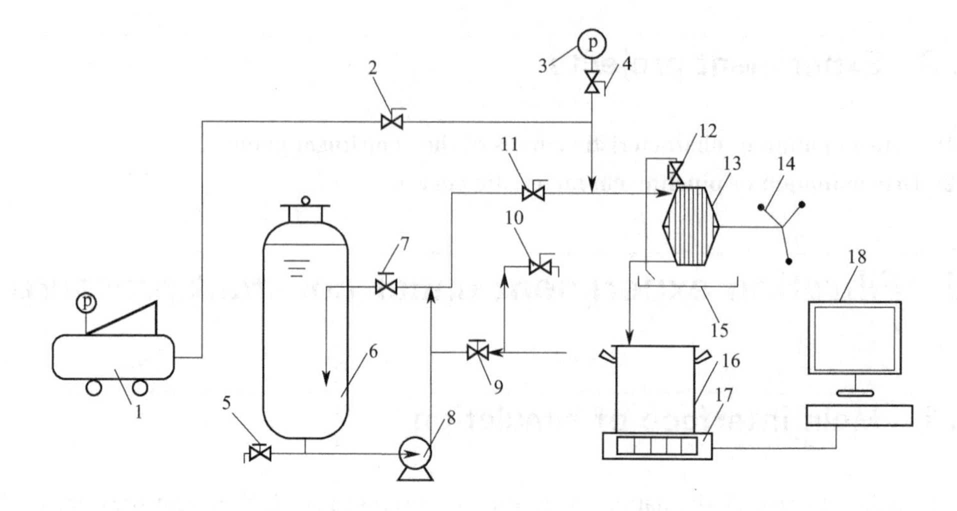

Figure 6-23 Diagram of the experimental device for filtration under constant pressure

1—Air compressor; 2—Compressed air valve; 3—Pressure gauge; 4—Pressure valve; 5—Exhaust valve; 6—Suspension tank; 7—Pressure regulating valve; 8—Feed pump; 9—Water inlet valve; 10—Faucet; 11—Feed valve; 12—Vent valve; 13—Plate filter; 14—Fastening screw; 15—Cake collection basin; 16—Clear liquid tank; 17—Electronic scale; 18—Computer

ter cake is blown dry with compressed air.

6. 5. 3 Experiment projects

Determination of filtration constant.

6. 5. 4 Adjustable experimental conditions

The experimental conditions can be changed (Table 6-1) to investigate the effects of different parameters on the results.

Table 6-1 Adjustable conditions for filtration experiment under constant pressure

Filter diameter/mm	120. 0	130. 0	140. 0
Density of suspension solution/ (kg/m^3)	1016. 0	1033. 0	1045. 0
Mass of clear liquid tank/kg	13. 2	8. 95	25. 5
Composition of suspension solution	$CaCO_3$	Starch	—

6. 5. 5 Notes for data processing

Since the solid particles did not form a filter cake on the filter media at the start of filtration, if the operation is performed at a constant pressure from the beginning of the experiment, clear liquid may not be obtained as some particles may pass through the filter media under the large filtration driving force. Therefore, at the onset of the experiment, a relatively low pressure is required for a small interval during operation. After the solid particles form a filter cake on the filter media, filtration can be performed under the set pressure until the end. Thus, the software design takes into account the quantity of filtrate q_1 that has passed through the filter during the period τ_1 before reaching the constant pressure, and transforms the constant pressure filtration equation into formula (6-1).

$$\frac{\tau-\tau_1}{q-q_1}=\frac{1}{K}(q-q_1)+\frac{2}{K}(q_1+q_e)=\frac{1}{K}(q+q_1)+\frac{2}{K}q_e \tag{6-1}$$

6.6 Comprehensive experiment of heat transfer

6.6.1 Main interface of simulation

The main interface of simulation for comprehensive experiment of heat transfer is shown in Figure 6-24.

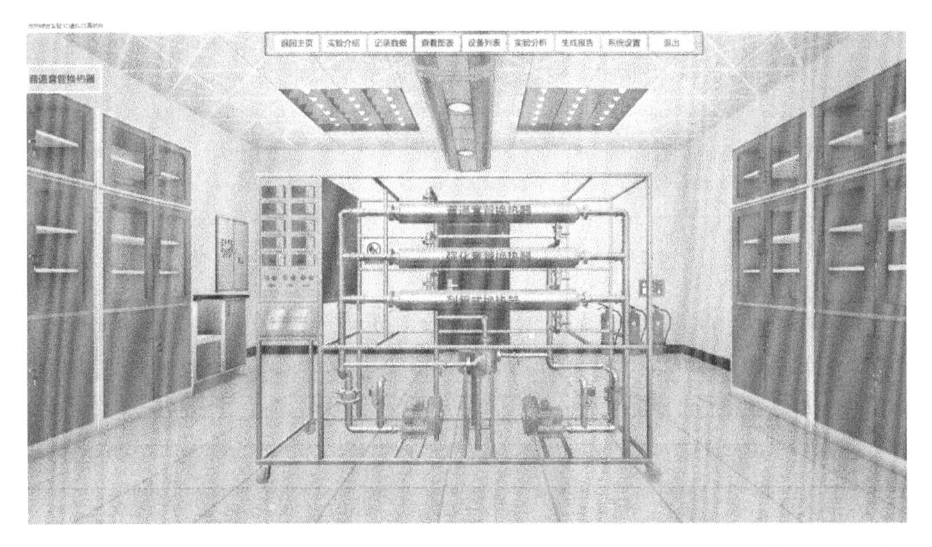

Figure 6-24 The main interface of simulation for comphensive experiment of heat transfer

6.6.2 Experimental device for comprehensive experiment of heat transfer

The diagram of the experimental device for the comprehensive experiment of heat transfer is shown in Figure 6-25.

For illustration, cold and hot air are used as media. The cold air is blown out by cold air blower 1 and regulated by cold air bypass regulating valve 2. It passes through orifice flowmeter 3, then enters the heat exchanger through different branches controlled by the branch control

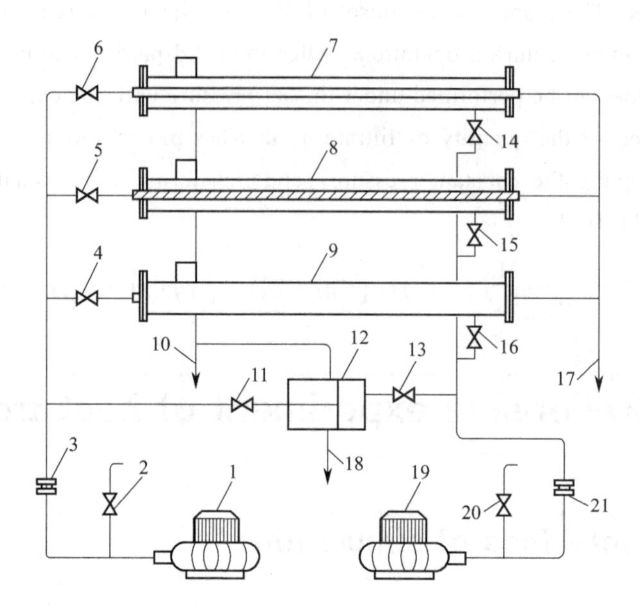

Figure 6-25 Diagram of the experimental device for the comprehensive experiment of heat transfer

1—Cold air blower; 2—Cold air bypass regulating valve; 3—Orifice flowmeter Ⅰ; 4—Tubular heat exchanger cold flow inlet valve; 5—Strengthened double pipe heat exchanger cold flow inlet valve; 6—Ordinary double pipe heat exchanger cold flow inlet valve; 7—Ordinary double pipe heat exchanger; 8—Strengthened double pipe heat exchanger; 9—Tubular heat exchanger; 10—Hot flow outlet; 11—Spiral plate heat exchanger cold flow inlet valve; 12—Spiral plate heat exchanger; 13—Spiral plate heat exchanger heat flow inlet valve; 14—Ordinary double pipe heat exchanger heat flow inlet valve; 15—Strengthened double pipe heat exchanger heat flow inlet valve; 16—Tubular heat exchanger heat flow inlet valve; 17, 18—Cold flow outlet; 19—Hot air blower; 20—Hot air bypass regulating valve; 21—Orifice flowmeter Ⅱ

valve. The tube-side hot air is blown out by the hot air blower, regulated by hot air bypass regulating valve 20, passes through orifice flowmeter Ⅱ 21, and enters the shell side of the heat exchanger following different branches controlled by the branch control valve. The hot air is naturally blown out of the exit at the other end to achieve the effect of counter-current heat exchange.

6.6.3 Experiment projects

(1) Ordinary double tube heat exchanger (cold air-hot air)

(2) Ordinary double tube heat exchanger (cold water-hot water)

(3) Ordinary double tube heat exchanger (toluene-steam)

(4) Strengthened double tube heat exchanger (cold air-hot air)

(5) Strengthened double tube heat exchanger (cold water-hot water)

(6) Strengthened double tube heat exchanger (toluene-steam)

(7) Tubular heat exchanger (cold air-hot air)

(8) Tubular heat exchanger (cold water-hot water)

(9) Tubular heat exchanger (toluene-steam)

(10) Spiral plate heat exchanger (cold air-hot air)

(11) Spiral plate heat exchanger (cold water-hot water)

(12) Spiral plate heat exchanger (toluene-steam)

6.6.4 Adjustable experimental conditions

The experimental conditions can be adjusted (Table 6-2) to investigate the effects of different experimental conditions on the experimental results.

Table 6-2 Adjustable conditions for heat transfer experiment

Items	Ordinary double pipe heat exchanger	Strengthened double tube heat exchanger	Tubular heat exchanger	Spiral plate heat exchanger
Tube length/m	1.90 2.00 2.10	1.90 2.00 2.10		
Inner diameter/mm	20 25 30	20 25 30		
Number of tubes			4 5 6	
Area of heat exchange/m^2				10 20 30

6.7 Carbon dioxide absorption and desorption experiment

6.7.1 Main interface of simulation

The main interface of simulation for carbon dioxide absorption and desorption experiment is shown in Figure 6-26.

6.7.2 Experimental device of carbon dioxide absorption and desorption

The diagram of the experimental device for carbon dioxide absorption and desorption is shown in Figure 6-27.

The main components of the experimental device include a CO_2 cylinder, fans, water

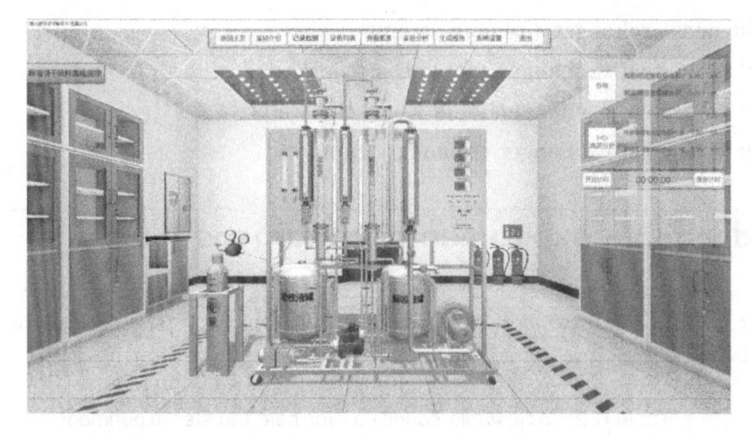

Figure 6-26 The main interface of simulation for carbon dioxide absorption and desorption experiment

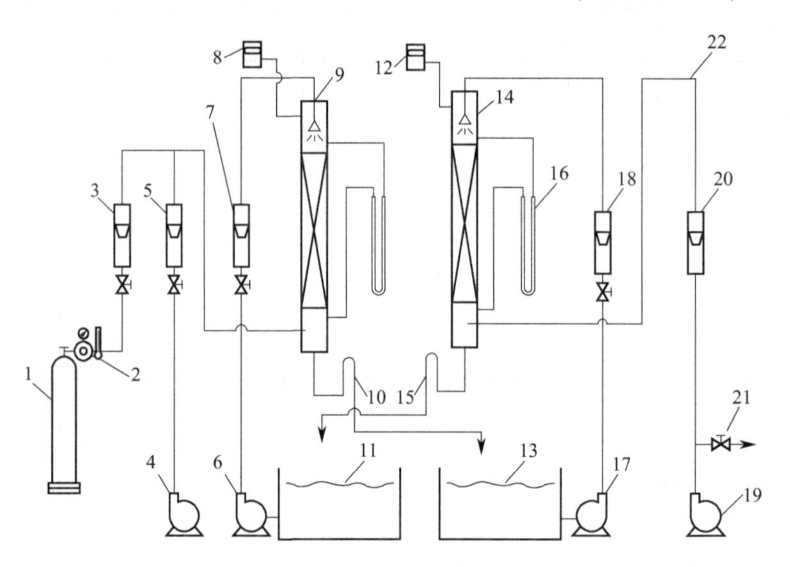

Figure 6-27 Diagram of the experimental device for carbon dioxide absorption and desorption

1—CO_2 cylinder; 2—Pressure relief valve; 3—CO_2 flowmeter; 4—Absorption fan; 5—Absorption tower air flowmeter; 6—Absorption water pump; 7—Absorption tower water flowmeter; 8—Absorption exhaust gas sensor; 9—Absorption tower; 10, 15—Liquid seal; 11—Desorption liquid tank; 12—Desorption exhaust gas sensor; 13—Absorption liquid tank; 14—Desorption tower; 16—Differential pressure meter; 17—Desorption pump; 18—Desorption tower water flowmeter; 19—Desorption fan; 20—Desorption tower air flowmeter; 21—Air bypass regulating valve; 22—π-shaped tube

pumps, an absorption tower, and a desorption tower. CO_2 is provided by the CO_2 cylinder 1. After decompression and flow adjustment, it is mixed with the regulated air from the absorption fan 4. The mixed gas enters the absorption tower 9, where the mass transfer of CO_2 from air to pure water occurs. Air is provided by the desorption fan 19 and regulated by the air bypass regulating valve 21 to desorb the absorption solution in the desorption tower.

6.7.3 Experiment projects

(1) Determination of performance curves for dry packing materials of the desorption tower

(2) Determination of performance curves for wet packing materials of the desorption tower

(3) Determination of the absorption mass transfer coefficient

6.7.4 Adjustable experimental conditions

The experimental conditions can be adjusted (Table 6-3) to investigate the effects of different parameters on the results.

Table 6-3 Adjustable conditions for absorption experiments

Packing height/m	0.65	0.78	0.90
Tower diameter/m	0.068	0.075	0.085
HCl concentration/(mol/L)	0.08	0.10	0.12

6.7.5 Precautions

It should be noted that the flowmeter readings from the absorption tower and desorption tower should be adjusted on a timely basis to ensure that their readings are consistent so that the operating conditions during the experiment remain unchanged.

6.8 Comprehensive experiment of rectification

6.8.1 Main interface of simulation

The main interface of simulation for comprehensive experiment of rectification is shown in Figure 6-28.

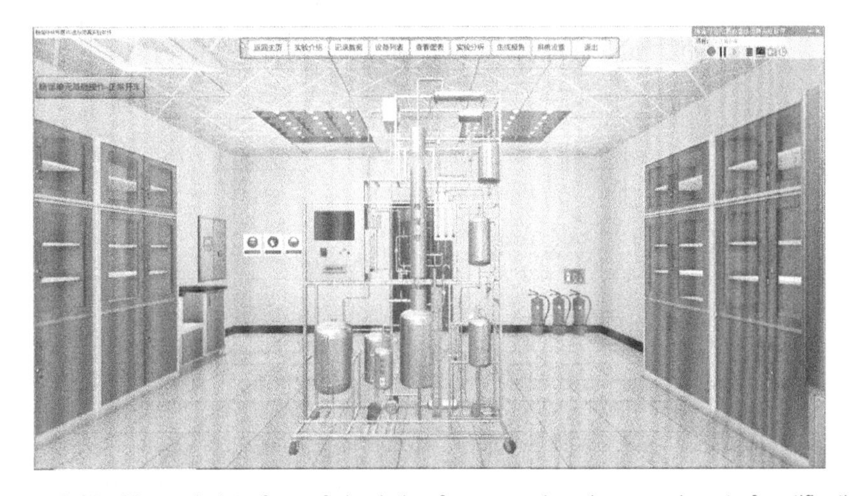

Figure 6-28 The main interface of simulation for comprehensive experiment of rectification

6.8.2 Experimental device for comprehensive experiment of rectification

The diagram of the experimental device for comprehensive experiment of rectification is shown in Figure 6-29. The names of main measurement points and operation control points are shown in Table 6-4.

The main components of the rectification process are a rectification tower, a reboiler, condensers, a reflux tank and a conveyer. After the feed solution at a certain temperature and pressure enters the rectification tower, the light components become gradually concentrated at the enriching section. After exiting the top of the tower, they are condensed and flow into the reflux tank, where a portion is collected as the top product (distillate), the rest is sent back to the reflux tank. The heavy components are concentrated in the stripping section. After that, a portion is collected as the tower-bottom product (residual liquid), the rest is returned to the tower after being heated by the reboiler.

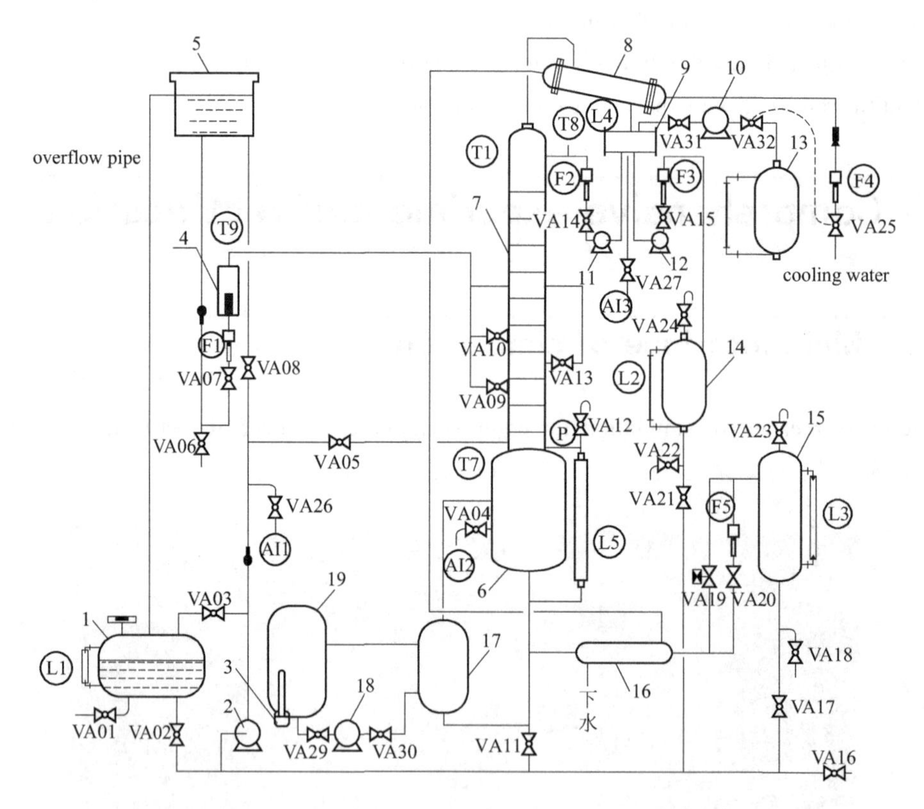

Figure 6-29 Diagram of the experimental device for comprehensive experiment of rectification

1—Storage tank; 2—Feed pump; 3—Heater; 4—Feed preheater; 5—Elevated tank; 6—Tower bottom; 7—Sieve plate rectification tower; 8—Condenser; 9—Reflux tank; 10—Vacuum pump; 11—Reflux pump; 12—Production pump; 13—Storage tank; 14—Tower-top product tank; 15—Tower-bottom product tank; 16—Tower-bottom condenser; 17—Reboiler; 18—Heat transfer oil pump; 19—Heat transfer oil tank

The control system of the rectification tower uses a DCS control system.

Table 6-4 Major measurement and operation control points

F1—Feed material flow rate	T7—Tower bottom temperature
F2—Reflux flow rate	T8—Reflux temperature
F3—Tower top product flow rate	T9—Feed temperature
F4—Coolant flow rate	L1—Liquid level of storage tank
F5—Tower bottom discharge flow rate	L2—Liquid level of the top product tank
AI1—Feed material concentration	L3—Liquid level of the bottom product tank
AI2—Tower bottom concentration	L4—Liquid level of the reflux tank
AI3—Tower top concentration	L5—Liquid level of tower bottom
T1—Tower top temperature	VA01—VA30 valve

6.8.3 Experiment projects

(1) Basic operation of rectification unit—guidance mode

(2) Basic operation of rectification unit—normal start-up

(3) Basic operation of rectification unit—normal shutdown

(4) Emergency management of abnormal situations and accidents—flooding

(5) Emergency management of abnormal conditions and accidents—entrainment

(6) Emergency management of abnormal situations and accidents—serious weeping

(7) Emergency management of abnormal situations and accidents—heat exchanger fouling

(8) Emergency management of abnormal conditions and accidents—cavitation

(9) Effect of changes in operating parameters of the atmospheric unit on the rectification process—reflux ratio

(10) Effect of changes in operating parameters of the atmospheric unit on the rectification process—feed temperature

(11) Effect of changes in operating parameters of the atmospheric unit on the rectification process—heating power of the heating transfer oil

(12) Effect of different pressures on the rectification process—normal pressure

(13) Effect of different pressures on the rectification process—pressurization

(14) Effect of different pressures on the rectification process—decompression

(15) Effect of equipment parameters on the rectification process—equipment parameters

(16) Effect of changes in the experimental system on the rectification processes—experimental system

6.8.4 Precautions

① Supply cooling water to the system before turning on the heating switch for the heat transfer oil. After the experiment is finished, turn off the heat and wait for the temperature at the top of the tower to drop below 70 °C before turning off the cooling water.

② Record the experimental data (temperature, pressure, liquid level and heating power) at certain times after the heating switch is turned on and observe the performance of the tower in real time.

③ During the operation, the heating power, tower temperature, bottom pressure, reflux rate, feed volume and discharge volume need to be continuously monitored to keep them in a steady state. Record data (temperature, pressure, liquid level, heating power and flow rate) at regular intervals. Take samples every 10 minutes to analyze the concentrations (top, bottom and feed) .

④ Maintain a constant liquid level in the reflux tank and record the reading of the reflux rotameter after the liquid level becomes constant.

6.9 Liquid-liquid extraction experiment

6.9.1 Main interface of simulation

The main interface of simulation for liquid-liquid extraction experiment is shown in Figure 6-30.

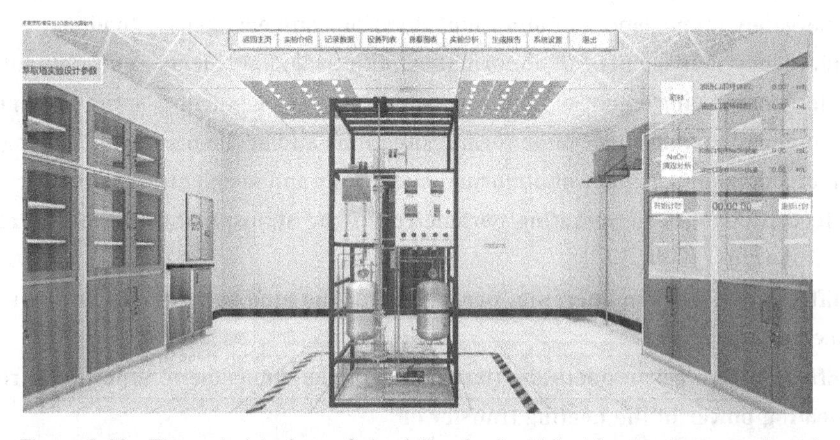

Figure 6-30 The main interface of simulation for liquid-liquid extraction experiment

6.9.2 Experimental device for liquid-liquid extraction

The diagram of the experimental device for liquid-liquid extraction is shown in Figure 6-31.

The experimental device for liquid-liquid extraction mainly consists of equipment and components such as an extraction tower, a raw material elevated tank, a water phase elevated tank, an extraction raffinate tank, a raw material delivery pump and a DC motor. During the operation of this device, the continuous phase (water) is injected into the extraction tower until the neck of its upper half, following which the dispersed phase (kerosene) is injected based on a phase ratio of 1 ∶ 1 (the corrected mass ratio of the extraction solvent to the raw material solution).

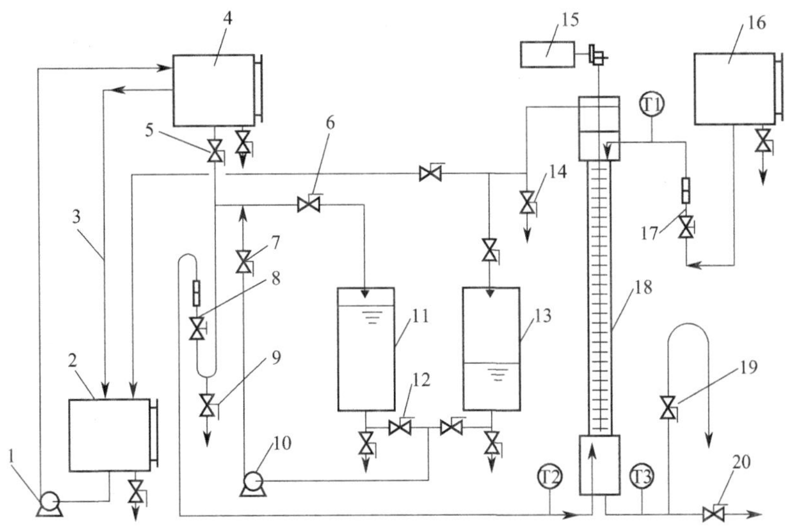

Figure 6-31 Diagram of the experimental device for liquid-liquid extraction

1—Raw material delivery pump Ⅰ; 2—Dosing tank; 3—Overflow pipe; 4—Raw material elevated tank; 5—Raw material inlet valve; 6—Bypass valve; 7—Pump outlet valve; 8—Raw material (oil) rotameter and valve; 9—Raw material (oil inlet) sampling valve; 10—Raw material delivery pump Ⅱ; 11—Raw material tank; 12—Pump inlet valve; 13—Extraction raffinate tank; 14—Extract (oil outlet) sampling valve; 15—DC motor; 16—Water phase elevated tank; 17—Water rotameter and valve; 18—Extraction tower; 19—π-shaped tube regulating valve; 20—Exhaust valve; T1—Water phase inlet thermometer; T2—Raw material inlet thermometer; T3—Extraction phase inlet thermometer

The dispersed phase moves upward in the tower, while the continuous phase flows downward. A counter-current contact mass transfer occurs between the two phases. The dispersed phase converges at the top of the tower and forms a light liquid layer with a certain thickness, forming a clear interface between the two phases in the upper part of the extraction tower. Through the regulating valve on the π-shaped tube at the continuous phase outlet, the two-phase interface is adjusted to the position in the middle of the continuous phase inlet and the water phase outlet. Kerosene is collected from the tower top and sent to the extraction raffinate tank and water discharges through the π-shaped tube at the bottom of the tower.

6.9.3 Experiment projects

Liquid-liquid extraction tower operation and determination of the height of an extraction mass transfer unit.

6.9.4 Adjustable experimental conditions

The experimental conditions can be adjusted (Table 6-5) to investigate the effects of different parameters on the experimental results.

Table 6-5 Adjustable conditions of the extraction experiment

Tower height/m	0.90	1.00	1.10
Tower diameter/m	0.025	0.030	0.035
Raw material oil concentration/(kg/kg)	0.00400	0.00425	0.00450
NaOH concentration/ (mol/L)	0.010	0.015	0.020
Sieve plate amplitude/mm	20.0	23.0	26.0
Dispersed phase	Kerosene	Water	
Continuous phase	Water	Kerosene	

6.9.5 Precautions

① When the position of the two-phase interface is adjusted using the regulating valve on the π-shaped tube, the water phase is not allowed to flow from the extraction tower into the extraction raffinate tank.

② Sampling and analysis must be carried out after the system becomes stable.

6.10 Tunnel drying experiment

6.10.1 Main interface of simulation

The main interface of simulation for tunnel drying experiment is shown in Figure 6-32.

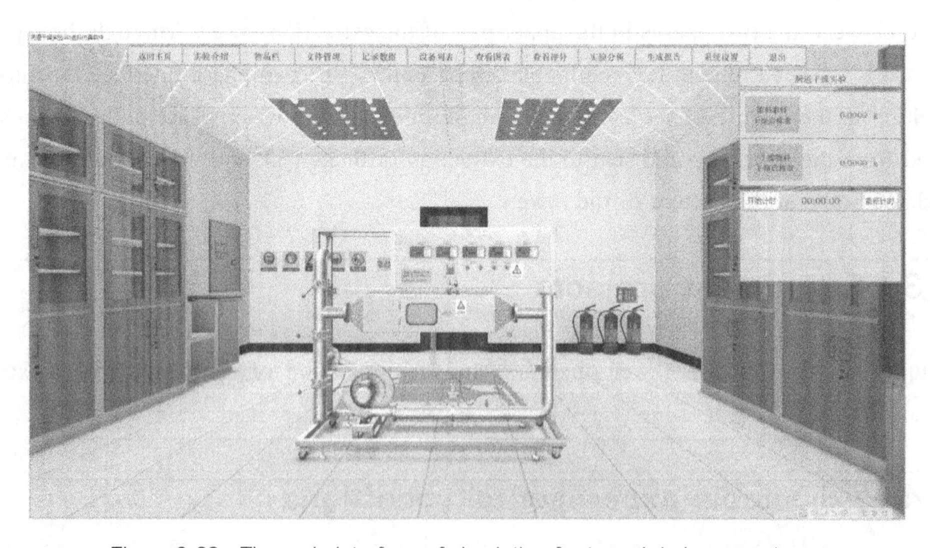

Figure 6-32 The main interface of simulation for tunnel drying experiment

6.10.2 Tunnel drying experimental device

The diagram of the tunnel drying experimental device is shown in Figure 6-33.

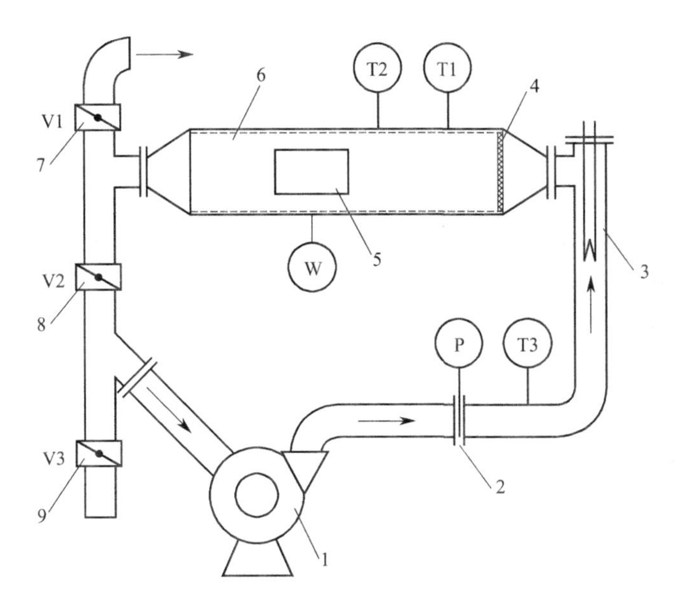

Figure 6-33 Diagram of the tunnel drying experimental device

1—Centrifugal fan; 2—Orifice flowmeter; 3—Preheater; 4—Air flow distributor; 5—Dryer door; 6—Tunnel dryer; 7—Fan outlet valve; 8—Circulation valve; 9—Fan inlet valve; T1—Dry bulb temperature; T2—Wet bulb temperature; T3—Air inlet temperature; P—Orifice pressure drop; W—Weight sensor display board; V1, V2, V3—Butterfly valve

Air is blown through the orifice flowmeter by the centrifugal fan for measurement, after that, it enters the preheater for heating. The heated air moves into the tunnel dryer and meets the wet material placed on the bracket of the weight sensor in the dryer for drying. The moisture on the surface of the wet material is transferred to the air and removed from the dryer. A portion of the exhaust gas is released through valve V1, and the rest is recycled through valve V2.

6.10.3 Experiment projects

Tunnel drying experiment.

6.10.4 Adjustable experimental conditions

The experimental conditions can be adjusted (Table 6-6) to investigate the effects of different experimental conditions on the experimental results.

Table 6-6 Adjustable conditions of the drying experiment

Length of drying frame/m	0.150	0.160	0.180
Width of drying frame/m	0.070	0.080	0.100
Weight of drying frame/g	79.5	88.6	93.8

6.10.5 Precautions

① It needs to be confirmed that air flows through the dryer at a constant flow rate before the heat switch is turned on to heat the air.

② After turning off the heat switch at the end of the experiment, wait until the dry bulb temperature drops to 45℃ to turn off the fan.

③ Remember to click "Dry and weigh the dried materials" after taking out the materials and to close the door at the end of the experiment.

Chapter 7

Demonstration Experiment of Chemical Engineering Principle

7.1 Reynolds experiment

7.1.1 Experimental objectives

① Understand the movement mode of fluid particles in a circular straight tube and the characteristics of different flow patterns; master the criteria for judging the flow patterns in the tube.

② Observe the flow pattern of the laminar flow turbulent flow and transition zone in the circular straight tube to determine the critical Reynolds number.

③ Observe the velocity distribution of laminar flow.

7.1.2 Experimental principles

In 1883, O. Reynolds of England observed that there are two different flow patterns of fluid in a circular tube, which including laminar flow (stagnation) and turbulent flow (turbulence). At the same time, the study found that the flow pattern of the fluid is related to the flow velocity u, viscosity μ, density ρ of the fluid and the diameter d of the tube through which the fluid flows, these four factors are expressed as Reynolds number as follows:

$$Re=\frac{du\rho}{\mu} \tag{7-1}$$

① $Re \leqslant 2000$, laminar flow, fluid particle motion is linear and parallel to each other. The velocity on the flow section in the tube is parabola distribution when laminar flow occurs.

② $Re \geqslant 4000$, there are irregular pulsations in other directions besides the main flow direction in turbulent flow.

③ $2000 < Re < 4000$, between the laminar flow and the turbulent flow, also known as the transition zone. In the transition zone, it may be laminar or turbulent, which is related to the external environment and the flow pattern is unstable.

7.1.3 Experimental device

(1) Experimental procedures

The Reynolds experimental device is shown in Figure 7-1.

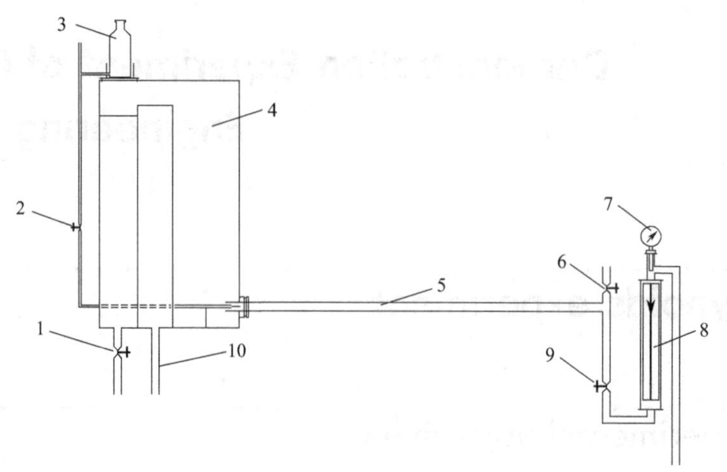

Figure 7-1 Flow chart of Reynolds experimental equipment

1—Water supply valve; 2—Regulating valve; 3—Tracer bottle;
4—High position water Tank; 5—Observation tube; 6—Exhaust valve;
7—Thermometer; 8—Flow meter; 9—Flow control valve; 10—Overflow pipe

(2) Main technical data of device

The effective length of the test pipe $L=1100$ mm, the outer diameter $D_o=30$mm, the inner diameter $D_i=24.2$mm.

7.1.4 Experimental procedures

(1) Preparations before the experiment

① Observe and properly adjust the position of the thin tube so that it is on the center line of the experimental observation tube before the experiment.

② Add the red ink diluted with water to the tracer bottle as the tracer for experiment.

③ Close the water flow regulating valve and exhaust valve. Open the water supply valve and drain valve, fill the high position water tank with water, make the water fill the water tank to produce overflow. Maintain a certain overflow flow to ensure the constant liquid level of the high position tank.

④ Gently open the water flow regulating valve to allow water to flow slowly through the experimental tube. Expel the bubbles from the red ink injection tube, fill the thin tube with red ink.

(2) Experimental procedures

① Adjust the water supply valve to maintain the minimum overflow flow based on the above preparations. Gently open the flow regulating valve and let the water flow slowly through the experimental pipeline (Figure 7-3).

② Slowly and controlledly open the red water flow regulating valve, then the red water stream will show different flow states. The red flow beam represents the current flow of water in the test tube (as shown in Figure 7-2) . Read the flow value and calculate the corresponding Reynolds number.

③ Vibration caused by water inflow and overflow may sometimes cause the red water stream in the experimental tube to deviate from the center line in the tube or swing to different degrees from left to the right. At this time, the water supply valve can be closed immediately, it can be stabilized for a period of time, and the red straight line that coincides with the center line of the tube can be seen in the test tube.

④ Gradually increase the opening of the upper water valve and the flow control valve, increase the flow of water in the experimental pipeline while maintaining the minimum overflow flow, then observe the flow of water in the experimental pipeline (Figure 7-3) .

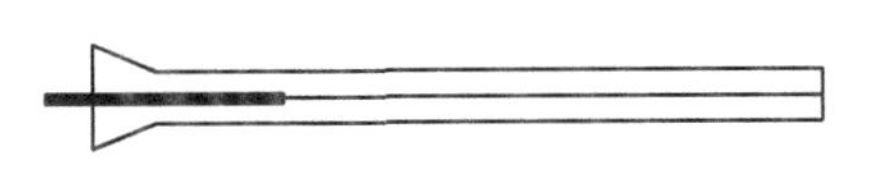

Figure 7-2 Diagram of laminar flow

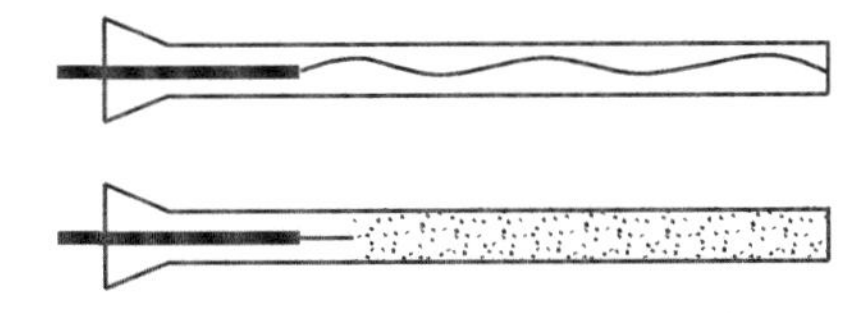

Figure 7-3 Diagram of transition flow and turbulent flow

(3) Demonstration experiment of fluid velocity distribution in a circular tube

① Open the water supply valve and close the water flow regulating valve.

② Open the red ink flow regulating valve and let a small amount of red ink to flow into the inlet end of the experimental pipeline.

③ Open the water flow regulating valve suddenly, the velocity distribution as shown in Figure 7-4 can be seen in the experimental pipeline.

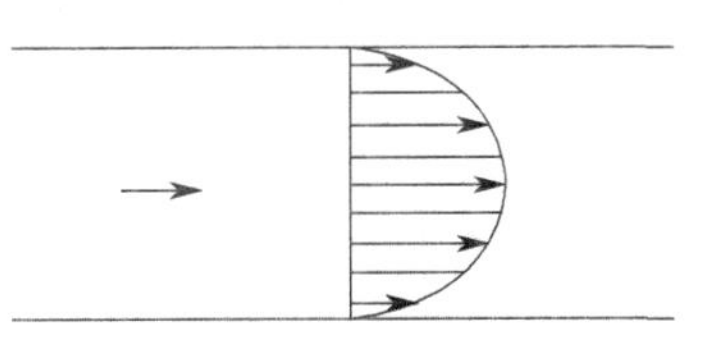

Figure 7-4 Diagram of velocity distribution

(4) End of experiment operation

① Close the red ink flow control valve to stop the flow of red ink.

② Close the water supply valve to stop the water flowing into the high-level tank.

③ Close the water flow control valve when the red color disappears after flushing the test tube.

④ If the device will not be used for a long time in the future, please keep all the water in the equipment clean.

7.1.5 Precautions

When demonstrating laminar flow, in order to make the stable laminar flow faster, please note the following points.

① The overflow flow of the water tank should be as small as possible, because the overflow is too large and the inflow flow is also large. The vibration caused by the inflow and overflow is relatively large, which will affect the experimental results.

② Try not to make the test stand vibrate artificially. In order to reduce the vibration and ensure the experimental effect, the bottom of the test stand can be fixed.

7.1.6 Questions

① If the red ink injection tube is not located in the center of the experimental pipeline, can the expected results of the experiment be obtained?

② How to calculate the Reynolds number under a certain flow? What is the criterion for judging the flow pattern by Reynolds number?

③ According to the experimental phenomena, explain the flow state of red ink in laminar and turbulent flow.

④ Imagine if the fluid is an ideal fluid, what should you see?

7.2 Demonstration experiment of energy conversion (fluid mechanical energy conversion)

7.2.1 Experimental objectives

① Observe and test the changes of kinetic energy, potential energy and static pressure energy when fluid flows through different tube diameters and positions at different flow rates.

② Master the relationship of energy conversion in fluid flow, and understand Bernoulli equation.

③ Understand the energy loss phenomenon when the fluid flows in the tube.

7.2.2 Experimental principles

The fluid has kinetic energy, potential energy and static pressure energy, which can be transformed into each other. In the flow process, some mechanical energy will be converted into heat energy due to friction and collision due to internal friction for the for actual fluids, this part of mechanical energy cannot be recovered in the process of flow. Therefore, for the actual fluid, the sum of mechanical energy on the upstream and downstream sections is not equal, the difference between them is energy (mechanical energy) loss.

Kinetic energy, potential energy and static pressure energy can be expressed by the height of the liquid column, which are called dynamic pressure head, potential pressure head and static pressure head. The difference of the sum of upper head, dynamic head and static head of any two sections is the head loss.

7.2.3 Experimental device

As shown in Figure 7-5, the test tube is connected by tubes of different diameters and heights. Several measuring points are selected at different positions of the test tube. Each measuring point is connected with two vertical measuring tubes, one of which is directly connected to the tube wall. Its liquid level height reflects the size of the static head at the measuring point, which is the static head measuring tube. Another piezometric tube opening is at the center of the tube, facing the water flow direction, and its liquid level height is the sum of static head and dynamic head, which is called punch head measuring tube. The measuring tube liquid level height can be read by the scale on the device. The water is returned to the water tank through the test tube from the high-level tank, the water in the tank is pumped to the high-level tank to ensure that the high-level tank is always in overflow state. The pipeline diagram of experimental test conduit is shown in Figure 7-6.

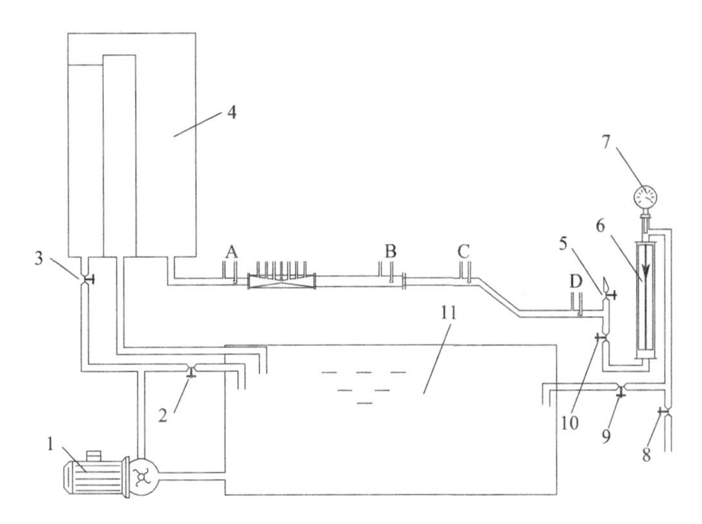

Figure 7-5 Diagram of energy conversion experiment flow

1—Centrifugal pump; 2—Circulating valve; 3—Water supply valve; 4—High position water tank; 5—Exhaust valve; 6—Flowmeter; 7—Thermometer; 8—Drain valve; 9—Backwater valve; 10—Flow control valve; 11—Water tank

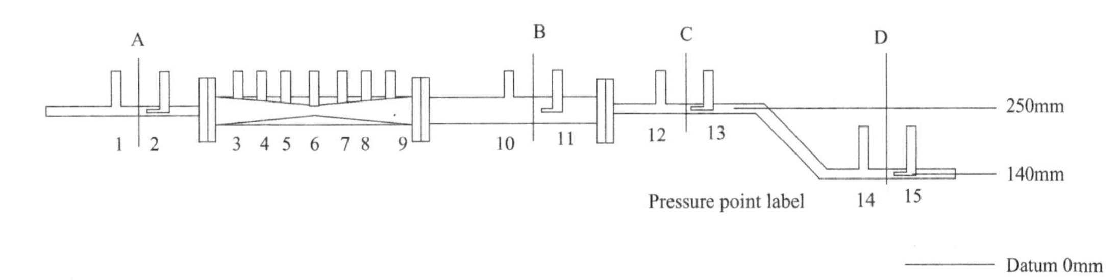

Figure 7-6 Pipeline diagram of experimental test conduit

7.2.4 Experimental procedures

① Add 3/4 volume of distilled water to the water tank，close the water supply valve，circulating water valve at the outlet of centrifugal pump，flow regulating valve，exhaust valve and drain valve at the outlet of experimental test conduit，open the return valve and start the centrifugal pump.

② Open the circulating water valve，fully open the outlet flow control valve of the pipeline，then gradually open the water outlet valve of the centrifugal pump to the overflow of the high-level tank overflow tube. Observe and read the level of each measuring tube after the flow is stable.

③ Gradually close the small flow control valve to change the flow rate，then observe the changes in the liquid level of each measuring tube at the same measurement point and different measurement points.

④ After closing the water supply valve，circulating water valve，outlet flow regulating valve and return valve of the centrifugal pump，the centrifugal pump is closed and the experiment is finished.

7.2.5 Precautions

① Do not open the water valve at the outlet of the centrifugal pump too much，to avoid the water flowing out or causing the liquid level of the high tank to be unstable.

② When the water flow increases，check whether the water surface in the high-level tank is stable. When the water surface drops，properly open the water supply valve to make up the water volume.

③ The flow control valve should be closed slowly when it is small，so as to avoid the flow rate to suddenly drop and cause the water in the measuring tube to overflow the tube.

④ The bubbles in the test tube and the measuring tube must be discharged when there are bubbles in the system.

7.2.6 Questions

① What energy is involved when fluid flows in a tube?

② How to measure the static head and total head on a certain section? How to measure the dynamic head on a certain section?

③ Is the static pressure difference between the two sections only caused by flow resistance if the height of the two pressure measuring surfaces is different from the reference surface?

④ Observe and compare the relative magnitude of each mechanical energy value at each pressure measurement point，then draw a conclusion.

7.3 Streamline demonstration experiment

7.3.1 Experimental objectives

① Students can further understand the trajectory of fluid flow and the basic characteristics of streamlines through demonstration experiments.

② Observe the flow phenomenon of liquid flowing through different solid boundaries, the region and shape of vortex, and enhance the perceptual understanding of fluid flow characteristics.

7.3.2 Experimental principles

When the actual fluid flows along the solid wall, due to the viscous effect, the friction between the stationary fluid layer and it's adjacent fluid layer will produce, which will slow down the flow speed of the adjacent fluid layer. Therefore, the velocity gradient du/dy is generated in the universal direction perpendicular to the fluid flow, the fluid layer with velocity gradient is called boundary layer.

Great attention is paid to the study of the boundary layer in the chemical engineering discipline. The significance of the concept of the boundary layer is to pay attention to the changes in the boundary layer when studying the flow of real fluid along the solid wall, which will directly affect the momentum transfer, energy transfer and mass transfer.

When the fluid flows through the curved surface, or the cross-section size or flow direction of the fluid changes, if the reverse pressure gradient dp/dx occurs at this time, the fluid boundary layer will be separated from the wall and form a vortex (or eddy current), which will intensify the collision between the particles of the fluid and cause the loss of fluid energy. The separation of the boundary layer from a solid wall is called separation or debonding of the boundary layer. From this, we can find the reason for the energy consumption of the fluid in the flow process. At the same time, the turbulent motion of the fluid micro clusters caused by the vortex and their collision and mixing will greatly enhance the transfer process. Therefore, the practical significance of fluid streamline research is that it can analyze and study the existing flow process and equipment, then strengthen the transfer process, provide a theoretical basis for the development of new high-efficiency equipment, guide the selection of appropriate operating control conditions.

This demonstration experiment uses the bubble tracing method, which can display the flow images of the fluid flow lines, the boundary layer separation phenomenon, the area and strength of the vortex when the fluid flows through the solids of different geometric shapes.

7.3.3 Experimental device

The flow of the experimental device, the shape of the demonstration board and the flow around the demonstration board are shown in Figure 7-7, Figure 7-8 and Figure 7-9 respectively. The water in the water tank is fed into the demonstrator by a centrifugal pump, and then returned to the water tank through the overflow device of the demonstrator. Water flows through the slit-type flow channel, a variety of water flow phenomena under different shape boundaries are demonstrated by the method of incorporating bubbles in the water flow, the corresponding flow lines are displayed. Each demonstrator in the device can be used as an independent unit or at the same time. To facilitate observation, the demonstrator is made of organic glass.

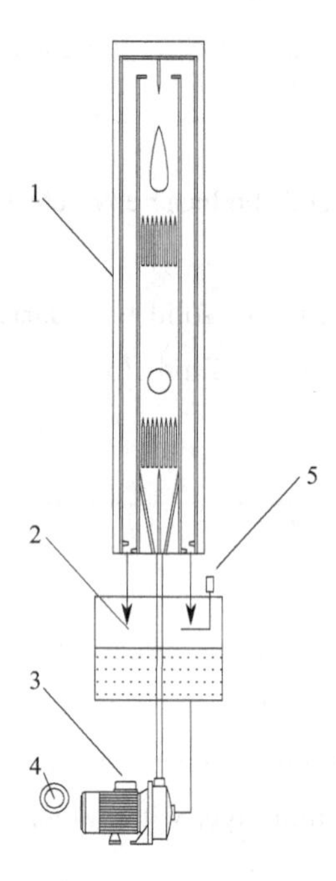

Figure 7-7 Device diagram of streamline demonstration experiment

1—Experimental panel; 2—Water tank; 3—Water pump; 4—Pressure control knob; 5—Aeration knob

Several flow demonstrators are described as follows:

Type (a): The fluid path with bubbles first flows through the gradually expanding, steady flow, single cylinder flow, steady flow, streamline body flow, right-angled bend, then flows into the circulating water tank.

Type (b): The fluid with bubbles flows into the circulating water tank after gradually narrowing, steady flow, rotameter and right-angle bend.

Type (c): The fluid with bubbles gradually flows into the circulating water tank after gradually expanding, steady flow, orifice flowmeter, steady

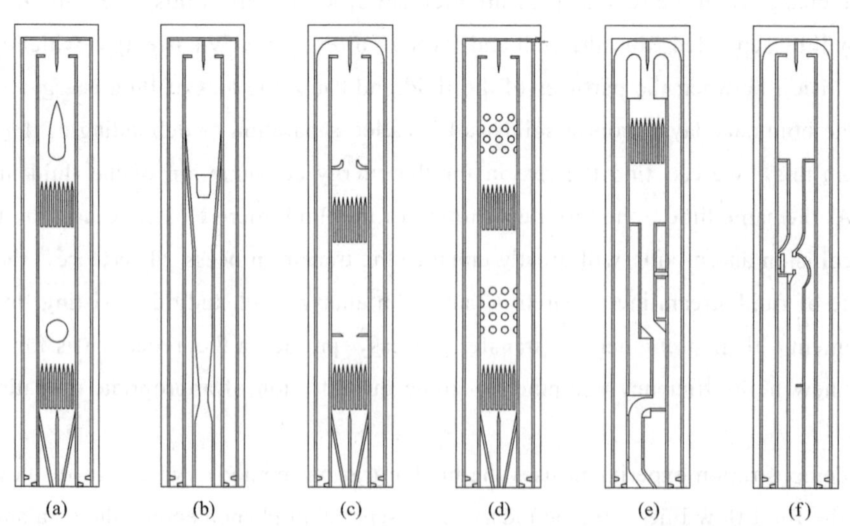

Figure 7-8 Outline drawing of the demonstration board

flow, nozzle flowmeter and right-angle bend.

Type (d): The fluid with bubbles gradually flows into the circulating water tank after gradually expanding, steady flow, multi-cylindrical flow (downstream), steady flow, multi-cylindrical flow (staggered flow), right-angled bend.

Type (e): The fluid with bubbles flows into the circulating water tank after passing through 45° angle bend, circular arc bend, right angle bend, sudden expansion, steady flow and sudden reduction.

Type (f): The fluid with bubbles passes through the shape of the valve, suddenly expands and flows into the circulating water tank after being bent at right angles.

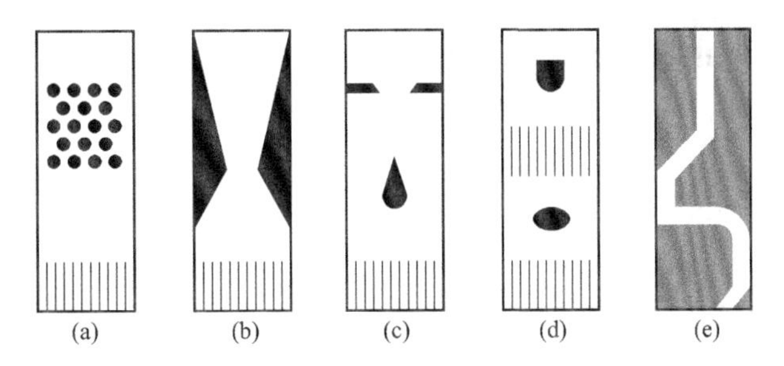

Figure 7-9 Circumferential flow demonstration board

(a) Shell and tube heat transfer simulation; (b) Venturi simulation; (c) Streamline body and orifice simulation; (d) simulation of circular and linear endings body; (e) Corner simulation

7.3.4 Experimental procedures

① Before the experiment, turn on the water adding switch and fill the water tank to 2/3.

② Turn on the speed control knob to make the drain on both sides of the display surface full of water at the maximum flow rate.

③ Adjust the amount of aeration to the best state and observe the experimental phenomenon.

④ The streamline demonstrations of several other experimental devices are operated according to the above requirements, then carefully observe the different streamlines.

⑤ Turn off the speed control knob and cut off the main power supply at the end of the experiment.

7.3.5 Questions

① Why should avoid vortices when transporting fluids?

② Why should proper vortices be formed in the process of heat and mass transfer?

③ Where does the boundary layer separation occur when the fluid flows around the cylinder? Is the separation point the same at different flow rates? What is the flow state of the fluid after the boundary layer is separated?

④ What is the difference between the flow state of the fluid through sudden expansion, sudden shrinkage and tapering and expanding?

⑤ What is the difference between the flow state of multi-cylinder in-line flow and multi-cylinder staggered flow? What are the implications for the arrangement of tube bundles of tube heat exchangers?

7.4 Demonstration experiment of heterogeneous gas-solid separation

7.4.1 Experimental objectives

① Understand the structure, characteristics and working principle of the settling chamber, cyclone separator and bag filter.

② Measure the static pressure distribution in the cyclone separator, recognize the necessity of a good seal between the dust outlet and the dust collection chamber.

③ Measure the influence of the inlet gas velocity on the separation performance of the cyclone separator, understand the calculation method of suitable operation gas velocity.

7.4.2 Experimental principles

(1) Dedusting principle of gravity settling

The process of gravity settling is that the dusty gas enters the gravity sedimentation equipment in the horizontal direction. Under the action of gravity, the dust particles gradually settle to the bottom of the equipment, while the gas continues to move forward along the horizontal direction and flows out of the settling equipment, so as to achieve the purpose of dust removal, which belongs to the coarse dust removal. The common gravity settling equipment used to separate gas-solid mixture has a settling tank.

In the gravity dedusting equipment, the lower the speed of gas flow, the more conducive to the deposition and separation of dust with small particle size, and improve the efficiency of dedusting. Therefore, the general control gas flow speed is 1～2m/s, and the dust removal efficiency is 40% ～ 60%. But the gas speed is too low, the equipment is relatively large, the investment cost is high. When the gas flow rate is basically fixed, the longer the gravity settling equipment is designed, the better the dust removal efficiency will be.

(2) Working principle of cyclone separator

The cyclone separator is shown in Figure 7-10. The dusty gas enters along the tangent direction from the air inlet on the cylinder part of the separator. It makes downward spiral movement under the restriction of the gas wall. The gas and dust particles are affected by centrifugal force at the same time in the cyclone separator. Because the density of dust particles is much higher than that of gas, the centrifugal force of dust particles is much greater than that of gas. Under the action of the centrifugal force, the dust particles rotate outward while performing radial movement outwards. As a result, the dust particles generate radial and centrifugal settling movements toward the walls. When they move to the wall, they lose kinetic energy and fall down and separate from the gas. Then, under the action of air friction and gravity, they make downward spiral movement along the wall surface, and finally fall into the ash outlet at the bottom of the cone. The dust gas is gradually purified in downward spiral motion. When reaching the conical part of the separator, the purified air flow changes from downward spiral movement in the range of space near the wall of the separator to upward spiral movement in the range of space near the central axis, finally it is discharged by the exhaust tube at the top of the separator. The downward spiral is outside, the upward spiral is inside, but the rotation direction of the two is the same. The upper part of downward spiral flow is the main dedusting area. The spiral track seen in the demonstration experiment is that the powder particles which have been thrown onto the wall of the device are swept down by the downward spiral air flow to move downward spirally on the surface of the device wall.

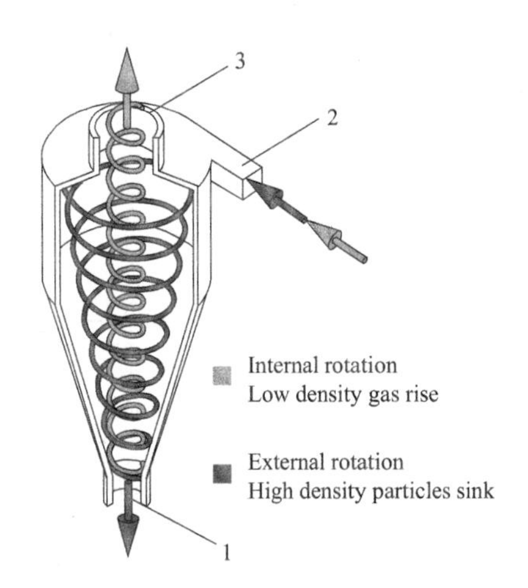

Figure 7-10 Diagram of cyclone separator

1—Ash outlet; 2—Air inlet; 3—Exhaust port

The pressure in the cyclone decreases gradually from the highest static pressure near the wall to the central static pressure. This is because the downward and upward spirals rotate in the same direction and the gas is pushed out by the inertial centrifugal force. One consequence of this static pressure intensity distribution is, part of the gas continuously flows from the downward swirl with higher pressure into the upward swirl with lower pressure in the radial direction, which is driven by the pressure difference. Therefore, the gas has a velocity in three directions at any position in the device, including tangential velocity, radial velocity and axial velocity.

From the wall to the center, the tangential velocity of the gas increases first and then decreases. The air flow within the circle with the maximum tangential velocity is called "air core", which has the following characteristics: a. The rising axial speed is quite large. b. The static pressure in the air core can be as small as the outlet pressure of the exhaust tube and the local atmospheric pressure. c. The low-pressure air core usually extends from the lower end of the exhaust pipe to the ash outlet of the cone bottom. Therefore, the dust outlet and the dust collection chamber under it should be well sealed, otherwise it is easy to leak into the air. The dust collected at the bottom of the cone is blown again, which seriously reduces the separation effect.

(3) Working principle of bag filter

As shown in Figure 7-11, after the dusty gas enters the dust remover from the inlet 2, the gas turns upward and enters the box. The dust is blocked on the outer surface of the cloth bag when passing through the cloth bag with metal framework inside. The purified gas enters the clean air chamber at the upper part of the cloth bag and is discharged from the clean air outlet.

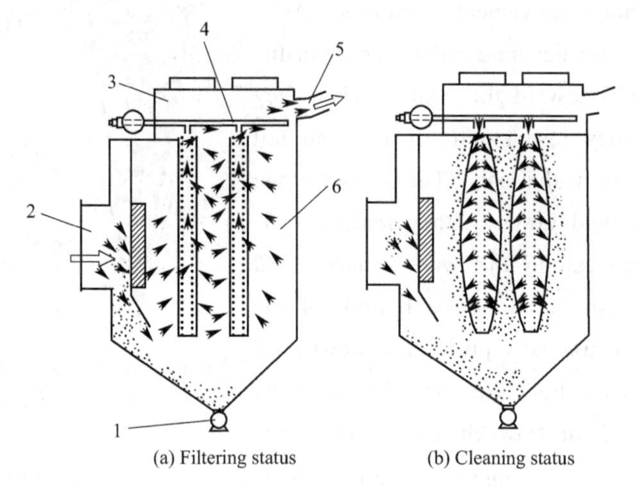

Figure 7-11 Diagram of bag filter

1—Ash outlet; 2—Dust-containing gas inlet; 3—Clean room; 4—Blowpipe; 5—Net gas outlet; 6—Casing

7.4.3 Experimental device

The heterogeneous gas-solid separation device is shown in Figure 7-12. The experimental

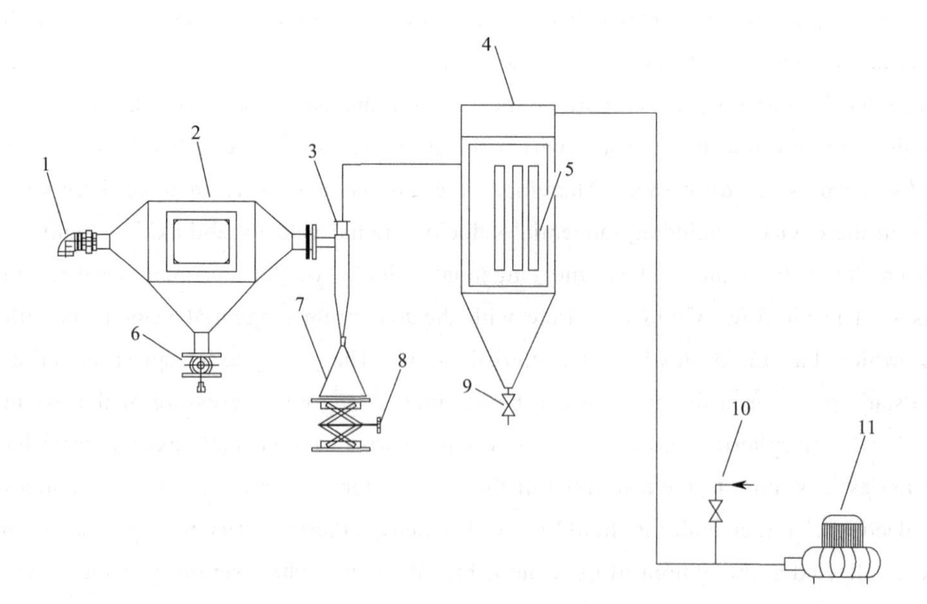

Figure 7-12 Diagram of the heterogeneous gas-solid separation device

1—Inlet; 2—Settling chamber; 3—Cyclone separator; 4—Bag filter; 5—Cloth bag; 6—Discharge valve; 7—Collection bottle; 8—Lifts; 9—Discharge valve; 10—Bypass valve; 11—Fan

device consists of a gravity settling chamber, cyclone separator, bag filter and fan. After the dusty gas passes through the gravity settling chamber, cyclone separator and bag filter, the particles are separated from the gas, then the purified gas is discharged into the atmosphere.

7.4.4 Experimental procedures

① Fully open the fan bypass valve 10. Turn on the power switch of the blower.

② Gradually close the bypass valve 10 and increase the air suction through the settling chamber and cyclone separator to understand the changing trend of gas flow.

③ Close the air flow regulating valve. Pour the solid materials (corn flour, washing powder, etc.) which used in the experiment into the feed container close to the material inlet, observe and analyze the dusty gas and the movement of the dust particles and gas in the separator. In order to observe the above situation continuously for a long period of time, the container can be gently moved by hand to push the dust particles to be continuously added. Although what the observer actually sees is the trajectory of the dust particles, the downward spiral movement of the dust particle wall is caused by the air flow. Therefore, it can be inferred from this that the flow path of the dusty gas flow and gas.

④ At the end of the experiment, first fully open the bypass valve 10, then cut off the power switch of the blower. If it is not used for a long time in the future, take out the solid powder from the dust collection room after parking.

7.4.5 Precautions

① During start-up and shut-down, the bypass valve shall be fully opened first, then the power switch of the blower shall be turned on or off.

② The connection between the ash discharge tube of the cyclone separator and the settling chamber should be relatively tight, so as to prevent the separated dust particles from being blown up and taken away by the air leaking into the negative pressure inside.

③ In the experiment, if the gas flow rate is small enough, the solid powder is wet, the solid powder will stick to the wall along the downward spiral motion path. If you want to remove the powder particles attached to the wall, you can use the high-speed rotating new powder particles separated from the dusty gas at the large flow rate to wash away the powder particles originally attached to the wall.

7.4.6 Questions

① Is the settling velocity constant in the process of particles settling towards the inner diameter of cyclone separator?

② What is the difference between centrifugal settling and gravity settling?

③ What are the main indicators for evaluating cyclones? What are the factors that affect its performance?

④ What is the difference between the separation principle of the settling chamber, cyclone separator and bag filter?

⑤ The sequence of three kinds of separation equipment in the experimental device is settling chamber, cyclone separator and bag filter. Why?

⑥ What is the difference between the size and quantity of solid particles retained by the settling chamber, cyclone separator and bag filter after the air volume is increased?

7.5 Demonstration experiment of hydrodynamic performance of plate column

7.5.1 Experimental objectives

① Understand the basic structure of tower equipment and tray (screen hole, float valve, bubble cap, tongue shape) through experiment.

② Observe the flow and contact conditions of gas and liquid phases on different types of trays. Observe the normal and abnormal operation phenomena in the experimental tower. Master the measurement method of pressure drop on the tray, and compare the hydrodynamic performance of different trays.

7.5.2 Experimental principles

Plate tower is an important gas-liquid mass transfer equipment, which is widely used in distillation and absorption operations. The tray is the core component of the plate tower, which determines the basic performance of the tower. In order to effectively realize the material and heat transfer between the gas and liquid phases, the tray requires the following two conditions: Good gas-liquid contact conditions must be created to form a larger contact area. The contact surface should be continuously updated to increase the driving force of mass transfer and heat transfer. The whole tower should ensure gas-liquid counter-current flow as a whole to avoid back-mixing and gas-liquid short circuit.

The tower achieves the purpose of mass transfer and heat transfer by contacting the bottom-up gas and top-down liquid when they flow on the tray. Therefore, the performance of mass and heat transfer of the tray mainly depends on the hydrodynamics of gas and liquid on the tray.

(1) Gas-liquid two-phase contact state on the tray

The gas and liquid phases are in bubbling contact state when the gas velocity is low. There is an obvious clear liquid layer on the tray. The gas is dispersed in the clear liquid layer in the form of bubble, the gas and liquid are transferred on the bubble surface. The gas and liquid phases are in foam contact when the gas velocity is high. At this time, the clear liquid layer on the tray is obviously thinned, the clear liquid can only be seen on the surface of the tray. The clear liquid layer decreases with the increase of the gas velocity. There is a large amount of foam on the tray, the liquid is mainly renewed The membrane morphology exists between very dense foams, the gas and liquid phases transfer mass on the surface of the liquid membrane. When the gas velocity is very high, the gas and liquid phases are in-jet contact, the liquid is dispersed in the gas phase in the form of continuously updated droplets, the gas and liquid phases transfer mass on the droplet surface.

(2) Abnormal flow on the tray

The normal gas-liquid load shall be maintained in the tower to avoid the following abnormal operation conditions during the operation of plate tower.

① Serious liquid leakage: When the rising gas speed is very low, the dynamic pressure of the gas through the riser of the tray is not enough to prevent the liquid layer on the tray from falling. And a large amount of liquid will leak down from the opening of the tray, which will cause serious liquid leakage.

② Entrainment: When the rising gas passes through the liquid layer of the tray, the liquid drops on the tray will be carried to the upper tray and resulting in the back-mixing of the liquid phase, which is called entrainment.

③ Flooding: The flow rate of one of the gas-liquid two phases in the tower increases, causing the liquid in the downcomer to flow down smoothly and the liquid in the downcomer accumulates. When the liquid in the tube crosses the top of the overflow weir, the liquid between the two plates is connected and rises in turn. This phenomenon is called flooding, also known as a flooded tower. At this time, the pressure drop of the tray rises, the whole tower operation is destroyed.

Therefore, the design of tray should strive to be simple in structure, good in mass transfer, large in gas-liquid capacity, low in pressure and flexible in operation.

7.5.3 Experimental device

The flow of the experimental device is shown in Figure 7-13. The flow includes four towers, which namely bubble tower, sieve plate tower, tongue tower, float valve tower, the four towers are connected in parallel. The air is measured by the vortex air pump through the orifice plate flowmeter and then sent to the bottom of each plate tower. It flows up through the plate and out from the top of the tower. The liquid is metered by the centrifugal pump through the rotameter, then enters the tower from the top and contacts with the air and flows back to the water tank from the bottom of the tower.

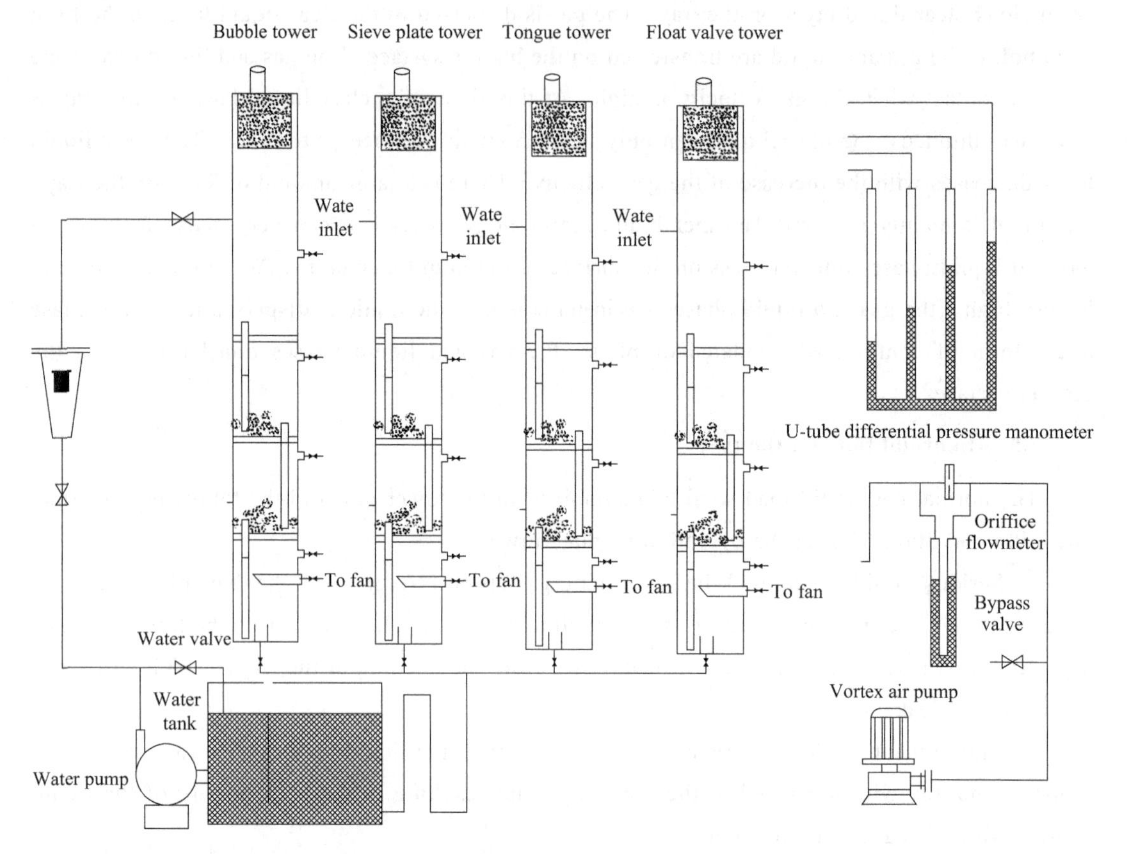

Figure 7-13 Diagram of plate tower test device

7.5.4 Experimental procedures

① Fill the water tank with distilled water, place the air flow adjustment valve in the fully open position, close the centrifugal pump flow adjustment valve.

② Turn on the vortex air pump and feed air into the tower. Turn on the centrifugal pump to deliver liquid to the tower, change the gas-liquid flow. Observe the gas-liquid flow and contact on the tray. Rrecord the tower pressure drop, air flow and liquid flow.

③ The pressure drop and gas-liquid flow and contact conditions of other towers were measured and observed by the same method.

④ After the experiment is completed, the regulating valve and the centrifugal pump are closed first, the vortex air pump is turned off when most of the liquid in the tower flows to the bottom of the tower to prevent water from entering the equipment and pipes.

7.5.5 Precautions

① Distilled water must be used in the experiment to protect the transparency of the organic glass tower.

② When driving, start the vortex air pump first, then start the centrifugal pump and vice versa. Shut down the other way, to prevent the liquid in the plate tower from pouring into the fan.

③ When changing the air flow or water flow during the experiment, you must observe the phenomenon and record the data after it stabilizes.

④ If the indicating liquid level of U-tube differential pressure gauge is too high, remove the pressure tube and suck out the indicating liquid with the ear ball.

⑤ The water tank must be filled with water, otherwise, the air pressure will be too high, it will be easy to short circuit and overflow from the water tank.

7.5.6 Questions

① Is the mass transfer area of the gas and liquid phases fixed in the plate tower?

② What are the indicators to evaluate the performance of the tray? Discuss the advantages and disadvantages of the four types of trays, such as sieve tray, float valve, blister and tongue tray.

③ The greater the degree of two-phase turbulence in the flow process, the lower the mass transfer resistance, which can be seen from the mass transfer theory. How to increase the degree of turbulence in two phases? Is the increase of turbulence unrestricted?

④ Qualitative analysis the affecting factors flooding.

Chapter 8

Chemical Engineering Principle "Three Type" Experiment

8.1 Fluid flow resistance experiment

8.1.1 Experimental objective

① Learn the engineering significance of determining friction coefficients and local resistance coefficients.

② Master how to measure the friction loss (Δp_f), friction coefficient (λ) and local resistance coefficient (ζ) and how to determine their relationships with Reynolds number (Re).

③ Learn how to use the logarithmic coordinate system.

④ Master the use of inverted U-tube differential nanometers and rotameters.

⑤ Understand the role of pipe fittings and valves in pipelines.

8.1.2 Experimental principles

The fluid will consume an amount of mechanical energy, because of the viscidity when it flows in the pipe. A pipeline is composed of straight pipe and pipe fittings (such as three-links, elbow, valve), etc. The straight pipe resistance is the loss of mechanical energy caused by the flow of fluid in a straight pipe. The local resistance is the loss of mechanical energy caused by the sudden changes in the flow path when the fluid flows through the local components.

(1) Measurement of friction resistance loss (Δp_f) and friction coefficient (λ) in a straight pipe

According to the basic theory of fluid mechanics, when fluid flows through a straight tube, the relationship between friction coefficient and friction resistance is consistent with the Fanning formula, whether in laminar or turbulent.

$$\Delta p_f=\lambda \frac{l}{d}\frac{\rho u^2}{2} \tag{8-1}$$

In a horizontal straight pipe with equal diameter, if there is no fluid transporting machine to do work, the friction resistance (Δp_f) is equal to the pressure drop in the pipe (p_1-p_2).

$$\Delta p_f=-\Delta p=p_1-p_2 \tag{8-2}$$

The pressure drop (Δp_f) in a horizontal cylindrical pipe can be measured at different Reynolds numbers. Then the friction coefficient (λ) can be calculated according to formula(8-1), finally the relationships between friction coefficient (λ) and Reynolds number (Re) can be obtained.

The theoretical formula, Hagen-Poiseuille equation, can give the friction resistance in laminar flow in a cylindrical pipe.

$$\Delta p_f = \frac{32\mu l u}{d^2} \tag{8-3}$$

where Δp_f is friction resistance, Pa; μ is fluid viscosity, Pa • s; l is pipe lengths, m; u is flow velocity, m/s; d is pipe diameter, m.

The relationships between λ and Re can be obtained under laminar flow by formulas (8-1) and (8-3):

$$\lambda = \frac{64}{Re} \tag{8-4}$$

Although the relationship between friction coefficient and Reynolds number is not theorized under turbulent flow due to the complexity, it can be built by the Dimensional Analysis Method The dimensional analysis shows that λ is a function of the Reynolds number (Re) and the relative roughness (ε/d) .

$$\lambda = \phi(Re, \varepsilon/d) \tag{8-5}$$

According to the experimental data, we can draw curves for relationships between λ and Re in laminar flow or turbulent flow. And the experimental results can be compared with the results calculated by known equations.

(2) Measurement of local resistance ($\Delta p_f'$) and coefficient (ζ)

The local resistance can be calculated based on the local resistance coefficient caused by the fluid flowing through the local parts such as the pipe fittings and valves.

$$\Delta p_f' = \zeta \frac{\rho u^2}{2} \tag{8-6}$$

where ζ is local resistance coefficient.

According to the experimental data and formula (8-6), we can calculate the local resistance coefficient (ζ) . However, the flow of the fluid before and after the pipe is unstable, the pressure measuring port cannot be placed close to the pipe, we can measure the resistance at a certain distance from the pipe fittings to avoid the instability. Therefore, the pressure measuring parts should be set as shown in Figure 8-1. The fluid flow resistance between a and a', or b and b' include the local resistance of the pipe and the flow resistance of the straight pipe.

The distances between different measuring locations were set as ab=bc and $a'b'=b'c'$.

Then $$\Delta p_{f,ab} = \Delta p_{f,bc};\ \Delta p_{f,a'b'} = \Delta p_{f,b'c'}$$

The Bernoulli equation between a and a' is as follows:

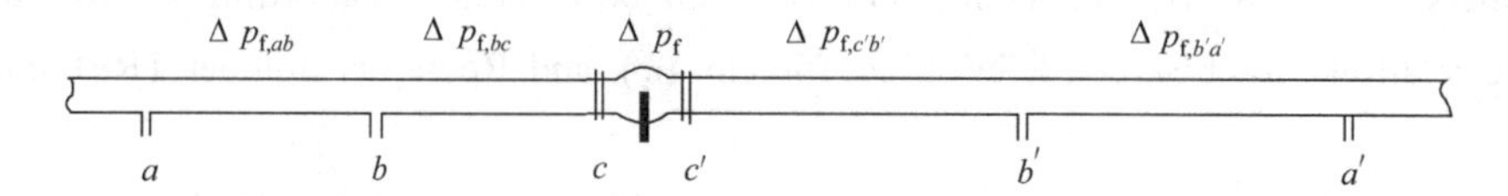

Figure 8-1 layout of measurement of local resistant pressure

$$p_a - p_{a'} = 2\Delta p_{f,ab} + 2\Delta p_{f,a'b'} + \Delta p_f' \tag{8-7}$$

The Bernoulli equation between b and b′ is as follows:

$$p_b - p_{b'} = \Delta p_{f,bc} + \Delta p_{f,b'c'} + \Delta p_f' = \Delta p_{f,ab} + \Delta p_{f,a'b'} + \Delta p_f' \tag{8-8}$$

Combining the formula (8-7) and formula (8-8) into the following equation, then

$$\Delta p_f' = 2(p_b - p_{b'}) - (p_a - p_{a'}) \tag{8-9}$$

where $(p_b - p_{b'})$ is named as proximal pressure difference, $(p_a - p_{a'})$ is named as distal pressure difference. They can be measured by a differential pressure sensor or inverted U-tube differential pressure gauge.

8.1.3 Experimental device

Main technical data of equipment are as follows.

① Straight pipe

Smooth pipe: ϕ10mm×1mm, pipe length l=1.70m, material stainless steel;

Rough pipe: ϕ12mm×1mm, pipe length l=1.70m, material stainless steel;

Local resistance pipe: ϕ25mm×1.5mm, material stainless steel ball valve.

② Glass rotameter:

Model LZB-25, Range of measurement 100～1000L/h;

Model VA10-15F, Range of measurement 10～100L/h.

③ Differential pressure transducer:

Model SM9320DP, Differential pressure range 0-200kPa.

④ Inverted U-Tube: Differential pressure range 0～900mmH_2O.

⑤ Centrifugal pump:

Model WB70/0.55, Rated flow 1.2～7.2m^3/h, Rated head 15～20.5m; Power 220V/380V, 0.55kW.

8.1.4 Experimental procedures and precautions

8.1.4.1 Experimental procedures

① Check the water level of the tank and take corresponding measures.

② Switch on the power.

③ Open the valves of 4, 10, 12, 13, 14, 16, 17, 19, 20, 21, 22, 23, 30, 31, 33, 34, 35 and close the other valves, then start the centrifugal pump.

④ Slowly open the valves 6 and 8, drive away bubbles in the pipeline under high flow rate.

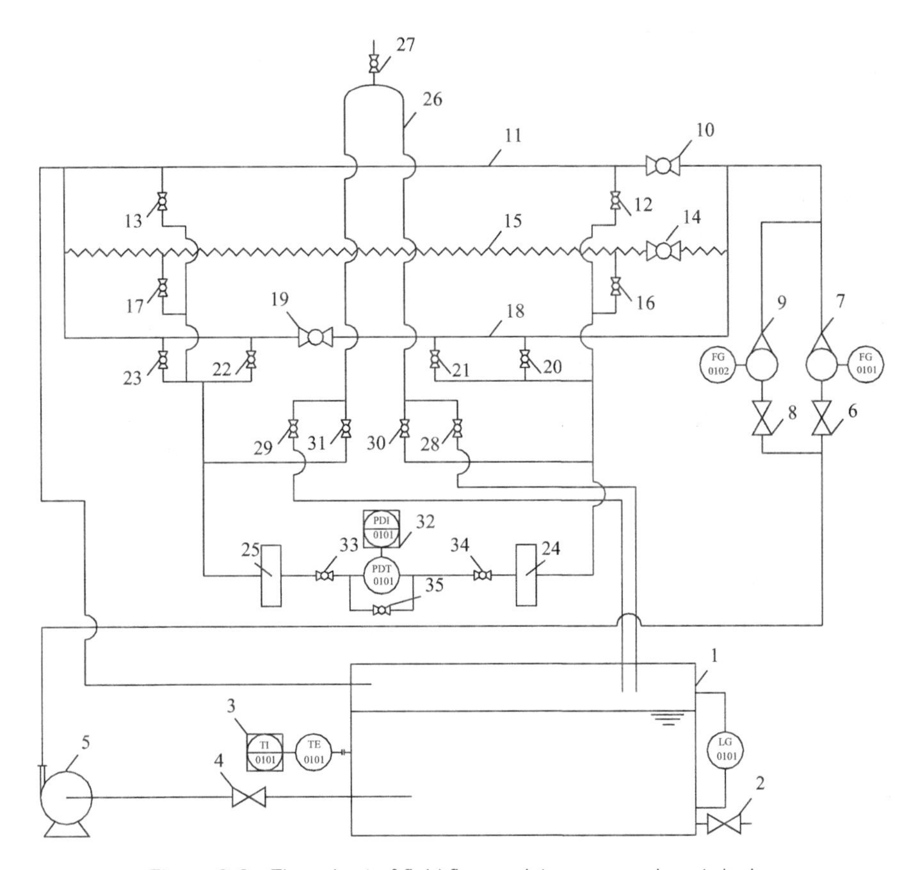

Figure 8-2 Flow chart of fluid flow resistance experiment device

1—Water tank; 2—Drain valve of water tank; 3—Thermometer; 4—Inlet valve of centrifugal pump; 5—Centrifugal pump; 6, 8—Flow control valve; 7, 9—Rotameter; 10—Control valve of smooth pipe; 11—Smooth pipe; 12, 13—Pressure measuring valve of smooth pipe; 14—Control valve of rough pipe; 15—Rough tube; 16, 17—Pressure measuring valve of rough tube; 18—Measured local resistance pipe; 19—Ball valve; 20, 21, 22, 23—Pressure measuring valve for local resistence; 24, 25—Buffer tank; 26—Inverted U-tube; 27—Venting valve of inverted U-tube; 28, 29—Drain valve of inverted U-tube; 30, 31—Outlet valve of inverted U-tube; 32—Differential pressure transducer; 33, 34—Differential pressure transducer valve; 35—Balance valve

⑤ Slowly open the valves above the two buffer tanks to drain the air from tank close them until a slight overflow of water occurs.

⑥ Adjust the two liquid levels in the inverted U-tube to the centre of the tube: Close valve 6 and valve 8 to reduce the flow rate to zero. Close valve 30 and valve 31, open valve 27, slowly open valve 28 and valve 29 to adjust the liquid level to half-height of the inverted U-tube differential pressure meter. Then close valve 27 and valve 35, open valve 30 and valve 31. If the liquid levels of the inverted U-tube differential pressure meter are equal at this moruent, it means that the air in the pipeline is exhausted, the experiment can be started. If the liquid level at both ends are not equal, repeat steps ④ and ⑤ to exhaust until the liquid levels of the inverted U-tube differential are equal.

⑦ Open the valves of the tested pipeline, close the valves of other unrelated pipelines, and the valves 28, 29 are also closed.

⑧ Measure the pressure drop for the straight pipe after the flux is stable. Measure 15～20 groups of data, the measurement sequence is from large flux to small flux. Measure the pressure drop by the differential pressure sensor which show the reading on the instrument panel when the flux is greater than 100L/h. The valves 30 and 31 are opened to measure the pressure drop by the inverted U-tube differential pressure gauge when the flux is lower than 100L/h.

⑨ After measuring a group of data of pipelines, close the flow regulating valve, check whether the liquid level of the inverted U-tube differential pressure gauge are equal, then repeat the above steps to measure the data of other pipelines.

It needs to measure the proximal pressure difference $(p_b - p_{b'})$ and distal pressure difference $(p_a - p_{a'})$ when measuring the local resistance coefficient of the ball valves, three groups of data should be measured at high flow rates; the inverted U-tube differential manometer is used to measure the proximal pressure difference and distal pressure difference when the ball valve is fully opened, while the differential pressure transducer is used to measure the proximal pressure difference and the distal pressure difference when the ball valve is half-open.

⑩ Close the flow control valve, and then turn off the pump and the instrument power. Finally, Switch off the total power of the equipment.

8.1.4.2 Precautions

① It is necessary to check whether all flow control valves are shut off before turning on the centrifugal pump.

② Drive away the bubbles in all tubes

③ In order to avoid damage to the meter, the switchs of the pressure gauge and vacuum gauge must be closed before turning on the centrifugal pump. The centrifugal pump inlet valve 4 is fully opened to avoid the occurence of cavitation.

④ When the differential pressure sensor is used to measure the pressure difference high flow rate, the valves 30 and 31 should be closed to prevent the formation of parallel pipelines from affecting the measurement results.

⑤ Measure and record data after the flux and the pressure are stable.

⑥ The other pressure measure valves need to be closed when using the inverted U-tube differential manometer or the differential pressure sensor to measure the differential pressure.

8.1.4.3 Physical properties of water

The density (ρ) and viscosity (μ) of the physical parameters of the water required in the experiment are checked by the manual or calculated by the following formula.

1) Density

$$\rho = -0.003589285t^2 - 0.0872501t + 1001.44 (\text{kg/m}^3) \tag{8-10}$$

where t is average temperature of water, ℃

2) Viscosity

$$\mu = 0.000001198\exp\left(\frac{1972.53}{273.15 + t}\right) (\text{Pa} \cdot \text{s}) \tag{8-11}$$

8.1.5 Questions

① What is the reason for the friction resistance generated in the straight tube when a fluid flows under a stable state? How is the friction resistance measured?

② What is the reason for the generation of local resistance when a fluid is flowing? How is the local resistance measured? How is the local resistance coefficient determined? How is the equivalent length determined based on the local resistance coefficient?

③ What is the measurement principle of the U-tube differential manometer? Can it be used to directly measure the absolute pressure?

④ How to choose indicator for the U-tube differential manometer?

⑤ Briefly describe the structure, working principle, characteristics, installation precautions and operation method of the orifice flow meters and rotameters.

⑥ Why does the air in the equipment need to be completely discharged before the measurement? How can air be quickly discharged?

⑦ During this experiment, if there is no water flowing out of the tube after the centrifugal pump is started, what are the possible causes?

⑧ When a horizontal straight pipe with equal-diameter was used in the experiment, the pressure difference between the upstream and downstream of the pipe section is equal to the friction resistance, that is, $\Delta p_f = p_1 - p_2$. If the pipe is placed at an angle, is the equation $\Delta p_f = p_1 - p_2$ still valid? Can the reading of the manometer reflect the friction resistance between the two pressure measuring points? Why?

⑨ When experiments are conducted in a rough tube, if the valve of the smooth tube is not completely closed, what effects will it have on the measurement result? On the friction coefficient graph, will the measurement points produce a positive or negative deviation?

⑩ During the experiment, if the two pressure measurement valves upstream and downstream of the other tubes are in an open state in addition to the tube being tested, how will it affect the measurement results?

⑪ Can the λ-Re data measured on different equipment (with the same relative roughness but different tube diameters) at different temperatures with Newtonian fluids be correlated into a single curve? If so, why?

⑫ By measuring various resistances in the pipeline, what measures do you think can be taken to reduce the flow resistance of the fluid in the pipeline?

⑬ Try to design a suitable experiment scheme to measure the local resistance coefficient ζ of a 90° elbow using water as the experimental fluid. Draw a flow chart of the experimental device, specify the data to be measured and the required instruments and meters, explain the data processing and calculation methods.

8.2 Centrifugal pump experiment

8.2.1 Experimental objective

① Familiar with the basic structure of the centrifugal pump. Master the operation methods of start and stop of the centrifugal pump.

② Master the measuring method of the characteristic curve of centrifugal pump and pipeline.

③ Master the collaboration of two pumps and the characteristics of two pumps in series or parallel.

8.2.2 Experimental principles

Centrifugal pump uses the high-speed rotation of the impeller to make the liquid filled in the pump obtain mechanical energy during the process of throwing from the centre of the impeller to the edge of the impeller under the action of centrifugal force, and improve the static pressure energy and kinetic energy. The liquid leaves the impeller and enters the pump casing (volute) . The special structure of the volute converts part of the kinetic energy into static pressure energy, and finally enters the discharge pipe with high-pressure liquid. The theoretical head of the centrifugal pump is obtain by analysing the movement of liquid particles in the centrifugal pump theoretically under ideal conditions, it represents the relationships between the head and flow of the centrifugal pump. Due to the performance of the centrifugal pump is affected by the internal structure, impeller form and speed of the pump and other factors, the flow of fluid in the pump will produce a variety of resistance losses in the actual work, the actual head is smaller than the theoretical head, the flow in the pump is more complex. Therefore, the characteristic parameters such as the head and efficiency of the centrifugal pump cannot be accurately calculated theoretically, which should be determined by experiments.

(1) The characteristic curve of centrifugal pump

The relationships between the head, power, efficiency and flow are called the characteristic curve of the centrifugal pump at a certain speed.

1) Flow Q The flow rate of the centrifugal pump is measured with a turbine flowmeter.

2) Head H The head of the pump can be calculated by the mechanical energy balance between the inlet and outlet of the pump. The Bernoulli equation between the inlet and outlet of the pump can be obtained as follows:

$$z_i+\frac{p_i}{\rho g}+\frac{u_i^2}{2g}+H=z_o+\frac{p_o}{\rho g}+\frac{u_o^2}{2g}+h_{f,i-o} \tag{8-12}$$

$$H=(z_o-z_i)+\frac{p_o-p_i}{\rho g}+\frac{u_o^2-u_i^2}{2g}+h_{f,i-o} \tag{8-13}$$

$$H=(z_o-z_i)+\frac{p_o-p_i}{\rho g}+\frac{u_o^2-u_i^2}{2g}=h_0+\frac{p_o-p_i}{\rho g}+\frac{u_o^2-u_i^2}{2g} \tag{8-14}$$

where H is the head of centrifugal pump, m; h_0 is the vertical distance between two pressure measurement ports, outlet pressure gauge and the inlet vacuum gauge, m; p_i, p_o are the pressure at the pump inlet and outlet respectively, Pa; u_i, u_o are the flow rate of the pump inlet and outlet respectively, m/s; ρ is fluid density, kg/m^3; g is acceleration of gravity, m/s^2.

It can be known from the above formula that as long as the pressure at the inlet and outlet of the pump and the height difference between the pressure gauge and vacuum gauge are measured, the flow rate can be calculated from the flow rate, the pump head can be obtained.

3) Shaft power N

$$N=N_p\times k \tag{8-15}$$

In formula, N_p is display value of electric power meter, W; k is motor transmission efficiency, $k=0.95$.

4) Efficiency η

The efficiency η of the pump is the ratio of the effective power N_e to the shaft power N. The effective power N_e is the actual work obtained when the fluid passes through the pump in unit time. Shaft power N is the work obtained from the motor in unit time. The difference between them reflects the magnitude of the hydraulic loss, volume loss, mechanical loss. The effective power N_e of the pump is calculated as follows:

$$N_e=HQ\rho g \tag{8-16}$$

Therefore, the pump efficiency is as follows:

$$\eta=\frac{HQ\rho g}{N}\times 100\% \tag{8-17}$$

In summary, the relationships between the head, power, efficiency and flow rate of the centrifugal pump in a certain speed are the characteristic curve, which can be measured by adjusting the opening of the pipeline valve to change flow rate.

(2) Characteristic curves of pipeline

The characteristic curve of the centrifugal pump is the characteristic of pump itself, which has nothing to do with the characteris tic of the pipeline. Centrifugal pump is always installed in a specific pipeline when it is used, which provides the mechanical energy required for the flow of liquid in the pipeline. Therefore, the actual working condition of the centrifugal pump should be determined by the characteristics of the pump and the characteristics of the pipeline. The characteristic curve of pipeline can be obtained from the Bernoulli equation:

$$H=\frac{\Delta p}{\rho g}+\Delta z+\frac{\Delta u^2}{2g}+\sum h_f=A+BQ^2 \tag{8-18}$$

where Δp is the differential pressure at both ends of pipeline, Pa; Δz is the vertical distance at both ends of the pipeline, m; u is flow rate, m/s; $\sum h_f$ is flow resistance, m; A is the total potential energy difference at both ends of pipeline, m; B is the coefficient of pipeline characteristic curve，h^2/m^5，Q is flow rate in pipeline，m^3/h.

The measurement method of head H and flow Q is the same as that of the characteristic curve of the centrifugal pump. A depends on the actual conditions at both ends of the pipeline，B depends on the pipeline conditions. B is constant when the liquid is in high turbulence. The measurement of the characteristic curve of pipeline cannot be changed by control valve of pipeline. Because the characteristic curve of the pipeline changes when the opening of the control valve changes，so the flow rate can only be adjusted by the frequency converter to change the speed of the centrifugal pump when measuring the characteristic curves of the pipeline. If the characteristic curves of the centrifugal pump and the pipeline can be plotted in the same coordinate，the intersection of the two curves is the working point of the centrifugal pump.

8. 2. 3 Experimental device

Main technical data of equipment are as follows.

① Turbine flowmeter：Model LWGY-40A05WSN，Measuring range $1\sim20m^3/h$.

② Centrifugal pump：Model MS100/0. 55，Rated flow $6m^3/h$，Rated head 14m；Power supply three-phase AC380V，0. 55kW.

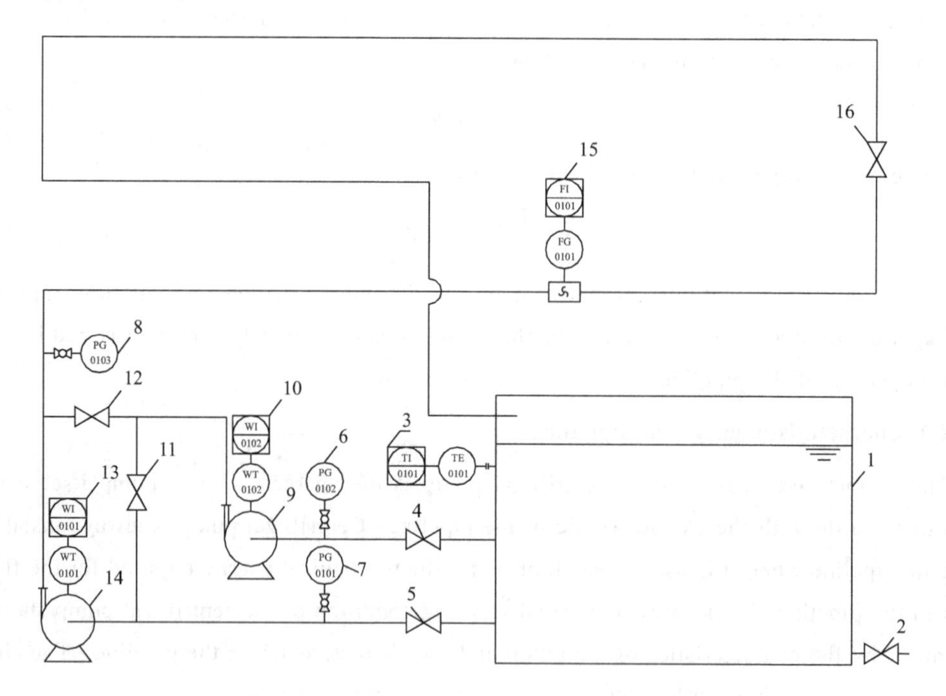

Figure 8-3 Flow chart of centrifugal pump experiment device

1—Water tank；2—Water tank drain valve；3—Temperature measuring point；4—Inlet valve of water pump 2＃；5—Inlet valve of water pump；6—Inlet vacuum gauge of pump 2＃；7—Inlet vacuum gauge of pump 1＃；8—Total outlet pressure gauge of pump；9—entrifugal pump 2＃；10，13— Power meter；11，12— Series and parallel control valve；14—Centrifugal pump 1＃；15— Turbine flowmeter；16— Flow control valve

③ Power meter: Model GPW201-V3-A3-F1-P2-O3, Accuracy 0.5%.

④ The vertical distance between two pressure measurement ports $h_0=0.29m$.

⑤ The inner diameter of inlet pipe of pump $d_1=0.035m$.

⑥ The inner diameter of outlet pipe of pump $d_2=0.042m$.

⑦ Vacuum meter: Model: Y-100, Measuring range −0.1 ~ 0MPa, Accuracy class 1.5.

⑧ Pressure gauge: Model Y-100, Measuring range 0~0.6MPa, Accuracy class 1.5.

⑨ Frequency converter: Mitsubishi frequency converter, Model FR-D740-0.75K-CHT, Power 0.75kW/380V.

⑩ Thermometer: platinum resistance, WZP-270501BX, measuring range 0 ~ 100℃, Accuracy class B.

8.2.4 Experimental procedures and precautions

8.2.4.1 Experimental procedures

(1) Determination the characteristic curve of centrifugal pump

① Check whether the water level of the water tank meets the experimental requirements.

② Switch on the main power and the instrument power.

③ Fully open the inlet valve 5 of the centrifugal pump 1# and close all other valves, then start the water pump 1#.

④ Slowly open the flow control valve 16 to fully open. When the fluid in the system is stable, and there is no gas in the system, then open the switch 7 of the inlet vacuum gauge of the centrifugal pump and the switch 8 of the outlet pressure gauge.

⑤ Open the flow control valve to the maximum to determine the flow measurement range. The order of measuring data can be from the maximum flow to zero or vice versa. Generally, 12 groups of data are measured.

⑥ The flow rate, inlet vacuum of pump, outlet pressure of pump, reading of power meter and fluid temperature shall be recorded at the same time for each measurement.

⑦ After the test, close the valve 16 and stop the centrifugal pump 1#.

(2) Determination of pipeline characteristic curve

① Switch on the main power and the instrument power.

② Fully open the inlet valve 5 of the centrifugal pump 1# and close all other valves, then start the water pump 1#. Adjust the flow control valve 16 to a certain state and make the pipeline with high resistance (the flow control valve 16 is adjusted to a flow rate of about $6m^3/h$) or low resistance (the flow control valve 16 is fully open).

③ The characteristic curve of the pipeline is measured by the frequency regulation of the centrifugal pump motor, and the regulation range is 50% ~100% of the full speed of the motor.

④ A set of data is recorded when the motor frequency is changed, including flow rate, vacuum at the inlet of the pump and pressure at the outlet of the pump. At least 6 sets of data should be recorded.

⑤ At the end of the test, restore the motor frequency to 100% of full speed, then close the flow control valve 16 of outlet and stop the water pump 1# .

(3) Measurement of characteristic curves of two pumps in parallel and in series

① Switch on the main power and the instrument power.

② In the parallel experiment, fully open valve 5, valve 4 and valve 12, all other valves are closed. Start water pump 1# and water pump 2# , turn on the vacuum gauge switchs 6 and 7 of inlet and the pressure gauge switch 8 of outlet. Record the inlet vacuum, outlet pressure, readings of two power meters and fluid temperature by adjusting the valve 16.

③ Fully open valve 4 and 11 and all other valves are closed. Start water pump 1# and water pump 2#, turn on the vacuum gauge switchs 6 and 7 inlet and the pressure gauge switch 8 outlet. Record the inlet vacuum, outlet pressure, readings of two power meters and fluid temperature by adjusting the valve 16.

④ At the end of the test, close the flow control valve 16 of outlet, stop the pump and cut off the power supply.

8.2.4.2 Precautions

① The centrifugal pump can only be started or shut down when the flow control valve of outlet is closed.

② The pipeline needs to be vented to ensure the continuous flow of fluid before the experiment.

③ Close the switches of pressure gauge and vacuum gauge before starting the centrifugal pump to avoid damage.

④ The flow control valve needs to be adjusted slowly to avoid water ingress into the sensor.

⑤ The inlet valve of the centrifugal pump must be fully opened to avoid cavitation.

8.2.5 Questions

① What are the principles for selecting a centrifugal pump?

② What is the difference between the characteristic curve of the centrifugal pump and the characteristic curve of the pipeline?

③ Why does the characteristic curve of the pipeline measured during this experiment have no obvious intercept and approximately pass through the origin?

④ Based on the measured characteristic curve, what measures can be taken to increase the flow range of the pipeline?

⑤ What are the characteristics of the double pump operations with a series connection and a parallel connection?

⑥ What kind of flow regulation methods are involved in the experiment? Explain their characteristics and differences.

⑦ When the flow control valve of the centrifugal pump is turned down, how do the flow rate in the pipeline and the display values of the inlet vacuum gauge and the outlet pressure gauge of the centrifugal pump change? Why?

⑧ In the experiment, why was the flow rate regulated by adjusting the motor frequency instead of flow control valve?

⑨ Based on the characteristic curves of the double-pump parallel and double-pump series operations, as well as the characteristic curves of the pipeline, try to explain the specific conditions suitable for the double pump parallel operation or double pump series operation.

⑩ In the measurement experiment of the characteristic curve of the centrifugal pump, as the flow rate decreases, the display values of the vacuum gauge gradually declines until it reaches zero. Why does this occur?

⑪ According to the experimental measurement results, please explain the following: a. Why should the outlet valve be closed when the centrifugal pump is starting. b. How can the appropriate working range of the centrifugal pump be determined? Why?

⑫ What is the most likely reasons for cavitation in the centrifugal pump used in this experimental device? What happens to the head of a centrifugal pump when it experiences a cavitation?

⑬ If the location of the centrifugal pump in this experiment is modified to be above the liquid level of the water tank, please design an experimental scheme for measuring the characteristic curve of the centrifugal pump. Please provide the experimental device flow chart, the experimental operation steps, and the record form for original data.

8.3 Filtration experiment

8.3.1 Experimental objective

① Be familiar with the basic process of plate and frame filter, the structure of plate and frame, master the operation method of plate and frame filter.

② Master the determination method of filtration parameters such as filtration constant and compressibility index under constant pressure filtration.

③ Master the engineering simplified treatment method and experimental research method of filtration problem.

8.3.2 Experimental principles

Filtration is a unit operation in which the solid particles in the suspension (solid-liquid mixture or filter slurry) are retained by the filter media to form filter cake (filter residue), while the liquid passed through the filter cake layer and filter media. Regardless of the filtering operation or the design of the filter, the filter constant is a very important basic data. The filtration constants of different suspensions are also different. Even for the same suspension, the filtration constants vary with the concentration of solid particles, the temperature of filtration slurry and the driving force of filtration, so it is necessary to measure the accurate and reliable filtration constants by experiments.

The constant pressure filtration equation is as follows:

$$q^2+2qq_e=K\theta \tag{8-19}$$

where q is volume of filtrate per unit filtration area, m^3/m^2; q_e is equivalent filtrate volume per unit filtration area, m^3/m^2; K is filter constant, m^2/s; θ is filtration time, s.

There are two main experimental methods to determine the filtration constant, differential method and integral method. Their principles are described as follows.

(1) Differential method for determination of filtration constant

Differentiate formula (8-19):

$$\frac{d\theta}{dq}=\frac{2}{K}q+\frac{2}{K}q_e \tag{8-20}$$

The incremental ratio $\frac{\Delta\theta}{\Delta q}$ can be used instead of $\frac{d\theta}{dq}$ when the time interval of each data point is not large, that is

$$\frac{\Delta\theta}{\Delta q}=\frac{2}{K}\bar{q}+\frac{2}{K}q_e \tag{8-21}$$

The above formula is a linear equation. Under the condition of constant filter driving force and temperature, a certain filter medium is used to filter the suspension of a certain concentration at a constant pressure, the filter time θ and filtrate accumulation q are measured, the relationship between the $\frac{\Delta\theta}{\Delta q}$ and $\bar{q}$ is plotted on the rectangular coordinate system. The slope of the straight line is $\frac{2}{K}$ and the intercept is $\frac{2}{K}q_e$. Then the filter constant can be obtained by the slope and intercept of the straight line.

(2) Integral method for determination of filtration constant

Formula (8-19) is the constant pressure filtration equation obtained by integrating the basic filtration equation, it can be rewritten as follows:

$$\frac{\theta}{q}=\frac{1}{K}q+\frac{2}{K}q_e \tag{8-22}$$

The above formula is also linear.

Filter the suspension to be measured under constant pressure, measure the data of the filtration time θ and the cumulative amount q of filtrate, and plot the relationship between $\frac{\theta}{q}$ and q on the rectangular paper. The slope of the obtained line is $\frac{1}{K}$, and the intercept is $\frac{2}{K}q_e$. The filter constants K and q_e can be obtained according to intercept.

(3) Determination of compressibility index

The compressibility index reflects the compression performance of the filter cake, which can be obtained from the relationship between the filter constant and the driving force of the filter. The definition of the filter constant is as follows:

$$K=\frac{2\Delta p}{\mu rc}=\frac{2\Delta p^{1-s}}{\mu r_0 c}=2k\Delta p^{1-s} \tag{8-23}$$

The empirical relationship between filter cake specific resistance and filtration driving force is as follows:

$$r=r_0\Delta p^{s} \tag{8-24}$$

where Δp is filtering power, Pa; μ is filtrate viscosity, Pa • s; r_0 is specific resistance of filter cake when Δp is 1Pa, $1/m^2$; r is filter cake specific resistance, $1/m^2$; c is filter cake volume formed when a unit volume of filtrate was obtained, m^3/m^3; s is compressibility index.

The equations of simultaneous formula (8-23) and formula (8-24) can be obtained as follows:

$$r=\frac{2\Delta p}{K\mu c} \tag{8-25}$$

The specific resistance of filter cake r can be obtained from formula (8-25) based on the experimental data of filter constant K measured under different filter driving forces Δp. The data of r and Δp are plotted on the double logarithm coordinate under the condition of different filter driving forces. From formula (8-24), the relationship between r and Δp should be a straight line. The compressibility index can be obtained from the slope s and intercept r_0 of the line. Or plot the data of K and Δp on the double logarithmic coordinates. The relationship between K and Δp is a straight line from formula (8-23), the slopes and intercept of which are s and $2k$ respectively, and then r_0 can be obtained according to the definition k in formula (8-23).

8.3.3 Experimental device

The flow chart of plate and frame filtration experimental device is shown in Figure 8-4, which is mainly composed of slurry tank, agitator, plate and frame filter, centrifugal pump, etc. The plate and frame filter is a small industrial filter, which is assembled from non-washed plates, frames and washed plates in a certain order. The diagram of the structure of the plate and frame is shown in Figure 8-5.

Main technical data of equipment are as follows.

① Stainless-steel horizontal centrifugal pump with single-stage: Model MS60/0.75, Rated flow 60L/min, Rated power 0.75kW.

② Electric agitator: Rated power 400W, Rotating speed 1400r/min.

③ Filter plate and frame: Stainless steel material, Frame diameter 0.09m, Frame thickness 0.02m.

④ Filter medium: Industrial filter cloth.

⑤ Electronic balance: Model LNW-15, Maximum weighing 15kg, Minimum weighing

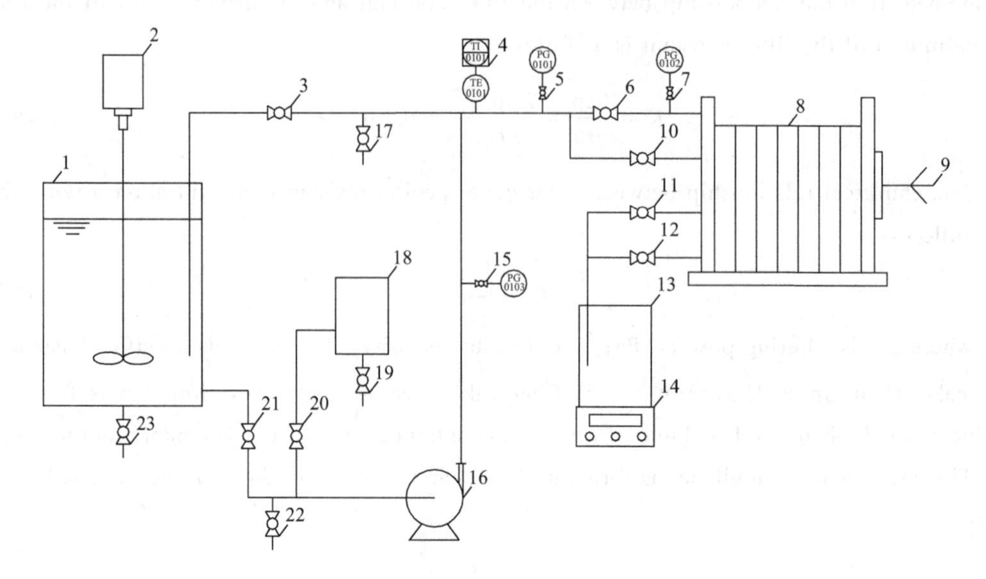

Figure 8-4 Flow chart of plate and frame filtration experiment device

1—Slurry tank; 2—Electric agitator; 3—Pressure control valve; 4—Temperature measuring point; 5—Pressure gauge before valve; 6—Inlet control valve of filter; 7—Pressure gauge behind valve; 8—Plate and frame filter; 9—Closing device; 10—Control valve of wash inlet; 11—Control valve of filtracte outlet; 12—Control valve of wash outlet; 13—Filtrate receiving tank; 14—Electronic balance; 15—Outlet pressure gauge of centrifugal pump; 16—Centrifugal pump; 17—Drain valve of pipe; 18—Water bucket; 19—Drain valve of water bucket; 20—Outlet valve of water bucket; 21—Outlet valve of filter barrel; 22—Drain valve; 23—Discharge valve of pulp drum

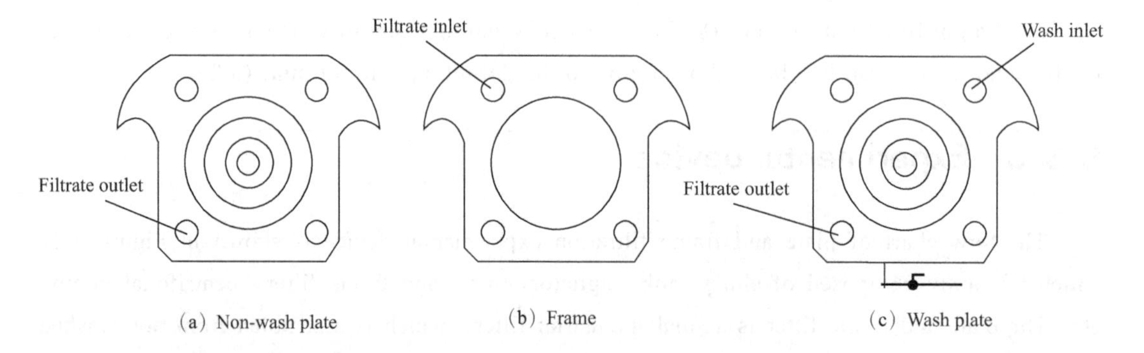

Figure 8-5 Diagram of the structure of the plate and frame

100g, Working environment temperature 0～40℃.

8.3.4 Experimental procedures and precautions

8.3.4.1 Experimental procedures

(1) Filtration

① Ingredients and mixing: Prepare a certain concentration of $CaCO_3$ aqueous suspension in the slurry tank. Switch on the main power, the instrument and the agitator in turn on the instrument control cabinet to make the $CaCO_3$ suspension stir evenly. The top cover of the batching tank shall be closed when mixing.

② Installation of plate and frame: The plates and frames shall be installed in the order of fixed head, non-wash plate, frame, wash plate, frame, non-wash plate and movable head. The filter cloth is soaked with water before use. The filter cloth must be tight and not wrinkled. The filter cloth shall be close to the filter plate, the sealing gasket shall be close to the filter cloth, then the plate and frame shall be compressed by the closing device for use. Note that the through-hole switch valve under the washing plate must be open during the filtration test.

③ Start centrifugal pump: Open valves 21, 11 and close all other valves. Turn on the centrifugal pump switch on the instrument control cabinet and start the centrifugal pump. Adjust the intelligent instrument to the interface of frequency, use the increase key to adjust the motor frequency of the centrifugal pump to 100% and fully open the pressure control valve 3.

④ Pressure setting: Use the pressure control valve 3 to adjust the filter pressure to the required value.

⑤ Filter: Open the control valve 6 of slurry inlet and start timing when you see the filtrate flowing out of the collecting tube of the filtrate outlet of the frame. It is recommended to record the corresponding filtering time θ or $\Delta\theta$ until the filter cake is close to the full frame when every 0. 5kg filtrate is collected.

⑥ After a constant pressure filtration experiment, close inlet control valve 6, fully open the pressure control valve 3, loosen the closing device, remove the frame, filter plate, filter cloth for cleaning. Note that in order to maintain the consistency of the slurry concentration, the filter cake should be returned to the slurry tank, the collected filtrate should be used to clean the frame, filter plate, and filter cloth. The cleaning liquid and the material in the collection tank under the plate and frame should be poured back into the slurry tank.

⑦ Repeat steps ④, ⑤ and ⑥ to determine the filtration constants at three different pressures to end the filtration experiment.

(2) Washing process

① Close the filtrate inlet valve 6 and outlet valve 11 of the plate and frame filter after a filtration experiment. Close the through-hole change-over valve under the washing plate.

② Close the pressure control valve 3; stop the feed pump; close the outlet valve 21 of the slurry tank. Then open the outlet valve 20 of the water bucket, start the pump on the instrument control cabinet, open drain valve 17 of the pipeline, clean the residual material liquid in the pipeline to the gutter.

③ Close drain valve 17, open the control valve 12 of wash outlet and the control valve 10 of wash inlet. At this time, the pressure gauge indicates the cleaning pressure. The cleaning solution flows out the collection tube.

④ After washing for about 2min, the end of the washing experiment can be judged according to the turbidity change of the washing liquid. General materials can not be cleaned.

(3) End of experiment

① Close the outlet valve of the centrifugal pump and stop the pump.

② Close the outlet valve 21 of the slurry tank and close the agitator motor.

③ Remove the frame, filter plate and filter cloth for cleaning.

④ Clean the feed pump and pipeline; fully open the pressure control valve 3; open the drain valve 22 at the bottom of the pipeline; drain the liquid in the pipeline. Close the pressure control valve 3, open the outlet valve 20 of the water bucket 18 and drain valve 17, start the feed pump, clean the feed pump and pipeline until the cleaning liquid is clear. Note that if the water in the water bucket is insufficient, a certain amount of water can be added.

8.3.4.2 Precautions

① Try not to fold the filter cloth and keep it flat when cleaning; properly place the sealing gasket between the filter plate and the frame. Press the plates and frames tightly with the handle to avoid leakage.

② The arrangement order and direction of plates and frames must be correct.

③ The handle of through-hole change-over valve under the wash plate is in the filtering state when it is parallel to the filter plate, and it is in the cleaning state when it is vertical to the filter plate.

④ The filter cake, cleaning fluid and the material in the collection tank under the plate and frame must be poured back into the slurry tank.

⑤ During the constant pressure filtration experiment, the pressure control valve cannot be adjusted.

8.3.5 Questions

① What are the filter slurry, filter cake, filtrate, filter media and filter aid?

② What are the advantages and disadvantages of the plate and frame filters? What are these filters used for?

③ Briefly describe the characteristics of constant pressure filtration.

④ What are the influencing factors of filtration constant?

⑤ Why is the filtrate often turbid at the beginning of the filtration, and later becomes clear? How does a variation in the filtration pressure impact this phenomenon?

⑥ During a constant pressure filtration, are the filter cake structures obtained at different filtration pressures the same? How does its porosity change with pressure?

⑦ Is the amount of filtrate the same when the filter is full of filter cake under different filtration pressure? Why?

⑧ Why does the suspension solution in the tank need to be stirred during the constant pressure filtration experiment? What is the function of the baffles used in the suspension tank?

⑨ Is the last data point of the filtration experiment near the full frame lower or higher than the other data? Why?

⑩ If the filtration rate is found to decrease significantly under the same filtration pressure during the filtration, what could be the reasons?

⑪ What are the measures used to speed up the filtration rate?

⑫ What is the difference in porosity at different positions in the filter cake? Why?

⑬ For a constant pressure filtration, is it feasible to increase the production capacity of the plate and frame filters by extending the filtration time? Why?

⑭ Please briefly describe the main factors affecting the production capacity of the intermittent filter and the ways to improve the production capacity of the intermittent filter .

⑮ What are the main factors affecting the filtration rate? When measuring K and q_e under constant pressure, how will K and q_e change with the filtration pressure increased?

8.4 Heat transfer experiment

8.4.1 Experimental objectives

① Understand the main structure of the heat transfer equipment. Master the operation method of the heat transfer equipment.

② Master the measurement methods of heat transfer coefficient K and convection heat supply coefficient α_1 in a tube, and understand of the concept and influencing factors.

③ Master the method of determining the constants in empirical correlation $Nu=ARe^m Pr^n$ by drawing or least squares method.

④ Compare the ordinary heat exchanger with the enhanced heat exchanger, and understand the measures of enhancing heat transfer in engineering.

8.4.2 Experimental principles

When the fluid is in forced turbulence in a straight tube, the correlation of the convective heat transfer coefficient can be written as:

$$Nu=ARe^m Pr^n \tag{8-26}$$

A and m need to be determined by experiments. In this experimental device, vapor is used to heat air. For the cold fluid in the tube, $n=0.4$. Therefore, formula (8-26) can be written as:

$$\frac{Nu}{Pr^{0.4}}=ARe^m \tag{8-27}$$

where $Nu=\frac{\alpha_1 d_1}{\lambda}$ $Re=\frac{d_1 u_1 \rho}{\mu}$ $Pr=\frac{c_p \mu}{\lambda}$

The flow rate u_1 in Re is calculated according to the flow rate measured by flow meter. The physical parameters such as ρ, μ, λ, c_p are obtained according to the qualitative temperature of the fluid by the parameter chart or the following correlation formula (t is the qualitative temperature, and the range of it is $0℃ \leqslant t \leqslant 100℃$).

Air density ρ (kg/m^3)

$$\rho = 1.2916 - 0.0045t + 1.05828 \times 10^{-5} t^2 \tag{8-28}$$

Specific heat of air　c_p [kJ/(kg · ℃)]

$$c_p = 1.00492 - 2.88378 \times 10^{-5} t + 8.88638 \times 10^{-7} t^2 - 1.36051 \times 10^{-9} t^3 + 9.38989 \times 10^{-13} t^4 - 2.57422 \times 10^{-16} t^5 \tag{8-29}$$

Viscosity of air μ (Pa · s):

$$\mu = 1.71692 \times 10^{-5} + 4.96573 \times 10^{-8} t - 1.74825 \times 10^{-11} t^2 \tag{8-30}$$

Thermal conductivity of air λ [W/(m · ℃)]

$$\lambda = 0.02437 + 7.83333 \times 10^{-5} t - 1.51515 \times 10^{-8} t^2 \tag{8-31}$$

Nu is obtained by α_1. For saturated steam heating the air on the other side, the heat transfer coefficient of the steam side $\alpha_2 \gg \alpha_1$, and the heat transfer tube is a metal one, with a large thermal conductivity and a thin tube wall. Therefore, the relationship between Nu and α_1 is as follows:

$$K \approx \frac{\alpha_1 d_1}{d_2} \tag{8-32}$$

Meanwhile

$$Q = m_{s2} c_{p2} (t_2 - t_1) = KA\Delta t_m \approx \alpha_1 \frac{d_1}{d_2} A\Delta t_m \tag{8-33}$$

As shown in formula (8-33), α_1 can be calculated by the mass flow of air, inlet and outlet temperature of air and steam, then Nu and Re under different flow rates can be obtained. The constant A and the index value m in correlation (8-27) are determined by the drawing or the linear regression method. The empirical correlation of the convective heat transfer coefficient is thus determined.

8.4.3 Experimental device

(1) Flowchart of equipment I

The equipment consists a double-tube heat exchanger and a shell-and-tube heat exchanger. The double-tube heat exchanger is composed of a common tube and a spiral coil reinforced tube. The steam generator is an electric heating kettle. The flow and pressure of the steam are adjusted by a steam control valve. The air is provided by a air blower. The bypass flow control valve is used to adjust the air flow. Steam and air enter each heat transfer device through seperate pipes.

(2) Technical data of equipment Ⅰ

① Double-tube heat exchanger

Specifications of ordinary tube: inner diameter 19mm, outer diameter 21 mm, heat exchange tube length L=960mm, red copper.

Specifications of reinforced tube: inner diameter 19mm, outer diameter 21 mm, heat exchanger tube length L=960mm, red copper; a spiral coil is added in the tube.

Outer tube specifications: ϕ 100mm×3mm, length L=1000mm, stainless steel.

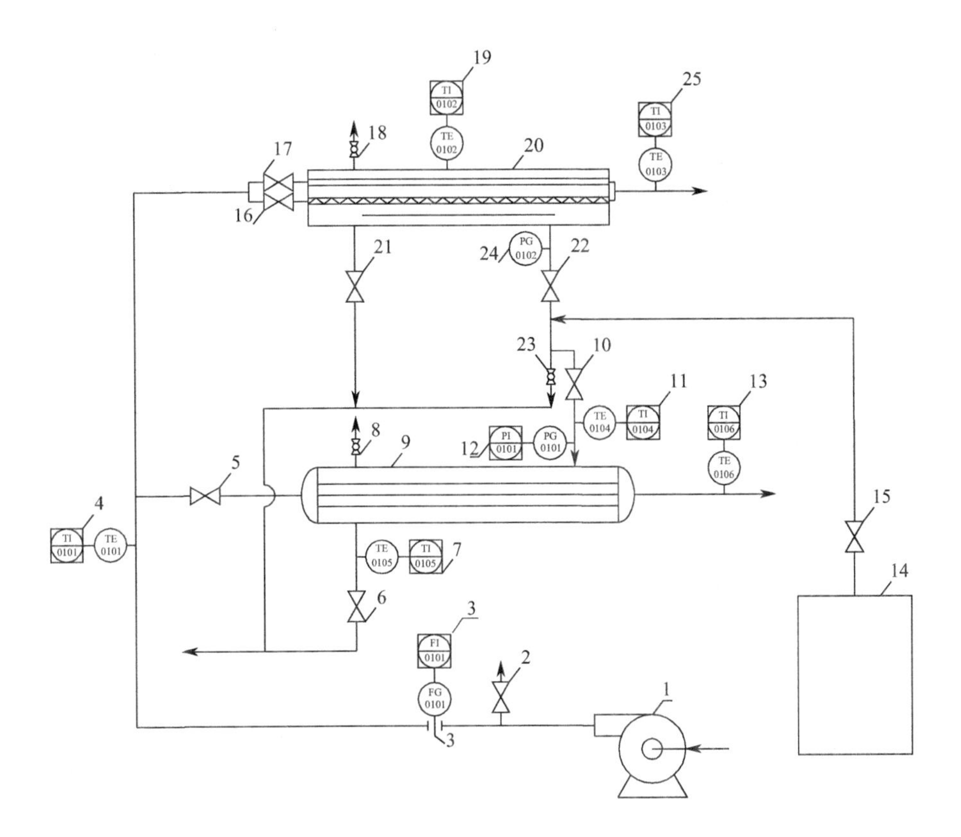

Figure 8-6 Flow chart of heat transfer experiment (equipment Ⅰ)

1—Air blower; 2—Air bypass control valve; 3—Orifice flowmeter; 4—Air inlet temperature; 5—Air inlet valve of shell-and-tube heat exchanger; 6—Condensate outlet valve for shell-and-tube heat exchanger; 7—Steam outlet temperature of shell-and-tube heat exchanger; 8—Vent valve of non-condensable gas in shell-side of shell-and-tube heat exchanger; 9—Shell-and-tube heat exchanger; 10—Steam inlet valve of shell-and-tube heat exchanger; 11—Steam inlet temperature of shell-and-tube heat exchanger; 12—Steam inlet pressure gauge of shell-and-tube heat exchanger; 13—Air outlet temperature of shell-and-tube heat exchanger; 14—Steam generator; 15—Steam control valve; 16—Air inlet valve of strengthening tube; 17—Air inlet valve of normal tube; 18—Vent valve of non-condensable gas in shell-side of double-tube heat exchanger; 19—Temperature of shell-side of double-tube heat exchanger; 20—Double-tube heat exchanger; 21—Condensate outlet valve of double-tube heat exchanger; 22—Steam inlet valve of double-tube heat exchanger; 23—Drain valve of steam tube of double-tube heat exchanger; 24—Shell-side steam pressure gauge of double-tube heat exchanger; 25—Air outlet temperature of double-tube heat exchanger

② Shell-and-tube heat exchanger

Tube side specifications: ϕ20mm × 1.5mm, length L = 800mm, quantity 7, stainless steel.

Shell side specification: ϕ139mm×2mm, stainless steel.

The control panel and instrument display are shown in Figure 8-7 and Figure 8-8.

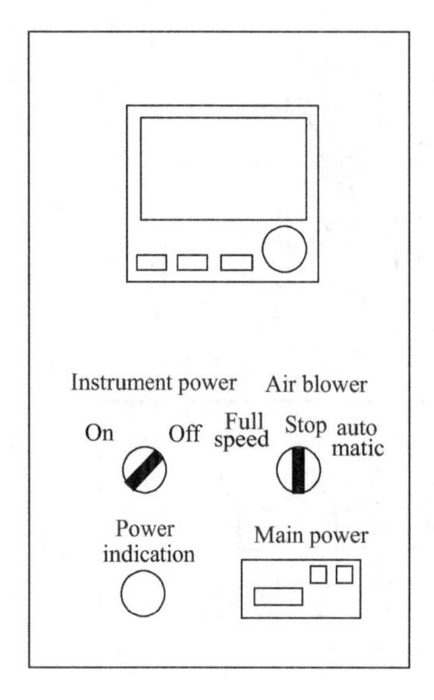

Figure 8-7 Control panel

TI101 ℃	TI104 ℃	TI105 ℃
TI106 ℃	TI102 ℃	TI103 ℃
PI103 kPa	FIC101 m^3/h	FIC101 Pa

Figure 8-8 Instrument display screan

TI101—Inlet temperature of Cold fluid; TI102—Steam term-perature of Double-tube heat exchanger; TI103—Cold fluid outlet temperature of double-tube heat exchanger; TI104—Steam inlet temperature of shell-and-tube heat exchanger; TI105—Steam outlet temperature of shell-and-tube heat exchanger; TI106—Cold fluid outlet temperature of shell-and-tube heat exchanger; PI103—Steam pressure; FIC101—Cold fluid flow

(3) Flow chart of equipment Ⅱ

The device is mainly composed of a double-tube heat exchanger and a shell-and-tube heat exchanger as shown in Figure 8-9. The double-tube heat exchanger can realize the conversion between common and enhanced double-tube through the internal disassembly and assembly of a spiral coil, and the shell-and-tube heat exchanger can change the heat transfer area using the tube plug. The steam generator is the electric heating kettle. The air is provided by the vortex air blower, and the air flow is controlled by the bypass control valve. Steam and air are input into the heat transfer equipment through the pipeline.

(4) Technical data of equipment Ⅱ

① Double-tube heat exchanger

Specification of inner tube: Inside diameter 20mm, outside diameter 22mm, length of heat exchange tube L=1200mm, red copper. The spiral coil can be added in the tube (Coil wire diameter 1mm, Pitch 40mm).

Specification of outer tube: Inside diameter 50mm, outside diameter 57mm, 6 tubes.

② Shell-and-tube heat exchanger

Specification of inner tube: Inside diameter 16mm, outside diameter 19mm, length of heat exchange tube L=1200mm, 6 tubes.

Specification of outer tube: Inside diameter 82mm, outside diameter 89mm.

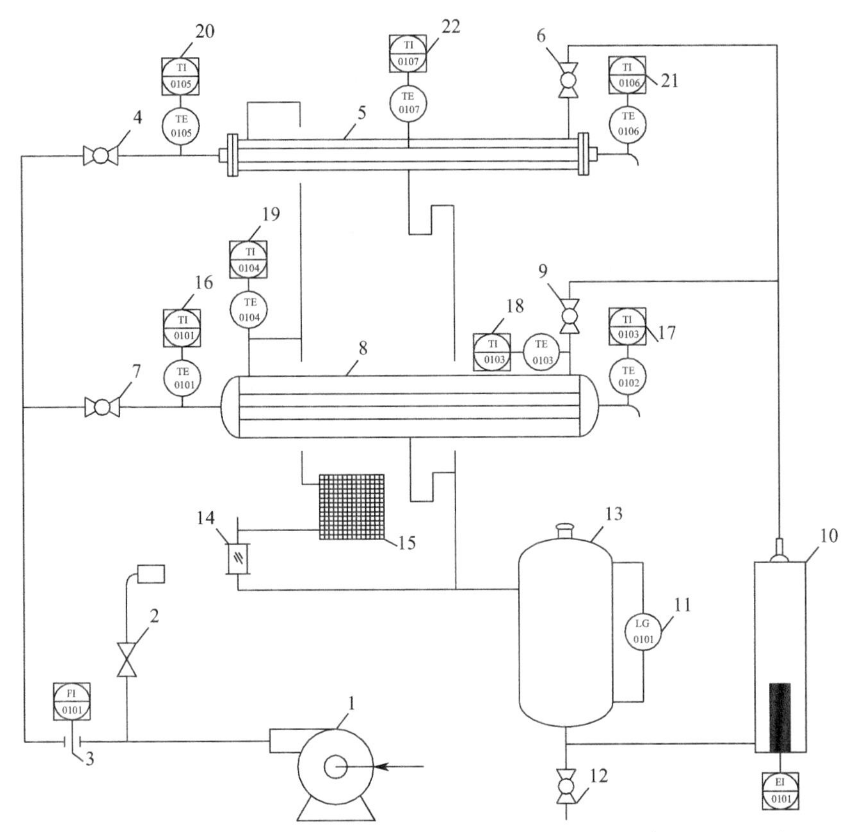

Figure 8-9 Flow chart of heat transfer experiment device (equipment Ⅱ)

1—Air blower; 2—Air bypass control valve; 3—Orifice flowmeter; 4—Air inlet valve of strengthening tube; 5—Double-tube heat exchanger; 6—Steam inlet valve of double-tube heat exchanger; 7—Steam inlet valve of shell-and-tube heat exchanger; 8—Shell-and-tube heat exchanger; 9—Steam inlet valve of shell-and-tube heat exchanger; 10—Steam generator; 11—Level gauge; 12—Drain valve; 13—Water tank; 14—Observation cup; 15—Radiator; 16—Air inlet temperature measuring point of shell-and-tube heat exchanger; 17—Air outlet temperature measuring point of shell-and-tube heat exchanger; 18—Steam inlet temperature measuring point of shell-and-tube heat exchanger; 19—Steam outlet temperature measuring point of shell-and-tube heat exchanger; 20—Air inlet temperature measuring point of double-tube heat exchanger; 21—Air outlet temperature measuring point of double-tube heat exchanger; 22—Wall temperature measurement point of double-tube heat exchanger

③ Orifice flowmeter

Pore flow coefficient 0. 65, Aperture 0. 017m.

The control panel is shown in Figure 8-10.

8. 4. 4 Experimental procedures and precautions

8. 4. 4. 1 Experimental procedures

(1) Experimental procedures of equipment Ⅰ

① Switch on the main power and the meter power.

② Open the steam inlet valve (valve 22 or valve 10) of the heat exchanger according to the

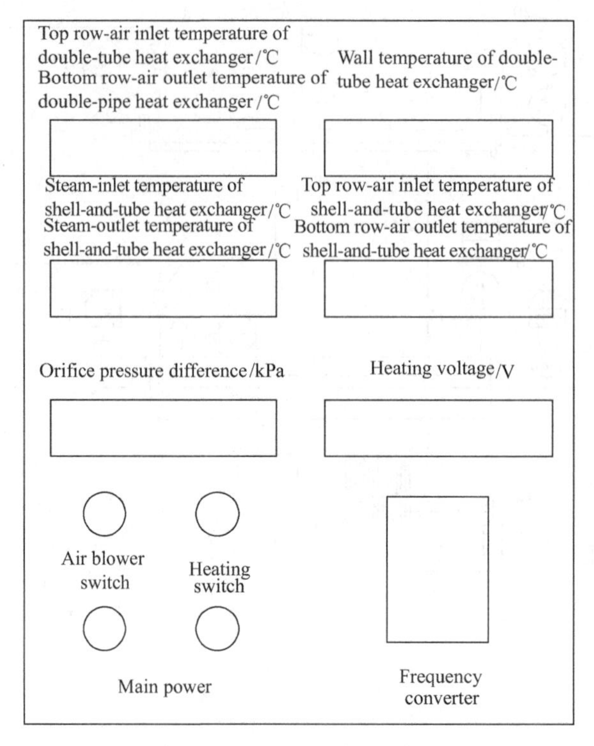

Figure 8-10 Control panel

heat exchanger used in the experiment, and open the water inlet valve of the steam generator to ensure that the steam generator can automatically replenish water; switch on the steam generator to start heating when the steam generator has been filled with water, then slightly open the steam control valve 15.

③ Fully open the air inlet valve of the corresponding heat exchanger (one of valve 16, valve 17, and valve 5) according to the heat exchanger used in the experiment.

④ Fully open the air bypass control valve 2, and switch on the air blower.

⑤ Open the condensate outlet valve 21 or valve 6 and close it after the condensate in the pipeline is discharged.

⑥ Open the non-condensable gas vent valve (valve 18 or valve 8) according to the heat exchanger used in the experiment, pay attention that the valve opening should not be too large; at the same time appropriately open the condensate drain valve (valve 21 or valve 6) .

⑦ Adjust the opening of the steam inlet valve 22 or 10 to make the steam slowly flow into the heat exchanger, and gradually heat the heat exchanger to change the heat exchanger from "cold" to "hot", and the heating time shall not be less than 10 min.

⑧ After preheating, adjust the steam pressure to about 0. 01 MPa using the steam inlet valve and maintain the temperature stable.

⑨ Manually adjust the air by pass control valve 2 to change the air flow (the flow can also be adjusted by controlling the frequency of fan frequency converter). Under each flow condition, the experimental values can only be recorded after the heat exchange process is stable. Generally, it takes at least

3～5 minutes for the heat exchange process to be stable under each flow condition. Change the flow, then record the experimental valuer under different flow conditions.

⑩ The air flow changes from small to large, and record 7 groups of experimental data to complete the experiment. Turn off the steam generator, and close the heat exchanger steam inlet valve after 15 minutes. After the heat exchanger temperature drops to room temperature, turn off the air blower and heat exchanger air inlet valve, drain the system condensate, close the steam generator water inlet valve, and switch off meter power and main power.

(2) Procedures of equipment Ⅱ

① Add distilled water to the water tank 13 to a height of 2/3 of the level gauge.

② Fully open the steam inlet valve 6 or valve 9 of the heat exchanger according to the heat exchanger used in the experiment. Switch on the main power and heating power to heat the water in the steam generator.

③ Fully open the air inlet valve 4 or 7 of the heat exchanger according to the heat exchanger used in the experiment, fully open the air bypass control valve 2, then start the air blower when the wall temperature of the double-tube heat exchanger or the inlet steam temperature of the shell-and-tube heat exchanger is close to 90℃.

④ Manually adjust the air bypass control valve 2 to change the air flow. Under each flow condition, the experimental value can only be recorded when the heat exchange process is stable. Generally, it takes at least 3～5 minutes for the heat exchange process to be stable under each flow condition. Record the experimental paramete values under different flow rate.

⑤ The air flow rate changes from small to large, and records 7 groups of experimental data to complete the experiment. Switch off the heating power of the steam generator. Switch off the air blower, close the steam inlet valve of the heat exchanger and the air inlet valve of the heat exchanger when the temperature of the heat exchanger drops to room temperature, then switch off the main power.

8.4.4.2 Precautions

(1) Precautions of equipment Ⅰ

① The steam valve can only be opened after a certain amount of air is input into the pipe in the heat exchanger. The steam can only be input into the double-tube heat exchanger after the condensate accumulated on the steam pipeline is discharged.

② The steam pressure should be controlled below 0.04MPa (gauge pressure) during operation.

③ Pay attention to the evacuation of inert gas and the adjustment of steam pressure or steam temperature all the time.

④ During the experiment or when switching operation is required between heat exchangers, the steam inlet valves of the double-tube heat exchanger and the shell-and-tube heat exchanger cannot be fully closed at the same time, and the air inlet valves of the double-tube heat exchanger and the shell-and-tube heat exchanger cannot be fully closed at the same time.

(2) Precautions of equipment Ⅱ

① Check whether the water level in the steam heating kettle is within the normal range, otherwise water should be added in time.

② During the experiment or when switching operation is required between heat exchangers, the steam inlet valves of the double-tube heat exchanger and the shell-and-tube heat exchanger cannot be fully closed at the same time, and the air inlet valves of the double-tube heat exchanger and the shell-and-tube heat exchanger cannot be fully closed at the same time.

③ Keep the rising steam flow stable, and the heating voltage should not be changed during the experiment.

④ When adding the spiral coil to the double-tube heat exchanger or installing a plug on the shell-and-tube heat exchanger, it is necessary to stop supplying steam to the heat exchanger under operation and cool it properly in advance, then stop supplying air and disassemble it.

8.4.5 Questions

① According to the experimental Nu-Re empirical correlation, what are the main factors that affect the convective heat transfer coefficient in the tube? How do these factors affect α?

② What conclusions can be drawn by comparing the Nu-Re correlations between ordinary tubes and reinforced tubes? What is the mechanism of the reinforced tubes to enhance heat transfer? What is the effect of heat transfer of spiral coil?

③ What other measures can be taken to enhance heat transfer, in addition to adding a spiral coil in a ordinary tube?

④ How should the air outlet temperature theoretically change with the increase of air flow during the experiment? Whether the facts are consistent with this conclusion, if not, why?

⑤ Is the wall temperature of the heat transfer tube close to the steam temperature or the air temperature during the experiment? Why?

⑥ How to calculate the Reynolds number Re of heat transfer experiment in the air?

⑦ What fluid should be discharged from the steam side during the experiment? Why?

⑧ How does the condensation and heat transfer of steam change when the air flow increases?

⑨ Why is the air flow in the reinforced tube smaller than that in the ordinary tube when the size of the ordinary tube and the reinforced tube are the same, the tube is the same and the same air blower is used?

⑩ What is the purpose of discharging noncondensable gas?

⑪ What is the function of the non-condensable gas discharge valve? How to use it correctly?

⑫ Can the experimentally determined Nu-Re correlation be used to calculate the convective heat transfer coefficient of water?

⑬ During the heat transfer experiment of the double-tube heat exchanger, if the inlet air valve of the tube heat exchanger is not completely closed, how will it affect the measurement result Nu-Re correlation? Why?

⑭ Are the Nu-Re empirical correlation obtained from the measurement of heat transfer between the shell-and-tube heat exchanger and the normal tube heat exchanger the same?

⑮ If the area of the heat exchanger is infinite, what will the air outlet temperature be?

⑯ What are the possible reasons for the increase of the steam pressure at the inlet of the heat exchanger when the steam generator is working normally? What measures should be taken to make the steam pressure at the inlet of the heat exchanger return to normal?

8.5 Absorption experiment

8.5.1 Experimental objectives

① Understand the basic structure of the packed absorption column and the process of the absorption device.

② Master the measurement method of the hydrodynamic performance of the packed absorption column, understand the influence of the gas and liquid flow rate on the pressure drop of the whole column under the conditions of the dry column and wet column.

③ Be familiar with the operation of the packed absorption column, the normal flow state of gas and liquid in the column and flooding phenomenon.

④ Master the measurement method of the volume total mass-transfer coefficient of packed absorption column.

8.5.2 Experimental principles

(1) Hydrodynamic Performance

The packed column is a gas-liquid mass transfer device. The gas and liquid phases usually flow counter-currently in the column. The function of packing is to provide contact area for mass transfer between the two phases as well as to form resistance to gas-liquid flow. Therefore, it is of great significance to know the hydrodynamic performance of packing layers, such as pressure drop, flooding and their changing rules in order to determine the suitable operating range of packing column and realize the operation and optimization of packed column.

In addition to the type, structure, size and material of the packing, the pressure drop of the packing layer is mainly related to the gas-liquid load in the packed column. Therefore, under a certain spray rate, the greater the gas velocity is, the greater the pressure drop of the packing layer will be. Under a certain gas velocity, the greater the spray rate is, the greater the pressure drop of the packing layer will be. As shown in Figure 8-11, the relationship between the pressure drop per unit packing layer height and the gas and liquid flow is plotted to obtain the pressure drop curve of the packing layer.

As shown in Figure 8-11, the flow state of gas in the packing layer is generally turbulent. When the liquid spray volume $L=0$, the pressure drop of the packing layer will increase proportionally with the increase of the gas flow rate, so that the pressure drop curve is a straight line with the slope of about 1.8 ~ 2.0, which is called the pressure drop curve of the dry packing layer. The pressure drop curve can be divided into three sections under a certain amount of liquid spray. When the gas velocity is small, the drag force of gas to the liquid film on the surface of packing is small, and increasing the gas velocity has little effect on the liquid flow and little change on the thickness of the liquid film on the surface of the packing. Therefore, the liquid holding capacity in the packing layer is basically the same, the relationship between pressure drop and gas velocity is a straight line, and the pressure drop line in the dry packing layer is parallel, so the area below point A is the constant liquid holding area. When the gas velocity exceeds point A, the drag force produced by the gas to the liquid is large, and the increase of the gas velocity will hinder the flow of the liquid, increasing the thickness and holding capacity of the liquid film. With the increase of liquid holding capacity, the channel of gas will be narrowed and the resistance of gas flow will be increased obviously. At this time, the slope of the pressure drop curve of the packing layer will be increased. With the increase of gas velocity, the liquid holding capacity of the packing layer continues to increase, so A is called loading point. The gas velocity at point A is the loading gas velocity, and the area between A and B is the loading liquid area. When the gas velocity reaches point B, the liquid holding capacity in the column increases sharply because the liquid can no longer flow down smoothly, and the liquid almost fills the whole packing layer. The liquid is transformed from the dispersed phase to the continuous phase, and the gas passes through the packing layer in the form of bubbles, transformed from the continuous phase to dispersed phase. At this time, a slight increase in gas velocity will cause a significant increase in the lamination of the packing, normal flow and mass transfer are severely damaged, a large amount of liquid is entrained by the gas out of the column top, and the operation of the absorption tower is extremely unstable. Therefore, the area above point B is called the flooding area, and point B is called flooding point.

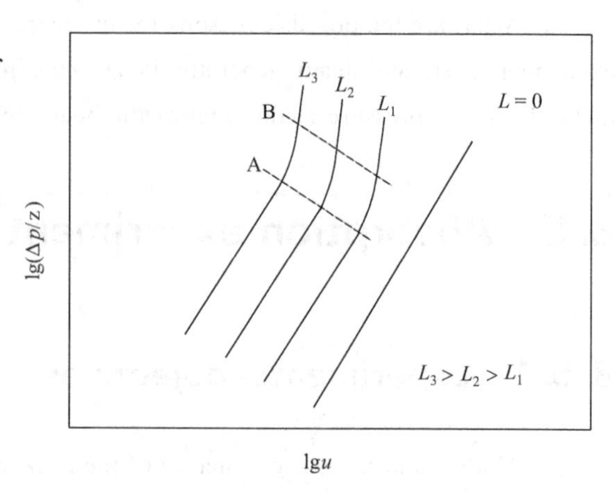

Figure 8-11 Pressure drop curve of packing layer

(2) Mass Transfer Performance

The mass transfer coefficient is one of the important parameters reflecting the mass transfer performance of the packed column. The system properties, operation conditions, gas-liquid flow state, packing type, structure and size will all affect the mass transfer coefficient. Since there are many complex factors affecting the mass transfer coefficient, the mass transfer coefficient is usually measured by experiments to provide accurate and reliable basic data for engineering design and operation. In this experiment, water is used to absorb CO_2 in the air. The absorption process is controued by liquid film. Because the solubility of CO_2 in water is small, the absorption process can be regarded as a low concentration gas absorption, and the temperature of the absorption process is constant. The phase equilibrium constant is also constant. In the experimental device with a certain height of packing layer, the number of mass transfer units can be calculated by measuring the liquid phase composition of the inlet and outlet gas of the absorption column. Then the mass transfer unit height can be calculated by the following formula, to obtain the mass transfer coefficient.

$$h_0=\int_0^{h_0} dh=\frac{L}{K_x a}\int_{x_a}^{x_b}\frac{dx}{x^*-x}=H_{OL}N_{OL}=\frac{L}{K_y a}\int_{y_a}^{y_b}\frac{dy}{y-y^*}=H_{OG}N_{OG} \tag{8-34}$$

where h_0 is packing height, m; L is liquid molar flow rate, kmol/(m^2 · h); G is gas phase molar flow rate, kmol/(m^2 · h); x_a, x_b are mole fraction of liquid inlet and outlet; y_a, y_b are mole fraction of gas phase outlet and inlet; H_{OL} is total mass transfer unit height of liquid phase, m; N_{OL} is total mass transfer unit number in liquid phase; H_{OG} is total mass transfer unit height of gas phase, m; N_{OG} is total mass transfer unit number of gas phase.

And

$$N_{OG}=\frac{1}{1-S}\ln\left[(1-S)\frac{y_b-mx_a}{y_a-mx_a}+S\right] \quad H_{OG}=\frac{G}{K_y a} \quad S=\frac{mG}{L}$$

$$N_{OL}=\frac{1}{1-A}\ln\left[(1-A)\frac{y_b-mx_a}{y_b-mx_b}+A\right] \quad H_{OL}=\frac{L}{K_x a} \quad A=\frac{L}{mG}$$

where, $K_y a$ is total mass transfer coefficient of gas volume, kmol/(m^3 · h); $K_x a$ is total mass transfer coefficient of liquid volume, kmol/(m^3 · h); S is desorption factor; A is absorption factor.

The phase equilibrium relationship of isothermal physical absorption at low concentration is in accordance with Henry law, the formula as follows:

$$y^*=mx \tag{8-35}$$

where m is phase equilibrium constant.

In the specific absorption column (column diameter, packing height, packing type, packing size, packing material, etc.), under certain operating conditions (such as temperature, pressure, gas-liquid ratio, etc.), the absorption effect can be evaluated by the absorptivity.

$$\eta=\frac{y_b-y_a}{y_b} \tag{8-36}$$

where η is absorptivity.

8.5.3 Experimental device

The experimental device includes absorption column and desorption column. The mixed gas of air and carbon dioxide enters the absorption column from the bottom of the column. The absorbent (water) is pumped in by the absorption liquid pump, being sprayed from the top of the column to contact with the mixed gas in the packing layer. After the absorption, the mixed gas is discharged from the top of the absorption column and the absorption liquid is discharged from the bottom of the absorption column. And enters the desorption liquid tank, then driven into the top of the desorption column by the desorption liquid pump. The ambient air, as the desorbent, is transported by a fan and input from the bottom of the desorption column, and contacts the liquid in the packing layer to transfer mass. The liquid is desorbed and regenerated in the desorption column, and discharged from the bottom of the desorption column, then enters the absorption liquid tank to be used as absorbent for recycling. The flow chart of absorption experiment device is shown in Figure 8-12.

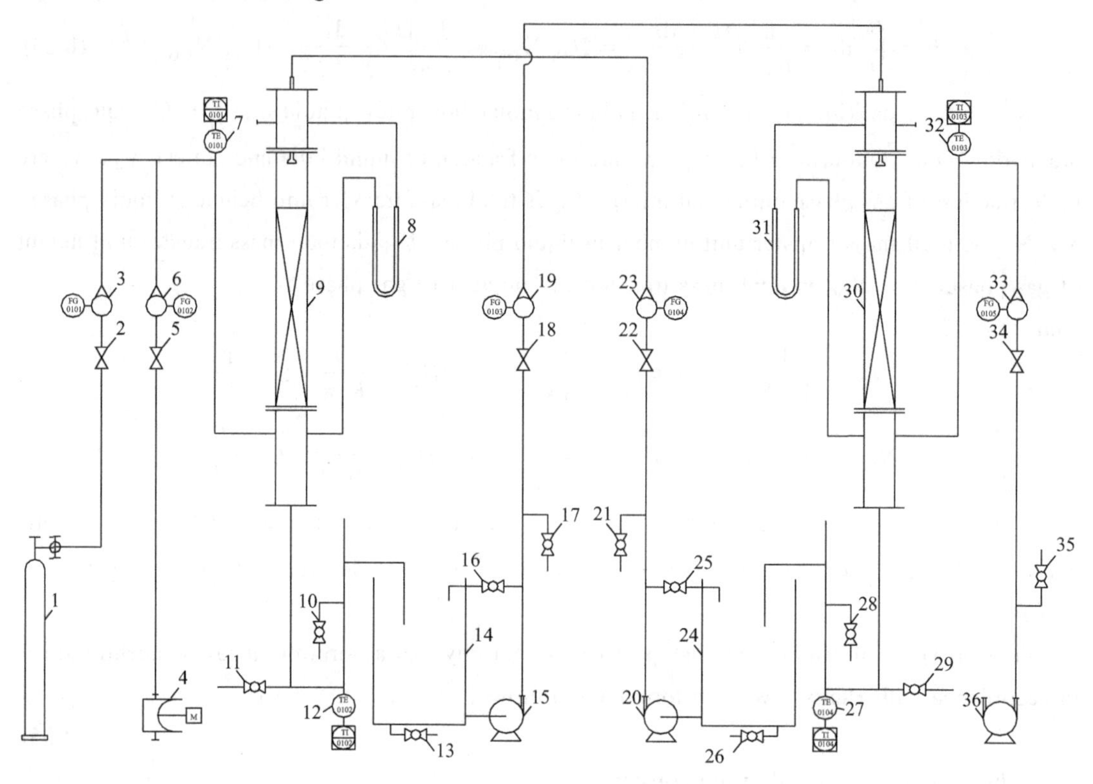

Figure 8-12 Flow chart of absorption experiment device

1—Carbon Dioxide Cylinder; 2—Flow control valve of carbon dioxide; 3—Carbon dioxide flowmeter; 4—Air pump of absorption column; 5—Air flow control valve of absorption column; 6—Air flowmeter of absorption column; 7—Mixed gas thermometer of absorption column; 8—U-tube differential pressure gauge of absorption column; 9—Absorption column; 10—Liquid outlet sampling valve of absorption column; 11—Drain valve of liquid outlet pipe of absorption column; 12—Liquid thermometer of absorption column; 13—Drain valve of desorption liquid tank; 14—Desorption liquid tank; 15—Desorption liquid pump; 16—Circulation valve of desorption liquid pump; 17—Liquid inlet sampling valve of desorption column; 18—Flow control valve of desorption liquid; 19—Desorption liquid flowmeter; 20—Absorption liquid pump; 21—Liquid inlet sampling valve of absorption column liquid; 22—Flow control valve of absorption liquid; 23—Absorption liquid flowmeter; 24—Absorption liquid tank; 25—Circulating valve of absorption liquid pump; 26—Drain valve of absorption liquid tank; 27—Thermometer of desorption liquid; 28—Liquid outlet sampling valve of desorption column; 29—Drain valve of liquid outlet pipe of desorption column; 30—Desorption column; 31—U-tube differential pressure gauge of desorption column; 32—Desorption thermometer; 33—Air flowmeter of desorption column; 34—Flow control valve of desorption air; 35—Bypass valve of desorption air fan; 36—Fan of desorption column

Main technical data of the device are as follows.

① Absorption column and desorption column: diameter ϕ76mm×3.5mm, height of packing layer 850mm.

② Packing: specific surface area of ceramic Raschig ring packing 440m^2/m^3; specific surface area of stainless steel Pall ring packing, 480m^2/m^3.

③ Flowmeter: CO_2 flowmeter, LZB-6 (0.06～0.6m^3/h); rotameter of desorption air

LZB-10 (0. 25～2. 5m^3/h); water rotameter LZB-15 (40～400L/h); rotameter of desorption column air, LZB-40 (4～40m^3/h).

④ Fan: HG-250-C vortex air pump.

⑤ Centrifugal pump: WB50/025.

⑥ Thermometer: PT100, AI501 digital display meter.

8. 5. 4 Experimental procedures and precautions

8. 5. 4. 1 Experimental procedures

(1) Experiment of hydrodynamic performance

The hydrodynamic performance experiments are carried out in the desorption column in the following steps.

① Check whether the device, valves, instrument, etc. are normal and whether the water level of the water tank is appropriate.

② Switch the main power and instrument power. The control panel of instrument is shown in Figure 8-13.

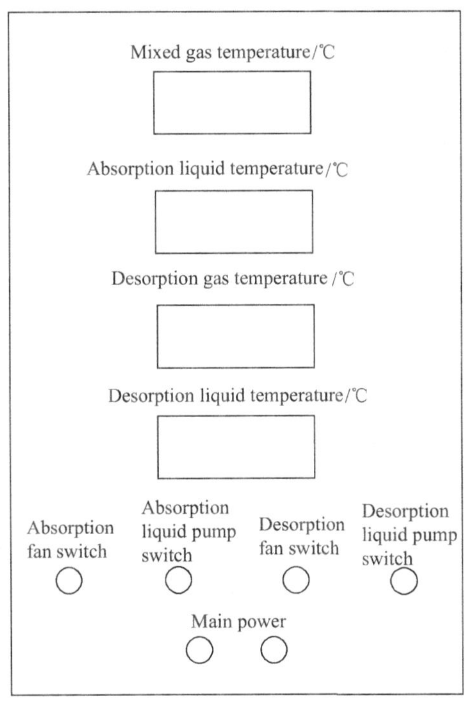

Figure 8-13 Control panel of the instrament

③ Open the desorption liquid pump 15 (close all the outlet valves of the centrifugal pump—pumb 16, 17 and 18, switch on the power of pump 15, then open the circulation valve 16. Open the flow control valve after stabilization; the subsequent start-up operations of the centrifugal pump will no longer be described), and fully wet the packing for 10～20min (this step could be omitted if the packing is already wetted before experiment).

④ Adjust the liquid flow control valve 18 to a certain flow (When the flow is 0, the pressure drop of the dry packing layer is measured; when the flow is not 0, the pressure drop of the wet packing layer is measured) .

⑤ Switch on the power of fan after fully opening the bypass valve of fan 35, use the control valve 34 (by pass 35 can be used for auxiliarx adjustment) to adjust the desorption flow from small to large, measure the pressure drop of the packed column and observe the flow state of gas and liquid in the desorption column, until flooding.

⑥ Close the desorbing air flow control valve 34, fully open the bypass valve 35, close the desorption air fan (the closing operations of fans will no longer be described afterwards); close the desorption liquid flow control valve 18, circulation valve 16 and switch of the centrifugal pump 15 (no longer be described afterwards) .

(2) Experiment of mass transfer performance

① Switch on the absorption liquid pump 20 and adjust the absorption liquid flow control valve 22 to a certain flow.

② Switch on the absorption air fan 4 and adjust the air flow to a certain value by control valve 5; open the main valve of carbon dioxide cylinder when the pressure reducing valve is closed, then slowly open the pressure reducing valve, and adjust carbon dioxide flow control valve 2 to a certain flow.

③ Switch on the power of desorption air fan after fully opening the bypass valve of desorption air fan, 35. Adjust the desorption air flow to a larger value by flow control valve 34 and by pass valve of desorption air fan 35 (no flooding blowed), ensuring the fully desorbed of desorption liquid) .

④ After the desorption liquid pump 15 is switched on, adjust the desorption liquid flow control valve 18 to make the desorption liquid flow consistent with the absorption liquid flow.

⑤ Maintain system temperature, flow liquid level and other parameters stable. After the system is stable, sample and analyze the liquid phase at the inlet and outlet of the absorption column.

⑥ After the experiment, close the main valve of carbon dioxide cylinder, and when the carbon dioxide flow is zero, close the carbon dioxide rotameter and cylinder pressure reducing valve, then switch off the air fan 4.

⑦ Close the flow control valve of absorption liquid 22, then shut down the absorption liquid pump 20; close the flow control valve of desorption liquid 18, then shut down the desorption liquid pump 15.

⑧ Fully open the bypass valve of desorption air fan 35, close the flow control valve of desorption air 34, then shut down the desorption liquid pump 15.

⑨ Switch off the instrument power and main power, clean up the experimental device and the site.

(3) Analysis

① The concentration of CO_2 in each sample is analyzed by acid-base titration. Take 20mL of liquid sample, add 5mL of $Ba(OH)_2$ (0. 1mol / L) solution inside. After full response, add a few drops of phenolphthalein indicator, then perform titration analysis using a hydrochloric acid .

② Calculate the concentration of CO_2 as follows:

$$c_{CO_2}=\frac{2c_{Ba(OH)_2}V_{Ba(OH)_2}-c_{HCl}V_{HCl}}{2V_{sample}}$$

③ Titrate blank sample to correct the analysis results of the samples.

④ Parallel samples shall be tested .

8.5.4.2 Precautions

① Pay attention to the control operation when the flooding state appears in the experiment of hydrodynamic performance control the flow to prevent a large amount of liquid foam from being entrained from the top of the column, or even overflowing the top of the column.

② When the flooding state is about to appear, quickly adjust the flow and record the relevant data.

③ After making sure that pressure reducing valve of the cylinder and the regulating valve of CO_2 gas flowmeter are closed, open the main valve of the cylinder; after opening the main valve, slowly open the pressure reducing valve and adjust the CO_2 flow control valve.

④ During the experiment of mass transfer performance water level of two tanks should be kept stable. In addition, the water level of the two tanks should also be seen to in the experiment of hydrodynamics performance.

⑤ Before sampling, the liquid in the pipeline should be replaced first; the sampling operation should be prompt and the lid should be placed on as soon as possible to reduce the contact between sample and the air.

8.5.5 Questions

① What is the significance of measuring the $\Delta p/z$-u curve of the packed column?

② What are the factors related to the flooding of the packed column through the experiments of hydrodynamic performance?

③ What is the engineering significance of measuring the total volume mass transfer coefficient?

④ Why is the CO_2 absorption process controlled by liquid film?

⑤ What temperature should be used to calculate Henry's coefficient when the temperatures of gas and liquid are different?

⑥ Why should the vent tube be set on the liquid outlet tube of column?

⑦ What is the function of the U-tube outlet pipeline of absorption column and desorption column? How to determine its height?

⑧ How does the pressure drop of packing layer change with other conditions unchanged and the liquid flow rate is increased? How does the gas velocity of flooding point change with the increase of liquid flow rate?

⑨ When the liquid flow rate of absorption column increases with other conditions unchanged, how does the composition of liquid outlet and absorptivity change?

⑩ What is the effect on the absorption process when the desorption air flow decreases?

⑪ What is the difference between the absorption experiment operation in winter and summer if other conditions are the same?

⑫ What will the deviation of the measured total mass transfer coefficient be when the bypass valve of the mixture flowmeter is not fully closed?

⑬ Which temperature should be used to calculate the phase equilibrium constant when the gas temperature and the absorbent temperature are different?

⑭ During the start-up prosedures of absorption column, why is it necessary to input the absorbent before the mixture gas?

⑮ What is the packing in the desorption column? Is the packing of this size suitable for the column? Why?

⑯ For bulk packing, the wetting rate of packing is generally required to be no less than 0. 08. Under the experimental conditions, please calculate whether the wetting rate of the absorption column packing can meet such a requirement? Knowm wetting tate= liquid spray density/ specific surface area of packing.

⑰ In the mass transfer process in absorption experiment, if the absorption rate is required to reach 10%, how to determine the flow rate of the absorbent given the known composition, flow rate of the mixed gas entering the column, as well as operating conditions?

⑱ What is the function of the regular packing above bulk packing?

⑲ Why is a circulation pipeline set between the centrifugal pump and water tank?

⑳ What are the typical characteristics of flooding phenomenon of packed column?

㉑ Which packing of absorption column and desorption column has larger pressure drop? Under the experimental conditions, the pressure drop of desorption column is much larger than that of absorption column, what is the main reason?

㉒ If the desorption temperature of your experimental device is significantly higher than that of other experimental devices under the same conditions, what is the possible reason the thermometer working normally?

㉓ In the absorption and desorption mass transfer experiment, if the liquid flow rate of the desorption column is lower than that of the absorption column, what effect will it have on the absorption process?

8. 6 Distillation experiment

8. 6. 1 Experimental objectives

① Understand the structure of the distillation tower and the operation process of distillation.

② Master the operation methods of start-up, shut-down and stable operation of distillation tower.

③ Master the measurement methods of overall efficiency and single tray efficiency.

④ Understand the influence of operation parameters on the mass transfer and separation performance and hydrodynamic performance of the distillation tower. Master the control and regulation methods for stable operation of the distillation tower.

8. 6. 2 Experimental principles

In the plate distillation tower, the steam generated by the tower kettle ascends plate by plate along the height of the tower, contacting with the reflux liquid descending plate by plate, so as to achieve multiple heat and mass tranfer between the two, from the top of the tower and the mixed liquid is seperated to a certain degree. The liquid-phase reflux at the top of the tower and the vapor-phase reflux at the bottom of the tower are the basis of the distillation operation. Reflux ratio is one of the important parameters of distillation operation. The reflux ratio, feed position, feed concentration and feed quantity affect the separation effect of distillation operation. In practical operation, due to the limit of mass transfer time, mass transfer area and other factors, it is impossible for vapor-liquid two-phase on the tray to reach equilibrium, so the separation effect of the actual plate is lower than that of the theoretical plate. Therefore, the overall efficiency (total plate efficiency) and the single plate efficiency are usually used to evaluate the separation effect of the whole tower and the single plate.

(1) Overall efficiency E

$$E=\frac{N}{N_e} \tag{8-37}$$

where E is overall efficiency; N is theoretical plate numbers (excluding tower kettle); N_e is actual number of trays.

(2) Single plate efficiency E_m

The single plate efficiency is known as Murphree efficiency. The single plate efficiency includes the liquid phase Murphree efficiency E_{ml} and vapor phase Murphree efficiency E_{mv}.

$$E_{ml,n}=\frac{x_{n-1}-x_n}{x_{n-1}-x_n^*} \tag{8-38}$$

where $E_{ml,n}$ is the liquid phase Murphree efficiency of the n practical plate; x_n, x_{n-1} are the liquid composition (mole fraction) of the n actual plate and the (n−1) actual plate, respectively; x_n^* is composition (mole fraction) of liquid phase in equilibrium with vapor phase concentration of the n actual plate.

$$E_{mv,n}=\frac{y_n-y_{n+1}}{y_n^*-y_{n+1}} \tag{8-39}$$

where $E_{mv,n}$ is the vapor phase Murphree efficiency of the n actual plate; y_n, y_{n+1} are the vapor phase composition (molar fraction) of the n actual plate and the (n−1) actual plate, respectively; y_n^* is composition (mole fraction) of gas phase in equilibrium with vapor phase concentration of the n actual plate.

Under the condition of total reflux, because the operating line is $y_{n+1}=x_n$, the measurement and calculation of Murphree efficiency can be simplified as:

$$E_{ml,n}=\frac{x_{n-1}-x_n}{x_{n-1}-x_n^*} \tag{8-40}$$

$$E_{mv,n}=\frac{y_n-y_{n+1}}{y_n^*-y_{n+1}}=\frac{x_{n-1}-x_n}{y_n^*-x_n} \tag{8-41}$$

where $x_n^*=f(y_n)=f(x_{n-1})$, $y_n^*=f(x_n)$, therefore, the Murphree efficiency can be measured by the liquid phase composition of two adjacent plates under the condition of total reflux.

(3) Theoretical plate numbers N

Theoretical plate numbers can be obtained by plate-by-plate calculation method or graphical method. Taking the ethanol-water system as an example, because the ethanol water system is a highly non-ideal system, the graphical method is used to calculate the number of theoretical plates. The basic principle is similar to the plate-by-plate calculation methods, and the graphical methods under the conditions of total reflux and partial reflux are shown in Figure 8-14 and Figure 8-15 respectively.

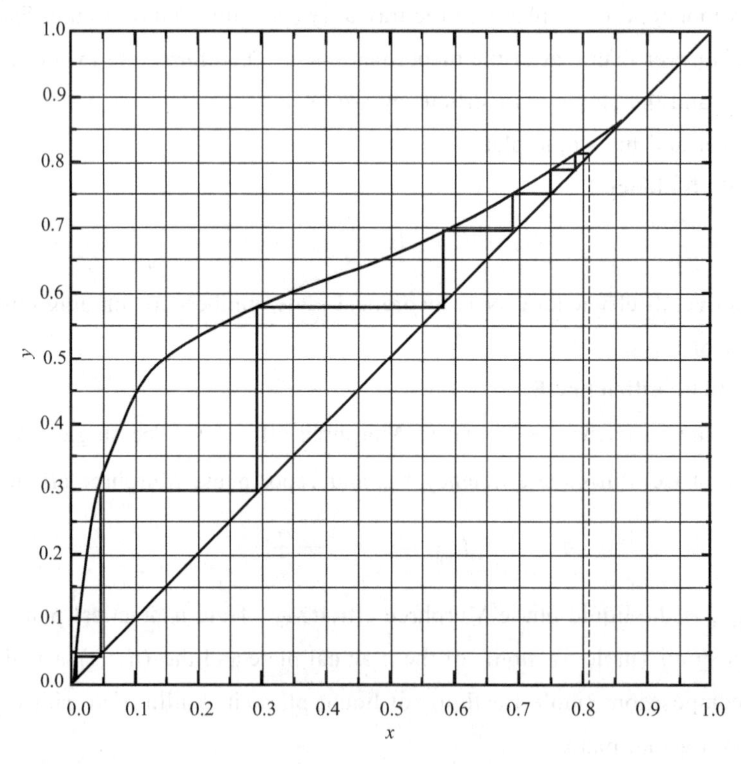

Figure 8-14 Calculation of theoretical plate numbers under total reflux

1) Calculation of theoretical plate numbers under total reflux The operation line coincides with the diagonal line on the x-y phase diagram under the condition of total reflux. Therefore, the number of theoretical plates can be obtained by plate calculation between the equilibrium line and the diagonal line on the x-y phase diagram only by determining the composition of the tower top and the tower kettle.

2) Calculation of theoretical plate numbers under partial reflux condition Under the condition of partial reflux, plate-by-plate calculation needs to be carried out between the equilibrium line and

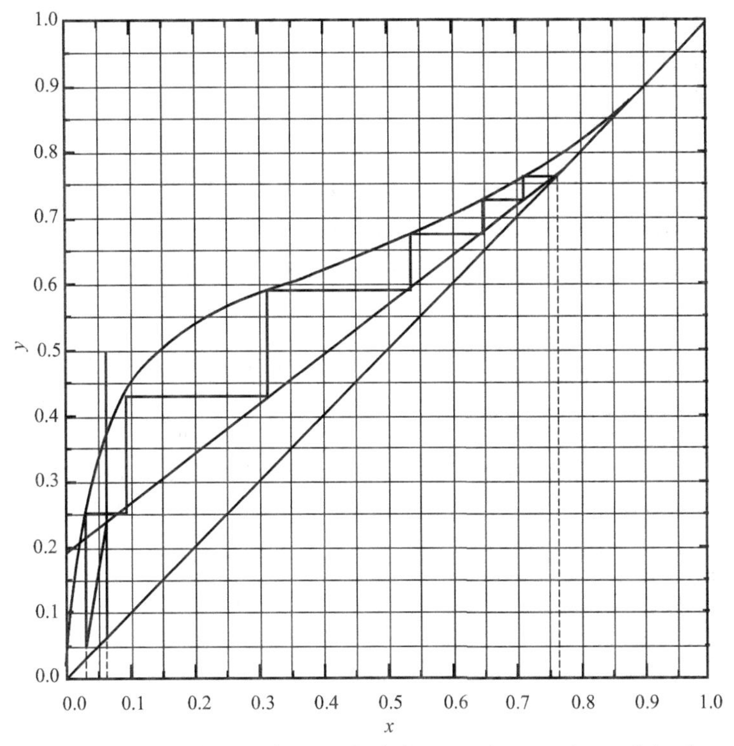

Figure 8-15 Calculation of theoretical plate numbers under partial reflux

the operation lines. Therefore, the operating line needs to be determined first. As shown in Figure 8-15, after the compositions of tower top and tower kettle are determined, the operation line equation and feed equation of the rectifying section could be drawn:

$$y=\frac{R}{R+1}x+\frac{x_D}{R+1} \tag{8-42}$$

$$y=\frac{q}{q-1}x-\frac{x_F}{q-1} \tag{8-43}$$

After the operation line of rectifying section and q line are determined on the phase diagram, the operation line of stripping section can be determined by connecting the intersection point of the two operation lines and the state point corresponding to the composition of the tower kettle x_W on the diagonal.

It can be seen that the key to the determination of the above operation line lies in the calculation of the reflux ratio R and the value q. Since the state of the feed liquid is cold liquid, it can be obtained according to the heat balance of the feed inlet.

$$q=1+\frac{c_{pF}(t_S-t_F)}{r_F} \tag{8-44}$$

where q is feed heat condition parameters; r_F is latent heat of vaporization with feed liquid composition, kJ/kmol; c_{pF} is specific heat capacity of feed liquid at average temperature $(t_S+t_F)/2$, kJ/(kmol · ℃); t_S is bubble point temperature of feed liquid, ℃; t_F is feed liquid temperature, ℃.

Similarly, when the cold liquid is refluxed, the supercooled liquid flowing back into the tower will partially condense the rising vapor in the tower, thereby increasing the amount of liquid actually flowing downward, and increasing the ratio of the vapor-liquid two-phase molar flow rate in the tower (the vapor and liquid flow rates in the tower are L and V respectively).

$$\frac{\overline{L}}{\overline{V}}=\frac{R'}{R'+1}$$

It is equivalent to changing the reflux ratio (represented by R' at this time, and in numerical value $R'> R= L/D$). Therefore, the reflux ratio R in the operation line of the rectifying section of formula (8-42) should be replaced with the actual reflux ratio R', and the actual reflux ratio in the tower is:

$$R'=\frac{L'}{D}=\frac{L}{D}\left[1+\frac{c_p(t_1-t_R)}{r}\right]=R\left[1+\frac{c_p(t_1-t_R)}{r}\right] \tag{8-45}$$

where R' is actual reflux ratio; R is reflux ratio in bubble point (ratio of reflux to produced liquid flow) L/D; c_p is average specific heat capacity of reflux at average temperature $(t_1+t_R)/2$, kJ/(kmol · ℃); t_1 is saturated steam temperature at the top of the tower, ℃; t_R is reflux temperature, ℃; r is vaporization latent heat of saturated steam at the top of the tower, kJ/kmol.

The main physical properties of the distillation system are as follows:

Specific heat of ethanol aqueous solution: [c_p, kJ/(kg · ℃); w, mass percentage, %; t, ℃]

$c_p=(1.0365-1.3485\times10^{-3}w-9.3326\times10^{-4}t-4.3944\times10^{-5}w^2+3.863\times10^{-5}wt+9.962\times10^{-6}t^2)\times4.187$

Vaporization heat of ethanol aqueous solution: (r, kJ/kg; w, mass percentage, %) $r=(4.745\times10^{-4}w^2-3.315w+5.3797\times10^{-2})\times4.187$

Vaporization heat of ethanol: $r=-0.0042t^2-1.5074t+985.14$ (r, kJ/kg; t, ℃)

Vaporization heat of n-propanol: $r=-0.0031t^2-1.1843t+839.79$ (r, kJ/kg; t, ℃)

Specific heat capacity of ethanol: $c_p=4.3357\times10^{-5}t^2+0.00621t+2.2332$ [c_p, kJ/(kg · ℃); t, ℃]

Specific heat capacity of n-propanol: $c_p=-8.3528\times10^{-7}t^3+1.2144\times10^{-5}t^2+0.00365t+2.222$ [c_p, kJ/(kg · ℃); t, ℃]

8.6.3 Experimental device

(1) Flow chart of equipment I

The ethanol aqueous solution in the raw material tank is pressurized by the feed pump and then enters the kettle liquid cooler, exchanging heat with the kettle liquid extracted from the tower kettle. The heated raw material liquid enters the distillation tower, and the liquid in the distillation tower kettle is heated by the electric heater and vaporized and then ascends plate by plate, performing mass and heat transfer with the liquid oneach plate. Finally, it enters the top of the distillation tower and be condensed by the coil condenser. The condensate flows out of the liquid collector, partly distilled on the top of the tower as the product and enters the product tank, and other part as refluxed liquid flows from top to kettle plate by plate, and finally enters the tower kettle for partial vaporization. The generated vaporization flows upward along the tower height, while the residual liquid at the kettle of the tower passes through the kettle liquid cooler and enters the kettle liquid tank. The specific experimental flow is shown in Figure 8-16.

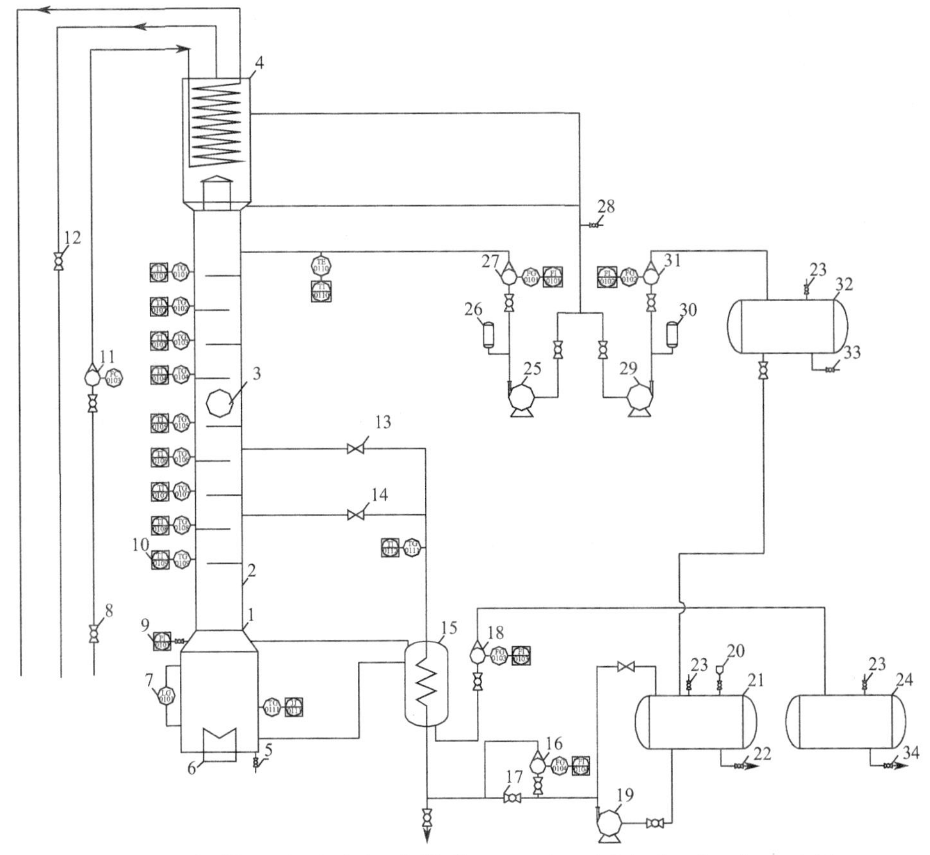

Figure 8-16 Flow chart of the distillation experimental device (equipment Ⅰ)

1—Tower kettle; 2—Dstillation tower; 3—Distillation tower visual cup; 4—Overhead condenser; 5—Tower kettle sampling valve; 6—Heating tube of tower kettle; 7—Level gauge of tower kettle; 8—Cooling water inlet valve; 9—Pressure gauge of tower kettle; 10—Thermometer; 11—Cooling water flowmeter; 12—Top vent valve; 13—Upper feed valve; 14—Lower feed valve; 15—Kettle liquid cooler; 16—Feed flowmeter; 17—Quick feed valve; 18—Kettle liquid flowmeter; 19—Feed pump; 20—Feeding port of raw material tank; 21—Raw material tank; 22—Outlet and sampling valves of raw material tank; 23—Vent valve; 24—Kettle tank; 25—Reflux pump; 26—Buffer tank of reflux pump; 27—Reflux liquid flowmeter; 28—Tower top sampling valve; 29—Product pump; 30—Buffer tank of product pump; 31—Product flowmeter; 32—Product tank; 33—Drain port of product tank; 34—Drain port of kettle liquid tank

(2) Main technical data of equipment I

The main structural parameters of the sieve plate tower: inner diameter of the tower $D=$ 68mm, thickness $\delta=2.5$mm, outer diameter is 73mm, thickness is 2. 5mm, number of plates N= 10, and the plate spacing $H_T=100$mm. The feeding positions are the 6th and 8th plate counting from top. The downcomer is bow-shaped and the height of the bottom gap of the downcomer is 4. 5mm. The overflow weir is a tooth shaped weir with a weir length of 56mm, height of 7. 3mm, and a tooth depth of 4. 6mm, and 9 teeth. Sieve hole diameter $d_0=1.5$mm, regular triangle distributed, hole spacing $t=5$mm, and the number of holes 54. The tower kettle is of internal electric heating type with heating power of 2. 5kW and effective volume of 10L. The tower top condenser and kettle liquid cooler are both coiled tube heat exchanger. The samples are taken from the 1st, the 2nd, the 9th and the 10th blocks counting from the top.

Raw material tank：ϕ250mm×540mm，horizontal；product tank：ϕ200mm×520mm，horizontal；kettle liquid tank：ϕ200mm×520mm，horizontal.

(3) Flow chart of equipment Ⅱ

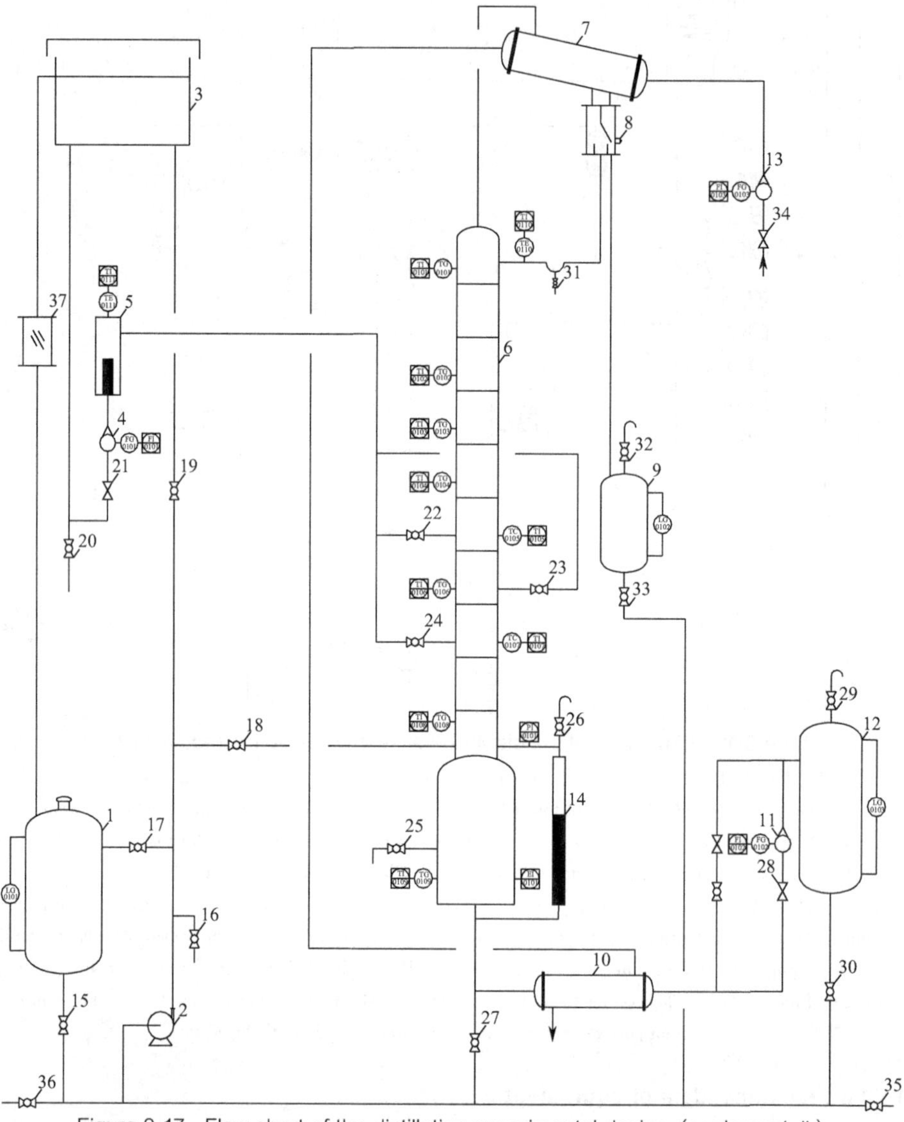

Figure 8-17　Flow chart of the distillation experimental device（equipment Ⅱ）

1—Raw material tank；2—Feed pump；3—High slot raw material；4—Feed flowmeter；5—Feed preheater；6—Distillation tower；7—Top condenser；8—Reflux ratio controller；9—Tower top product tank；10—Tower kettle liquid cooler；11—Tower kettle residual liquid flowmeter；12—Tower kettle residual liquid tank；13—Cooling water flowmeter；14—Magnetic flap level gauge；15—Bottom valve of raw material tank；16—Valve；17—Circulation valve of feed pump；18—Quick feed valve of tower kettle；19—Feed valve of high level tank；20—Raw material sampling valve；21—Main feed valve；22—Feed valve 1；23—Feed valve 2；24—Feed valve 3；25—Tower kettle sampling valve；26—Tower kettle vent valve；27—Tower kettle bottom valve；28—Residual liquid discharge valve of tower kettle；29—Vent valve of residual liquid tank；30—Bottom valve of residual liquid tank；31—Tower top product sampling valve；32—Vent valve of top product tank；33—Bottom valve of tower kettle product tank；34—Cooling water valve；35，36—Drain valve；37—Glass sight cup

(4) Main technical data of equipment II

Main structural parameters of sieve plate tower: Inner diameter of tower D=50mm, Number of plates N=9, Plate spacing H_T=120mm. The downcomer is round and the bottom gap height of the downcomer is 4.5mm. The tower kettle is electrically heated with a heating power of 2.5kW. The top condenser and kettle liquid cooler are both tube heat exchangers.

Raw material tank: ϕ300mm × 400mm, vertical; High-level tank: 200mm × 100mm × 200mm; Product tank: ϕ150mm × 260mm, vertical; Kettle liquid tank: ϕ150mm × 260mm, vertical; Overhead condenser: ϕ89mm × 600mm, horizontal; Tower cooler: ϕ76mm × 200mm, horizontal.

8.6.4 Experimental procedures and precautions

8.6.4.1 Experimental procedures

(1) Experimental procedures of equipment I

1) Total reflux

① Prepare 10% ~ 25% ethanol aqueous solution and add it into the raw material tank 21.

② Switch on the main power and the instrument power. Start the feed pump 19 after opening the top vent valve 12. Open valves 14 and 17 and drive the feed into the distillation tower kettle until the scale of the kettle level gauge is approximately 30~35 cm. Close the feed valves 14 and 17 and turn off the feed pump.

③ Switch on the power of the electric heating tube after closing the discharge tube at the top of the tower and the kettle of the tower. Adjust the heating voltage appropriately to slowly increase the temperature of the tower kettle.

④ Before the steam ascends to the tower top open the cooling water inlet valve 8 and adjust the flow meter 11 to an appropriate cooling water flow.

⑤ When condensate appears at the top of the tower, fully open valve 27. Then start reflux pump 25 to make the distillation tower in the state of total reflux.

⑥ Adjust the heating power of the tower kettle so that the return flow is in a proper state through observing the vaper-liquid contact status on the plate. When the temperatures of the tower top and kettle are stable, record the experimental data and take samples from sampling valves 5, 28 respectively to analyze the concentration at the tower top x_D and tower kettle x_W.

⑦ Take samples from the 1th, 2th, 9th and 10th plates from the top to measure the Murphree efficiency. Use syringes to slowly extract liquid from the sampling port of the corresponding plate. Take about 1mL and inject it into the sample bottle for washing and drying. Try to take samples simultaneously.

2) Partial reflux

① When the total reflux operation of the distillation tower is stable, open the feed valve 13 or 14. Ajust the feed rate to an appropriate value by feed flowmeter 16.

② Control the stroke of overhead reflux pump 25 and product pump 29 and regulate the flow of reflux liquid and overhead distillate through flow meters 27 and 31 to keep the reflux ratio at a certain value.

③ Adjust the flowmeter 18 to control the discharge of residual liquid in the tower kettle, and maintain the balance of materials in and out of the whole tower.

④ When the temperatures at the top of the tower and each plate are stable, record the experimental data and take samples from the sampling valves 28, 5 and 22 to analyze the concentration of the tower top x_D, tower kettle x_W and feed x_F.

⑤ Stop the heating of the tower kettle, close the feed pump, feed valve, discharge pump and discharge valve after the experiment.

(2) Experimental procedures of equipment Ⅱ

1) Total reflux

① Prepare a certain concentration of ethanol-n-propanol solution and add it into the raw material tank 1.

② Switch on the main power and instrument power. Start the feed pump 2 according to the start-up procedures of centrifugal pump, open the circulation valve 17, the quick feed valve 18, the vent valve 26 of tower kettle, then drive the feed liquid into the distillation tower kettle to the liquid level 2/3 of tower kettle, close the valves 17, 18, 26 and close the feed pump.

③ Close the tower top discharge tube and the tower kettle discharge tube; start the tower kettle heating power; adjust the heating voltage appropriately and make the tower kettle temperature rise slowly; open the pressure gauge valve at the same time.

④ Before the steam adscends to the tower top open the cooling water valve 34 and adjust the flowmeter 13 to an appropriate cooling water flow (80～120L/h).

⑤ After condensate appears on the top of the tower, adjust the heating power of the tower kettle to make the reflux flow in an appropriate state. When the temperature of the tower top and the temperature of the tower kettle are stable, record the experimental data and take samples from the top sampling valve 31 and the tower sampling valve 25 respectively, to analyze the composition of the tower top of the x_D and tower kettle x_W.

⑥ Sample at certain plate of the rectification tower to determine the efficiency of the single plate based on needs. Try to take sample simultaneously.

2) Partial reflow

① When the total reflux operation of the distillation tower is stable, start the feed pump 2, open the valves 17 and 19, pump the feed liquid into the high slot 3. When the stable fluid is observed to pass through the glass sight cup, open one of the feed valves 22, 23 and 24, then open the main feed valve 21, and adjust the feed flowmeter 4 to an appropriate value.

② Turn on the reflux ratio controller to keep the reflux ratio at a certain value.

③ Open the vent valve 32 of the top product tank 9 and the vent valve of the tower kettle residual liquid tank 29; open the residual liquid discharge valve 28 and adjust the residual liquid flowmeter 11 to maintain material balance in and out of the whole tower and a stable liquid level of the magnetic level indicator 14.

④ When the temperatures of the tower top and each plate are stable, record the experimental data and sample from the top product sampling valve 31, the tower kettle sampling valve 25 and the raw material sampling valve 20 to analyze the tower top concentration x_D, tower kettle concentration x_W and Feed concentration x_F.

⑤ When the experiment is completed, stop the heating of the tower kettle, close the feed pump, and feed valve and the discharge valve. When there is no steam rising in the tower and the tower temperature drops to a proper temperature, close the cooling water flowmeter valve and water inlet valve, then close other valves.

8. 6. 4. 2 Precautions

① Make sure the vent valve at the top of the tower is opened. Otherwise, the distillation tower may not be operated under normal pressure condition and the pressure in the tower may be dangerously high.

② The heating power of the tower kettle can only be turned on after the feed liquid is added to the 2/3 of total liquid level, otherwise the electric heating wire will be damaged due to dry burning of low liquid level.

③ During the operation of the feed pump, the valve of the circulation pipeline between the feed pump and the raw material tank should be opened to maintain the uniform composition in the raw material tank.

④ The feed flow, return flow, product flow and discharge flow of kettle liquid should be adjusted appropriately to keep the material balance in and out of the tower.

⑤ Pay attention to keep a slow adjustment of the heating power of the tower kettle to avoid boiling during the experiment.

⑥ The cooling water should be connected before heating the tower kettle, or the cooling water should be connected before the steam enters the tower top during start-up. When shutting down, the heating power of the tower kettle should be closed first, and the cooling water on the tower top could only be closed after the system is cooled.

8. 6. 5 Questions

① What are the flow characteristics of vapor-liquid two-phase flow in plate tower?

② What factors will affect plate efficiency?

③ Why is tower kettle pressure an important operating parameter in distillation tower operation? What factors are related to the pressure of the tower kettle?

④ What is reflux ratio? Why is the reflux ratio necessary in distillation? Try to explain the role of reflux, and how to determine and control the reflux ratio according to the experimental equipment?

⑤ How to increase reflux ratio in operation? Can it be achieved by reducing tower top output D ?

⑥ What is the effect on the separation performance of the distillation tower if the reflux ratio is changed while other conditions remain the same?

⑦ Is the feed position optional? What effect does it have on the separation performance of distillation tower?

⑧ If the product is unqualified due to the excessive recovery rate at the top of the tower during the operation，what is the fastest and most effective way to correct it?

⑨ Why is normal pressure operation used for the distillation of ethanol-water system instead of pressure distillation or vacuum distillation?

⑩ How to realize the normal pressure operation of the distillation tower? How to achieve pressurization or decompression operation?

⑪ How many parameters should be measured when determining the overall efficiency and Murphree efficiency? Where are the sampling locations?

⑫ How to get x_n^* after measuring the liquid phase composition of the N and (N−1) plate in total reflux and partial reflux?

⑬ In total reflux，if the liquid phase composition of plate n and n-1 is already measured，can the vapor phase Murphree efficiency of plate n be obtained?

⑭ How to reasonably determine the amount of distillate at the top of the distillation tower? If the actual volume of distillate from the top of the tower does not meet the requirements，how to adjust the reflux and the heating power of the tower?

⑮ What is the sensitive plate and it's temperature? What is the significance of sensitive plate temperature to the operation of the distillation tower?

⑯ What are the possible reasons for the rise of tower temperature during the experiment? What measures should be taken?

⑰ What are the possible reasons for the continuous decline of the liquid level in the tower kettle during operation? What measures should be taken to correct it?

⑱ If flooding or serious leakage occur in the process of operation，what measures should be taken to correct it?

⑲ When the amount and composition of feed are fixed，how to determine the amount of distillate from the top of the distillation tower to meet the criteria that the composition of the top of the distillation tower is no less than x_D^0?

8.7 Drying experiment

8.7.1 Experimental objectives

① Understand the structure and process of the tunnel dryer.

② Master the operation methods of start-up，shut-down and stable operation of drying equipment.

③ Master the measurement methods and influencing factors of drying curve and drying rate curve.

④ Understand the significance of the drying rate curve in industrial dryers designing.

8.7.2 Experimental principles

Drying is a unit operation that uses heat to dehumidify (water is the commonly seen moist component). It involves not only the heat and mass transfer between gas and solid phases, but also the mass transfer mechanism of moisture from the inside to the surface in the form of gas or liquid. Dwe to different moisture content, shape and structure of the material and different drying conditions, the drying rate in materials varies greatly. Generally speaking, the drying rate is affected by the nature, structure and water content of the material, the state temperature and humidity and velocity of the drying medium humid air usually, and the contact mode between the drying medium and the material, etc. Since the influencing factors of the drying rate are complicated, the drying rate is usually determined by experiments.

Under constant drying conditions (that is, the temperature, humidity and flow rate of the drying medium; the contact mode of the drying medium and the wet material are all determined and unchanged), put the wet material in the drying medium to measure the change of moisture content, temperature and drying rate of the material with the drying time, and the drying curve and drying rate curve as shown in Figure 8-18 and Figure 8-19 can be obtained respectively. Under the constant drying conditions, the drying process can be divided into three stages: preheating, constant speed drying, and reduced speed drying (heating). In the preheating stage (A→B), the wet material enters the dryer and comes into contact with the wet air, and the temperature and drying rate gradually increase. This stage has a relatively short time in the actual drying process, so it is often incorporated into the constant-speed drying stage in engineering calculations. If the surface of the material can be kept wet, constant-rate drying stage (B→C) will occur after the preheating stage. In this stage, the temperature of the wet material is constant and equal to the wet-bulb temperature corresponding to the air state. At this time, the driving force of heat and mass transfer is constant, and the heat transferred from the wet air to the wet material is all used for evaporation of the moisture in the wet material. With a large and constant drying rate, this stage is called constant speed drying stage. The drying rate of the constant speed drying stage is related to the nature and flow state of the wet air and the contact mode with the materials. Therefore, this stage is also known as the "surface evaporation control stage". When the rate of moisture inside spreads to the surface is lower than the evaporation rate of water on the surface of the material, the wet material will not be able to remain moist, and there will be some "dry areas" on the surface of the material with drying rate decreased. At this time, the heat from the air to the material is more than the latent heat required for water vaporization, causing the material to be heated, and the drying process enters the reduced speed drying stage (C→D). When the surface moisture completely evaporates, the drying rate will further decrease significantly (D→E) to zero, when the moisture content of the material reaches the equilibrium moisture content. The drying rate of this stage mainly depends on the nature, size and shape of the material. Therefore, the reduced drying speed stage is also called "the stage controlled by internal diffusion".

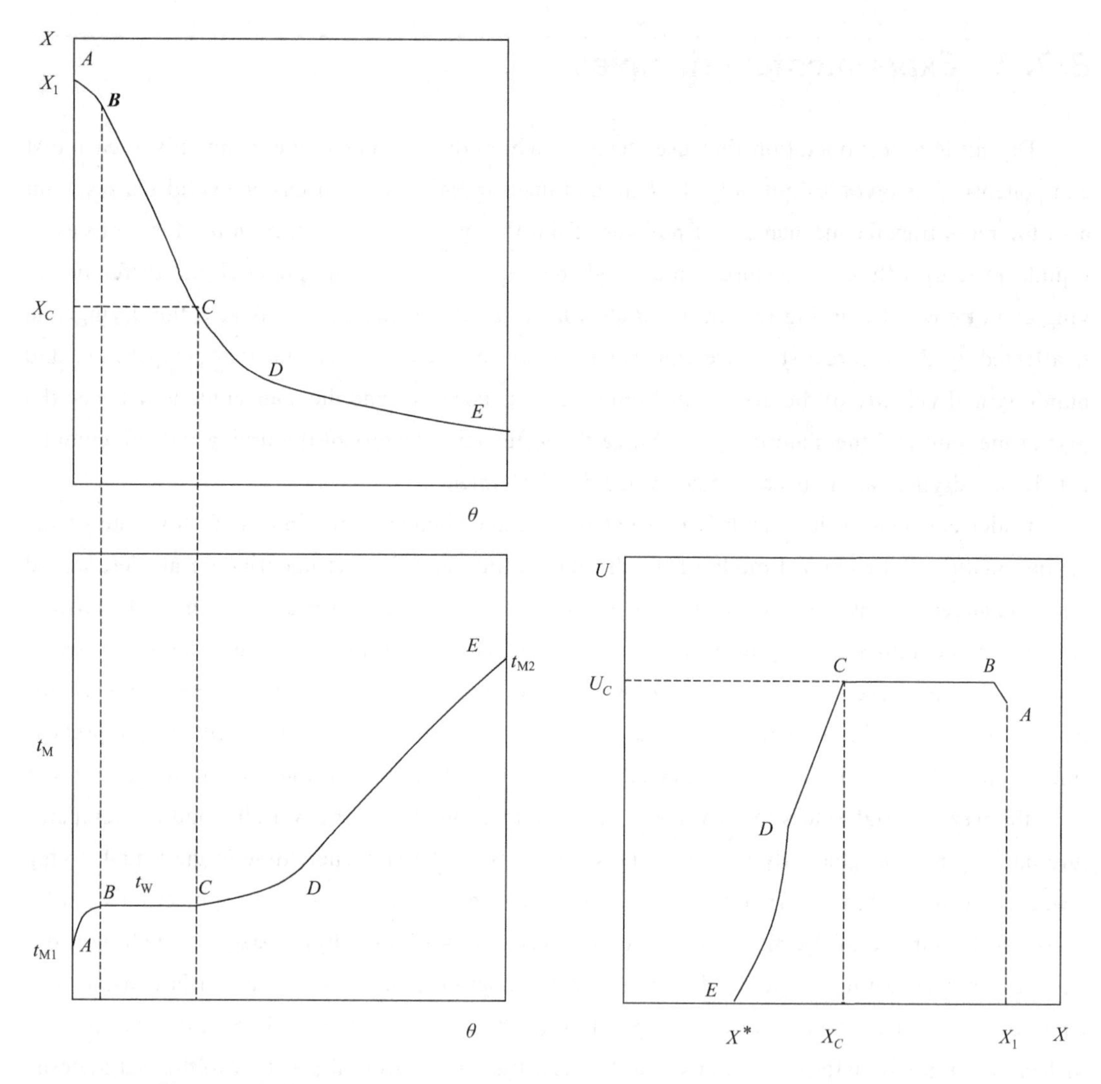

Figure 8-18　Drying curve　　　Figure 8-19　Drying rate curve

Drying rate is water evaporated per unit of material surface area per unit time, as follows:

$$U=\frac{dW}{Ad\theta} \tag{8-46}$$

where U is drying rate, $kg/(m^2 \cdot h)$; W is the mass of evaporated water during drying, kg; A is drying area, m^2; θ is drying time, s.

Since $dW=-G_C dX$, and considering the operability of the experimental measurement, the difference quotient $\Delta X/\Delta\theta$ isused instead of derivative $dX/d\theta$. Therefore:

$$U=-\frac{G_C dX}{Ad\theta}=-\frac{G_C \Delta X}{A\Delta\theta} \tag{8-47}$$

where G_C is absolute dry material mass, kg; X is dry basis moisture content, mass of water (kg) /mass of absolute dry material (kg) .

In the process of drying, the average drying rate in a certain time interval $\Delta\theta$ can be determined by measuring the mass of the wet materials at the beginning and end time respectively.

$$U=-\frac{G_C\Delta X}{A\Delta\theta}=\frac{G_C}{A\Delta\theta}\left(\frac{G_{S,i}-G_C}{G_C}-\frac{G_{S,i+1}-G_C}{G_C}\right)=\frac{G_{S,i+1}-G_{S,i+1}}{A\Delta\theta} \tag{8-48}$$

where $G_{S,\ i}$ is the mass of wet material at the beginning of time interval $\Delta\theta$, kg; $G_{S,\ i+1}$ is the mass of wet material at the end of time interval $\Delta\theta$, kg.

In the time interval $\Delta\theta$, the drying rate of the constant speed drying stage is constant, while that of the reduced speed drying stage decreases with time. Therefore, the dry basis water content of wet materials should be expressed by the average value of dry basis water content in the time interval $\Delta\theta$.

$$\bar{X}=\frac{X_i+X_{i+1}}{2}=\frac{1}{2}\left(\frac{G_{S,i}-G_C}{G_C}+\frac{G_{S,i+1}-G_C}{G_C}\right)=\frac{G_{S,i+1}+G_{S,i+1}}{2G_C}-1 \tag{8-49}$$

According to the drying rate U and the average dry basis water content $\bar{X}$ calculated by formula (8-48) and (8-49), the drying rate curve can be obtained. If the drying time θ_i at the beginning and end of each time interval $\Delta\theta$ is plotted against its dry basis moisture content X_i, the drying curve can be obtained.

8.7.3 Experimental device

The drying equipment used in this experiment is tunnel type circulating dryer, which can dry bulk materials under constant drying condition. The flow chart of the device is shown in Figure 8-20. The air is fed into the pipeline by the fan 1, theflow measured by the orifice flowmeter 3 and heated by the electric heater 4, then sent in to the drying chamber. After heating the wet materials placed on the sample rack in the drying chamber. In the end, the air partly returns to the inlet of the fan, and partly into the atmosphere. Dry bulb thermometer and wet bulb thermometer are installed in front of the drying chamber. As the drying process goes on, the mass of water evaporated from the material is measured by weight sensor and transformed into an electrical signal, then displayed in the intelligent digital display instrument. Air flow and temperature are regulated, controlled and displayed by this intelligent digital display instrument.

Main technical data of equipment are as follows.

① Centrifugal low noise medium pressure air blower: CZT-75, 370W;

② Electric heater: rated power 4.5kW;

③ Drying room: 180mm×180mm×1250mm;

④ Dry material: cardboard;

⑤ Weight Sensor: L6J8, 0~500g.

8.7.4 Experimental procedures and precautions

8.7.4.1 Experimental procedures

① Measure the sample size before the experiment to determine the dry area, and measure

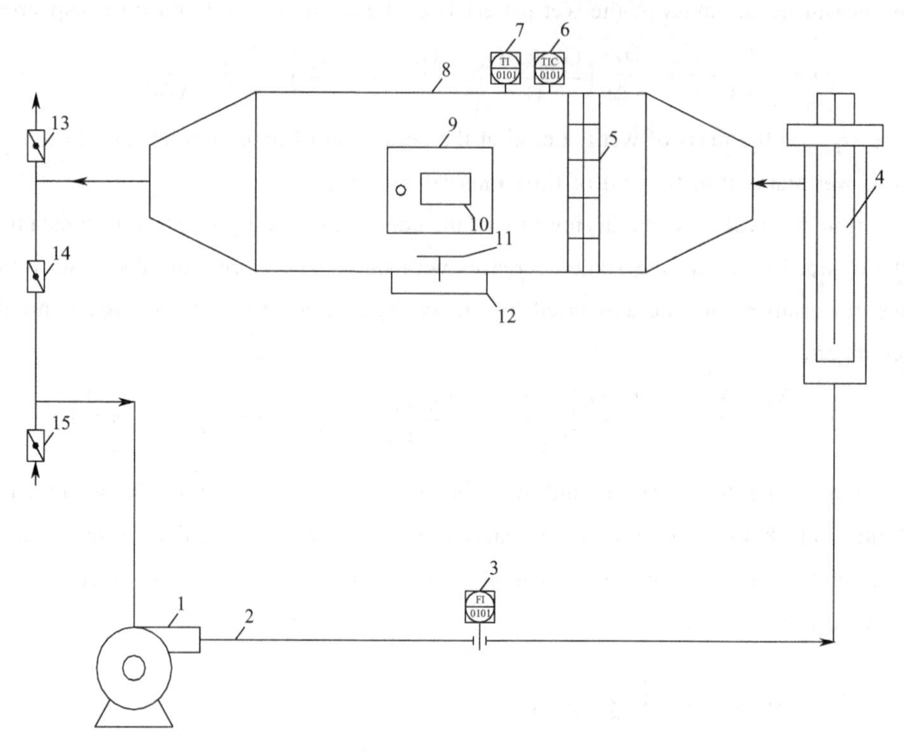

Figure 8-20 Flow chart of drying experimental device

1—Fan; 2—Pipeline; 3—Orifice flowmeter; 4—Electric heater; 5—Air distribution; 6—Dry-bulb thermometer; 7—Wry-bulb thermometer; 8—Tunnel dryer; 9—Desiccation room door; 10—Glass sight; 11—Sample holder; 12—Weight sensor; 13～15—Butterfly valve

the mass of absolute dry sample.

② Switch on the fan after turning on the main power and meter power. The air flow can be adjusted, controlled and displayed by intelligent digital display instrument.

③ Switch on the heater, and the temperature can be adjusted, controlled and displayed through intelligent digital display instruments.

④ Add some water into the U-shaped funnel on the back of the device.

⑤ Adjust the openings of the butterfly valves 13～15 to keep the velocity, temperature and humidity of the air constant.

⑥ Pay attention to the dry bulb temperature, and only carry out the experiment after the predetermined dry temperature (70℃, e. g.) is reached and kept stable.

⑦ Immerse the sample in water for a while to wet it entirely. Note that the water content should not be too much or too little.

⑧ Open the drying chamber door when the temperature is constant. Carefully place the sample holder adding the wet sample onto the weight sensor, close the drying chamber door, and record the initial mass of the wet sample.

⑨ Record the drying time required for each evaporation of a certain mass of moisture, until the mass of the sample is close to the equilibrium value.

⑩ At the end of the experiment, switch off the heating power to cool down the system.

Then switch off the fan, instrument power and main power in order. Carefully remove the sample holder and clean up the experimental site.

8.7.4.2 Precautions

① The heater must be turned on only after the fan is on, otherwise the heating tube may be burned out; the heater must be firstly turned off at the end of the experiment, and the fan should be turned off after the system cools down.

② When placing the sample holder, pay special attention not to press down forcefully. Because the upper limit of the measurement of the weight sensor is only 500g, damage to the sensor could be caused due to too much force.

③ Do not knock or hit the equipment during the experiment, or it will cause the sample holder to shake and affect the measurement results.

8.7.5 Questions

① What is the significance of measuring the drying rate curve? What help does it have to design dryers or guide production?

② What is convective drying? What is the function of the drying medium in the convective drying process?

③ Why is the drying process a heat and mass transfer process?

④ How to judge whether the drying condition is stable?

⑤ Why the fan should be turned on before the electric heater?

⑥ Will different initial water temperature affect the measurement results when measuring wet bulb temperature? Why?

⑦ Did the dry bulb and wet bulb temperatures change during the experiment? Why?

⑧ During the experiment, of the water in the wet bulb thermometer is dried, will the reading of the wet bulb thermometer be influenced? Why?

⑨ If the air flow or temperature is increased, how will the constant drying rate, critical moisture content, and equilibrium moisture content change? why?

⑩ What are the factors that affect the drying rate? What measures should be taken to improve the drying strength (take the device into consideration) ?

⑪ What is the critical moisture content? What factors can affect the critical moisture content in drying?

⑫ Is there exhaust gas circulation in this experiment? What is the effect of exhaust gas circulation on the drying process? Why is exhaust gas circulation often used when drying heat-sensitive materials or easily deformed and cracked materials?

⑬ Will absolutely dry material be gained after drying in 70～80℃ air flow for a long time? Why?

⑭Some materials are dried in hot air flow with low relative humidity, while some materils are dried in hot air flow with higher relative humidity. Why?

⑮ How to judge the end of the experiment?

⑯ Try to analyze the mechanism of constant speed drying stage and reduced speed drying stage.

⑰ What are the constant drying conditions? Please explain why the drying rate in the constant-rate drying stage is constant under constant drying conditions, considering the drying rate $U=(\alpha/r_w)(t-t_w)$.

⑱ If the opening of the butterfly valve on the pipeline becomes smaller, what effect will it have on the measured drying rate curve?

⑲ What are the differences in the drying process (drying rate, critical content, equilibrium content) of samples with different thickness according to the experimental results?

Chapter 9

Self-designed Experiments of Chemical Engineering Principle

9.1 Purpose of self-designed experiments

As introduced before, the experiments of chemical engineering principle include demonstration, design, research, comprehensive and advanced experiments are required to be completed on certain devices. Moreover, the experimental tasks, content, and procedures are predetermined, and even the data analysis and discussions in the experiment report are pre-arranged. To some extent, it reduces the motivation of students and may even prevent the development of their hands-on ability, innovation and engineering thinking. With the emergence of new engineering subjects and improved courses, especially the rapid development of the economy as well as science and technology in China, qualified and talented personels are in great needs. Therefore, four self-designed experiments are presented here involving fluid resistance, pump performance, enhanced heat transfer and tubular heat exchanger. The characteristics and purposes of these experiments are as follows.

(1) No designated experimental topics

According to their own learning wills, students may independently determine topics for experimental research. Students are encouraged to be more motivated in self-learning.

(2) No pre-set experimental devices

The laboratory provides a basic framework of nesessary experimental equipment, as well as the transportation machinery, tubes and fittings, measuring instruments, and other acessories and installation tools required . Students need to assemble the experimental device on their own based on the requirements of the research topics, which will enhance their in-depth understanding of chemical equipment, instruments and meters, and significantly improve their hands-on ability.

(3) No detailed teaching notes

Teaching notes usually provide information such as the objective of the experiment, experimental device, principles, procedures and precautions. With teaching notes, students conduct experiments following the notes step by step, in which process no much opportunity for independent thinking would be provided, and the experimental process cannot evaluate the quality and ability of students. Through self-designed experiments, students can independently complete tasks such as the selection of experimental topics, design of devices, formulation of experimental plan, implementation of experimental steps, solution of unexpected problems rising from the experimental process and writing experimental reports. It provides comprehensive training to enhance the quality of engineering research and the ability of the students.

9.2 Requirements of self-designed experiments

① Self-designed experiments should be scheduled only after completing the experiments required by the syllabus.

② Before the experiment, students need to fully understand the laboratory settings (including laboratory equipment and related acessories), carefully read the laboratory's safety procedures and precautions, and learn how to use various tools safely.

③ Students need to determine the experimental topics and objectives, complete the design of the experimental setup, develop the experimental plan, prepare possible solutions to potential problems occured in the experimental process and coordinate tasks as a team. Only after obtaining approval from an instructor may students enter the laboratory to conduct experiments.

④ Students should strictly follow the regulations and rules of the laboratory during the process, and special attention should be paid to the safety issues in the device assembly and disassembly process.

⑤ After the experiment is completed, students should submit a complete experimental report.

9.3 Equipment for self-designed experiments

9.3.1 Self-designed experiment of fluid flow resistance

The main equipment, instrument and acessories for self-designed fluid resistance experiment are shown in Table 9-1.

Table 9-1 Main equipment, instrument and acessories for self-designed experiment of fluid flow resistance

Name of equipment, instrument and acessories	Specifications
Stainless steel frame	2.50m×0.55m×1.85 m
Centrifugal pump	MS100/0.55SSC model, stainless steel, 0.55 kW, 7.2 m^3/h, $H=12.2$ m
Smooth tube	Stainless steel, $DN8$, 1.7 m
	Stainless steel, $DN10$, 1.7 m
Rough tube	Stainless steel, $DN15$, 1.7 m
	Stainless steel, $DN25$, 1.7 m
	Stainless steel, $DN32$, 1.7 m
	Stainless steel, $DN40$, 1.7 m

continued

Name of equipment, instrument and acessories	Specifications
Gate valve	Stainless steel, $DN15$
Ball valve	Stainless steel, $DN15$
Shut-off valve	Stainless steel, $DN15$
Needle valve	Stainless steel, $DN15$
Butterfly valve	Stainless steel, $DN15$
Rotameter	LZB-40
	LZB-25
	VA10-15F
Turbine flowmeter	LWGY model, $DN25$
Orifice flowmeter	$d_o=15$ mm
Venturi flowmeter	$d_o=15$ mm
Frequency converter	0~50 Hz
Glass inverted U-shaped differential pressure gauge	—
Differential pressure sensor	WNK3051 model, 0~200.0 kPa
Digital temperature display	—
Digital flow display	—
Digital differential pressure display	—
Glass tube level gauge	—
Resistance thermometer	Pt100
Water tank	400mm×420mm×600 mm
Water tank	780mm×420mm×500 mm
Pipeline	Various length and I. D.
Valves	Pressure valves etc.
Other parts	Direct, elbow, tee, hose, gasket, etc.

9.3.2 Self-designed experiment of pump performance testing

The main equipment, instrument and acessories for the self-designed experimet of pump performance test are shown in Table 9-2.

Table 9-2 Main equipment, instrument and acessories for the self-designed experiment of pump performance test

Name of equipment, instrument and acessories	Specifications
Stainless steel frame	2.50m×0.55m×1.85m
Clean water pump	MS100/0.55SSC model, stainless steel, 0.55 kW, 6.0 m^3/h, $H=14$ m
	MS60/0.55SSC model, Stainless steel, 0.55 kW, 3.6 m^3/h, $H=19.5$ m
	1/2DB70 model, Stainless steel, 0.55 kW, 3.0 m^3/h, $H=70$ m

continued

Name of equipment, instrument and acessories	Specifications
Irrigation device	—
Rough tube	Stainless steel, DN15
	Stainless steel, DN25
	Stainless steel, DN32
	Stainless steel, DN40
Pressure gauge	0～0.25 MPa, precision grade 1.5
Vacuum gauge	−0.1～0 MPa precision grade 1.5
Power transducer	PS-139 model
Turbine flowmeter	LWGY model, DN40
Frequency converter	0～50 Hz
Electric power digital display	—
Digital temperature display	—
Digital flow display	—
Resistance thermometer	Pt100
Water tank	780mm×420mm×500 mm
Baffle	—
Pipeline	Various length and I. D.
Valves	Gate valves, ball valves, check valves, etc.
Other parts	Direct, elbow, tee, hose, gasket, etc.

9.3.3 Self-designed experiment of enhanced heat transfer

The main equipment, instrument and acessories for the self-designed experimet of enhanced heat transfer are shown in Table 9-3.

Table 9-3 Main equipment, instrument and acessories for the self-designed experiment of enhanced heat transfer

Name of equipment, instrument and acessories	Specifications
Stainless steel frame	2.2m×0.55m×1.77 m
Heating kettle	DZF-3 model, 3.0 kW, stainless steel
Ordinary heat transfer tube	DN15, 1.2 m
Fluted tube	DN15, 1.2 m
Transverse corrugated tube	DN15, 1.2 m

continued

Name of equipment, instrument and accessories	Specifications
Converging-diverging tube	*DN*15, 1.2 m
Finned tube	*DN*15, 1.2 m
Spiral coil heat transfer tube	*DN*15, 1.2 m(6 sizes of pitch and wire diameter available)
Vortex fan	XGB-12 model, 100 m^3/h
Orifice flowmeter	—
Resistance thermometer	Pt100
Thermocouple thermometer	Copper-constantan
Differential pressure sensor	WNK3051 model, 0～10.00 kPa
Resistance temperature display	—
Thermocouple temperature display	—
Differential pressure digital display	—
Water tank	240mm×240mm×240 mm
Pipeline	Various length and I. D.
Valves	Gate valves, ball valves, relief valves, etc.
Other parts	Direct, elbow, tee, hose, gasket, etc.

9.3.4 Self-designed experiment of shell-and-tube heat exchanger

The main equipment, instrument and acessories for the self-designed experiment of shell-and-tube heat exchanger are shown in Table 9-4.

Table 9-4 Main equipment, instrument and accessories for the self-designed experiment of shell-and-tube heat exchanger

Name of equipment, instrument and acessories	Specifications
Stainless steel frame	2.2m×0.55m×1.77 m
Tubular heat exchanger	Stainless steel, *DN*159, 1.5 m
Hot water pump	Stainless steel
Cold water pump	Stainless steel
Water tank	1000mm×500mm×600 mm
Heater	ES60V-U1(E) model, 3 kW
Rotameter	VA10-15F
Resistance thermometer	Pt100
Differential pressure sensor	WNK3051 model, 0 ～ 10.00 kPa

continued

Name of equipment, instrument and acessories	Specifications
Head	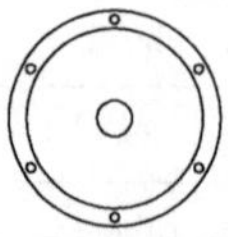
	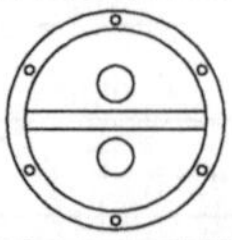
	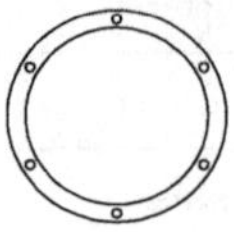
	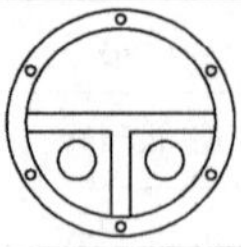
	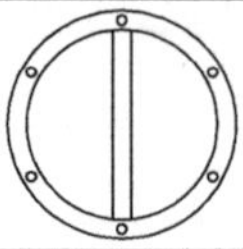
Pressure gauge	—
Pipeline	Various length and I. D.
Valves	Gate valves, ball valves, check valves, etc.
Other parts	Direct, elbow, tee, hose, gasket, etc.

Chapter 10
Advanced Experiment and Practical Training

10.1 Advanced experiment of liquid-liquid extraction

10.1.1 Experimental objectives

① To learn the structure, process and operation method of stirred extraction column, packed extraction column and spray extraction column, as well as the differences on hydrodynamic and mass transfer performance between different columns.

② To learn the main factors affecting extraction process, and to explore the effect of operating conditions on the process.

③ To master the measuring methods of hydrodynamic performance parameters such as extraction column retention fraction and flood point.

④ To master the experimental determination method of the number of mass transfer units, the height of mass transfer unit, and total mass transfer coefficient of a centain extraction column.

10.1.2 Experimental principles

Extraction is an important unit operation for separating and purifying substances. The component separation achieved by extraction is based on the difference in solubility of each component in the extractant. Two liquids flow countercurrently in the column, one of which acts as a dispersed phase and passes through the other continuously liquid the continuous phase in the form of droplets. The concentration of the two liquid phases continuously changes along the height of the column. The density difference between the two liquid phases causes the separation of phases at two ends of the column. When the light phase is used as the dispersed phase, the phase interface appears at the upper end of the tower column. Conversely, when the heavy phase is used as the dispersed phase, the phase interface appears at the lower end of the column. For the column height of a differential countercurrent extraction column, a calculation method similar to the method to obtain the height of the packing layer in absorption operation can also be used, which is, to use the number and height of mass transfer unit to calculate the height of extraction column. The number of mass transfer unit indicates the degree of difficulty of extraction, and the height of mass transfer unit indicates the performance of the equipment.

$$h_0 = H_{OE} N_{OE} \tag{10-1}$$

where h_0 is effective mass transfer height of the extraction column, m; H_{OE} is total mass transfer unit height based on extraction phase, m; N_{OE} is total mass transfer unit based on extraction phase, dimensionless.

The number of mass transfer unit is:

$$N_{OE}=\int_{y_a}^{y_b}\frac{dy}{y^*-y} \tag{10-2}$$

where y is composition of the extraction phase at a section in the column, kg BA/kg water; y^* is composition of the extraction phase in equilibrium with the raffinate phase at a section in the column, kg BA/kg water; y_a is the composition of the extractant when it enters the column, kg BA/kg water; y_b is the composition of the extraction phase as it leaves the tower, kg BA/kg water.

The y^* in the number of total mass transfer unit N_{OE} can be calculated by the extraction equilibrium relationship (10-3), and the raffinate phase composition in formula (10-3) can be obtained by the operation line equation of the column. In this experiment, water is used as the extractant to extract benzoic acid (BA) in kerosene. The solubility of benzoic acid in the solvent kerosene and the extractant water are both rather low, and the solubility between water and kerosene is low too. Therefore, the mass flow of the two liquid phases in the extraction column can be considered constant. However, according to liquid-liquid equilibrium equation, the composition between extraction phase and raffinate phase shows a non-linear relationship. For example, the phase equilibrium equation of benzoic acid in water (extractant) and kerosene (original solvent) at 25℃ is:

$$y^*=4.443060\times10^{-6}+1.247730x-4.440948\times10^{2}x^{2}+5.048579\times10^{4}x^{3}+6.414260\times10^{6}x^{4} \tag{10-3}$$

The formula above reflects the relationship between the equilibrium composition of the extraction phase y^* and the composition of the raffinate phase x, while x and y conform to the material balance relationship. A material balance calculation for the entire extraction column could be performed as follows:

$$Fx_b+Sy_a=Ey_b+Rx_a \tag{10-4}$$

where F is material flow, kg/h; S is extractant flow, kg/h; E is extraction phase flow, kg/h; R is raffinate phase flow, kg/h; x_a is composition of raffinate phase as it leaves the column, kg BA/kg kerosene; x_b is composition of the mixed liquid when it enters the column, kg BA/kg kerosene.

For dilute solution extraction $F\approx R$, $E\approx S$, while $y_a=0$, therefore

$$y_b=\frac{F}{S}(x_b-x_a) \tag{10-5}$$

Similarly, The operation line equation of the extraction column can be obtained by performing material balance calculation at any section in the extraction column:

$$y=\frac{F}{S}(x-x_a) \tag{10-6}$$

According to formula (10-3) and (10-6), y^* in formula (10-2) can be expressed as a function of y, then N_{OE} is obtained by numerical integration. Finally, H_{OE} and the total mass transfer coefficient can be gained according to the effective mass transfer height of the column h_0.

$$H_{OE}=\frac{S}{K_y aA} \tag{10-7}$$

where A is cross-sectional area of the column, m^2; $K_y a$ is total volume mass transfer coefficient of extraction phase, kg/ (m^3 · h · kg/kg) .

10.1.3 Experimental device

The flowcharts of the experimental equipments are shown in Figure 10-1 (pulse packed extraction column, spray extraction column if without packing) and Figure 10-2 (liquid-liquid stirred extraction column). The equipment is mainly composed of extraction column, light phase storage tank, heavy phase storage tank, magnetic pump, stirring motor, air compressor and pulse generator. In operation, the water in the heavy phase tank should be firstly transported to the extraction column by the heavy phase pump to form a continuous phase in the column. Then the light phase pump should be turned on to transport the kerosene in the light phase tank into the column to be dispersed into droplets to form dispersed phase. The dispersed phase (light phase) moves upward in the column, while the continuous phase (heavy phase) flows downwards. The two phases are in countercurrent contact for mass transfer. The dispersed phase will gather at the top of the column and form a light liquid layer with a certain thickness, an obvious interface between the light and the heavy phase forming in the upper part of the column. The position of the interface can be adjusted by the π-tube valve at the continuous phase outlet to be stabilized at a certain height. The light phase is extracted from the top of the column into the raffinate phase recovery tank, and the heavy phase is extracted from the bottom of the column and led out through the π-tube. When the light phase is used as the continuous phase and the heavy phase as the dispersed phase, the flow is similar as described above, except that the interface between the light and heavy phases is in the lower end of the column.

10.1.4 Experimental procedures and precautions

(1) Experimental procedures

① Dissolve benzoic acid in kerosene (the mass fraction of benzoic acid is about 0.0015 ~ 0.002), then pour the solution into the light phase tank. Open the outlet and inlet valve of the light phase magnetic pump to fill the pump, open the light phase tank drain valve to empty the pipeline air; and close the drain valve of light phase tank and the outlet valve of light phase magnetic pump.

② Connect the water pipes, pour water into the heavy phase tank, open the inlet and outlet valves of the heavy phase magnetic pump to fill the pump, open the drain valve of heavy phase tank to drain the pipeline air; and close the drain valve of heavy phase tank and the outlet valve

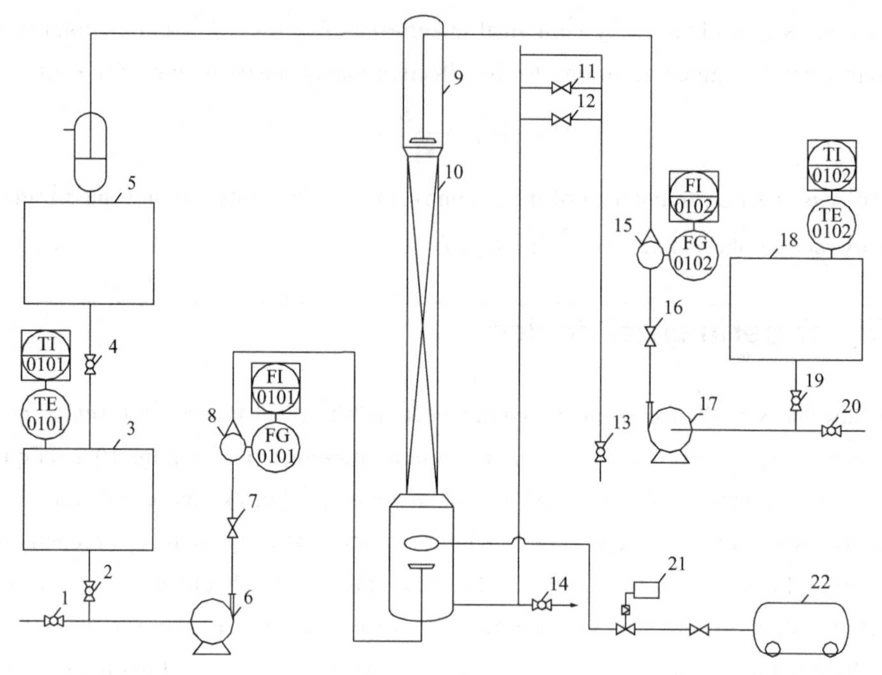

Figure 10-1 Flow chart of pulsed packed extraction experimental device

1—Drain valve of light phase tank; 2—Inlet valve of light phase magnetic pump; 3—Light phase tank; 4—Link valve for raffinate phase recovery tank and light phase tank; 5—Raffinate recovery tank; 6—Light phase magnetic pump; 7—Outlet valve of light phase magnetic pump ; 8—Light phase float flowmeter; 9—Packed extraction column; 10—Packings; 11—π-tube gate valve ; 12—π-tube gate valve 1 ; 13—Heavy phase outlet valve 2; 14—Drain valve of extraction column; 15—Heavy phase float flowmeter; 16—Outlet valve of heavy phase magnetic pump; 17—Heavy phase magnetic pump; 18—Heavy phase tank; 19—Inlet valve of heavy phase magnetic pump; 20—Tank drain valve of heavy phase; 21—Pulse regulator; 22—Air compressor

of heavy phase magnetic pump outlet valve.

③ As shown in Figure 10-5 or Figure 10-6, switch on the main power and instrument power.

④ Start the heavy phase magnetic pump, slowly open the outlet valve of the magnetic pump to send water into the column until it is almost full, then open the π-tube gate valve and outlet valve of the extraction phase.

⑤ Switch on the light phase magnetic pump, and slowly open outlet valve of the magnetic pump to transport kerosene into the column. The kerosene will be dispersed in the extraction column and flow upward, then converging at the top of the column.

⑥ For a stirred extraction column, turn on the stirrer to an appropriate speed; for pulsed extraction column, switch on the pulse power.

⑦ Adjust the flow of the two phases to about 10～15L/h, then pay attention to the correction of the flow measured by the kerosene float flowmeter.

Main technical parameters of the equipment are as follows.

ⅰ. Extraction column: total height 1280mm, the inner diameter 50mm.

ⅱ. Stirring paddle: 8 types as shown in Figure 10-3.

ⅲ. Packings: 11 types as shown in Figure 10-4.

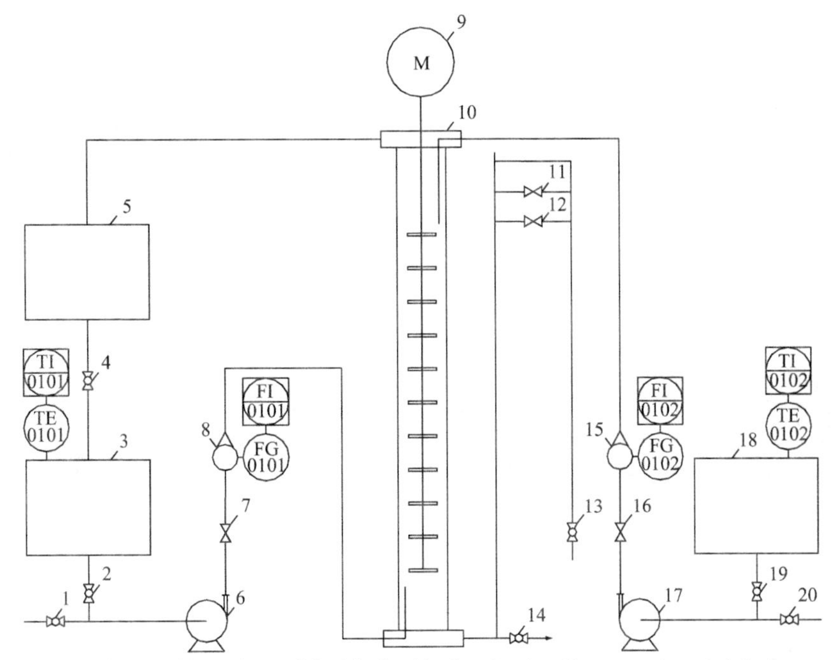

Figure 10-2 Flow chart of liquid- liquid stirred extraction experimental device

1— Drain valve of the light phase tank; 2—Inlet valve of light phase magnetic pump; 3—Light phase tank; 4—Link valve for raffinate phase recovery tank and light phase tank; 5—Raffinate phase recovery tank; 6—Light phase magnetic pump; 7—outlet valve of light phase magnetic pump; 8—Light phase float flowmeter; 9—Stirring motor; 10—Mechanical stirring extraction column; 11—π-tube gate valve 1; 12—π-tube gate valve 2; 13—Outlet valve of heavy phase; 14—Drain valve of extraction column; 15—Heavy phase float flowmeter; 16—Outlet valve of heavy phase magnetic pump; 17—Heavy phase magnetic pump; 18—Heavy phase tank; 19—Inlet valve of heavy phase magnetic pump; 20—Drain valve of heavy phase tank

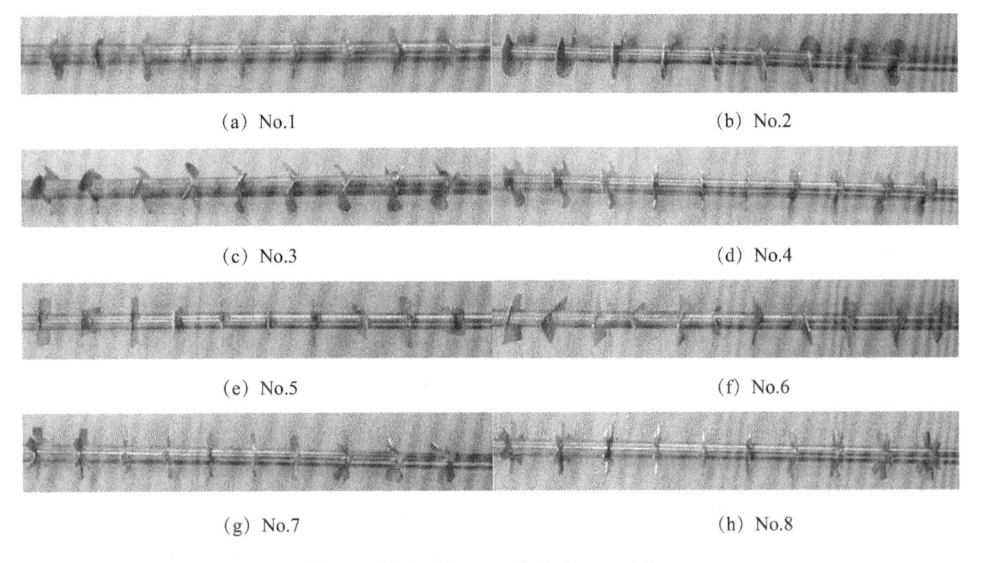

(a) No.1 (b) No.2

(c) No.3 (d) No.4

(e) No.5 (f) No.6

(g) No.7 (h) No.8

Figure 10-3 Types of stirring paddles

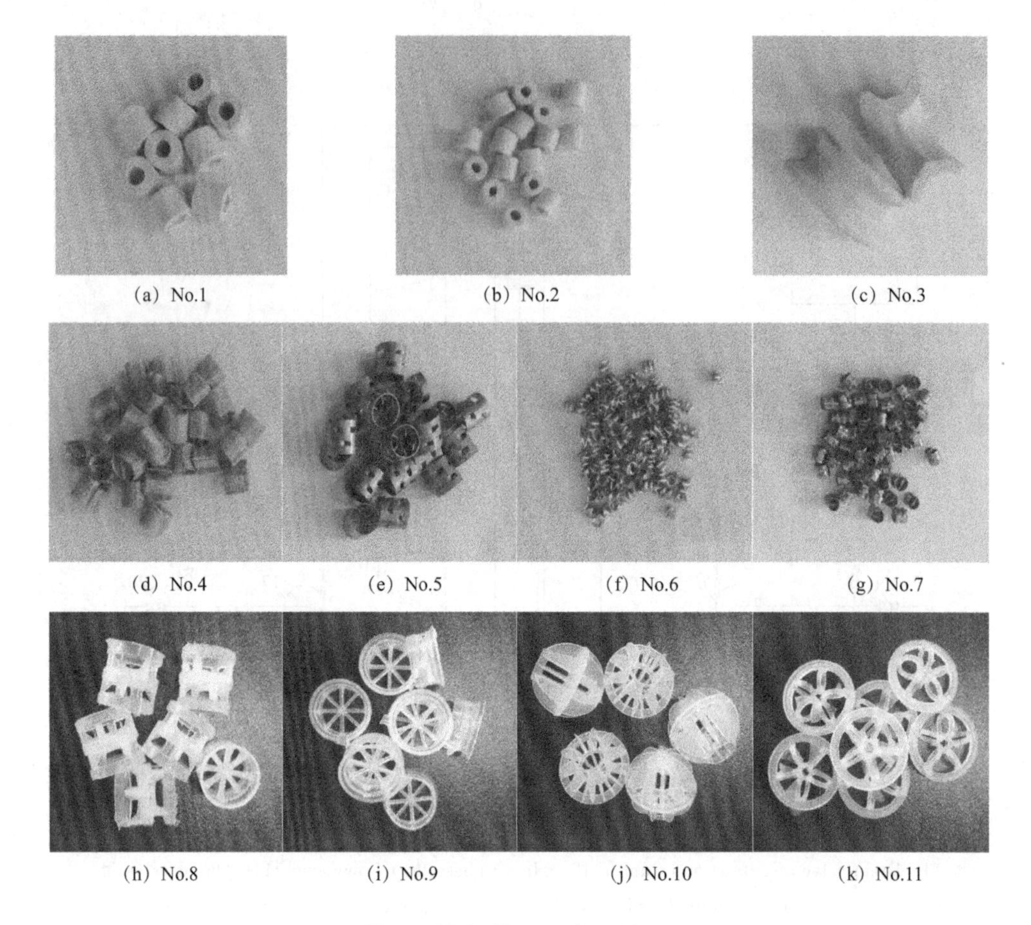

(a) No.1 (b) No.2 (c) No.3

(d) No.4 (e) No.5 (f) No.6 (g) No.7

(h) No.8 (i) No.9 (j) No.10 (k) No.11

Figure 10-4 Types of packings

$$V_{\text{h-actual}}=V_{\text{h-measured}}\sqrt{\frac{(\rho_{\text{f}}-\rho_{\text{actual}})\rho_{\text{calibrated}}}{\rho_{\text{actual}}(\rho_{\text{f}}-\rho_{\text{calibrated}})}}$$

where $V_{\text{h-actual}}$ is actual volume flow of kerosene, L/h; $V_{\text{h-measured}}$ is flow read by kerosene float meter, L/h; ρ_{f} is float density of kerosene float flowmeter, 7800kg/m^3; ρ_{actual} is actual fluid density, kg/m^3; $\rho_{\text{calibrated}}$ is the density of the calibrated fluid (water) under the conditions of the float meter calibration, $1000\ \text{kg/m}^3$.

⑧ By controlling the opening of the gate valve on the heavy phase outlet π-tube, the interface between the light and heavy phases in the column should be adjusted to be stable and between the light phase outlet and the heavy phase inlet at the top of the column.

⑨ When the working state of the column is stable, the take 60～80mL samples from kerosene inlet (raw material liquid), the kerosene outlet (extract phase) and the water outlet (extraction phase) for analysis according to the following method.

The composition of benzoic acid in samples should be analyzed by acid-base titration with ethanol solution of NaOH and phenolphthalein as the indicator. The specific steps are as follows.

ⅰ. Rinse the burette with the calibrated NaOH ethanol solution, and then fill the burette with an appropriate amount of NaOH ethanol solution.

ⅱ. Use a pipette to take 10 mL or 20 mL of sample into a conical flask, and add 1 to 3 drops of phenolphthalein reagent.

ⅲ. Titrate to the end point with NaOH ethanol solution, then the measured composition is

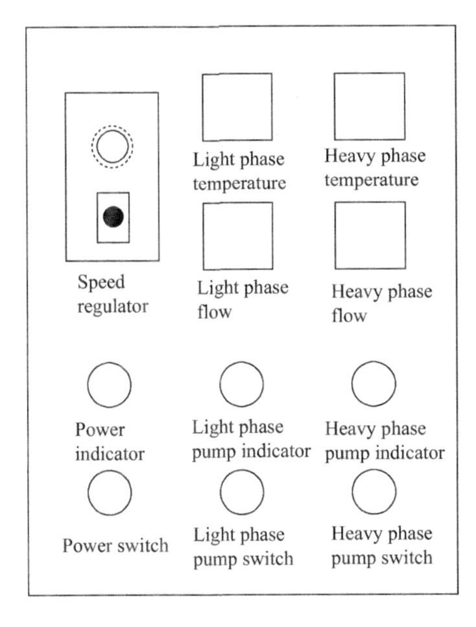

Figure 10-5 Control panel of stirred extraction column

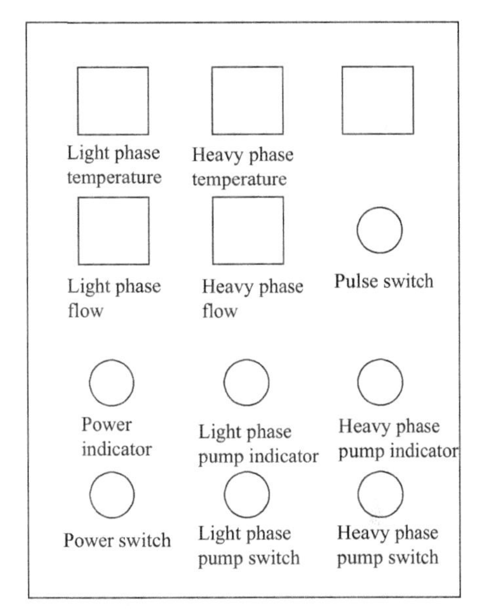

Figure 10-6 Control panel of pulse packing / spray extraction column

$$x=\frac{122.12N\Delta V}{800V_{\text{sample}}}$$

where N is equivalent concentration of NaOH ethanol solution, mol/L; ΔV is volume of NaOH ethanol solution consumed by titration, mL; V_{sample} is sample volume, mL.

⑩ Measure the mass transfer height of the column (which is the vertical distance between the light phase entrance and the interface of two phases) . Meanwhile, close the inlet and outlet valves of the light and heavy phases, switch off the magnetic pumps of two phases and the stirrer or pulse. After the light and heavy phases are completely separated, measure the mass tranfer height again. The retention fraction will thus be obtained.

⑪ Switch off the main power of the equipment, open the drain valve of the column to empty the liqui, empty the liquid in the raffinate recovery tank, light phase tank and heavy phase tank, and finally clean up the equipment and the site.

(2) Precautions

① Do not prepare kerosene-benzoic acid solution directly in the light phase tank, otherwise undissolved solid particles may block the inlet pipeline of the pump.

② Never operate the magnetic pump without load.

③ The outlet valve of magnetic pump should be opened slowly, otherwise the fluid may impact the electromagnetic float flowmeter and cause the float to stuck and cannot be reset.

④ During the acid-base titration analysis, the sample to be tested (especially the kerosene sample) should be fully shaken after NaOH salution is drippde, to ensure accurate determination of the end point of the titration.

10.1.5 Questions

① Analyze and compare the similarities and differences among extraction, absorption and distillation.

② What are the factors affecting the extraction process?

③ How does the external energy such as pulse and mechanical stirring affect the extraction process (such as mass transfer coefficient and extraction rate) ?

④ How will an increase in the extractant amount affect the compositions of the raffinate and extraction phases?

⑤ Analyze the performance differences among mechanically stirred extraction column, spray extraction column and packed extraction column according to the experimental data.

⑥ How is the retention fraction measured? How does it afffect the extraction process?

⑦ Is a smaller dispersed phase droplet more beneficial to the extraction process? Why?

⑧ Try to explain the flooding phenomenon of the extraction column.

⑨ What methods can be used to determine the composition of the raw material liquid, extraction phase and raffinate phase? During acid-base neutralization titration, why is the NaOH ethanol solution chosen as the standard base instead of NaOH aqueous solution?

⑩ If kerosene is selected as the continuous phase and water as the dispersion phase, how should the extraction column be started?

⑪ What is the role inπ-tube in the extraction phase outlet?

⑫ The pipeline through which kerosene enters the extraction column is an inverted U-tube. Why is it set up this way?

⑬ Analyze the qualitative relationship between the retention fraction and the droplet size of dispersed phase of the spray column comparing to the packed column.

⑭ If the phase interface moves down, how should it be adjusted back up?

⑮ How does the retention fraction of dispersed phase change if the dispersed phase or continuous phase flow increases?

10.2 Multi-effect evaporation comprehensive training

10.2.1 Training objectives

① To understand the basic structure and working principle of the evaporator, as well as to master the process and operation method of multi-effect evaporation.

② To master the control of instrument parameters, ensuring a normal and stable operation of the equipment.

③ To practice and master manual and automatic operation of multi-effect evaporation equipment, and to be familiar with the DCS control system.

④ To develop the ability to analyze and determine the types and causes of abnormal phenomena and take appropriate measures to deal with them.

⑤ To master the experimental research method of the multi-effect evaporation process and the determination method of related performance parameters.

⑥ To develop a production awareness of safety, standard, environmentally friendliness and energy-saving, a work ethic in strict compliance with operating procedures and a spirit of teamwork.

10.2.2 Basic principles

Evaporation is a unit operation to boil a dilute solution of non-volatile solutes to vaporize part of the solvent, thus achieve concentration. The equipment with such function is called an evaporator. In order to ensure a continuous and stable evaporation process, the evaporator must be provided with a continuous heat source to supply the heat required for solvent evaporation. Meanwhile, the vaporized steam must be continuously discharged from the evaporator. Therefore, the evaporator is generally composed of a heating chamber as well as an evaporation chamber.

Requiring a large amount of heat to vaporize solvent, evaporation is a high energy-consuming unit operation. Energy-saving methods such as increasing the utilization of heating steam are important to improve the economics of evaporation operations. Multi-effect evaporation consists of multiple evaporators connected in series. The secondary steam generated by the pre-effect evaporator is introduced into the post-effect evaporator as heating steam, thereby improving the utilization rate of the generated steam, achieving the purpose of energy saving.

This training equipment is a double-effect evaporation equipment, and the evaporator is an external-heated one. It consists of a heating chamber and an evaporation chamber. The heating chamber is located at the lower part of the evaporator and consists of a bundle of heating tubes. The heating steam outside the tube causes the solution inside the tube to boil and vaporize. The upper part of the evaporator is an evaporation chamber, in which the vapor generated is separated from the entrained liquid foam, then enters the second effect evaporator as a heat source or enters the condenser. Part of the concentrated solution is recycled back to the heating chamber to continue evaporation and concentration, part of it is discharged from the bottom of the evaporator.

10. 2. 3 Process, main devices and instruments

The raw material liquid is transported from the raw material tank V0101 to the first-effect heater E0101 by raw material pump P0101. The raw steam generated by the steam generator R0101 is condensed on the shell side of the heater E0101 to heat the raw material on the tube side. The liquid circulates naturally between the first-effect heater E0101 and the first-effect evaporation chamber E0102. The secondary steam generated by vaporization is separated from the entrained liquid foam in the first-effect evaporation chamber E0102. The second-effect evaporator operates under negative pressure. So tohe completion liquid from the first-effect evaporator automatically flows into the second-effect evaporator. The secondary steam from the first-effect evaporation chamber E0102 enters the shell of second-effect heater E0103 as the heating source for the second-effect evaporator, heating the liquid in the tube side. The liquid is circulated and concentrated between the second-effect heater E0103 and the second-effect evaporation chamber E0104. The completion liquid of the second-effect evaporator is driven into the product tank V0102 for storage by the completion liquid pump P0102. The secondary steam separated from the second-effect evaporation chamber E0104 then enters the condenser E0105 for removal. The condensed water in the first- and second-effect heaters is removed by stream traps. The specific process is shown in Figure 10-7.

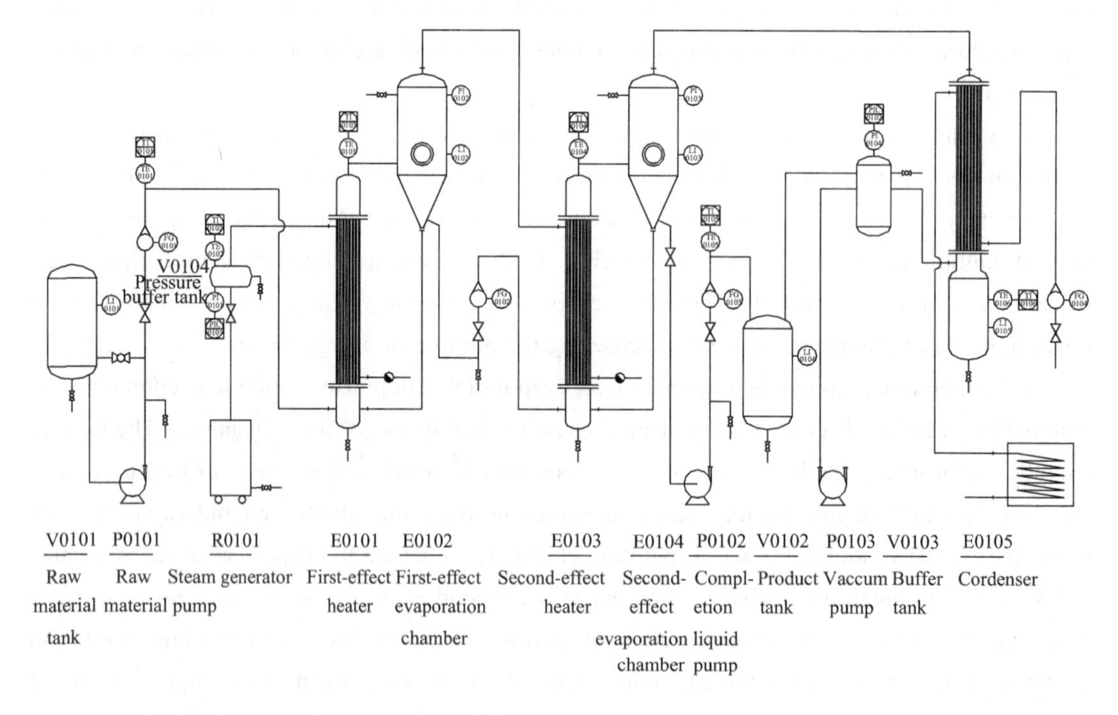

Figure 10-7 Flow chart of multi-effect evaporation comprehensive training device

The control panel is shown in Figure 10-8.

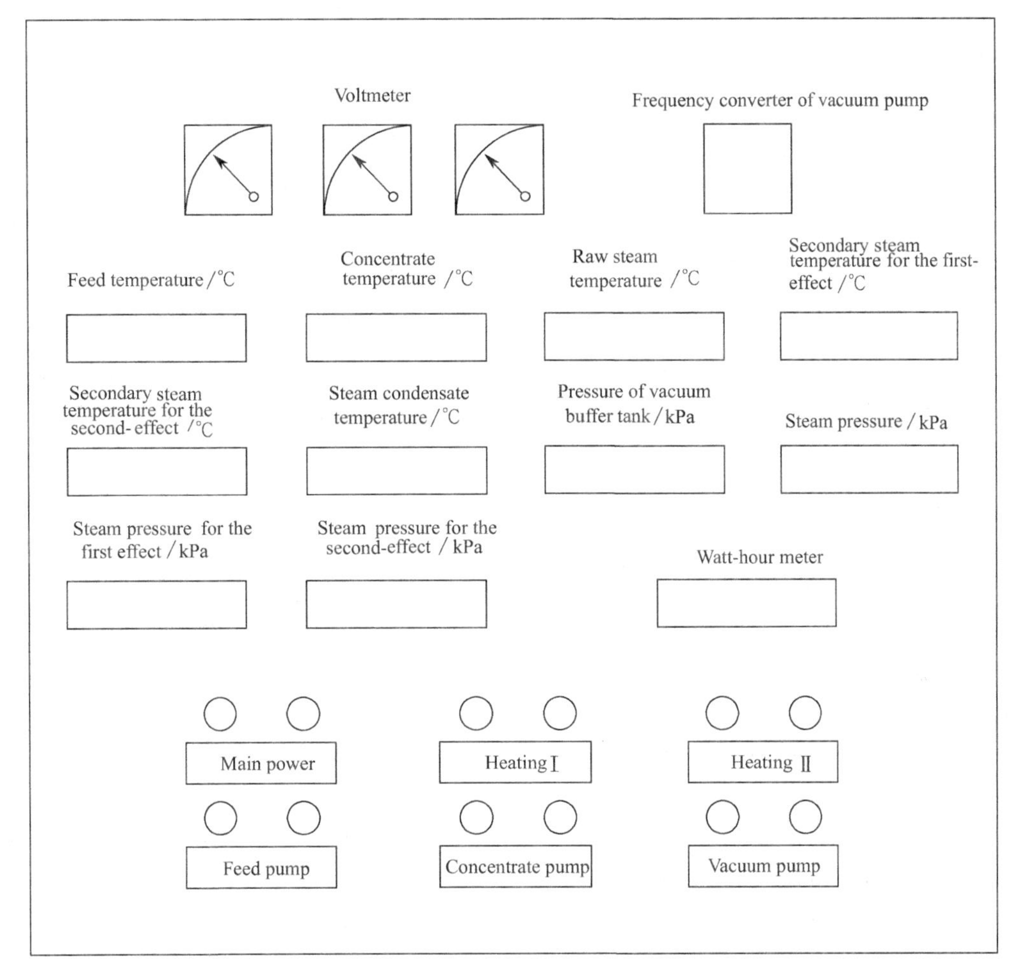

Figure 10-8 Control panel

10.2.4 Training projects

(1) Project 1: Chart reading and equipment recognition

① Read the flow chart of the device, independently and fluently detail the process in contrast with the equipment, and specifically defined the roles of each component and valve in the equipment.

② Identify the measurement and control points of the device, be familiar with the measurement points and the instrument position, and master the metering operation and parameter control of the equipment.

③ Formulate a detailed operating procedures of the multi-effect evaporation according to the actual situation of the equipment.

(2) Project 2: Dynamic and static inspection before starting

① Check whether each device is in perfect state.

② Check that the tubes, tube fittings and valves are well intact and that the valves are flexible and in correct position.

③ Inspect the entire equipment for leaks.

④ Check whether the fluid conveying devices are in perfect state, including understanding the nameplate of the centrifugal pump, checking whether the installation height of the centrifugal pump is appropriate, whether the centrifugal pump needs to be filled, and whether the inlet and outlet valves of pumps are in a correct state to be turned on.

(3) Project 3: Inspection of raw materials, water, electricity, gas and other instruments

① Check the liquid level of the storage tanks. Drain the remaining liquid in the heater, evaporation chamber, product tank, condenser, etc., and add raw material liquid to the raw material tank to keep the liquid level at about 4/5.

② Identify all measuring instruments, such as flow measuring instruments (rotor flowmeter, orifice flowmeter, venturi flowmeter, turbine flowmeter), pressure and liquid level measuring instruments (pressure gauge, vacuum gauge, magnetic turnover level gauge, glass tube level gauge), temperature measuring instruments (thermal resistance thermometers) as well as control instruments and related components (electric regulating valves, sensors, inverters), etc.

③ Check the power connection of the instrument cabinet. suitch on the main control, check whether the voltmeter shown on the meter cabinet indicates 380V, and whether the red indicator on the main power is on. Swith on the main power. After 3 minutes of stability, check whether the displayed values are in normal range.

④ Master the correct operation and monitoring method of the DCS system and field console instrument.

(4) Project 4: Dynamic equipment test run

① Centrifugal pumps, vacuum pumps and other equipments need to be barred over before power on. Before driving, check the inlet and outlet pipelines of the pump, valves and pressure gauge joints for leaks and the ground screws and other connections for looseness.

② Bar over to check whether the rotor is flexible, whether there is a metal collision sound inside the pump.

③ Open the inlet valve of the centrifugal pump and the outlet vent valve to fill the pump with liquid. Close the vent valve after removing the accumulated air in the pump and prepare for the start.

(5) Project 5: The normal operations of turning on and off of the centrifugal pump

Startup operation:

① Open the inlet valve of the centrifugal pump completely, close all the outlet valves, then start the motor. After the pump is running, check the working condition of it.

② Check that the motor and pump rotate in the correct direction.

③ Check the motor and pump for noise, abnormal vibration or leakage.

④ Check whether the motor current is less than the rated value. When the overload is displayed, stop the motor immediately and check.

⑤ Adjust the regulating valve of outlet flow so that the working point of the pump is in the state required by the process.

Shutdown operation:

① Gradually close the outlet valve of the centrifugal pump to full close.

② When all the outlet valves are completely closed, shut down the pump.

③ When the pump stops running, close the inlet valve of the centrifugal pump.

(6) Project 6: Multi-effect evaporation operation

① Feeding: Close all valves in the equipment, open the vent valve of first-effect evaporator, start the raw material pump according to the startup procedure and feed the liquid from the raw material tank to the first-effect evaporator. When the level of the raw material liquid approaches the middle position of the observation hole of the first-effect evaporator, open the regulating valve of buffer tank pressure, start the vacuum pump, open the liquid flow regulating valve between the two evaporators, and adjust the pressure regulating valve of buffer tank . Under negative pressure, the liquid in the first-effect evaporation chamber enters the second-effect evaporator. When the liquid level in the second-effect evaporator reaches the middle position of the observation hole, close the outlet valve of material pump to stop feeding, close the flow adjustment valve between the two evaporators, open the buffer tank pressure adjustment valve to reduce the negative pressure, and turn off the vacuum pump and material pump, then close the vent valve of first-effect evaporator.

② Preheating: Check whether the liquid level of the steam generator is normal, open the outlet valve of steam generator, then turn on the steam generator heating switch. The secondary steam generated by the first-effect evaporator enters the second-effect heater. When the liquid temperature reaches a certain value, turn on the vacuum pump, and adjust the buffer tank pressure and the cooling water flow to a certain value.

③ Evaporation: When secondary steam is generated in the second-effect evaporator, start the raw material pump to feed at a certain flow rate, and adjust the flow rate by the adjustment valve at the same time to be at a certain value (below the amount of feeding flow) . Start the completed liquid pump and adjust the finished liquid outlet valve so that the flow is a certain value (below the flow between effects) . Adjust the pressure, flow and other parameters to keep the system temperature, pressure and liquid level stable. Monitor and record the system temperature, pressure, liquid level and other data. After stable operation for a certain period of time, sample and analyze the composition.

④ Shutdown: Turn off the heating power of the steam generator; close the feed flow regulating valve, flow regulating valve and the completion liquid flow regulating valve, feeding and discharging thus stopped. Fully open regulating valve of the buffer tank pressure, and shut down the vacuum pump after the system returns to normal pressure. Stop the cooling water when the system cools down. Switch off the main power, and when the temperature of the system liquid drops to near room temperature, drain the liquid in the system and clean up the site.

NOTE: Before the practice training, students should have a comprehensive understanding of the device, be familiar with all the equipment, pipelines, valves and instruments in the equipment, and clarify their working principles and functions. The detailed operating procedures should be formulated in advance, and the training operations can be carried out only after the operating procedures are reviewed and qpproved. The multi-effect evaporation operation project should be performed on the premise that all other training projects have been completed.

10.3 Absorption-desorption comprehensive training

10.3.1 Training objectives

① To understand the basic structure and working principle of absorption column and desorption column, and to master the process and operation method of absorption and desorption.

② To master the control of instrument parameters, ensuring a normal and stable operation of the equipment.

③ To practice and master manual and automatic operation of absorption-desorption equipment, and to be familiar with the use of DCS control system.

④ To develop the ability to analyze and determine the types and causes of abnormal phenomena and take appropriate measures to deal with them.

⑤ To master the experimental research method of absorption-desorption process and the determination method of related performance parameters.

⑥ To develop a production awareness of safety, standard, environmentally friendliness and energy-saving, a work ethic in strict compliance with operating procedures and a spirit of teamwork.

10.3.2 Basic principles

Gas absorption, abbreviated as absorption, is a unit operation in which a mixed gas is brought into contact with a liquid for mass transfer, to dissolve soluble components in the liquid thus separate it from the mixed gas. Gas absorption is mainly used for separating mixed gas, removing harmful or unwanted components, recovering useful components or preparing gas-liquid reaction products. The operation of removing dissolved gas from liquid is called desorption, which is the reverse process of absorption, with an opposite mass transfer direction. To achieve reuse of the absorbent or recovery and purification of the absorbing components, the absorption and desorption processes are usually combined. There are two types of absorption and desorption equipment: packed column and plate column.

This absorption-desorption training equipment includes absorption column and desorption column, both of which use Ball ring packing. Water in the absorption column absorbs carbon dioxide in the air, while in the desorption column air is used to desorb carbon dioxide in the absorption liquid. The solubility of carbon dioxide in water is very small, and its absorption and desorption processes can be treated at a low concentration. The mass transfer process is controlled by a liquid film.

10.3.3 Process, main equipment and instruments

As shown in Figure 10-9, air is provided by fan P0101, and carbon dioxide (solute) is provided by cylinder X0101. After air and carbon dioxide are mixed, the mixed air enter the bottom of the absorption column T0101 through a π-tube, flow through the packing layer from bottom to top, and come into contact with the absorbent (from the desorption column) in countercurrent. Carbon dioxide is absorbed by the water, and the gas after absorption will be vented. The absorbent after absorption enters the storage tank V0101 from the bottom of the absorption column, and is transported by the centrifugal pump P0103. After being preheated by the heater E0101, it enters the top of the desorption column T0102, and it is desorbed countercurrently with the fresh air from the fan P0104. The desorbed and regenerated liquid is cooled by the cooler E0102, then enters the storage tank V0102, and conveyed by the centrifugal pump P0102 into the top of the absorption column as an absorbent. The desorption gas is vented from the top of the desorption column.

The control panel is shown in Figure 10-10.

10.3.4 Training projects

(1) Project 1: Chart reading and equipment recognition training

① Read the flow chart of the device, detail the process independently and fluently according to the equipment, and clarify the role of each device and valve in the equipment.

② Identify the measurement and control points of the device, be familiar with the measurement points and the instrument position, and master the metering operation and parameter control of the equipment.

③ Formulate a detailed operating procedures of the absorption-desorption according to the actual situation of the equipment.

(2) Project 2: Dynamic and static inspection before starting

① Check whether each device is in perfect state.

② Check that the tubes, fittings and valves are well intact and that the valves are flexible and in the correct position.

③ Inspect the entire instrument for leaks.

④ Check whether the fluid conveying equipment are in perfect state, including understand-

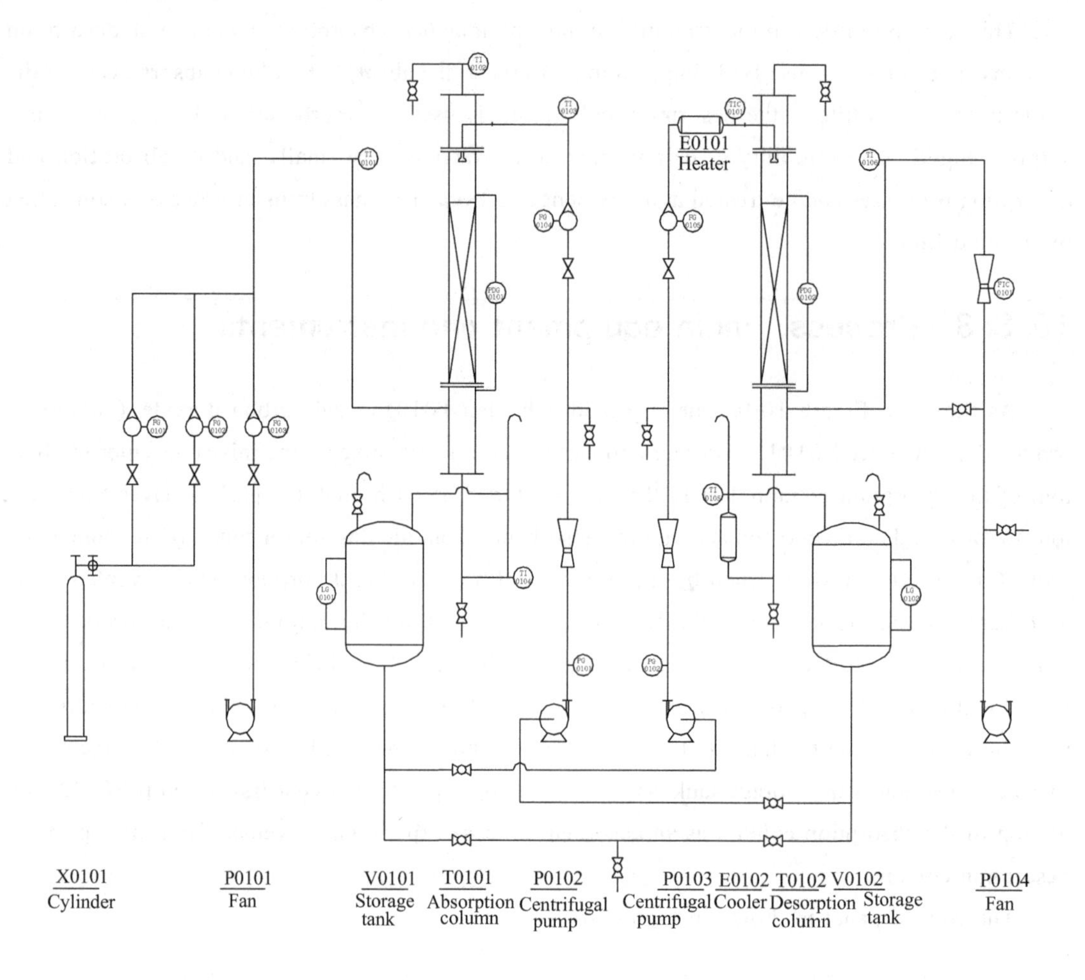

Figure 10-9　Flow chart of absorption-desorption comprehensive training device

ing the nameplate of the centrifugal pump and the fan，checking whether the installation height of the centrifugal pump is appropriate，whether the centrifugal pump needs to be filled，and whether the inlet and outlet valves of pumps and fans are in a correct state to be turned on.

(3) Project 3：Inspection of raw materials and water，electricity，gas and other instruments

① Check the liquid level in the storage tank of the absorption liquid and desorption liquid，whether the storage capacity of the carbon dioxide cylinder is appropriate，and whether the pressure reducing valve is normal.

② Identify all measuring instruments，such as flow measuring instruments (rotor flow meters，venturi flow meters)，pressure and level measuring instruments (pressure gauges，glass tube level gauges)，temperature measuring instruments (thermal resistance thermometers) as well as control instruments and related components (electrically controlled valves，sensors，frequency converters)，etc.

③ Check the power connection of the instrument cabinet，swich on the main control，check whether the voltmeter shown on the meter cabinet indicates 380V，and whether the red indicator

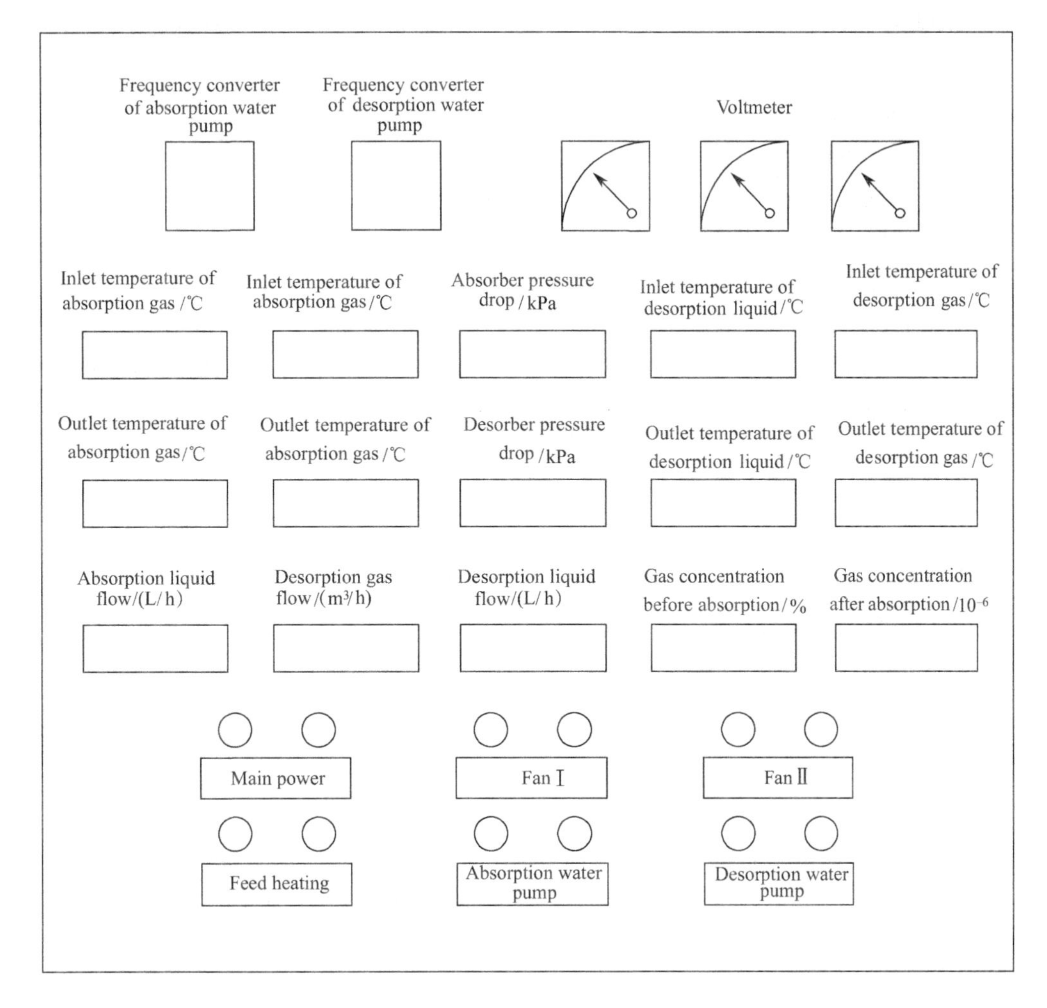

Figure 10-10 Control panel

of the main power is on. Switch main power on. After 3 minutes of stability, check whether the displayed values are in normal range.

④ Master the correct operation and monitoring method of the DCS system and field console instrument.

(4) Project 4: Dynamic equipment test run

① Centrifugal pumps and fans need to be barred over before power on. Check the pump inlet and outlet pipelines, valves, pressure gauge joints for leaks, and the ground screws and other connections for looseness.

② Bar over to check whether the rotor is flexible, whether there is a metal collision sound inside the pump.

③ Open the inlet valve of the centrifugal pump and open the outlet vent valve to fill the pump with liquid. Close the vent valve after removing the accumulated air in the pump and prepare for the start.

(5) Project 5: The normal operations of turning on and off of the centrifugal pump

Startup operation:

① Open the inlet valve of the centrifugal pump completely, close all the outlet valves, then start the motor. After the pump is running, check the working condition of it.

② Check that the motor and pump rotate in the correct direction.

③ Check the motor and pump for noise, abnormal vibration or leakage.

④ Check whether the motor current is less than the rated value. When the overload is displayed, stop the motor immediately and check.

⑤ Adjust the regulating valve of outlet flow so that the working point of the pump is in the state required by the process.

Shutdown operation:

① Gradually close the outlet valve of the centrifugal pump to full close.

② When all the outlet valves are completely closed, shut down the pump.

③ When the pump stops running, close the inlet valve of the centrifugal pump.

(6) Project 6: The normal operations turning on and off of vortex fan

Startup operation:

① Fully open the bypass valve of the vortex fan, start the motor, check the working condition after the fan is running.

② Check that the direction of rotation of the motor.

③ Check the motor and fan for noise and abnormal vibration.

④ Adjust the fan bypass valve so that the working point of the fan is in the state required by the process.

Shutdown operation:

① Gradually open the bypass valve of the fan to full open.

② Shut down the motor.

(7) Project 7: Absorption-desorption operation

① Startup: Start the absorption liquid centrifugal pump according to the operation regulations, feed the absorbent into the absorption column, fully wet the packing. Turn on the mixed gas air fan and open the carbon dioxide cylinder to adjust the carbon dioxide flow so that the mixed gas composition at the inlet of the absorption column reaches a certain value. Start the desorption fan in accordance with the operating procedures, start the desorption liquid centrifugal pump to send the desorption liquid to the desorption column for contact with the desorption air, and appropriately turn on the absorption liquid cooler and the desorption liquid heater when necessary.

② Stable operation: Adjust the flow of mixed gas, liquid and desorbed gas, keep the parameters stable, Monitor and record the data of flow, temperature, pressure drop, liquid level, etc. After the system is stable, sample and analyze the composition. Change the gas and liquid flow, regain the stability of the system, and measure various parameters of the system under different conditions to obtain the hydrodynamics and mass transfer performance of the absorption and desorption processes.

③ Shutdown: Close the main valve of the carbon dioxide cylinder, close the pressure reducing valve, turn off the air fan, and turn off the absorption liquid centrifugal pump, the desorption liquid centrifugal pump and the desorption gas fan according to the shutdown procedures. Switch off the main power, vent the fluid in the system, and clean up the site.

NOTE: Before the training, students should have a comprehensive understanding of the device, be familiar with all the equipment, pipelines, valves and instruments in the equipment, and clarify their principles and functions. The detailed operating procedures should be formulated in advance, and the training operations can be carried out only after the operating procedures are reviewed and approved. The absorption-desorption operation project should be performed on the premise that all other training projects have been completed.

附　录

附表 1　二氧化碳在水中的亨利系数（$E\times10^{-5}$，kPa）

气体	温度/℃											
	0	5	10	15	20	25	30	35	40	45	50	60
CO_2	0.738	0.888	1.05	1.24	1.44	1.66	1.88	2.12	2.36	2.60	2.87	3.46

注：$E=24.1419t^2+3.097\times10^3t+7.283\times10^4$。

附表 2　空气密度随温度变化表

t/℃	T/K	ρ/(kg/m³)	t/℃	T/K	ρ/(kg/m³)
20	293.15	1.2050	46	319.15	1.1070
21	294.15	1.2010	47	320.15	1.1035
22	295.15	1.1970	48	321.15	1.1000
23	296.15	1.1930	49	322.15	1.0965
24	297.15	1.1890	50	323.15	1.0930
25	298.15	1.1850	51	324.15	1.0897
26	299.15	1.1810	52	325.15	1.0864
27	300.15	1.1770	53	326.15	1.0831
28	301.15	1.1730	54	327.15	1.0798
29	302.15	1.1690	55	328.15	1.0765
30	303.15	1.1650	56	329.15	1.0732
31	304.15	1.1613	57	330.15	1.0699
32	305.15	1.1576	58	331.15	1.0666
33	306.15	1.1539	59	332.15	1.0633
34	307.15	1.1502	60	333.15	1.0600
35	308.15	1.1465	61	334.15	1.0569
36	309.15	1.1428	62	335.15	1.0538
37	310.15	1.1391	63	336.15	1.0507
38	311.15	1.1354	64	337.15	1.0476
39	312.15	1.1317	65	338.15	1.0445
40	313.15	1.1280	66	339.15	1.0414
41	314.15	1.1245	67	340.15	1.0383
42	315.15	1.1210	68	341.15	1.0352
43	316.15	1.1175	69	342.15	1.0321
44	317.15	1.1140	70	343.15	1.0290
45	318.15	1.1105			

注：$\rho=1.07346\times10^{-5}t^2-0.00448t+1.29009$。

附表 3 二氧化碳密度随温度变化表

t/℃	T/K	ρ/(kg/m³)	t/℃	T/K	ρ/(kg/m³)
20	293.15	1.8303	46	319.15	1.6812
21	294.15	1.8241	47	320.15	1.6759
22	295.15	1.8179	48	321.15	1.6707
23	296.15	1.8117	49	322.15	1.6655
24	297.15	1.8056	50	323.15	1.6604
25	298.15	1.7996	51	324.15	1.6552
26	299.15	1.7936	52	325.15	1.6501
27	300.15	1.7876	53	326.15	1.6451
28	301.15	1.7817	54	327.15	1.6401
29	302.15	1.7758	55	328.15	1.6351
30	303.15	1.7699	56	329.15	1.6301
31	304.15	1.7641	57	330.15	1.6252
32	305.15	1.7583	58	331.15	1.6202
33	306.15	1.7526	59	332.15	1.6154
34	307.15	1.7468	60	333.15	1.6105
35	308.15	1.7412	61	334.15	1.6057
36	309.15	1.7355	62	335.15	1.6009
37	310.15	1.7300	63	336.15	1.5961
38	311.15	1.7244	64	337.15	1.5914
39	312.15	1.7189	65	338.15	1.5867
40	313.15	1.7134	66	339.15	1.5820
41	314.15	1.7079	67	340.15	1.5774
42	315.15	1.7025	68	341.15	1.5728
43	316.15	1.6971	69	342.15	1.5682
44	317.15	1.6918	70	343.15	1.5636
45	318.15	1.6865			

注：$\rho=1.67705\times10^{-5}t^2-0.00683t+1.95988$。

附表 4 乙醇-水汽液相平衡数据（组成以乙醇表示）

t/℃	摩尔分数		t/℃	质量分数	
	液相	气相		液相	气相
100.02	0.0000	0.0000	100.02	0.0000	0.0000
93.93	0.0250	0.2169	97.17	0.0250	0.2330
90.31	0.0500	0.3302	94.86	0.0500	0.3692
87.94	0.0750	0.3994	92.95	0.0750	0.4585
86.29	0.1000	0.4458	91.35	0.1000	0.5216
85.09	0.1250	0.4790	89.99	0.1250	0.5684

续表

t/℃	摩尔分数		t/℃	质量分数	
	液相	气相		液相	气相
84.18	0.1500	0.5041	88.82	0.1500	0.6045
83.48	0.1750	0.5237	87.82	0.1750	0.6331
82.92	0.2000	0.5397	86.94	0.2000	0.6563
82.46	0.2250	0.5531	86.18	0.2250	0.6755
82.08	0.2500	0.5649	85.51	0.2500	0.6917
81.74	0.2750	0.5754	84.92	0.2750	0.7055
81.45	0.3000	0.5851	84.39	0.3000	0.7174
81.19	0.3250	0.5943	83.92	0.3250	0.7278
80.94	0.3500	0.6032	83.51	0.3500	0.7371
80.72	0.3750	0.6120	83.13	0.3750	0.7453
80.50	0.4000	0.6207	82.79	0.4000	0.7528
80.30	0.4250	0.6296	82.48	0.4250	0.7596
80.11	0.4500	0.6387	82.19	0.4500	0.7660
79.92	0.4750	0.6481	81.92	0.4750	0.7720
79.75	0.5000	0.6578	81.67	0.5000	0.7778
79.57	0.5250	0.6679	81.43	0.5250	0.7834
79.41	0.5500	0.6785	81.20	0.5500	0.7889
79.25	0.5750	0.6896	80.98	0.5750	0.7944
79.11	0.6000	0.7012	80.76	0.6000	0.8001
78.97	0.6250	0.7134	80.55	0.6250	0.8059
78.83	0.6500	0.7263	80.34	0.6500	0.8120
78.71	0.6750	0.7397	80.12	0.6750	0.8184
78.60	0.7000	0.7540	79.91	0.7000	0.8254
78.50	0.7250	0.7689	79.69	0.7250	0.8328
78.41	0.7500	0.7847	79.48	0.7500	0.8410
78.33	0.7750	0.8014	79.26	0.7750	0.8500
78.26	0.8000	0.8189	79.05	0.8000	0.8600
78.21	0.8250	0.8375	78.84	0.8250	0.8710
78.18	0.8500	0.8570	78.65	0.8500	0.8834
78.16	0.8750	0.8777	78.47	0.8750	0.8973
78.15	0.9000	0.8995	78.32	0.9000	0.9129
78.16	0.9250	0.9225	78.21	0.9250	0.9307
78.19	0.9500	0.9469	78.15	0.9500	0.9508
78.24	0.9750	0.9727	78.18	0.9750	0.9737
78.31	1.0000	1.0000	78.31	1.0000	1.0000

附表5 体积分数、质量分数、密度对照表（20℃，组成以乙醇表示）

体积分数	质量分数	密度/(g/mL)	体积分数	质量分数	密度/(g/mL)
0	0	1.00020			
0.0020	0.0016	0.99793	0.2200	0.1787	0.97145
0.0040	0.0032	0.99764	0.2300	0.1871	0.97036
0.0060	0.0047	0.99734	0.2400	0.1954	0.96925
0.0080	0.0063	0.99705	0.2500	0.2038	0.96812
0.0100	0.0079	0.99675	0.2600	0.2122	0.96699
0.0120	0.0095	0.99646	0.2700	0.2206	0.96583
0.0140	0.0111	0.99617	0.2800	0.2291	0.96466
0.0160	0.0127	0.99587	0.2900	0.2376	0.96346
0.0180	0.0143	0.99558	0.3000	0.2461	0.96224
0.0200	0.0159	0.99529	0.4000	0.3330	0.94806
0.0220	0.0175	0.99500	0.5000	0.4243	0.93019
0.0240	0.0190	0.99471	0.6000	0.5209	0.90916
0.0260	0.0206	0.99443	0.7000	0.6239	0.88551
0.0280	0.0222	0.99414	0.8000	0.7348	0.85932
0.0300	0.0238	0.99385	0.9000	0.8566	0.82926
0.0320	0.0254	0.99357	0.9100	0.8696	0.82590
0.0340	0.0270	0.99329	0.9200	0.8829	0.82247
0.0360	0.0286	0.99300	0.9300	0.8963	0.81893
0.0380	0.0302	0.99272	0.9400	0.9100	0.81526
0.0400	0.0318	0.99244	0.9420	0.9128	0.81450
0.0420	0.0334	0.99216	0.9440	0.9156	0.81373
0.0440	0.0350	0.99189	0.9460	0.9184	0.81297
0.0460	0.0366	0.99161	0.9480	0.9212	0.81220
0.0480	0.0382	0.99134	0.9500	0.9240	0.81144
0.0500	0.0398	0.99106	0.9520	0.9269	0.81065
0.0600	0.0478	0.98973	0.9540	0.9297	0.80986
0.0700	0.0559	0.98845	0.9560	0.9326	0.80906
0.0800	0.0640	0.98719	0.9580	0.9355	0.80827
0.0900	0.0720	0.98596	0.9600	0.9384	0.80748
0.1000	0.0801	0.98476	0.9620	0.9413	0.80665
0.1100	0.0883	0.98356	0.9640	0.9442	0.80582
0.1200	0.0964	0.98239	0.9660	0.9471	0.80500
0.1300	0.1046	0.98123	0.9680	0.9501	0.80417
0.1400	0.1127	0.98009	0.9700	0.9530	0.80334
0.1500	0.1209	0.97897	0.9720	0.9560	0.80247
0.1600	0.1291	0.97787	0.9740	0.9590	0.80159
0.1700	0.1374	0.97678	0.9760	0.9620	0.80072
0.1800	0.1456	0.98571	0.9780	0.9651	0.79984
0.1900	0.1539	0.97465	0.9800	0.9681	0.79897
0.2000	0.1621	0.97360	0.9900	0.9837	0.79431
0.2100	0.1704	0.97253	1.0000	1.0000	0.78927

附表 6　乙醇-正丙醇汽液相平衡数据（组成以乙醇表示）

t/℃	摩尔分数		t/℃	质量分数	
	液相	气相		液相	气相
97.16	0.0000	0.0000	85.40	0.5098	0.6852
96.20	0.0392	0.0748	84.70	0.5490	0.7171
95.20	0.0784	0.1459	84.03	0.5882	0.7472
94.19	0.1176	0.2132	82.17	0.7059	0.8290
91.25	0.2353	0.3911	81.60	0.7451	0.8540
90.32	0.2745	0.4426	81.05	0.7843	0.8781
89.41	0.3137	0.4904	80.51	0.8235	0.9014
88.54	0.3529	0.5350	80.00	0.8627	0.9241
86.90	0.4314	0.6151	79.01	0.9412	0.9680
86.14	0.4706	0.6513	78.38	1.0000	1.0000

附表 7　乙醇-正丙醇体系折光指数（n_D）与液相乙醇质量分数（w）之间的关系

w	n_D		
	25℃	30℃	35℃
0	1.3827	1.3809	1.3790
0.05052	1.3815	1.3796	1.3775
0.09985	1.3797	1.3784	1.3762
0.1974	1.3770	1.3759	1.3740
0.2950	1.3750	1.3755	1.3719
0.3977	1.3730	1.3712	1.3692
0.4970	1.3705	1.3690	1.3670
0.5990	1.3680	1.3668	1.3650
0.6445	1.3607	1.3657	1.3634
0.7101	1.3658	1.3640	1.3620
0.7983	1.3640	1.3620	1.3600
0.8442	1.3628	1.3607	1.3590
0.9064	1.3618	1.3593	1.3573
0.9509	1.3606	1.3584	1.3653
1.0000	1.3589	1.3574	1.3551

参考文献

[1] 谭天恩，窦梅．化工原理［M］．4版．北京：化学工业出版社，2013.
[2] 施小芳，李微，林述英．化工原理实验［M］．2版．福州：福建科学技术出版社，2010.
[3] 李玲，叶长燊，施小芳，等．化工原理实验［M］．北京：经济科学出版社，2012.
[4] 张金利，郭翠梨，胡瑞杰，等．化工原理实验［M］．2版．天津：天津大学出版社，2016.
[5] 史贤林，张秋香，周文勇，等．化工原理实验［M］．北京：化学工业出版社，2019.
[6] 居沈贵，夏毅，武文良．化工原理实验［M］．北京：化学工业出版社，2016.
[7] 都健，王瑶，王刚．化工原理实验［M］．北京：化学工业出版社，2017.
[8] 杨祖荣．化工原理实验［M］．2版．北京：化学工业出版社，2014.
[9] 徐伟，鞠彩霞，刘书银．化工原理实验［M］．北京：化学工业出版社，2017.
[10] 马江权，魏科年，韶晖，等．化工原理实验［M］．上海：华东理工大学出版社，2016.
[11] 程振平，赵宜江，钱运华，等．化工原理实验［M］．2版．南京：南京大学出版社，2017.
[12] 方安平，叶卫平．Origin 8．0实用指南［M］．北京：机械工业出版社，2010.
[13] 潘祖亭．分析化学教程［M］．北京：科学出版社，2020.
[14] 厉玉鸣．化工仪表及自动化［M］．6版．北京：化学工业出版社，2019.